# Guanine Quartets
Structure and Application

# Guanine Quartets
## Structure and Application

Edited by

**Wolfgang Fritzsche**
*Institute of Photonic Technology (IPHT), Nano Biophotonics Department, PO Box 100 239, 07702 Jena, Germany*
*E-Mail: wolfgang.fritzsche@ipht-jena.de*

**Lea Spindler**
*Faculty of Mechanical Engineering, University of Maribor, Smetanova 17, SI-2000 Maribor, also J. Stefan Institute, Jamova 39, SI-1000, Ljubljana, Slovenia*
*E-Mail: lea.spindler@uni-mb.si*

RSCPublishing

ISBN: 978-1-84973-460-8

A catalogue record for this book is available from the British Library

© The Royal Society of Chemistry 2013

*All rights reserved*

*Apart from any fair dealing for the purpose of research or private study for non-commercial purposes, or criticism or review as permitted under the terms of the UK Copyright, Designs and Patents Act, 1988 and the Copyright and Related Rights Regulations 2003, this publication may not be reproduced, stored or transmitted, in any form or by any means, without the prior permission in writing of The Royal Society of Chemistry or the copyright owner, or in the case of reprographic reproduction only in accordance with the terms of the licences issued by the Copyright Licensing Agency in the UK, or in accordance with the terms of the licences issued by the appropriate Reproduction Rights Organization outside the UK. Enquiries concerning reproduction outside the terms stated here should be sent to The Royal Society of Chemistry at the address printed on this page.*

The RSC is not responsible for individual opinions expressed in this work.

Published by The Royal Society of Chemistry,
Thomas Graham House, Science Park, Milton Road,
Cambridge CB4 0WF, UK

Registered Charity Number 207890

Visit our website at www.rsc.org/books

# PREFACE

Guanosine (G) molecules show a remarkable ability to self-assemble into highly complex patterns. The most common structural motif is the G-quartet, a hydrogen-bonded planar arrangement of four guanosine molecules, that is formed in a variety of guanosine derivatives but also in G-rich DNA and RNA sequences. Stacking of G-quartets leads to the formation of G-quadruplexes; complex and highly ordered helical structures. The extent of stacking, and consequently the length of these supramolecular structures, can be controlled by temperature, pH value, solution concentration and by cations added to the solution. Other self-assembling motifs, like G-ribbons, were also identified in lipophilic guanosine derivatives.

Guanine-rich regions were found to abound in the human genome and these G-rich DNA sequences have the propensity to fold into G-quadruplexes. Theoretical models predict the folding of these putative G-quadruplex folding sequences into different topologies and methods like high-resolution NMR spectroscopy and X-ray crystallography are used for their experimental determination. Recent findings indicate the possible involvement of DNA-quadruplex structures in the regulation of gene transcription and explore quadruplex DNA as a viable therapeutic target.

G-quadruplexes as well as the planar G-ribbons manifest themselves as self-organised wires of controllable length. This property if revealed by atomic force microscopy and scanning tunnelling microscopy when such structures are deposited onto solid surfaces. These so-called G-wires are characterized by high stiffness, heat resistivity and mechanical stability, with respect to natural double-stranded DNA, which makes them prosperous candidates for molecular wires.

This book summarises recent advances in the field, with an emphasis on those resulting from European researchers brought together by Cooperation in Science and Technology (COST) Action MP0802 "*Self-assembled guanosine structures for molecular electronic devices*". This network was established in an effort to explore the basic principles of guanosine-assembly, to understand and predict the folding of G-rich oligonucleotides into G-quadruplexes, to synthesize new optimised materials, and to explore guanosine-based materials as possible nanoarchitectures for molecular electronic devices. Between November 2008 and November 2012 more than 60 research groups from 19 different countries participated in the network, which resulted in an intense exchange of ideas and knowledge, numerous joint scientific publications and, finally, in the publication of this book. It presents in 5 chapters the various aspects of G-quartet structures, ranging from design, simulation and synthesis over characterization to applications in bioanalytics, therapy as well as nanoelectronics. This documentation of the state of the art will, without doubt, further strengthen the G-quadruplex community and allow for the identification of emerging trends in this fascinating, truly interdisciplinary field.

We would like to thank all the authors for their contributions, without their efforts the publication of the book would not have been possible. We are also grateful to the chapter editors for all their work in preparing and organising their chapters. And finally, we gratefully acknowledge the encouragement and support from the COST Office, especially Science Officer Dr. Caroline Whelan, Administrative Officer Ms. Milena Stoyanova, and Rapporteur of our Action, Dr. Anthony Flambard.

Lea Spindler and Wolfgang Fritzsche

# Chapter Editors

Lea Spindler
Faculty of Mechanical Engineering
University of Maribor
Smetanova 17, SI-2000 Maribor
also with J. Stefan Institute
Jamova 39, SI-1000 Ljubljana, Slovenia
E-Mail: lea.spindler@uni-mb.si

Gian Piero Spada
Professor of Organic Chemistry
Alma Mater Studiorum – Università di Bologna
Dipartimento di Chimica Organica "A. Mangini"
Via San Giacomo 11 – 40126 Bologna, Italy
E-Mail: gianpiero.spada@unibo.it

Shozeb Haider
Senior Lecturer in Drug Discovery
Centre for Cancer Research and Cell Biology
Queen's University Belfast
97 Lisburn Road, Belfast BT9 7BL, UK
E-Mail: s.haider@qub.ac.uk

Mateus Webba da Silva
Reader in Pharmaceutical Chemistry
School of Biomedical Sciences
University of Ulster
Coleraine, BT51 1SA, UK
E-Mail: mateus@webbas.org

Wolfgang Fritzsche
Institute of Photonic Technology (IPHT)
Nano Biophotonics Department
PO Box 100 239, 07702 Jena, Germany
E-Mail: wolfgang.fritzsche@ipht-jena.de

# Contents

## Chapter 1

Introduction: From G-Quartet to G-Quadruplex and its Nanoarchitectures ... 1
   *Lea Spindler*

Guanylic acid Self-Assembly: 100 years on ... 3
   *Gang Wu*

Functional Assemblies made from Supramolecular G-Quadruplexes ... 15
   *J.M Rivera*

Self-Assembly of Lipophilic Guanosines: Switching between different Assemblies ... 28
   *Stefano Masiero, Silvia Pieraccini and Gian Piero Spada*

Nanopatterning the Surface with ordered Supramolecular Architectures: Controlling the Self-Assembly of Guanine-based Hydrogen-Bonded Motifs ... 40
   *Artur Ciesielski, Mathieu Surin, Gian Piero Spada and Paolo Samor*

Morphological Heterogeneity of Supramolecular G-DNA Polymers derived from Guanine Rich Oligonucleotides ... 48
   *T.C. Marsh, Z.M. Henseler and M.A. Klimstra*

Thermondynamics of G-Quadruplexes ... 63
   *C Giancola*

G-Quadruplex Nanostructures Probed at the Single Molecular Level by Force-Based Methods ... 73
   *Soma Dhakal, Hanbin Mao, Arivazhagan Rajendran, Masayuki Endo, and Hiroshi Sugiyama*

## Chapter 2

Introduction: Synthesis and Characterization ... 87
   *Gian Piero Spada*

Synthesis and Properties of Oligonucleotides Forming G-Quadruplexes ... 89
   *A. Aviñó, and R. Eritja*

Electrochemical Characterization of Guanine Quadruplexes ... 100
   *A.–M. Chiorcea–Paquim, P. Santos, V.C. Diculescu, R. Eritja and A.M. Oliveira-Brett*

AFM of Guanine Rich Oligonucleotide Surface Structures ... 110
   *James Vesenka*

Solution Dynamics and Structure of G-Quadruplexes Studied by Dynamic Light Scattering  121
*Lea Spindler*

GMP-Quadruplex Structures in Dilute Solutions and in Condensed Phases: An X-Ray Scattering Analysis  135
*E. Jr. Baldassarri, M. G. Ortore, A. Gonnelli, M. Marcinekova, S. Mazzoni, M. L. Travaglini and P. Mariani*

Temperature-Gradient Gel Electrophoresis: Unfolding of G-Quadruplexes  147
*Viktor Víglaský, Katarína Tlučková, Petra Tóthová and Ľuboš Bauer*

Specific Behaviour of Guanosine in Liponucleoside Thin Films  154
*L. Coga, M. Devetak, S. Masiero, G.P. Spada and I. Drevenšek-Olenik*

## Chapter 3

Theoretical Modelling, Analysis and Prediction  165
*Shozeb Haider*

Fundamentals and Applications of the Geometric Formalism of Quadruplex Folding  167
*Mateus Webba da Silva and Andreas Ioannis Karsisiotis*

Guanine, Xanthine and Uric Acid Assemblies: Comparative Theoretical and Experimental Studies  179
*Gábor Paragi, János Szolomájer, Zoltán Kupihár, Gyula Batta, Zoltán Kele, Petra Pádár, Botond Penke, Hester Zijlstra, Célia Fonseca Guerra, F. Matthias Bickelhaup, and Lajos Kovács*

Computational Methods for Studying G-Quadruplex Nuceic Acids  194
*B. Islam, V. D'Atri, M. Sgobba, J. Husby and S. Haider*

## Chapter 4

Introduction: Recognition of Quadruplexes  213
*Mateus Webba da Silva*

Biological Functions of G-Quadruplexes  215
*Nancy Maizels*

Regulation of Gene Transcription by DNA G-Quadruplexes  223
*Michael Fry*

The Reality of Quadruplex Nucleic Acids as a Therapeutic Target  237
*G. N. Parkinson*

Contents

| | |
|---|---|
| Screening for Quadruplex Binding Ligands: A Game of Chance?<br>　*E. Largy and M.-P. Teulade-Fichou* | 248 |
| Recognition of G-Quadruplexes by Metal Complexes<br>　*Kogularamanan Suntharalingam and Ramon Vilar* | 263 |

# Chapter 5

| | |
|---|---|
| Introduction: Applications in Bioanalytics, Therapy and Molecular Electronics<br>　*Wolfgang Fritzsche* | 275 |
| Catalytic G-Quadruplexes<br>　*Dipankar Sen* | 277 |
| Catalytic G-Quadruplexes for the Detection of Telomerase Activity<br>　*Joanna Kosman, Bernard Juskowiak* | 285 |
| G-Quadruplex forming Oligonucleotides with Tailor-made Modifications as effective aptamers for Potential Therapeutic Applications<br>　*Domenica Musumeci and Daniela Montesarchio* | 292 |
| Dep-Based Integration of G-Quadruplex Structures<br>　*Christian Leiterer, Andreas Kopielski, Irit Lubitz, Alexander Kotlyar, Antti-Pekka Eskelinen, Päivi Törmä and Wolfgang Fritzsche* | 306 |
| Conductive Behaviour of G4-DNA-Silver Nanoparticle Structures<br>　*T. Parviainen, G. Eidelshten, A. Kotlyar, and J.J. Toppari* | 314 |
| Novel Materials for Molecular Electronics – Synthese and Characterization of Long G4-DNA<br>　*Dvir Rotem, Gennady Eidelshtein, Alexander Kotlyar and Danny Porath* | 324 |

**Subject Index**     **337**

# Chapter 1

## From G-Quartet to G-Quadruplex and its Nanoarchitectures

Chapter Editor: **Lea Spindler**

In 1910 Ivar Bang reported that concentrated solutions of guanylic acid formed a gel. It took half a century before Gellert *et al.* in 1962 discovered the structural motif, a guanine-quartet, to be the basis for guanylic acid gelation. This chapter starts with a historical overview and gives credit to Ivar Bang and several other pioneers in the field. Then the basic principles of guanine self-assembly are explored, either for individual molecules like guanosine and its derivatives, or for guanine-rich oligonucleotides and G-rich DNA-sequences. Finally, we show how these guanine-based nanoarchitectures are visualised by surface techniques like scanning tunnelling microscopy or atomic force microscopy.

# GUANYLIC ACID SELF-ASSEMBLY: 100 YEARS LATER

Gang Wu

Department of Chemistry, Queen's University, Kingston, Ontario, Canada K7L 3N6

## 1 INTRODUCTION

It is commonly accepted that the work of Gellert, Lipsett, and Davies[1] published in 1962 marks the beginning of modern research on G-quartet related molecular systems. Interestingly, the opening sentence of this seminal paper stated, "In 1910, Bang reported that concentrated solutions of guanylic acid formed a gel." This referred to a paper published half of a century earlier.[2] They then described two structural models used to explain the experimental X-ray diffraction patterns obtained for guanylic acid fibers. The fibers were drawn from two different types of guanylic acid gels: one from concentrated aqueous solutions of guanosine 3'-monophosphate (3'-GMP) under an acidic condition (pH 5) and the other from the 5' isomer, 5'-GMP, under a similar condition. The central structural building block proposed for the 3'-GMP gel consists of four hydrogen bonded guanine bases in a planar fashion, which is now known as a G-quartet or G-tetrad. The structural model proposed for the 5'-GMP gel was initially ambiguous, due to the limited quality of the X-ray data. Later, Davies and co-workers[3] showed that the correct structural model for 5'-GMP gels obtained at pH 5 consists of hydrogen bonded guanine bases that form a continuous helix similar to the shape of a lock-washer. These two models are illustrated in Figure 1.

In the past 20 years, research activities in the field of G-quartet related molecular systems have grown exponentially. Recent discoveries of the existence of the G-quartet motif in many biologically important systems such as telomeres, promoters of many genes, and sequences related to various human diseases have triggered tremendous research interest in this unusual type of nucleic acid structures. Now G-quartet structures can be found in such diverse areas as molecular biology, medicinal chemistry, supramolecular chemistry, and nanotechnology. It is interesting to note that, as the field of G-quartet research expands in recent years, one can also find increasingly more references to the aforementioned 1910 article by Bang. Unfortunately, many citations are either inaccurate in terms of what exactly Bang reported in the 1910 paper or completely wrong in terms of who Bang was. For example, he was sometimes misidentified as being a German biochemist (obviously his 1910 paper was published in *Biochemische Zeitschrift* and indeed in German). One may wonder how many modern researchers in the field have

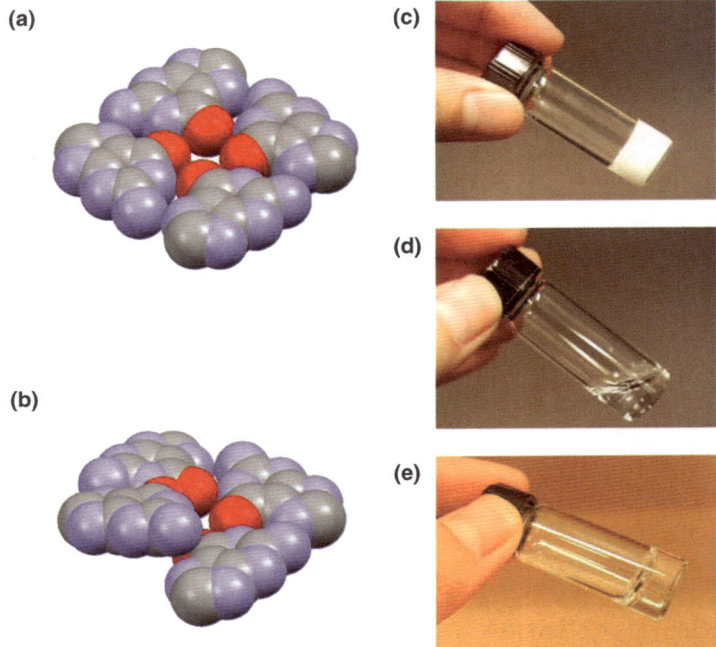

**Figure 1**  (a) The planar G-quartet model. (b) The lock-washer G-quartet model. (c) A gel formed from concentrated $Na_2(5'\text{-}GMP)$ solution at pH 5. (d) Concentrated $Na_2(5'\text{-}GMP)$ solution at pH 8. (e) A gel formed on cooling of guanosine aqueous solution in the presence of KCl.

actually read Bang's 1910 article. It seems, however, that all we can learn from the modern literature is that Bang somehow noticed gelation of guanylic acid under certain conditions, but we cannot find any clue as to the context he made this important observation. So who was Bang and why did he study guanylic acid in the first place? In this article, we set out to examine briefly the history centered on the discovery of guanylic acid (1894-1912), then discuss some early results on its self-assembly properties/structures (1962-1990), and finally describe some recent progress related to guanylic acid (since 2000). The main objective of this article is to provide the reader with a sense of history and a big picture about what we have learned about this fascinating molecule and its self-assembly properties over a period of 100 years.

Here we should credit the title of this article to several previous workers in the field. In particular, Guschlbauer, Chantot, and Thiele[4] published the first review article on the subject in 1990 which was entitled "Four-stranded nucleic acid structures 25 years later: From guanosine gels to telomer DNA". In 2004, Davis[5] wrote the most popular review in the field with a similar title: "G-quartet 40 years later: From 5'-GMP to molecular biology and supramolecular chemistry". Thus it seems appropriate for us to follow this good tradition, except that, as the reader may have already noticed, we extend the beginning of our time line to Bang's 1910 discovery.

## 2 THE BEGINNING

In 1894, Olof Hammarsten (1841-1932) isolated from ox pancreas a substance that he called β-nucleoprotein.[6] The non-protein portion of β-nucleoprotein was given the name of guanylic acid, because on hydrolysis it releases an excessive amount of guanine but no trace of other bases such as thymine and cytosine. This appears to be very different from other nucleic acids known at the time. From 1897-99, Ivar Christian Bang (1869-1918, a Norwegian physician and clinical chemist), who had obtained his medical degree two years earlier from Oslo, joined Hammarsten's Physiological Chemistry Laboratory in Uppsala, Sweden, as a trainee (perhaps modern equivalent of a postdoctoral fellow).[7] In the next decade, with a two-year interruption (1900-1902) when he practiced medicine in Oslo, Bang attempted to elucidate the chemical structure of guanylic acid.[2,8-13] He spent a great deal of effort trying to obtain a reasonably large quantity of pure guanylic acid for chemical analysis, which turned out to be particularly difficult. In the meantime, several other groups were also studying guanylic acid.[14-18] Because of the difficulties involved in producing pure, crystalline guanylic acid for proper chemical analysis, different researchers obtained different results, triggering a lively debate that lasted more than 10 years.

Bang's 1910 paper[2] was the last and also most comprehensive one of the series of papers that he published on the subject of guanylic acid (afterwards he turned his attention to some other problems in clinical chemistry and died suddenly from coronary occlusion in 1918 at age of 49). In this lengthy paper (18 pages), he correctly identified for the first time the chemical compositions of guanylic acid as being composed of guanine, pentose, and phosphoric acid in equal molar amounts. However, because his elemental analysis was inaccurate, he was unable to account for an unknown residue of $C_4H_{10}O_2$. In this work, he also described an observation that, after a concentrated guanylic acid solution was neutralized with KOH, re-acidifying the solution with acetic acid led to gel formation. This kind of gel formation was commonly known for thymus nucleic acids at that time. For this reason, Bang argued forcefully (sometimes filled with personal anger) but *incorrectly* against the view of other contemporary researchers who considered guanylic acid to be a simple nucleotide just like inosinic acid, a mononucleotide discovered by Liebig in 1848 from beef booth. This latter opinion was clearly expressed by Albrecht Kossel (1853-1927, a German biochemist) in his Nobel lecture presented in December of 1910, in which, after describing the complex nature of nucleic acids, he stated, "The composition of *inosinic* and *guanylic* acids is still simpler."[19]

The controversy around the chemical nature of guanylic acid was finally settled in 1912 by an American biochemist, Phebus Aaron Levene (1869-1940) who had worked on guanylic acid since 1901.[14] In 1909, Levene successfully obtained crystalline guanosine, a guanine nucleoside, through hydrolysis of guanylic acid.[20] So he was well aware of the fact that, upon cooling, a hot guanosine aqueous solution will form a gel in the presence of a small amount of $K^+$ ions; see Figure 1(e). In a paper published in 1912,[21] he argued, "However, guanosine—a simple guanine-pentoside—shares with guanylic acid the property of gelatinizing when it contains only slight proportion of mineral impurity." Therefore, gel formation alone was insufficient to disprove the simple nature of guanylic acid, as Bang had attempted to do. To address the chemical composition problem, Levene performed a more accurate elemental analysis that showed unambiguously the correct chemical formula for guanylic acid to be $C_{10}H_{14}N_5PO_8$ with a structure shown in Figure 2.[21] Now we know that this particular structure corresponds to a slightly different isomer, 3'-GMP, rather than the most common one, 5'-GMP. However, this work was a truly remarkable achievement, considering the following three facts: (1) X-ray was discovered by Röntgen in 1895; (2) Its diffraction effect on crystals was discovered by Friedrich, Knipping, and Laue in 1912, the same year as the publication of Levene's work; and (3) Of course, nuclear magnetic resonance (NMR) was not discovered until 1945. It is perhaps very hard for those students in my 2$^{nd}$ year organic spectroscopy course to grasp the concept of chemical analysis or structural elucidation without modern spectroscopic techniques. It is also amazing to see how science progresses over time. Thus, as for the chemical nature of guanylic acid, Bang was unfortunately incorrect. However, very few people in the field of G-quartet research are aware of the fact that Bang is actually best known for his pioneering contributions to clinical chemistry; he was considered to be the founder of modern clinical chemistry.[22] For example, he invented the technique for analyzing blood sugar levels on a microscale. To return to the subject on hand, now what is the structural basis for guanylic acid gelation? The answer to this question would have to wait for another 50 years.

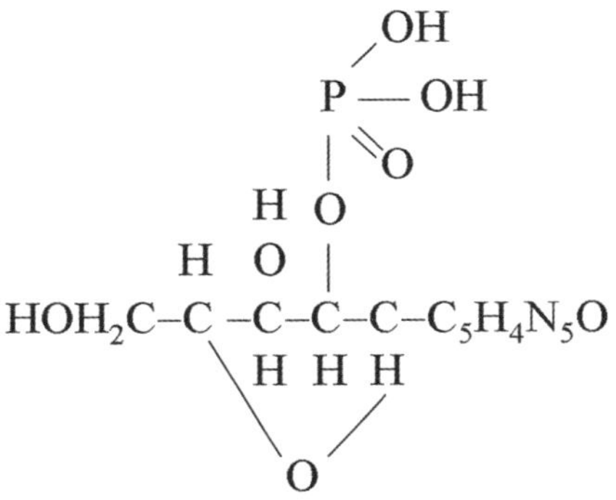

**Figure 2** *The molecular structure of guanylic acid proposed by Levene in 1912.*

### 3 EARLY STUDIES

Since 1955, David Davies at the National Institute of Health had been studying RNA structures using X-ray fiber diffraction. During an effort to prepare polyguanylic acid, poly G, he and his co-workers accidentally discovered the G-quartet motif in 1962. Davies described this serendipitous discovery in his autobiography published in 2005: "Marie [Lipsett] originally thought that she had been able to make poly G but was then disappointed to discover that what she had was unpolymerized GMP that was forming a viscous solution that looked just like DNA."[23] Recall that this was exactly what Bang had seen 50 years earlier! Then Davies continued, "As soon as she told me this I rushed over and pulled some fibers that gave diffraction patterns that could be explained by the formation of G-quartets." Soon after, Iball, Morgan, and Wilson reported that deoxyguanosine nucleosides and nucleotides also form similar helices.[24] Many guanine nucleosides and nucleotides were subsequently found to be able to form gels containing similar helices.[25-27]

In 1972, based on infra-red (IR) spectroscopic evidence, Todd Miles at the NIH reported a new helical structure formed by 5′-GMP in neutral or slightly basic solution (pH 7~8) in the presence of $Na^+$ ions.[28] As 5′-GMP behaves as a regular liquid under such conditions, it is suitable for solution NMR studies. Thus he approached his NMR colleague at the NIH, Ted Becker. At that time, Becker' lab was equipped with one of the early high-field 220 MHz NMR spectrometers. Together, they established that the planar G-quartet motif is responsible for the 5′-GMP helix formation in neutral solution.[29-31] Years later, Becker[32] described this discovery in *Encyclopedia of Nuclear Magnetic Resonance*: "Tom Pinnavaia came from Michigan State University to spend his sabbatical leave in my lab in 1974-75. He discovered that guanosine-5′-monophosphate (GMP) in neutral or basic solution self-associates by hydrogen bonding to give a stable structure in which the H-bonds exchange only slowly on the NMR timescale. We were able to interpret the NMR, along with ancillary infrared spectra data, in terms of an unexpected tetrameric structure." Remarkably, the 5′-GMP self-assembly process in neutral solution exhibits great sensitivity towards the type of alkali metal ions present in solution. As shown in Figure 3, while $Li^+$ and $Cs^+$ ions are inactive, $Na^+$, $K^+$ and $Rb^+$ appear to promote different self-assembled structures. The very clean four-peak pattern ($\alpha$, $\beta$, $\gamma$, and $\delta$ signals for $H_8$) observed for $Na_2$(5′-GMP) strongly suggests the presence of well-defined aggregates, yet the proper interpretation of this spectral pattern had remained illusive for more than 30 years until recently (*vide infra*). In 1976, Zimmerman confirmed the helix formation of $Na_2$(5′-GMP) under neutral conditions with X-ray fiber diffraction data.[33] In the several years that followed, $^1H$, $^{13}C$ and $^{31}P$ NMR techniques were used to gain insights into the various aspects of this new type of 5′-GMP self-assembly.[34-42] In the meantime, Laszlo and co-workers[43-46] used $^{23}Na$ NMR to probe ion binding in the 5′-GMP self-assembly. In the 1990s, Gottarelli, Spada, Mariani and their co-workers[47-53] reported extensive studies on the various liquid crystal phases formed from guanylic acid and related derivatives.

By late 1980s it became clear that the helical structure formed by 5′-GMP at an acidic pH consists of a simple continuous helix whereas the helices that utilize stacking of planar G-quartets are far more complex. How can this be? In the early 1990s, suddenly several discoveries suggested that the G-quartet motif has some profound biological implications and thus stirred great interest in G-quadruplex DNA. Meanwhile the subject of guanylic acid self-assembly seemed to have been totally forgotten.

## 4 RECENT PROGRESS

Since 2000, several occasions prompted us to revisit the helical structure of 5′-GMP self-assembly in neutral solution. As a result, there has been a renewed interest in fully understanding the structural details of the 5′-GMP self-assembly. Because of space limitation, we will focus only on recent results primarily from our laboratory in two areas: (1) ion binding to G-quadruplexes and (2) structural details of the $Na_2$(5′-GMP) self-assembled helix in neutral solution. For other recent studies related to guanylic acid, the reader should consult the literature.[54-63]

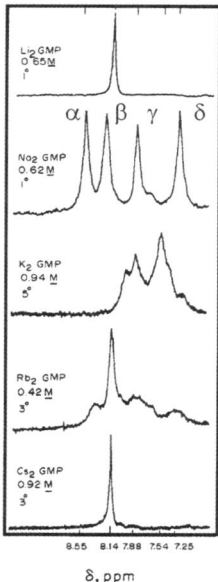

**Figure 3** *The $H_8$ regions of $^1H$ NMR spectra obtained for different salts of 5′-GMP in $D_2O$ at pH 8. Reproduced with permission from ref. 30.*

## 4.1 5′-GMP as a model system for studying ion binding to G-quadruplex

In 1996, when I was a postdoctoral fellow with Prof. Bob Griffin at MIT, Bob suggested that we used solid-state $^{23}Na$ NMR to probe $Na^+$ ion binding to G-quadruplex DNA. In collaboration with Prof. Juli Feigon of UCLA, we started with a simple DNA sequence, d(TG$_4$T). At that point, the beautiful crystal structure of parallel stranded [d(TG$_4$T)]$_4$ G-quadruplex was just published.[64] So it seemed logical to choose this simple DNA sequence for exploring the potential of solid-state $^{23}Na$ NMR. Dr. Nick Hud (a postdoc from the Feigon group) prepared several DNA samples and David Rovnyak (a graduate student in the Griffin group) and I carried out the NMR measurements. After getting some very exciting preliminary $^{23}Na$ NMR results, I left MIT in June of 1997 taking up a faculty position at Queen's University. Later, David and Dr. Marc Baldus (another postdoc in the Griffin group) did more NMR experiments. The paper was finally published in 2000.[65] However, in this study, we had some troubles assigning the observed $^{23}Na$ NMR signals to the proper $Na^+$ binding sites in DNA. It was clear that we would need additional information to help with spectral assignment. After consulting the literature, we decided to use 5′-GMP as a model,[66] as the ion binding properties of 5′-GMP had been known from several earlier studies as mentioned in the previous section. As shown in Figure 4, besides the very sharp signal at 7 ppm which can be readily identified as due to residual NaCl, there are essentially two $^{23}Na$ NMR signals observed for 5′-GMP: one at −19 ppm and the other at about −1 ppm. By examining the signal changes upon addition of other alkali metal ions such as K[+] and Cs[+] ions, we were able to assign unambiguously the signal at − 19 ppm to be due to the Na[+] ions inside the G-quadruplex channel. Subsequently, in collaboration with Jeff Davis, we further confirmed this spectral assignment by examining both solid-state $^{23}$Na NMR and crystal structure of a liphophilic guanosine nucleoside.[67] Later we were able to use the solid-state NMR approach to determine relative binding affinity of ions to the 5′-GMP helical structure.[68] Then we showed that NMR can be used to detect Na[+] ions in G-quadruplex DNA.[69-71] In addition, we also obtained NMR spectral signatures for K[+], Rb[+], and Ca[2+] ions bound to G-quadruplexes.[72-76] In all these studies, the 5′-GMP helix turns out to be as an excellent model for G-quadruplex nucleic acids.

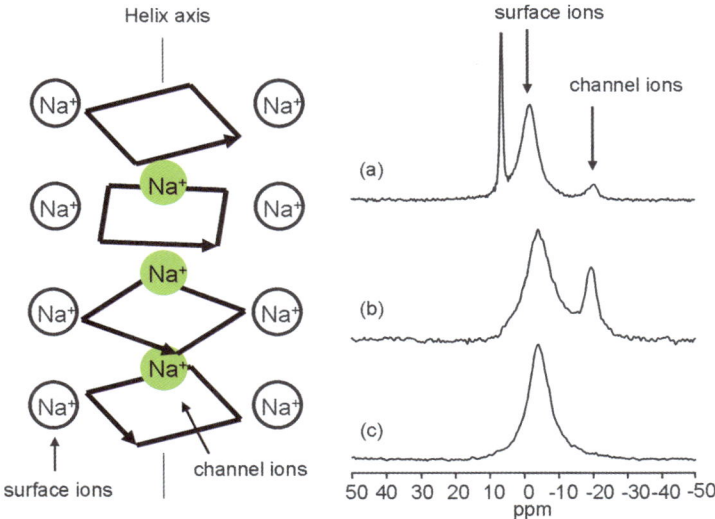

**Figure 4**  (Left) Illustration of two types of Na[+] binding sites to the self-assembled 5′-GMP helix formed under the neutral condition. (Right) Solid-state $^{23}$Na NMR spectra of (a) Na$_2$(5′-GMP), (b) Na$_2$(5′-GMP) in the presence of Cs[+], and (c) in the presence of K[+]. All spectra were obtained at 11.75 T under the magic-angle spinning condition.

## 4.2 Structural details of the Na$_2$(5′-GMP) helix in neutral solution

As mentioned earlier, Na$_2$(5′-GMP) self-associates into an order helical structure in neutral solution, giving rise to a complex but well-defined four peak pattern for each resonance in its $^1$H NMR spectrum; see Figure 3. However, the structural details for this helix have never been established. In the pioneering studies carried out in the 1980s, only rather simple 1D NMR techniques were utilized. It was surprisingly that no attempt had ever been made to fully characterize this helix. In 2004, we decided to investigate this problem using modern NMR techniques. The first thing we did was to determine the aggregate size using an NMR method called diffusion-order spectroscopy (DOSY), in collaboration with Lea Spindler who provided dynamic light scattering (DLS) data.[77] In this study, we discovered two things. First, as shown in Figure 5, the so-called β signal corresponds to a different type of 5′-GMP aggregates, which has a smaller size than those giving rise to the α and δ signals. Later we proved that the β signal is not related to G-quartet formation as previously thought but due to a centrosymmetric GG dimer.[78] Second, we demonstrated that the size of $Na_2$(5′-GMP) aggregates is on the order of 8-30 nm, much larger than those (i.e., octamers, dodecamers, or hexadecamers) considered exclusively by previous workers.

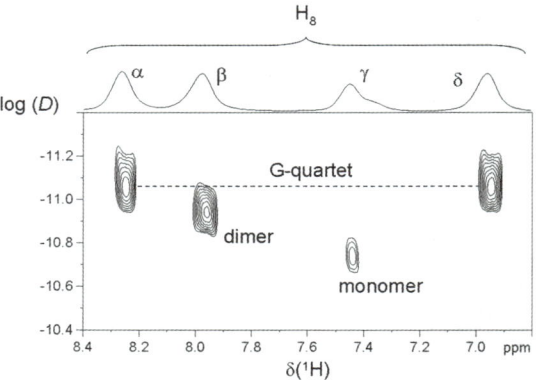

**Figure 5**   *The $H_8$ region of a 2D $^1H$ DOSY spectrum obtained for 1.0 M $Na_2$(5′-GMP) in $D_2O$ (pH 8) at 273 K. Reproduced with permission from ref. 78.*

The DOSY study provided us with an opportunity to look at an old problem with fresh eyes.  To solve the spectral overcrowding problem, we designed a DOSY-NOESY experiment in which a regular NOESY experiment is preceded with a DOSY preparation period.  This combination is equivalent to running a chromatography followed by a NOESY experiment, resulting in the "filtering out" of unwanted signals.  This method makes it possible to obtain structural information from overcrowded spectral regions, as illustrated in Figure 6.

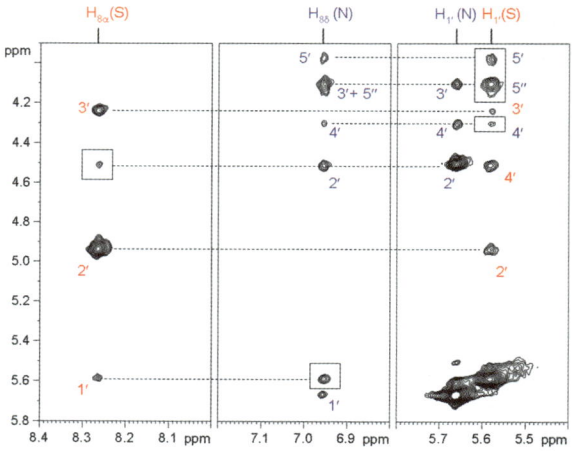

**Figure 6**   *Regions of a 2D DOSY-NOESY $^1H$ NMR spectrum of 1.0 M $Na_2$(5′-GMP) in $D_2O$ at pH 8 and 298 K. Reproduced with permission from ref. 78.*

The final structural determination for the Na$_2$(5'-GMP) helix was published in 2009 and the result was a total surprise.[78] The key feature of the Na$_2$(5'-GMP) helix is that the two distinct sugar pucker conformations, C2'-*endo* (S) and C3'-*endo* (N), are present simultaneously in an equal amount, which explains the α and δ signals in a 1:1 ratio. Along the Na$_2$(5'-GMP) quadruple helix, we found alternating G$_4$(N) and G$_4$(S) G-quartets that are stacked on top of one another with a rotation of 30° forming a right-handed helix. Between G$_4$(N) and G$_4$(S) is a Na$^+$ ion coordinating to the eight carbonyl oxygen atoms in a square anti-prism fashion. So the basic repeating unit of the helix is a G-octamer, G$_4$(N)-Na$^+$-G$_4$(S). Interestingly, in the 1962 paper by Gellert, Lipsett, and Davies,[1] they also proposed an octamer model as the basic repeating unit of the 3'-GMP helix. It is possible that the 3'-GMP helix also consists of mixed C2'-*endo* and C3'-*endo* sugar puckers. Figure 7(a) shows a single "strand" of the Na$_2$(5'-GMP) quadruple helix to highlight how individual 5'-GMP molecules are "stitched" together via P–O$^-$···H–O hydrogen bonds with C2'-*endo* and C3'-*endo* sugar puckers alternating along the helical strand. Moreover, an additional [P(S)–O$^-$]$_i$···[H–O$_3$'(N)]$_{i+3}$ hydrogen bond inter-locks the helical structure. Figure 7(b) displays the hydrogen bond linkage along the 5'-GMP helix in a conventional fashion used for polynucleotides. It is striking to notice that the arrangement of adjacent 5'-GMP molecules is such that they are perfectly positioned for phosphodiester bond formation. This is the first time that we see how individual mononucleotides utilize weak molecular forces as well as their intrinsic flexibility in the sugar pucker conformation to self-associate into a quadruple helix.

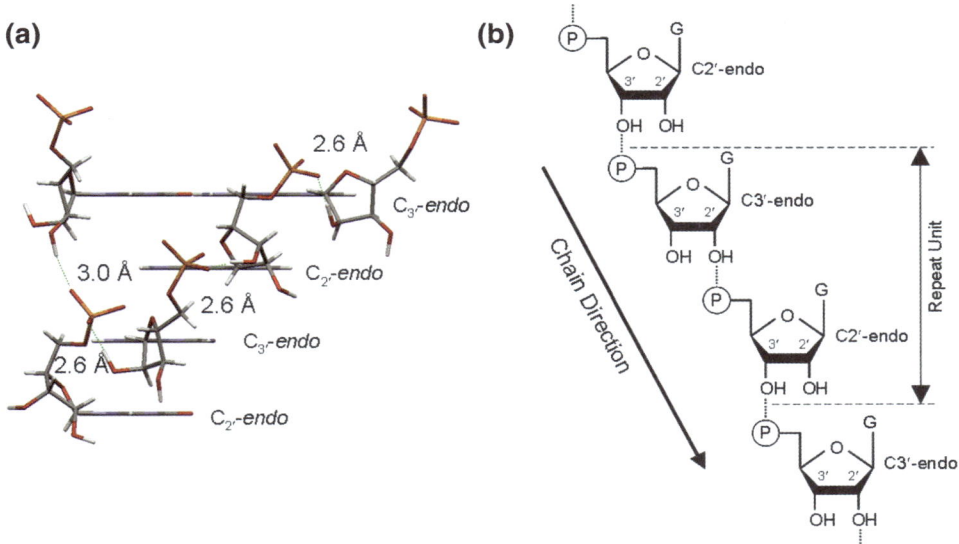

**Figure 7**  (a) A partial structure of the Na$_2$(5'-GMP) helix showing key hydrogen bonds. The O···O distances are given. (b) Scheme of the hydrogen bond linkage along the Na$_2$(5'-GMP) helix following the convention used for polynucleotides. Reproduced with permission from ref. 78.

## 5 CONCLUDING REMARKS

Over a span of 100 years, the story of guanylic acid has unfolded from the discovery of its chemical identity, to the observation of its gel formation, and finally to the elucidation of its remarkable helical structures. To date, there are only two fully characterized helical structures within the guanylic acid family: the continuous helix of 5'-GMP formed at pH 5 and the $Na_2$(5'-GMP) helix formed at pH 8 in aqueous solution. The two helices, although related, exhibit very different structural features, all simply due to the slightly different ionization states of their phosphate groups. Meanwhile helical structural details from other members of the guanylic acid family await to be uncovered. For example, what is the exact helical structure of $K_2$(5'-GMP) in neutral solution? Even though guanylic acid has been studied more than 100 years, it is perhaps safe to predict that this fascinating molecule will continue to surprise us.

### Acknowledgements

Our research was supported by the Natural Sciences and Engineering Research Council (NSERC) of Canada. I wish to thank my graduate students (Alan Wong, Ramsey Ida, Sanela Martic, Irene Kwan, Parisa Akhshi) for their contributions. I am also grateful to Dr. Lea Spindler (University of Maribor and J. Stefan Institute), Professor Jeffery T. Davis (University of Maryland) and Dr. Zhehong Gan (National High Magnetic Field Laboratory) for collaborations.

### References

1. M. Gellert, M.N. Lipsett and D.R. Davies, *Proc. Natl. Acad. Sci. U.S.A.*, 1962, **48**, 2013.
2. I. Bang, *Biochem. Z.*, 1910, **26**, 293.
3. V. Sasisekharan, S. Zimmerman and D.R. Davies, *J. Mol. Biol.*, 1975, **92**, 171.
4. W. Guschlbauer, J.-F. Chantot and D. Thiele, *J. Biomol. Struct. Dyn.*, 1990, **8**, 491.
5. J.T. Davis, *Angew. Chem. Intl. Ed.*, 2004, **43**, 668.
6. O. Hammarsten, *Z. Physiol. Chem.*, 1894, **19**, 19.
7. V. Schmidt, *Clin. Chem.*, 1986, **32**, 213.
8. I. Bang, *Z. Physiol. Chem.*, 1898, **26**, 133.
9. I. Bang, *Z. Physiol. Chem.*, 1901, **31**, 411.
10. I. Bang, *Z. Physiol. Chem.*, 1901, **32**, 201.
11. I. Bang and C.A. Raaschou, *Beitr. Z. Chem. Physiol. Pathol.*, 1903, **4**, 175.
12. I. Bang, *Beitr. Z. Chem. Physiol. Pathol.*, 1908, **11**, 76.
13. I. Bang, *Z. Physiol. Chem.*, 1910, **69**, 167.
14. P.A. Levene, *Z. Physiol. Chem.*, 1901, **32**, 541.
15. O. von Fürth and E. Jerusalem, *Beitr. Z. Chem. Physiol. Pathol.*, 1907, **10**, 174.
16. H. Steudel, *Z. Physiol. Chem.*, 1908, **53**, 539.
17. P.A. Levene and J.A. Mandel, *Biochem. Z.*, 1909, **10**, 215; 1909, **10**, 221.
18. W. Jones and L.G. Rowntree, *J. Biol. Chem.*, 1908, **4**, 289.
19. From *Nobel Lectures, Physiology or Medicine 1901-1921*, Elsevier, Amsterdam, 1967.
20. P.A. Levene and W.A. Jacobs, *Ber. Dtsch. Chem. Ges.*, 1909, **42**, 2469.

21  P.A. Levene and W.A. Jacobs, *J. Biol. Chem.*, 1912, **12**, 421.
22  D.D. Van Slyke, *Scand. J. Clin. Lab. Invest.*, 1957, **10** (Suppl. 31), 18.
23  D.R. Davies, *Annu. Rev. Biophys. Biochem. Struct.*, 2005, **34**, 1.
24  J. Iball, C.H. Morgan and H.R. Wilson, *Nature*, 1963, **199**, 688.
25  J.F. Chantot and W. Guschlbauer, *FEBS Lett.*, 1969, **4**, 173.
26  J.F. Chantot, *Arch. Biochem. Biophys.*, 1972, **153**, 347.
27  P. Tougard, J.F. Chantot and W. Guschlbauer, *Biochim. Biophys. Acta*, 1973, **308**, 9.
28  H.T. Miles and J. Frazier, *Biochem. Biophys. Res. Commun.*, 1972, **49**, 199.
29  T.J. Pinnavaia, H.T. Miles and E.D. Becker, *J. Am. Chem. Soc.*, 1975, **97**, 7198.
30  T.J. Pinnavaia, C.L. Marshall, C.M. Mettler, C.L. Fisk, H.T. Miles and E.D. Becker, *J. Am. Chem. Soc.*, 1978, **100**, 3625.
31  C.L. Fisk, E.D. Becker, H.T. Miles and T.J. Pinnavaia, *J. Am Chem. Soc.*, 1982, **104**, 3307.
32  E.D. Becker, *Encyclopedia of Nuclear Magnetic Resonance*, John Wiley and Sons, Ltd., 1996. Vol. 1, p. 207.
33  S.B. Zimmerman, *J. Mol. Biol.*, 1976, **106**, 663.
34  S.B. Oetersen, J.J. Led, E.R. Johnston and D.M. Grant, *J. Am. Chem. Soc.*, 1982, **104**, 5007.
35  E. Bouhoutsos-Brown, C.L. Marshall and T.J. Pinnavaia, *J. Am. Chem. Soc.*, 1982, **104**, 6576.
36  J.A. Walmsley and T.J. Pinnavaia, *Biophys. J.*, 1982, **38**, 315.
37  J.A. Walmsley, R.G. Barr, E. Bouhoutsos-Brown and T.J. Pinnavaia, *J. Phys. Chem.*, 1984, **88**, 2599.
38  J.J. Led and H. Gesmar, *J. Phys. Chem.*, 1985, **89**, 583.
39  R.G. Barr and T.J. Pinnavaia, *J. Phys. Chem.*, 1986, **90**, 328.
40  J.A. walmsley and B.L. Sagan, *Biopolymers*, 1986, **25**, 2149.
41  J.A. Walmsley and J.F. Burnett, *Biochemistry*, 1999, **38**, 14063.
42  F.S. Jiang and M.W. Makinen, *Inorg. Chem.*, 1995, **34**, 1736.
43  C. Detellier and P. Laszlo, *Helv. Chim. Acta*, 1979, **62**, 1559.
44  A. Delville, C. Detellier and P Laszlo, *J. Magn. Reson.*, 1979, **34**, 301.
45  M. Borzo, C. Detellier, P. Laszlo and A. Paris, *J. Am. Chem. Soc.*, 1980, **102**, 1124.
46  C. Detellier and P. Laszlo, *J. Am. Chem. Soc.*, 1980, **102**, 1135.
47  P. Mariani, C. Mazabard, A. Garbesi and G.P. Spada, *J. Am. Chem. Soc.*, 1989, **111**, 6369.
48  S. Bonazzi, M. Capobianco, M.M. De Morais, A. Garbesi, G. Gottarelli, P. Mariani, M.G. Ponzi, Bossi, G.P. Spada and L. Tondelli, *J. Am. Chem. Soc.*, 1991, **113**, 5809.
49  L.Q. Amaral, R. Itri, P. Mariani and R. Micheletto, *Liq. Cryst.*, 1992, **12**, 913.
50  H. Franz, F. Ciuchi, G. Di Nicola, M.M. De Morais and P. Miriani, *Phys. Rev. E*, 1994, **50**, 395.
51  G. Gottarelli, G. Proni and G.P. Spada, *Liq. Cryst.*, 1997, **22**, 563.
52  P. Mariani and L. Saturni, *Biophys. J.*, 1996, **70**, 2867.
53  P. Mariani, F. Ciuchi and L. Saturni, *Biophys. J.*, 1998, **74**, 430.
54  L. Spindler, I. Drevenšek-Olenik, M. Čopič, R. Romih, J. Cerar, J. Škerjanc and P. Mariani, *Eur. Phys. J. E*, 2002, **7**, 95.
55  L. Spindler, I. Drevenšek-Olenik, M. Čopič and P. Mariani, *Liq. Cryst.*, 2003, **395**, 317.
56  L. Spindler, I. Drevenšek-Olenik, M. Čopič, J. Cerar, J. Škerjanc and P. Mariani, *Eur. Phys. J. E*, 2004, **13**, 27.
57  K. Kunstelj, L. Spindler, F. Federiconi, M. Bonn, I. Drevenšek-Olenik and M. Čopič, *Chem. Phys. Lett.*, 2008, **467**, 159.

58   Y. Yu, D. Nakamura, K. DeBoyace, A.W. Neisius and L.B. McGown, *J. Phys. Chem. B*, 2008, **112**, 1130.
59   P. Mariani, F. Spinozzi, F. Federiconi, M.G. Ortore, H. Amenitsch, L. Spindler and I. Drevenšek-Olenik, *J. Phys. Chem. B*, 2009, **113**, 7934.
60   J.B. Hightower, D.R. Olmos and J.A. Walmsley, *J. Phys. Chem. B*, 2009, **113**, 12214.
61   K. Loo, N. Degtyareva, J. Park, B. Sengupta, M. Reddish, C.C. Rogers, A. Bryant and J.T. Petty, *J. Phys. Chem. B*, 2010, **114**, 4320.
62   M. Panda and J.A. Walmsley, *J. Phys. Chem. B*, 2011, **115**, 6377.
63   D. Hu, J. Ren and X. Qu, *Chem. Sic.*, 2011, **2**, 1356.
64   G. Laughlan, A.I.H. Murchie, D.G. Norman, M.H. Moore, P.C.E. Moody, D.M.J. Lilley and B. Luisi, *Science*, 1994, **265**, 520.
65   D. Rovnyak, M. Baldus, G. Wu, N.V. Hud, J. Feigon and R.G. Griffin, *J. Am. Chem. Soc.*, 2000, **122**, 11423.
66   G. Wu and A. Wong, *Chem. Commun.*, 2001, 2658.
67   A. Wong, J.C. Fettinger, S.L. Forman, J.T. Davis and G. Wu, *J. Am. Chem. Soc.*, 2002, **124**, 742.
68   A. Wong and G. Wu, *J. Am. Chem. Soc.*, 2003, **125**, 13895.
69   G. Wu and A. Wong, *Biochem. Biophys. Res. Commun.*, 2004, **323**, 1139.
70   A. Wong, R. Ida and G. Wu, *Biochem. Biophys. Res. Commun.*, 2005, **337**, 363.
71   R. Ida and G. Wu, *J. Am. Chem. Soc.*, 2008, **130**, 3590.
72   G. Wu, A. Wong, Z. Gan and J.T. Davis, *J. Am. Chem. Soc.*, 2003, **125**, 7182.
73   R. Ida and G. Wu, *Chem. Commun.*, 2005, 4294.
74   R. Ida, I.C.M. Kwan and G. Wu, *Chem. Commun.*, 2007, 795.
75   I.C.M. Kwan, A. Wong, Y.M. She, M.E. Smith and G. Wu, *Chem. Commun.*, 2008, 682.
76   G. Wu, Z. Gan, I.C.M. Kwan, J.C. Fettinger and J.T. Davis, *J. Am. Chem. Soc.*, 2011, **133**, 19570.
77   A. Wong, R. Ida, L. Spindler and G. Wu, *J. Am. Chem. Soc.*, 2005, **127**, 6990.
78   G. Wu and I.C.M. Kwan, *J. Am. Chem. Soc.*, 2009, **131**, 3180.

# FUNCTIONAL ASSEMBLIES MADE FROM SUPRAMOLECULAR G-QUADRUPLEXES

J.M. Rivera

Department of Chemistry, College of Natural Sciences, University of Puerto Rico, Río Piedras Campus, San Juan, PR, 00931

## 1 INTRODUCTION

Guanosine derivatives can self-assemble into ribbon-like supramolecular polymers or into supramolecular G-quadruplexes (SGQs), which are stacks of G-tetrads or quartets stabilized by cations such as potassium among others (Figure 1.0.1). The subject of G-quadruplex has been reviewed in various reviews and monographs in the last decade.[1-4] The last major review on the subject of SGQs (as defined below) was that of Spada and co-workers.[1] We therefore aim to review selected articles on the particular subject for the years 2009-2012 (June). The main focus of this review will be on discrete systems (precise assemblies of well-defined size and shape), other assemblies of guanosine such as ribbon-like supramolecular polymers (Figure 1.1b,c) will not be discussed. Systems based on intramolecularly constrained guanosine self-assembly (e.g., Template-Assembled Synthetic G-Quartets and similar structures), and those based on G-rich oligonucleotides (OGQs) are also out of the scope of this manuscript. Emphasis will be given to the development of functional assemblies based on SGQs as divided in two main topics: (1) inherent functionality of SGQs (e.g., ionophores); (2) systems where the G-quadruplex serves as a scaffold for the functional elements (e.g., dendrimers, polymers). Before the conclusion a brief mention is made of systems were the dynamics of the SGQs are controlled by external stimuli (e.g., switches).

### 1.1 Guanosine as a versatile scaffold

*Guanine is a privileged small-molecule recognition motif.* Almost a century ago Bang[5] reported the gelating properties of concentrated solutions of 5'-GMP. Five decades later, Davies[6] proposed that the stacking of planar tetramers of guanines (G-tetrads, Figure 1.1) was responsible for the formation of columnar aggregates that formed Bang's gels (stacks of G-tetrads are collectively known as G-quadruplexes).[4, 7-11] Guanine self-assembles into quadruplexes because of the high density of molecular information encoded in its many hydrogen-bonding sites and its relatively large π-surface area. A comparison of the guanine base with the other naturally occurring nucleobases reveals that not only does guanine posses the greater total number of hydrogen-bonding donating and accepting sites (seven vs. five), but they are also relatively evenly divided with three donors and four acceptors (Figure 1.1).

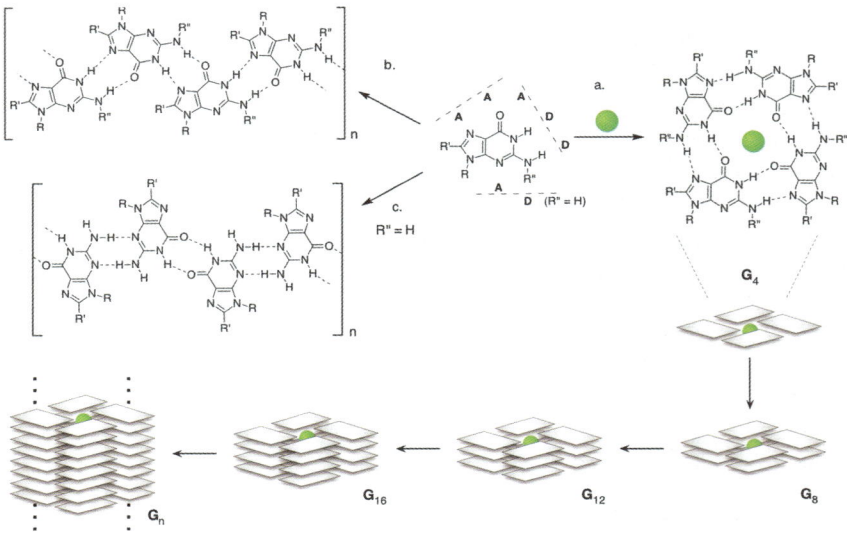

**Figure 1.1**   *The guanine base has up to four H-bond acceptor sites (A) and three HB donor (D) sites. This enables its self-assembly into (a) closed cyclic tetramers (G-tetrads or quartets) or into two (b, c) alternative linear supramolecular polymers (ribbons). The formation of either type of assembly can be biased by the nature of the substituents attached to the guanine moiety (R, R', R'') and other extrinsic parameters such as solvent, temperature or the addition of cations. The formation of G-tetrads, in particular, is promoted by the addition of a wide variety of cations of an appropriate size. Both types of supramolecular structures G-tetrads and ribbons can further assemble into higher order supramolecules, with the former leading to the formation of G-quadruplex structures.*

The structural characteristics of guanine (and related derivatives) enables it to self-assemble into G-quadruplexes as individual subunits (e.g., as the free base, or as part of nucleosides, or nucleotides) into supramolecular G-quadruplexes (SGQs) (Figure 1.2a) or from G-rich oligomeric systems into oligomeric G-quadruplexes (OGQs). Historically, OGQs have been developed mostly from G-rich oligonucleotides (Figure 1.2b) and other variations of the ribophosphate backbone such as PNAs, were G-subunits are arranged in a linear fashion. A more recent variant is to covalently constrain G-subunits in cyclic oligomer by using platforms such as calixarenes, porphyrins and other macrocycles (e.g., TASQs).[12-19] Although functional nanostructures have been developed using systems based on both of these two broad categories, this article will focus exclusively on the recent reports on the developments of SGQs for the construction of functional assemblies. or oligonucleotides (OGQs).

Guanine is a versatile motif from which to develop organized structures. As explained in detail in other parts of this book, this is due to the array and relative orientation of H-bond donors and acceptors in the guanine moiety. The relative arrangement (Figure 1.1) of 90° enables the guanine subunits to form both linear and cyclic arrangements. Whereas the latter lead to supramolecular polymers with two different types of ribbon arrangements (Figure 1.1), the latter is constrained to discrete quartets or tetrads. These quartets are promoted by cations that engage in dipole-ion interactions with the O6 in a process that greatly stabilizes the structure, and promotes the subsequent coaxial stacking of multiple G4 units in a hierarchical process.

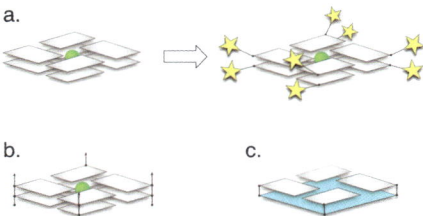

**Figure 1.2**  *Modes of classification of G-Quadruplexes: (a) supramolecular (SGQs) (b, c) oligomeric (OGQs). The latter are constrained by covalent bonds into (b) linear G-oligomers and (c) cyclic G-oligomers.*

In the late 1980s Gottarelli and Spada pioneered studies on the liquid crystalline (LC) phases formed by guanosine nucleotides like guanosine monophosphates (e.g. 5'GMP), dinucleotides (d(GpG)) and G- rich oligonucleotides.[20-23] Later on, concurrently, the groups of Gottarelli/Spada and Davis demonstrated that lipophilic G-derivatives could also self-assemble in organic media.[1, 3, 4, 24-27]

Gottarelli and Spada have focused their attention on the use of lipophilic G-derivatives for the development functional supramolecules such as liquid crystals and other applications in nanotechnology.[2, 20, 28-31] On the other hand, Davis and coworkers have focused their attention on the development of self-assembled ionophores based on lipophilic guanosine and iso-guanosine derivatives.[32-35] For example, they have shown that radioactive cations (e.g. $^{226}Ra^{2+}$) could be extracted with high efficiency, making them attractive for environmental applications.[36] In addition, they have also studied the dynamics of cation exchange and have demonstrated how the lipophilic quadruplexes have cation binding sites with different affinities.[37]

Gottarelli, and Spada[38] have shown that their lipophilic 2'-deoxyguanosine derivatives have a preference to form octamers at low cation concentration but will form polymeric columnar aggregates with an excess salt. In contrast, Davis uses guanosine derivatives that, due to the increased steric crowding imparted by the additional 2'-substituent, will only form octamers (even with excess salt).[39] If, however, the counter anion is a 2,6-dinitrophenolate such as picrate two octamers dimerize to form a hexadecamer.[40] The dynamics of Davis' lipophilic hexadecamers are intimately related to the nature of the anion as demonstrated by the X-ray structure in which four picrates form an "anionic belt" associated via H-bonds with the exocyclic N2H moieties (R" in Figure 1.1). They have taken advantage of this property to modulate its supramolecular properties.[4, 34, 41]

*Enhancing the properties of quadruplexes via chemical modifications.*[42-44] The way in which G-subunits are arranged in GQs leave three positions open to derivatization (N9, C8, N2) without disturbing the recognition elements. Substitutions at the guanine base occur most commonly at N9, which is usually connected to a sugar (ribose or deoxyribose), which can in turn be further derivatized at the 2', 3' or 5' or the sugar can be replaced by a non-natural group.[45] Alternatively, the guanine base can be connected to other groups (e.g., amino acids in PNAs) or other backbones to obtain linear oligomers.[46] In 1998, He reported that hydrophobic substituents located at the N2 and/or C8 of the *syn*-dG residues, stabilized the structure and increased the activity of the QDNA-containing thrombin-binding aptamer (TBA is a 15-mer oligonucleotides that forms a two- tetrad QDNA with a chair topology).[47] Shortly after, Sessler and coworkers described how a lipophilic G-derivative with an 8-(*p*-dimethylaminophenyl) group could form a G-tetrad in the solid state and in solution, even in the absence of a templating metal cation.[48] More recently, Davis and coworkers described the synthesis of a lipophilic guanosine analogue with alkene groups attached to the nucleobase's N2 and C8 positions. Such G-analogue formed a stable octamer in CDCl$_3$ but attempts to covalently capture an individual G-tetrad using olefin cross-metathesis were unsuccessful.[49] In a somewhat related report, Wu and coworkers reported the G-quadruplex formation from N2-modified lipophilic guanosine analogues that self-assembled into $D_4$-symmetric octameric SGQs in the presence of either K$^+$ or Na$^+$.[50-52] The Rivera group has embarked in the systematic study on how intrinsic parameters (particularly C8 and N9 substitutions) such as structural changes in the guanine moiety affect the formation of SGQs of various molecularities (octamers, dodecamers, hexadecamers) in both organic and aqueous media.[53-56]

The following section provides a discussion of selected examples showing how such modifications can be used to fine tune the inherent properties of the SGQs (Section 2.1) and how can the SGQs can be used as scaffolds to organize a wide variety of functional elements (Section 2.2).

## 2 FUNCTIONAL STRUCTURES BASED ON SGQS

**2.1 Inherent functionality**

In this section we discuss the "built in" functionality of SGQs, whereas in the next one we focus on the use of SGQs as scaffolds for organizing multiple functional elements.

*2.1.1 Ionophores (transporters, channels, etc.)*

The affinity of SGQs for a wide variety of cations have been discussed extensively,[57-59] and has motivated many studies aimed at evaluating their properties as ionophores.[1-4] Following we will discuss three articles that report the use of SGQs as carriers[60] or ion channels[61] to transport cations across lipophilic membranes as well as a host systems to perform $^{23}$Na or $^{39}$K NMR studies.[62, 63] Wu et al. reported for the first time the detection of different K$^+$ ions by high-resolution $^{39}$K NMR residing inside an SGQ channel using the multiple-quantum magic-angle-spinning method.[63] Specifically, they relied on the use of Davis' hexadecamer formed by a lipophilic guanosine derivative (5'-*t*-butyldimethylsilyl-2',3'-O-isopropylidene guanosine).[4] These studies followed up on a report from the previous year of using a combination of solid-state $^{23}$Na NMR and quantum chemical calculations to determine the detailed coordination geometry of a Na$^+$ ion complexed to a dimer of calix[4]arene **1.1** held together by the formation of a single G-tetrad.[62, 64] They determined that the Na$^+$ ion was not co-planar with the G-tetrad, but instead, was slightly shifted above the plane to in order to simultaneously coordinate (in a square-pyramidal geometry) to a water molecule co-encapsulated in the calix[4]arene cavity (Figure 2.1.1).

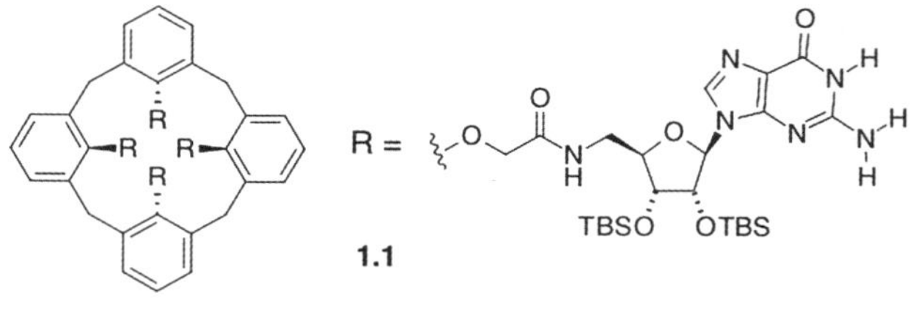

**Figure 2.1.1** *Structural depiction of Davis' calix[4]arene-guanosine conjugate.*[63]

Another aspect of SGQs is their potential as cation transporters (either as carriers and/or channels) as pioneered by the groups of Gotarelli, Spada and Davis.[1-4] Martín-Hidalgo et al. reported the use of SGQs formed from 8-phenyl-2'-deoxyguanosine derivatives (8PhGs) to transport metal cations across a bulk lipophilic liquid membrane.[60] Such results built on a previous report where they had demonstrated that 8PhGs (**1.2b,c**; Figure 2.1.2) were effective extracting metal cations from an aqueous phase into an organic phase by forming self-assembled ionophores (SAIs).[65] In most cases, when compared to the parent *d*G derivative **1.2a** and covalent ionophores (e.g., 18-crown-6 ether derivative and a [2•2•2] cryptand) such SAIs demonstrated transport efficiencies (after 24 hrs) at least equal, but most often superior, for cations such as $K^+$, $Na^+$ and $Sr^{2+}$. Similar experiments with $Li^+$ showed all the tested ionphores to be equally poor transporters, interestingly, however, the parent *d*G derivative **1.2a** was the least poor of all. The authors attributed such differences to a combination of factors such as the stability and the different molecularities (e.g., octamers, hexadecamers) of the various SGQs.

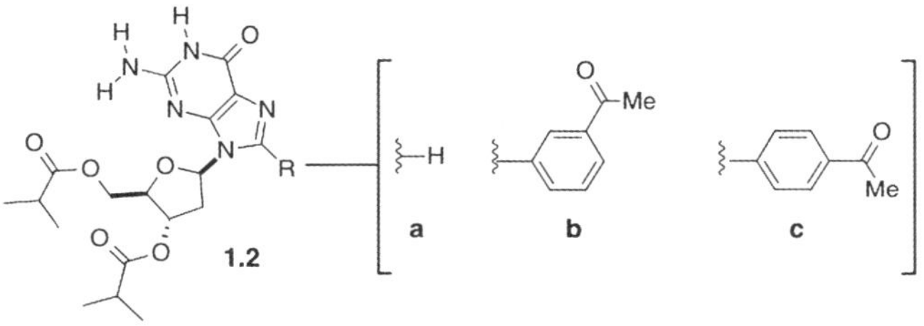

**Figure 2.1.2** *G-derivatives used to make self-assembled ionophores.*[60]

An additional example of a SAI, this time acting as an ion channel incorporated into lipid bilayers was examined by Ma et al.[61] Following up on their earlier discovery that SGQs formed by attaching guanosines to a membrane-spanning lithocholate dimer.[66] They showed that replacing the linkers' original *bis*-carbamate (**1.3a**) with bis-urea moieties (**1.3b**) could stabilize the resulting self-assembled ion channels by enhancing the number of H-bonds holding the supramolecule together (Figure 3.1.3b). Furthermore, although they had anticipated the G-quadruplex core to serve as the channel for ion transport, the large conductances measured (in the 1-5 nS range) were consistent with channels with much larger diameters (>1 nm).

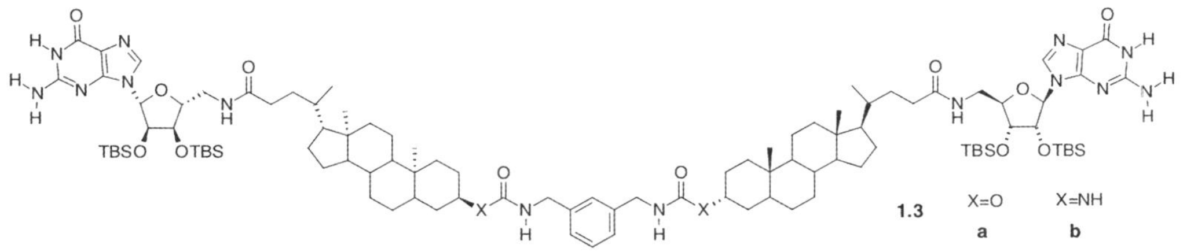

**Figure 2.1.3** *Structural depiction of Davis' bis-guanosine-sterol carbamate (**1.3a**) and urea (**1.3a**) derivatives.*[61]

## 2.2 Scaffolded functionality

In this section we discuss the "built in" functionality of SGQs, whereas in the next one we focus on the use of SGQs as scaffolds for organizing multiple functional elements.

### 2.2.1 Self-Assembled dendrimers

The Rivera group reported the construction of two hydrophilic self-assembled dendrimers (SADs) formed relying on the 8-(m-acetylphenyl)-2'-deoxyguanosine (mAG) scaffold.[67-69] These developments built on their development of analogous SADs made of sixteen lipophilic mAG derivatives ((**2.1b**)$_{16}$, (**2.1d**)$_{16}$.[70] Some of the attractive characteristics of these SADs are their accessible synthesis (when compared with covalent dendrimers of a similar size), the fact that they are chiral and optically pure, their well-defined (precise) structure, their thermodynamic stability and their functional core. Their discovery that hydrophilic mAG derivative self-assembles isostructurally into hexadecameric supramolecular G-quadruplexes in aqueous media[55] prompted them to developed analogous hydrophilic SADs. (This led to the discovery of a thermosensitive SAD that showed the lower critical solution temperature (LCST) phenomenon.)

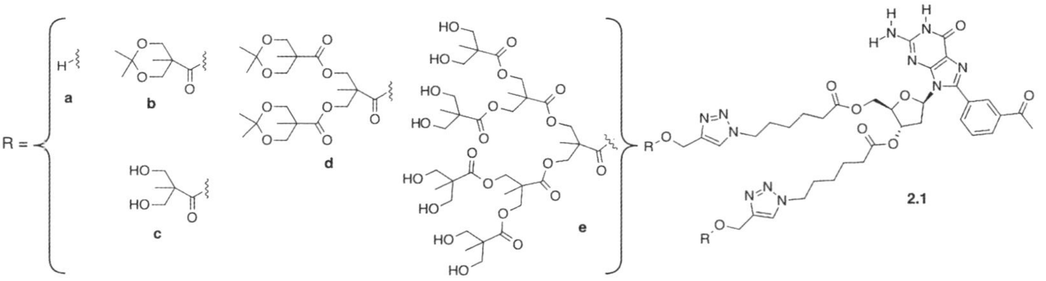

**Figure 2.2.1** *G-derivatives used to make self-assembled dendrimers.*[67-69]

The largest member of the family **2.1e** leads to a fourth-generation hydrophilic self-assembled hexadecameric dendrimer with a size (hydrodynamic diameter of 5.0 nm as measured by DLS) and shape similar to those of globular proteins (with 256 hydroxyl end groups). By comparison, the smallest, first generation congener has a hydrodynamic diameter of 4.3 nm, hinting that (**2.1e**)$_{16}$ have reached a so-called fractal geometry leading to an increase in surface irregularity, and the formation of void spaces suitable for encapsulating smaller guest molecules. The thermal stability of these systems (**2.1a**)$_{16}$, (**2.1c**)$_{16}$, (**2.1e**)$_{16}$ decreases slightly with an increase in generation number with $T_m$ values of 77 °C, 72 °C, and 68 °C, respectively. Since these SADs are held together by a variety non-covalent interactions it is possible to fine-tune their structure and dynamics via a variety of external stimuli. These systems have potential as molecular containers and one example of their use for the encapsulation of the anticancer drug doxorubicin will be discussed in Section 3.[68]

### 2.2.2 Self-assembled star polymers

The self-assembly of guanosine derivatives can lead to supramolecular polymers where the basic subunits are individual G subunits or G-tetrads. Alternatively, polymer chains can be attached to guanosine subunits that produce supramolecular star polymers upon assembly.[71] reported the preparation of supramolecular star polymers held together by octameric SGQs made of **2.2a** using atom-transfer radical polymerization. They prepared the supramolecular star methyl methacrylate polymers (PMMA) via atom-transfer radical polymerization (ATRP) and tested three different synthetic routes: (1) a so-called "core-first" approach where the polymerization is performed on the surface of the resulting octameric SGQ (**2.2a**)$_8$; (2) a the convergent "arm-first" approach where polymerization starts from individual subunits that subsequently assembled; and (3) a "one pot" procedure where both the self-assembly and the polymerization were supposed to occur simultaneously. While the first two methods resulted in supramolecular star polymers of similar molecular weights and distributions (4-7 KDa as determined by GPC; PDI = 1.3) the third method produced polymers with molecular weights about three times larger and of broader distributions.

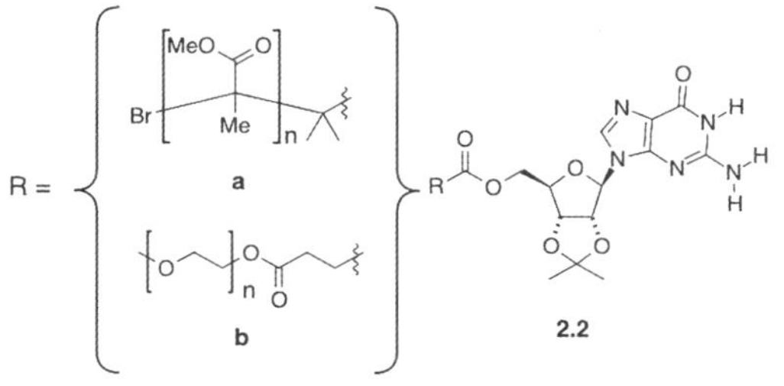

**Figure 2.2.2** *G-derivatives used to make supramolecular star polymers.*[71, 72]

The main focus of the study by Likhitsup et al.[71] was to determine the scope and limitations of the synthetic strategy, however, the issue of how does the nature of the polymer chain affect the supramolecular properties of the resulting assemblies were not addressed. Filling out this gap, however, Gadwal et. al.[72] recently reported the synthesis and study of guanosine derivatives with polyethylene glycol (PEG) chains attached to the 5'-oxygen of the ribose moiety (**2.2b**). Using a combination of techniques (NMR, UV/Vis, CD) they observed a direct correlation between the polymer chain length and the thermal stability of the resulting octameric SGQs (**2.2b**)$_8$, where the thermodynamic stability of the assemblies decreased concomitantly with as the polymer chain-length increased due to the increased steric repulsions between the attached polymer chains. They also found that the PEG-based assemblies were more susceptible to higher temperatures, dilution, and the presence of polar solvents than their PMMA counterparts. They rationalized this higher sensitivity to the potential competition for hydrogen-bonds by the glycol units, although it is possible that the PEG chains are also competing with the SGQs for cation complexation.

### 2.2.3 Exotic functionalities

*Organogold moieties.* Hirao and coworkers published a trilogy of papers describing the self-assembly of organogold guanosine derivatives containing a variety of alkynyl-Au(I/III)-Ligand appendages.[73-75] They relied on functionalization of a guanosine derivatives at C8 with a 4-alkynylphenyl moiety used to linked a gold(I/III)-(capping ligand) complex (Figure 2.3.3). The size and nature of the capping ligand increased from the first to the last two reports, starting with phenyl isonitrile, followed by 2,6-diphenylpyridyl and triphenyl phosphine (Figure 2.3.3). The potassium cation (KPF$_6$) promoted formation of octamer (**2.3a**)$_8$ in chloroform leads to an enhancement of the luminescence due Au(I)–Au(I) interactions.

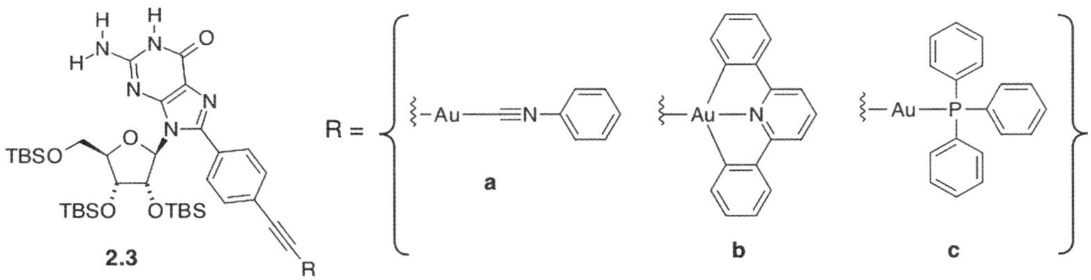

**Figure 2.2.3** *G-derivatives used to make organogold containing octameric SGQs.*[73-75]

In a follow up study,[74] they reported the synthesis and self assembly studies of the cyclometalated gold(III) derivative **2.3b**. The rigidity and enhanced steric bulk of **2.3b** might be responsible for the apparent lower stability of the resulting assemblies. While the formation of an octameric SGQ in $CD_2Cl_2$ seems viable at 25 °C, in THF-$d_8$ it appears to be the main species only at low temperatures (<-30 °C). Although the authors claim this species to be a columnar aggregate, the symmetry apparent from the signals as well as the chemical shift of the N1H seems more consistent with the formation of a putative $D_4$–symmetric octamer. At room temperature, it appears that the system is a single G-tetrad with a loosely bound cation. It seems that the capping ligand imposes too much steric hindrance to favor the formation of an octameric SGQ as stable as $(2.3a)_8$ as evident also by the lower enhancement in the luminescence of **2.3b** upon addition of $KPF_6$, when compared to **2.3a**.

Derivative **2.3c** was evaluated in a more recent report[73] where, not surprisingly and similar to **2.3b**, it shows difficulties to self-assemble at room temperature even after the addition of $KPF_6$. Although, lowering the temperature once again seemed to promote the formation of an octameric SGQ, the authors relied on changing the promoting cation from $K^+$ to $La^{3+}$ as it has been suggested to promote the formation of more stable SGQs.[76,77] The strategy seemed effective as evidenced by a variety of methods such as 1H NMR, UV-vis, CD, and fluorescence spectroscopies. Upon addition of $La(OTf)_3$ the emission intensity enhancement, ascribed to the putative Au(I)–Au(I) interactions, appears to be larger than the analogous measurements of **2.3b**, but still smaller than those of **2.3a**.

*Paramagnetic moieties.* In their ongoing efforts to develop (supra)molecular-scale magnetic devices Spada and coworkers[78] reported the first example of a guanosine derivative containing a persistent radical by installing a 4-carbonyl-2,2,6,6-tetramethylpiperidine-1-oxyl (TEMPO) moiety at the 5'-postion (**2.4**). In their first communication they demonstrated the viability of using a $D_4$-symmetric octameric SGQ to organize eight TEMPO subunits that showed through space spin-spin interactions as evidenced by using EPR measurements. Furthermore, they showed that these interactions could be switched ON and OFF, by the reversible assembly/disassembly of $(2.4)_8$, induced by the sequestration of the $K^+$ template using the potent [2.2.2] cryptand ionophore.

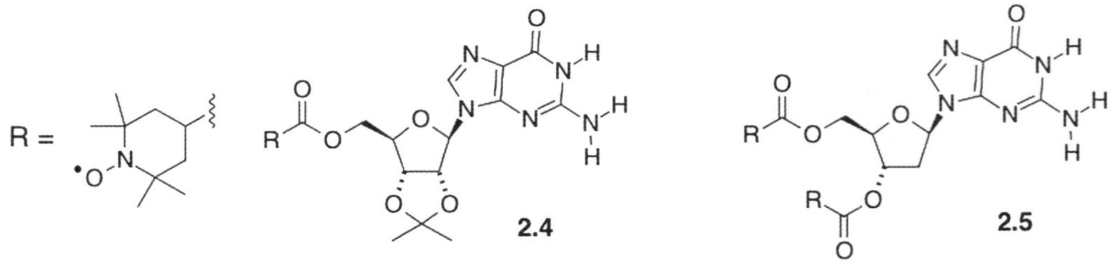

**Figure 2.2.4** *G-derivatives used to make paramagnetic octameric SGQs.*[78,79]

In order to improve the strength of the spin-spin interaction, they developed the analogous 5'-dG derivative **2.5**, which carries double (sixteen) the number of TEMPO moieties.[79] The resulting octameric SGQ $(2.5)_8$ showed similarities to the one formed by its guanosine (**2.4**) congener, yet, the symmetry of the assembly shifted to $C_4$-symmetric, as it has been reported previously for related lipophilic dG derivatives.[39] It also showed more drastic magnetic changes when induced to disassembly via a similar sequestration of the $K^+$ by the [2.2.2] cryptand ionophore. Molecular modeling studies provided insight into some possible relative configurations of the TEMPO moieties. The authors noticed that the sixteen TEMPO groups are divided into two main groupings, four tri-radical modules diverging from the sides of the SGQ and one tetra-radical module located above one of the tetrads.

*Pi-conjugated systems.* SGQs can also be used to provide greater organization to molecules that are normally used as the components in (supra)molecularly precise materials for the development of OLEDs, photovoltaic solar cells, luminescent probes and other advanced materials.[80] Along these lines, González-Rodríguez et al.[81] reported the synthesis and self-assembly studies of guanosine (**2.6a,b**) and guanine (**2.6c**) derivatives with oligo(p-phenylene-vinylene) (OPV) moieties attached at the C8 (Figure 2.2.5). Their selection of OPVs was based on their wide use as components in organic semiconducting and luminescent materials.[82] Similar to most lipophilic guanosine derivatives, in the absence of a cation promoter (or template), these compounds self-assemble (both in solution and on solid surfaces) into a mixture of supramolecules such as G-ribbons and empty G-tetrads. The latter are favored as the steric hindrance around the guanine moiety increases. The addition of $Na^+$ or $K^+$ promotes the formation of disk-shaped $D_4$-symmetric octameric SGQs. Numerous attempts to induce the formation of SGQs with larger molecularities (e.g., by changing the solvent, cations, anions) failed to produce the desired results. On the other hand, the high thermal and dilution stability of these SGQs allowed the researchers to transfer and image the intact assemblies over solid substrates with little dissociation or further aggregation. Self-assembly of these G-derivatives also resulted in up to three-fold enhancement of the fluorescence emission, underscoring the potential of similar derivatives in optoelectronics applications and the development of fluorescent probes.

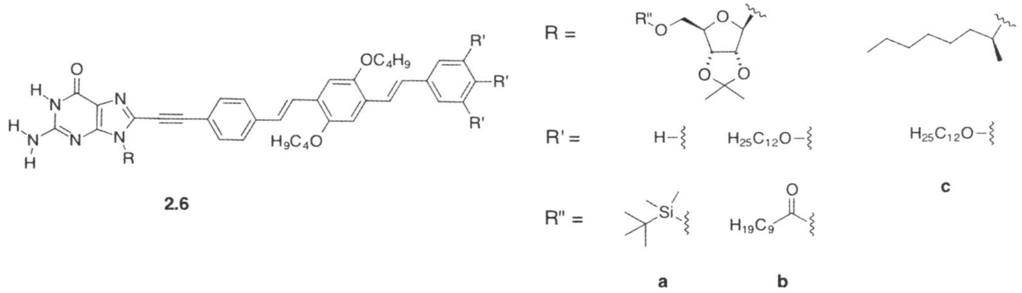

**Figure 2.2.5** *G-derivatives used to make pi-conjugated octameric SGQs.*[81]

*2.2.4 Controlling the dynamics of SGQs*

Because of space constraints, it is not feasible to provide a detailed discussion of articles describing the control over the dynamics of SGQs, using a variety of external stimuli. Before concluding this review, however, a brief mention of such studies should provide a good starting point for those readers interested in the subject. Martín-Hidalgo et al. recently reported the reversible interconversion between hexadecameric and octameric SGQs.[83] The process occurred with high fidelity and was triggered by alternating the salt used to promote the assembly from $K_2SO_4$ to $SrI_2$, respectively. The Rivera group reported the discovery of thermosensitive amphiphilic SGQs that exhibited the phenomenon of Lower Critical Solution Temperature (LCST).[69] This phenomenon allows the reversible thermally induced self-assembly of SGQs into microglobules capable of encapsulating small molecules such as the anticancer drug doxorubicin.[68] The interplay between the solvent, the SGQs and the salts that promote their formation has been reported as a means to trigger changes in the molecularity of the resulting assemblies.[54, 84, 85] Last but not least, was the report by Lena et al.[86] of the reversible photo-control over the formation of $D_4$-symmetric octameric SGQ made from a lipophilic 8-phenylvinyl-2'-dG derivative.

## 3 CONCLUSION

It is evident that SGQs have become extremely useful in the development of functional supramolecular assemblies. Although there a good variety of functional elements have been grafted to the guanosine moiety, many more functional elements remained unexplored. Some of the challenges in this area are still the poor solubility of some guanosine derivatives; although there are some examples there is still the need to develop reliable ways of constructing SGQs of molecularities larger than octamers with high fidelity; and finally there is also the need to integrate these assemblies within larger systems (e.g., biological). All the examples discussed are merely a scratch on the surface in this developing field of functional SGQs where the added option of "real-time" control over the dynamics will certainly open the door to new and exciting research opportunities and applications.

**References**

1. S. Lena, S. Masiero, S. Pieraccini and G. P. Spada, *Chem. Eur. J.*, 2009, **15**, 7792-7806.
2. S. Lena, S. Masiero, S. Pieraccini and G. P. Spada, *Mini-Reviews in Organic Chemistry*, 2008, **5**, 262-273.
3. J. T. Davis and G. P. Spada, *Chem. Soc. Rev.*, 2007, **36**, 296-313.
4. J. T. Davis, *Angew. Chem. Int. Ed.*, 2004, **43**, 668-698.
5. I. Bang, *Biochem. Z.*, 1910, **26**, 293.
6. M. Gellert, M. N. Lipsett and D. R. Davies, *Proc. Natl. Acad. Sci. U.S.A.*, 1962, **48**, 2013-2018.
7. G. Wu and I. Kwan, *J. Am. Chem. Soc.*, 2009, **131**, 3180-3182.
8. A. Wong, R. Ida and G. Wu, *Biochem. Biophys. Res. Commun.*, 2005, **337**, 363-366.
9. C. L. Fisk, E. D. Becker, H. T. Miles and T. J. Pinnavaia, *J. Am. Chem. Soc.*, 1982, **104**, 3307 - 3314.
10. E. Bouhoutsos-Brown, C. L. Marshall and T. J. Pinnavaia, *J. Am. Chem. Soc.*, 1982, **104**, 6576-6584.
11. C. F. Chantot and W. Guschlbauer, The Purines; Theory and Experiment, Jerusalem, 1972.
12. M. Nikan, G. A. L. Bare and J. C. Sherman, *Tetrahedron Lett.*, 2011, **52**, 1791-1793.
13. M. Nikan and J. C. Sherman, *J. Org. Chem.*, 2009, **74**, 5211-5218.
14. M. Nikan and J. C. Sherman, *Angew. Chem. Int. Ed.*, 2008, **47**, 4900-4902.
15. H.-J. Xu, L. Stefan, R. Haudecoeur, S. Vuong, P. Richard, F. Denat, J.-M. Barbe, C. P. Gros and D. Monchaud, *Org. Biomol. Chem.*, 2012.
16. L. Stefan, H.-J. Xu, C. P. Gros, F. Denat and D. Monchaud, *Chem. Eur. J.*, 2011, **17**, 10857-10862.
17. L. Stefan, A. Guédin, S. Amrane, N. Smith, F. Denat, J.-L. Mergny and D. Monchaud, *Chem. Commun.*, 2011, **47**, 4992-4994.
18. L. Stefan, F. Denat and D. Monchaud, *J. Am. Chem. Soc.*, 2011, **133**, 20405-20415.
19. P. Murat, R. Bonnet, A. Van der Heyden, N. Spinelli, P. Labbé, D. Monchaud, M.-P. Teulade Fichou, P. Dumy and E. Defrancq, *Chem. Eur. J.*, 2010, **16**, 6106-6114.

20. G. P. Spada and G. Gottarelli, *Synlett*, 2004, 596–602.
21. T. Giorgi, S. Lena, P. Mariani, M. A. Cremonini, S. Masiero, S. Pieraccini, J. P. Rabe, P. Samori, G. P. Spada and G. Gottarelli, *J. Am. Chem. Soc.*, 2003, **125**, 14741-14749.
22. S. Bonazzi, M. Capobianco, M. M. De Morais, A. Garbesi, G. Gottarelli, P. Mariani, M. G. Ponzi Bossi, G. P. Spada and L. Tondelli, *J. Am. Chem. Soc.*, 1991, **113**, 5809-5816.
23. P. Mariani, C. Mazabard, A. Garbesi and G. P. Spada, *J. Am. Chem. Soc.*, 1989, **111**, 6369-6373.
24. G. Gottarelli, S. Masiero, E. Mezzina, S. Pieraccini, J. P. Rabe, P. Samori and G. P. Spada, *Chem. Eur. J.*, 2000, **6**, 3242-3248.
25. S. Tirumala and J. T. Davis, *J. Am. Chem. Soc.*, 1997, **119**, 2769-2776.
26. G. Gottarelli, S. Masiero and G. P. Spada, *J. Chem. Soc., Chem. Comm.*, 1995, 2555-2557.
27. J. T. Davis, S. Tirumala, J. R. Jenssen, E. Radler and D. Fabris, *J. Org. Chem.*, 1995, **60**, 4167-4176.
28. G. P. Spada, S. Lena, S. Masiero, S. Pieraccini, M. Surin and P. Samori, *Adv. Mater.*, 2008, **20**, 2433-2438.
29. S. Pieraccini, S. Masiero, O. Pandoli, P. Samorì and G. P. Spada, *Org. Lett.*, 2006, **8**, 3125-3128.
30. S. Pieraccini, T. Giorgi, G. Gottarelli, S. Masiero and G. P. Spada, *Mol. Cryst. Liq. Cryst.*, 2003, **398**, 57-73.
31. G. Gottarelli, S. Masiero, E. Mezzina, G. P. Spada, P. Mariani and M. Recanatini, *Helv. Chim. Acta*, 1998, **81**, 2078-2092.
32. F. W. v. Leeuwen, J. T. Davis, W. Verboom and D. N. Reinhoudt, *Inorg. Chim. Acta*, 2006, **359**, 1779-1785.
33. A. Wong, J. C. Fettinger, S. L. Forman, J. T. Davis and G. Wu, *J. Am. Chem. Soc.*, 2002, **124**, 742-743.
34. X. Shi, J. C. Fettinger and J. T. Davis, *J. Am. Chem. Soc.*, 2001, **123**, 6738-6739.
35. F. W. Kotch, J. C. Fettinger and J. T. Davis, *Org. Lett.*, 2000, **2**, 3277-3280.
36. F. W. van Leeuwen, W. Verboom, X. Shi, J. T. Davis and D. N. Reinhoudt, *J. Am. Chem. Soc.*, 2004, **126**, 16575-16581.
37. L. Ma, M. Iezzi, M. S. Kaucher, Y. F. Lam and J. T. Davis, *J. Am. Chem. Soc.*, 2006, **128**, 15269-15277.
38. E. Mezzina, P. Mariani, R. Itri, S. Masiero, S. Pieraccini, G. P. Spada, F. Spinozzi, J. T. Davis and G. Gottarelli, *Chem. Eur. J.*, 2001, **7**, 388-395.
39. A. L. Marlow, E. Mezzina, G. P. Spada, S. Masiero, J. T. Davis and G. Gottarelli, *J. Org. Chem.*, 1999, **64**, 5116-5123.
40. S. L. Forman, J. C. Fettinger, S. Pieraccini, G. Gottarelli and J. T. Davis, *J. Am. Chem. Soc.*, 2000, **122**, 4060-4067.
41. J. T. Davis, M. S. Kaucher, F. W. Kotch, M. A. Iezzi, B. C. Clover and K. M. Mullaugh, *Org. Lett.*, 2004, **6**, 4265-4268.
42. S. Sivakova and S. J. Rowan, *Chem. Soc. Rev.*, 2005, **34**, 9-21.
43. J. L. Sessler and J. Jayawickramarajah, *Chem. Commun.*, 2005, 1939-1949.
44. J. Wengel, *Organic & Biomolecular Chemistry*, 2004, **2**, 277-280.
45. J. McCallum, S. Amare and R. Nolan, *Nucleos. Nucleot. Nucl.*, 2010, **29**, 801-808.
46. D. J. Hill, D. J. Hill, M. J. Mio, M. J. Mio, R. B. Prince, R. B. Prince, T. S. Hughes, T. S. Hughes and J. S. Jeffrey S Moore, *Chem. Rev.*, 2001, **101**, 3893-4012.

47. G. X. He, S. H. Krawczyk, S. Swaminathan, R. G. Shea, J. P. Dougherty, T. Terhorst, V. S. Law, L. C. Griffin, S. Coutre and N. Bischofberger, *J. Med. Chem.*, 1998, **41**, 2234-2242.
48. J. L. Sessler, M. Sathiosatham, K. Doerr, V. Lynch and K. A. Abboud, *Angew. Chem. Int. Ed.*, 2000, **39**, 1300-1303.
49. M. S. Kaucher and J. T. Davis, *Tetrahedron Lett.*, 2006, **47**, 6381-6384.
50. X. Liu, I. C. M. Kwan, S. Wang and G. Wu, *Org. Lett.*, 2006, **8**, 3685-3688.
51. S. Martic, G. Wu and S. Wang, *Can. J. Chem.*, 2010, **88**, 524-532.
52. S. Martic, X. Liu, S. Wang and G. Wu, *Chem. Eur. J.*, 2008, **14**, 1196-1204.
53. M. a. d. C. Rivera-Sánchez, I. Andújar-de-Sanctis, M. García-Arriaga, V. Gubala, G. Hobley and J. M. Rivera, *J. Am. Chem. Soc.*, 2009, **131**, 10403-10405.
54. J. E. Betancourt, M. Martín-Hidalgo, V. Gubala and J. M. Rivera, *J. Am. Chem. Soc.*, 2009, **131**, 3186-3188.
55. M. García-Arriaga, G. Hobley and J. M. Rivera, *J. Am. Chem. Soc.*, 2008, **130**, 10492-10493.
56. V. Gubala, J. E. Betancourt and J. M. Rivera, *Org. Lett.*, 2004, **6**, 4735-4738.
57. J. Plavec, in *Metal Complex-DNA Interactions*, eds. N. Hadjiliadis and E. Sletten, John Wiley & Sons, Ltd, Chichester, UK, Editon edn., 2009, pp. 55-93.
58. A. E. Engelhart, J. Plavec, O. Persil and N. V. Hud, in *Nucleic Acid-Metal Ion Interactions*, Royal Society of Chemistry, Editon edn., 2009, pp. 118-147.
59. N. V. Hud and J. Plavec, in *Quadruplex Nucleic Acids*, Royal Society of Chemistry, Cambridge, Editon edn., 2006, pp. 100-130.
60. M. Martín-Hidalgo, K. Camacho-Soto, V. Gubala and J. M. Rivera, *Supramol. Chem.*, 2010, **22**, 862-869.
61. L. Ma, W. A. Harrell and J. T. Davis, *Org. Lett.*, 2009, **11**, 1599-1602.
62. A. Wong, F. W. Kotch, I. C. M. Kwan, J. T. Davis and G. Wu, *Chem. Commun.*, 2009, 2154-2156.
63. G. Wu, Z. Gan, I. C. M. Kwan, J. C. Fettinger and J. T. Davis, *J. Am. Chem. Soc.*, 2011, **133**, 19570-19573.
64. F. W. Kotch, V. Sidorov, Y.-F. Lam, K. J. Kayser, H. Li, M. S. Kaucher and J. T. Davis, *J. Am. Chem. Soc.*, 2003, **125**, 15140-15150.
65. V. Gubala, D. De Jesus and J. M. Rivera, *Tetrahedron Lett.*, 2006, **47**, 1413-1416.
66. L. Ma, M. Melegari, M. Colombini and J. T. Davis, *J. Am. Chem. Soc.*, 2008, **130**, 2938–2939.
67. L. R. Rivera, J. E. Betancourt and J. M. Rivera, *Langmuir*, 2011, **27**, 1409-1414.
68. J. E. Betancourt, C. Subramani, J. L. Serrano-Velez, E. Rosa-Molinar, V. M. Rotello and J. M. Rivera, *Chem. Commun.*, 2010, **46**, 8537-8539.
69. J. E. Betancourt and J. M. Rivera, *J. Am. Chem. Soc.*, 2009, **131**, 16666-16668.
70. J. E. Betancourt and J. M. Rivera, *Org. Lett.*, 2008, **10**, 2287-2290.
71. A. Likhitsup, S. Yu, Y.-H. Ng, C. L. L. Chai and E. K. W. Tam, *Chem. Commun.*, 2009, 4070-4072.
72. I. Gadwal, S. De, M. C. Stuparu, S. G. Jang, R. J. Amir and A. Khan, *Journal of Polymer Science Part A Polymer Chemistry*, 2012, **50**, 2415-2420.
73. X. Meng, T. Moriuchi, Y. Sakamoto, M. Kawahata, K. Yamaguchi and T. Hirao, *RSC Adv.*, 2012, **2**, 4359-4363.
74. X. Meng, T. Moriuchi, N. Tohnai, M. Miyata, M. Kawahata, K. Yamaguchi and T. Hirao, *Org. Biomol. Chem.*, 2011, **9**, 5633-5636.
75. X. Meng, T. Moriuchi, M. Kawahata, K. Yamaguchi and T. Hirao, *Chem. Commun.*, 2011, **47**, 4682-4684.

76. S. Shinoda, T. Noguchi, M. Ikeda, Y. Habata and H. Tsukube, *J Incl Phenom Macrocycl Chem*, 2011, **71**, 523-527.
77. I. C. M. Kwan, Y.-M. She and G. Wu, *Chem. Commun.*, 2007, 4286-4288.
78. C. Graziano, S. Masiero, S. Pieraccini, M. Lucarini and G. P. Spada, *Org. Lett.*, 2008, **10**, 1739-1742.
79. P. Neviani, E. Mileo, S. Masiero, S. Pieraccini, M. Lucarini and G. P. Spada, *Org. Lett.*, 2009, **11**, 3004-3007.
80. P. M. Beaujuge and J. M. J. Fréchet, *J. Am. Chem. Soc.*, 2011.
81. D. González-Rodríguez, P. G. A. Janssen, R. Martín-Rapún, I. De Cat, S. De Feyter, A. P. H. J. Schenning and E. W. Meijer, *J. Am. Chem. Soc.*, 2010, **132**, 4710-4719.
82. A. Facchetti, *Chem. Mater.*, 2011, **23**, 733-758.
83. M. Martín-Hidalgo and J. M. Rivera, *Chem. Commun.*, 2011, **47**, 12485–12487.
84. D. González-Rodríguez, J. L. J. van Dongen, M. Lutz, A. L. Spek, A. P. H. J. Schenning and E. W. Meijer, *Nature Chem.*, 2009, **1**, 151-155.
85. S. Pieraccini, S. Bonacchi, S. Lena, S. Masiero, M. Montalti, N. Zaccheroni and G. P. Spada, *Org. Biomol. Chem.*, 2010, **8**, 774-781.
86. S. Lena, P. Neviani, S. Masiero, S. Pieraccini and G. P. Spada, *Angew. Chem. Int. Ed.*, 2010, **49**, 3657-3660.

# SELF-ASSEMBLY OF LIPOPHILIC GUANOSINES: SWITCHING BETWEEN DIFFERENT ASSEMBLIES

Stefano Masiero, Silvia Pieraccini and Gian Piero Spada*

Alma Mater Studiorum – Università di Bologna, Dipartimento di Chimica Organica "A. Mangini", Via San Giacomo 11, I-40126 Bologna, Italy (*gianpiero.spada@unibo.it*)

## 1 INTRODUCTION

Cyclic discrete G-quartet and "infinite" tape-like G-ribbon are the main different H-bonded supramolecular motifs that lipophilic guanosines (LipoGs) can form under different experimental conditions. The switching between these supramolecular motifs has been obtained by a variety of external stimuli. After a general presentation of LipoG self-assembly, examples involving three different stimuli will be discussed. A first example is represented by chemical stimuli: addition of an alkali metal ion stabilizes the G-quartet while its removal shifts the equilibrium towards the G-ribbon. In the second case, a lipoG armed with a terthiophene unit undergoes a pronounced variation of its supramolecular organisation by changing the polarity of the solvent: in chloroform the derivative assembles via H-bonding in a guanosine driven structure, while in the more polar (and H-bond competing) acetonitrile different aggregates are observed, where the terthiophene chains are $\pi$-$\pi$ stacked in a helicoidal arrangement. Finally, a third type of stimulus is represented by light: the self-assembly of a modified guanosine nucleobase with a photoactive unit at C8 can be photocontrolled by selecting the appropriate wavelength.

Besides the importance of self-assembled hydrophilic guanosine derivatives in biological systems,[1] LipoGs have been shown to exhibit a rich supramolecular chemistry in organic solutions, on surfaces and in the solid state.[2-4]

Specifically, in 1995 we showed that G-quartet formation occurs in organic solvents in the presence of metal ions for the deoxyguanosine derivative $dG(C10)_2$ **1** (see Figure 2),[5] while Davis et al. demonstrated analogous behaviour for an isoguanosine derivative.[6] Quartet is indeed one of the possible motifs that guanine, a multiple H-bonding unit, is able to form. The AADD homocoupling of four guanines exposing to the observer the same face (the two faces of guanine are heterotopic) leads to the formation of this discrete cyclic H-bonded motif (Figure 1).

For **1**, well-organised G-quartet based, yet different, supramolecular assemblies were observed depending on the relative guanosine to metal ion molar ratio, as characterised by solution-state NMR, notably NOE experiments:[7,8] while at low ion content the dominant species is an octamer composed of two stacked G-quartets, at higher ion concentration a pseudo-polymeric supramolecular assembly composed of several stacked G-quartes is formed (Figure 2).

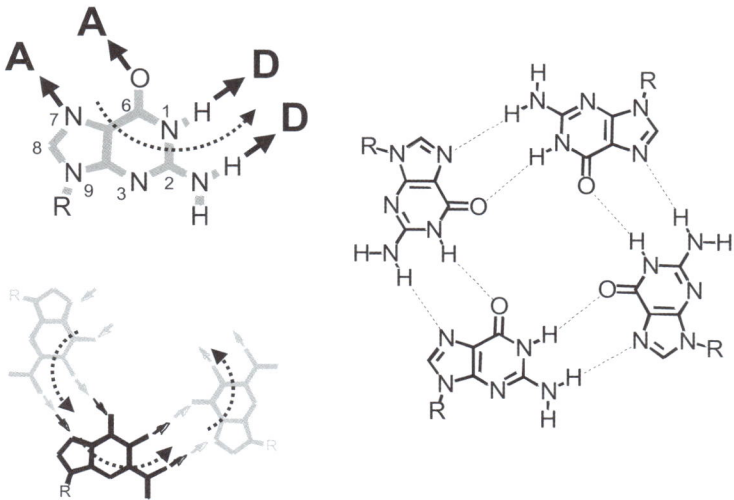

**Figure 1** *The G-quartet as composed by the AADD self-assembly of four guanines showing the same side (A: H-bond acceptor, D: H-bond donor).*

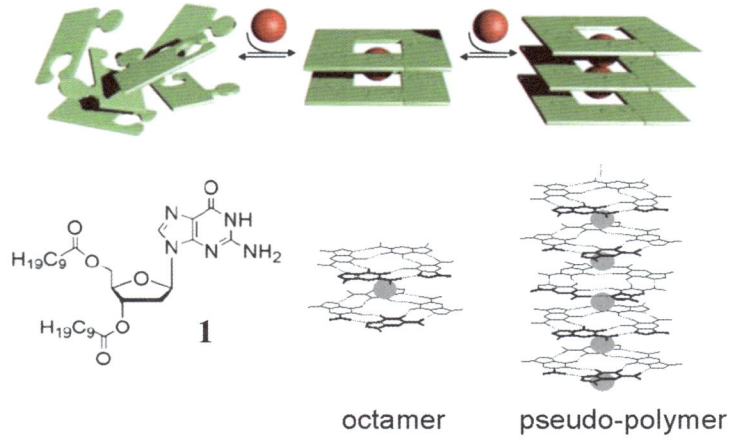

**Figure 2** *The cation-templated self assembly of Lipo-G **1** from the unassembled molecule to an octameric species and finally to a pseudo-polymeric aggregate (the spheres represent cations).*

While the cation-driven formation of octameric assemblies seems a common behavior of all LipoGs so far investigated, the formation of higher aggregates is not always observed and the reasons for that are not yet fully understood. Recently, Rivera and co-workers have shown that fine control over the specific assembly of quartets can be achieved by tuning or changing the solvent composition for guanosines with a phenyl substituent at the C8 position;[9-11] furthermore Meijer and co-workers have shown that Coulombic interactions between the separated cation and anion and solvent polarity are also important in determining the self-assembly of guanosine quartets into 8-, 12-, 16- or 24-mer structures.[12] Moreover, Sreenivasachary and Lehn have shown that different propensity for self-assembly and thus gelation can be exploited in a combinatorial chemistry approach for component selection in the generation of constitutional dynamic hydrogels.[13]

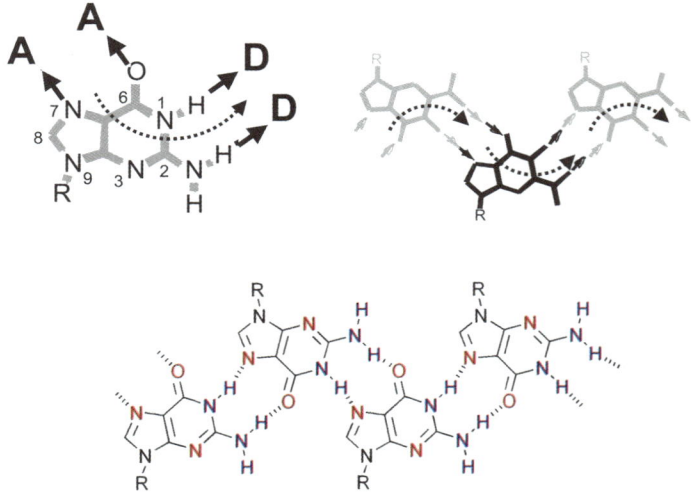

**Figure 3** *The non-centrosymmetric ribbon supramolecular structure by guanine, characterized by N2–H...O6, as composed by AADD homocoupling of guanines ("ribbon A").*

In the absence of metal ions, **1** has been shown to self-assemble into ribbon-like structures in organic solvents.[14] The same AADD homocoupling at the basis of the G-quartet formation may indeed drive to the formation of a different supramolecular motif (see Figure 3): in fact, when a couple of guanines exposes to the observer their opposite sides, an infinite H-bonded motif is obtained ("ribbon A"). Furthermore, a different homocoupling (ADDA) in which different H-bonding sites of the guanine are involved, leads to the formation of a further kind of H-bonded ribbon ("ribbon B", see Figure 4). The transformation over time from "ribbon A" (non centrosymmetric) to "ribbon B" (centrosymmetric) has been observed in CDCl$_3$ by solution-state NMR.[15]

These long anisometric supramolecular ribbons can form liquid-crystalline phases in organic solvents.[16] While it has not been possible to obtain a diffraction structure for the longer-chain derivative **1**, an X-ray single-crystal diffraction structure for the three-carbon atom tail derivative dG(C3)$_2$ reveals the "ribbon A" type self-assembly,[17] analogously to what observed in the crystal structure of guanosine monohydrate.[18] Similar ribbon-like self-assembly is observed in crystal structures presented by Araki and co-workers for a guanosine derivative with three Me$_2$t-BuSi- substituents.[19]

Pham et al. have shown that $^{15}$N refocused INADEQUATE[20-22] solid-state NMR spectra of $^{15}$N-labelled **1** and dG(C3)$_2$ allow the unambiguous identification of the distinct intermolecular hydrogen bonds that are characteristic of self-assembly into quartet-like or ribbon-like arrangement.[23,24]

On surfaces, guanosine ribbon-like and G-quartet self assemblies have been observed by SFM and STM.[15,25-28]

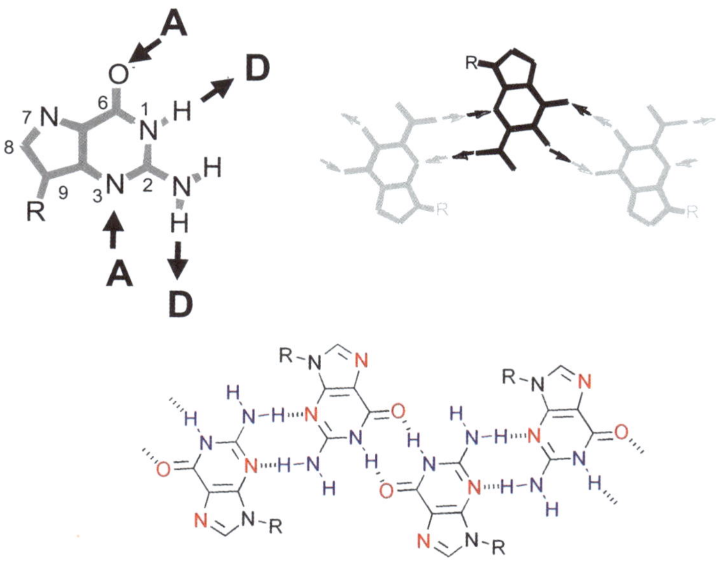

**Figure 4**  *The centrosymmetric ribbon supramolecular structures by guanine, characterized by N1–H...O6, as composed by ADDA homocoupling of guanines ("ribbon B").*

The rich self-assembly exhibited by guanosines has led to the preparation of a remarkable range of nanostructures. In the presence of metal ions to template guanosine quartet formation, architectures such as nanotubes formed by a calix[4]arene guanosine,[29] anion-bridged nanosheets,[30] a membrane film,[31] and a star polymer[32] have been observed, while Barboiu and coworkers have shown how an organic-inorganic hybrid exhibiting an ion-channel-like columnar architecture can be formed by the self organisation of silicon substituted guanosine quadruplexes and ureidocrown ethers.[33] Moreover, Davis and coworkers have synthesized a guanosine derivative that is able to function as a transmembrane transporter of sodium ions,[34] while the formation of a conjugate with bile acid or a subsequent modification of carbamate to urea leads to large and stable pores that could allow transport of larger biomolecules.[35,36]

Self-assembled nanoribbons of LipoG **1** have been used to interconnect gold nanoelectrodes fabricated by electron beam lithography, so as to produce devices with interesting electrical properties, e.g. a photoconductive device,[37] a rectifier,[38] or a field-effect transistor (FET).[39] Derivative **1** has also been used for biophotonic applications[40] and to produce a molecular electronic device with rectifying properties when conjugated to a wide band gap GaN semiconductor.[41]

## 2 SWITCHING BETWEEN DIFFERENT ASSEMBLIES

The control of molecular assembly into well-defined structures on the nanoscale is a key step to improve the performances of materials[42-47] to be used, for example, as components in electronic nanodevices, such as solar cells, light-emitting diodes (LEDs), and field effect transistors (FETs). This control has enormous potential for materials science due to the possibility of bridging the gap between the molecular scale and the macroscopic one in terms of structural order, when precise control of such self-assembly processes is achieved.

Among weak interactions, π-stacking has been the first to be employed to drive the self-assembly of conjugated (macro)-molecular systems into well-defined nanoscale assemblies that feature a high degree of order at the supramolecular level.[48-50] Further control of nanoarchitectures might be possible by incorporating more specific non-covalent interaction sites in the building blocks.[43-47] Among the various non-covalent interactions, multiple hydrogen bonds have been widely adopted because of their directionality and selectivity.[51] Many examples of bottom-up nanostructurization of π-conjugated oligomers assisted by multiple hydrogen-bonding interactions have been reported.[43]

As mentioned in the Introduction, guanine moiety is a versatile hydrogen bonding building block. In particular, lipophilic guanosines can undergo different self-assembly pathways originating diverse nanoarchitectures, and two typical assemblies are the ribbons and the cyclic-quartet system previously described. Furthermore, the easy fictionalization of guanosine in the sugar hydroxyl groups or in the aromatic base (in particular in C8 position) makes it a promising building block for the fabrication of complex architectures with functional units located in pre-programmed positions (Figure 5).[4] Finally, as it will be demonstrated in the following sections, by use of external stimuli the equilibrium between the different nanoarchitectures can be controlled and accordingly some physical properties, possibly relevant for molecular electronics, organic photovoltaics, photonics and spintronics, can be tuned. Three types of stimuli will be considered: a chemical stimulus (namely, addition or removal of cations), variation of solvent polarity and light (UV-vis).

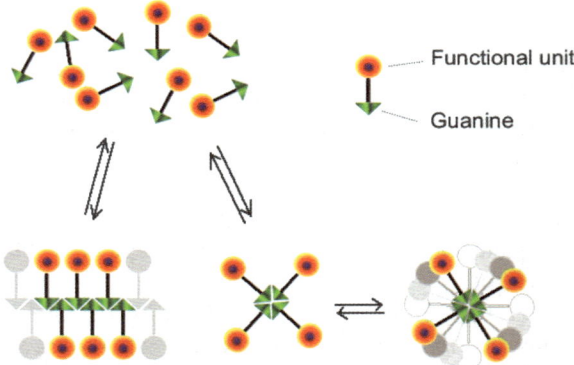

**Figure 5** *The different spatial distribution of the functional units in the interconvertible different assemblies leads to different properties*

## 2.1 Switching by addition/removal of cations

Considering that the supramolecular motifs obtained in the presence or in the absence of cations are different, an obvious chemical stimulus is represented by the addition and removal of potassium ions to/from a solution of a LipoG, e.g. **1**. We could control the addition/removal of $K^+$ ions by means of cryptand [2.2.2] (see Figure 6).[52] In fact this cryptand has a high affinity for $K^+$ and allows its removal from the system. However, the ability of [2.2.2] to capture the K ion is pH-dependent and in its protonated form this macrocycle is no longer active as cryptand. This fact can be exploited to switch reversibly from one supramolecular motif to the other by successive addition of acid and base.[53]

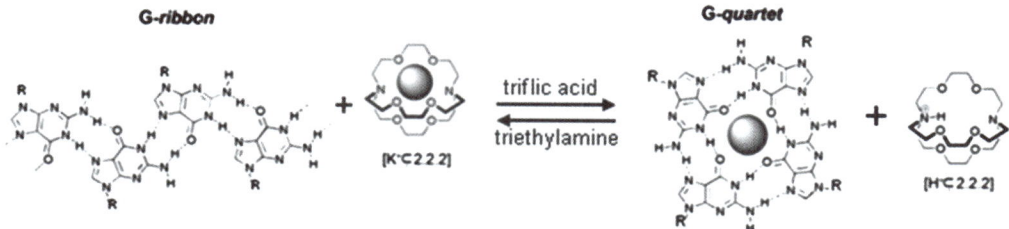

**Figure 6**  *The acid/base controlled interconversion between G-ribbons and G-quartet base structures.*

More in detail, the addition of potassium picrate to a chloroform solution of LipoG **1** transforms the supramolecular ribbon into the octameric complex based on the G-quartet motif. Upon subsequent addition to the quadruplex solution of the [2.2.2] cryptand, potassium is captured by the cryptand (hence the cryptate is formed) and the system reverts to the original G-ribbon. At this point upon addition of an acid (namely, triflic acid), $K^+$ is released from the cryptate and the G-quartet based system is regenerated. Finally, adding thereafter a base (namely triethylamine) the protonated cryptand deprotonates, the free cryptand recaptures $K^+$ and the G-ribbon is formed again. The acid/base addition steps can be repeated several times.

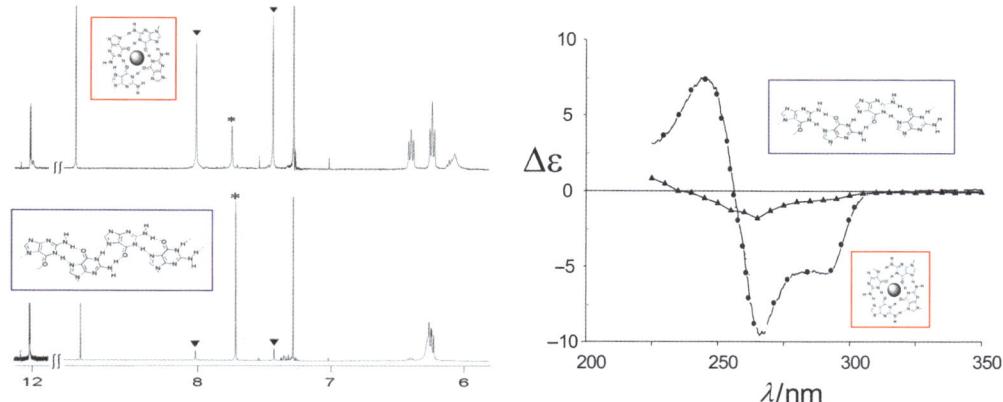

**Figure 7**  *$^1$H-NMR (left) and Circular Dichroism (right) spectral variation upon reversible quartet-to-ribbon transformation by acid/base addition.*

$^1$H-NMR and Circular Dichroism (CD) can be both exploited to monitor the ribbon-quadruplex interconversion (see Figure 7). Without entering into details, NMR spectra of octamer and ribbon are definitely different: for example, the former has a double set of signals (a few selected signals are marked with triangles in Figure 7) arising from its $C_4$-symmetry, while the latter shows a single set (marked with a star in Figure 7). Furthermore circular dichroism is diagnostic of the formation of stacked G-quartet-based assemblies.[54-56] In fact, the tetramers do not stack in register, but are rotated with respect to each other to give, in the 230-300 nm region, characteristic of the $\pi$-$\pi$* transitions of guanine chromophore, a double signed exciton-like CD signal.[57,58]

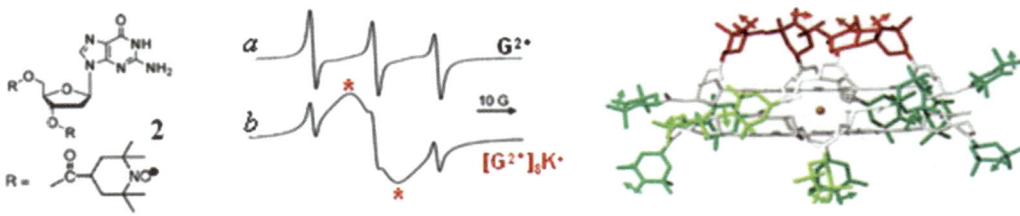

**Figure 8** *The EPR spectrum of LipoG **2** before (trace a) and after (trace b) $K^+$-directed formation of an octameric G-quartet based species. Stars mark the signals due to intermolecular spin-spin exchange. In the left part: a molecular model of the assembled species.*

This same chemical-stimulus-controlled self-assembly has been exploited with guanosine derivatives armed with one or two persistent paramagnetic units, 4-carbonyl-2,2,6,6-tetramethylpiperidine-1-oxyl (TEMPO).[59,60] As shown in the ESR spectra, in the absence of metal cations the spectrum of LipoG **2** (Figure 8, trace a) is characterised by three equally spaced lines (with a broadening between them indicating that intramolecular spin exchange is occurring). In sharp contrast, the ESR spectrum recorded after solid-liquid extraction of potassium picrate shows mainly one very broad signal whose integrated intensity corresponds to the initial amount of radicals (Figure 8, trace b). The broadening of the signal is independent of concentration and temperature, and thus inter-assembly interactions and motional broadening can be discounted. This spectrum is reminiscent of those obtained from very concentrated nitroxide solutions (>0.05 M).[61] Since the spectrum was obtained at 0.5 mM concentration, the signal broadening is ascribed to the proximity of spin centers of LipoG **2** within the framework of the octamer. The octameric assembly allows the confinement of 16 paramagnetic units in a small volume giving rise a drastic change of magnetic properties upon reversible formation. Since the relative geometry of the radical units is the outcome of $K^+$-directed self-assembly, the spin-spin interaction is suppressed by removing the alkaline ion by means of the cryptand/acid-base system described above.

The metal templated reversible assembly/re-assembly into quartets and ribbons has been described also at the solid-liquid interface by means sub-molecularly resolved STM.[62] The molecule used in this study was a $N^9$-alkylguanine, **3**, and the substrate was a highly oriented pyrolitic graphite (HOPG) surface. The presence of a long aliphatic side-chain and the absence of the sugar are expected to promote the molecular physisorption on HOPG. The self-assembly on HOPG has been studied for neat **3** and upon sub-sequent addition of [2.2.2] cryptand, potassium picrate and trifluoromethanesulfonic acid to trigger the reversible interconversion between two different highly ordered supramolecular motifs. For a full account on this subject the reader is referred to the chapter by P. Samorì and coworkers in the present volume.

## 2.2 Switching by variation of solvent polarity

In favourable conditions also a variation of solvent properties may control the type of supramolecular organisation of LipoGs.[10]

An interesting case is represented by LipoG **4**, armed with a terthiophene unit, that can form in THF either a ribbon-like motif or a G-quartet based columnar structure in the absence or presence of alkali metal ions, respectively,[63] thus allowing the control of the inter-oligothiophene interactions. Interestingly, the supramolecular organization of LipoG **4** undergoes a pronounced variation by changing the polarity of the solvent.[64] In chloroform guanosine derivative **4**, when templated by alkali metal ions, assembles via H-bonding in G-quartet based $D_4$-symmetric octamers where the polar guanine bases are located into the inner part of the assembly and act as a scaffold for the terthienyl pendants. On the other hand, in the more polar (and H-bond competing) acetonitrile (ACN) different aggregates are observed: the terthiophene chains are π-π stacked in a helicoidal (left-handed) arrangement in the central core and the guanine bases, free from hydrogen bonding, are located at the periphery and exposed to the solvent. The system can be switched from one state (guanine-directed) to the other (thiophene-directed) by subsequent addition of chloroform and acetonitrile. The solvent-induced switching can be easily followed by Circular Dichroism spectroscopy (see Figure 9): the CD exciton-couplet in the guanine chromophore absorption region observed in chloroform disappears after the addition of acetonitrile, indicating the disassembly of the G quartet-based octameric structure, while an intense quasi-conservative exciton splitting in the 300-450 nm spectral region becomes predominant in the CD spectrum. This latter strong bisignate optical activity can be ascribed to the helical packing of conjugated terthiophene moieties stabilized by π-π interactions. Both NMR spectra and photophysical investigations confirm the structures of the guanine-directed and thiophene-directed assemblies in chloroform and acetonitrile, respectively.

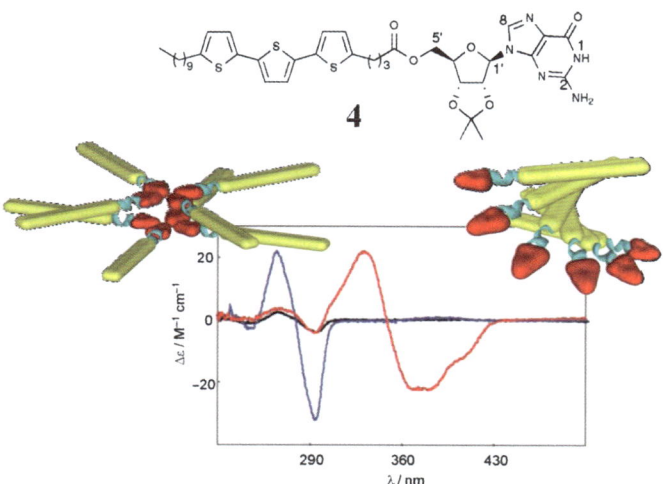

**Figure 9** *CD spectra of **4**-KPic in CHCl$_3$ (blue line) and in ACN/CHCl$_3$ 9/1 (red line). The idealised models represent the supramolecular structures in the two solvent conditions, respectively (in red the guanine, in light blue the sugar and in yellow the terthienyl moiety)*

### 2.3 Switching by UV-Vis irradiation

We investigated the behavior of LipoG ***E*-5**,[64] whose photoresponsive structure was inspired by previous work of Ogasawara and coworkers on oligonucleotides containing a modified guanosine.[66,67]

When a weighted amount of KI is added to a ACN solution of *E*-5, $^1$H-NMR and CD spectra show the formation of stacked G-quartets templated by the cation. In particular, an octameric species composed of two stacked G-quartets arranged in a $D_4$-symmetry is formed. Although no detailed information on the electronic transitions are available so far for 8-styrylguanine chromophore, the CD spectral changes observed upon addition of potassium ion closely resemble those reported for other unmodified lipophilic guanosines.[3,4] The strong increase of the CD signal associated with the formation of the *E*-5/$K^+$ aggregate can analogously be attributed to interchromophore couplings taking place in the stacked complex (see Figure 10, dashed line in the CD spectra). When samples of the *E*-5/$K^+$ octameric complex are irradiated at 365 nm, photoconversion to the *Z* isomer takes place and the *Z*-PSS is reached. The photoisomerization has a dramatic effect on the assembled species. The CD spectrum of the solution of 5/KI recorded at the Z-PSS shows very weak signals (Figure 10, dotted line in the CD spectra): this spectrum is practically superimposable to the CD spectrum of *Z*-5 prior to KI addition (Figure 10, solid line in the CD spectra) and it is similar to that of uncomplexed *E*-5. The disappearance of the strong CD bands at 255 and 350 nm is an evidence of the complex decomposition: stacked G-quartets no longer exist in solution. The absence of stacked G-quartets in the case of *Z*-5/$K^+$ mixtures is likely due to the fact that in the *Z* form the phenyl group of the styryl unit is twisted with respect to the G-quartet plane. The consequent steric hindrance could force quartets away from van der Waals contact or it could produce a conformational change around the glycosydic bond, which, in turn, would hamper the stacking. Additionally, in the *Z* form the N7 is probably shielded by the styryl unit and is no longer available for H-bonding.

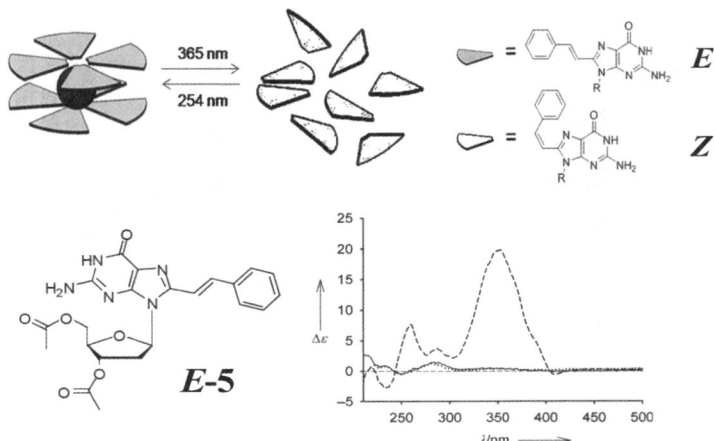

**Figure 10** *Top: cartoon of the photo-triggering of self-assembly of 5. Bottom: CD spectra of a solution of E-5/KI (dashed line), of 5/KI at the Z-PSS (dotted line) and 5 at the Z-PSS (solid line) in acetonitrile.*

The *Z* form can be converted back to the *E* form either photochemically, by irradiating at 254 nm, or thermally. Retroisomerization to the *E* isomer determines, at the supramolecular level, the recreation of the octameric complex: the CD spectrum of the solution at this point perfectly overlaps to the starting (*E*-5)$_8$$K^+$ trace. Thus, the G-quartet based complex can be cyclically assembled and disassembled by light.

## 3 CONCLUSIONS AND PERSPECTIVES

The self-aggregating behaviour of a given system is the outcome of a combination of several effects including ππ-stacking, van der Waals interactions, H-bonding and amphiphilicity. Any stimulus acting on one (or more) of the above mentioned effects, will, in principle, affect the self-aggregation behaviour and hence the properties of the system. The ability to control, in a predictable way, the switching between different aggregates, and thus between different properties, enormously expands the possible applications of these organized systems.

**Acknowledgements**

We gratefully acknowledge the European Commission through the COST Action MP0802 ("Self-assembled guanosine structures for molecular electronic devices"), for offering a stimulating environment for highly qualified discussions on the topic. This work was supported by MIUR through the National Interest Research Programme (PRIN 2009, grant 2009N5JH4F_002).

**References**

1. The Nobel prize for Physiology or Medicine 2009 was awarded to E. H. Blackburn, C. W. Greider and J. W. Szostak, for "the discovery of how chromosomes are protected by telomeres and the enzyme telomerase"; see: E. H. Blackburn, Nobel lecture, http://nobelprize.org/nobel_prizes/medicine/laureates/2009/blackburn-lecture.html.
2. J. T. Davis, *Angew. Chem, Int. Ed. Engl.*, 2004, **43**, 668.
3. J. T. Davis and G. P. Spada, *Chem. Soc. Rev.*, 2007, **36**, 296.
4. S. Lena, S. Masiero, S. Pieraccini and G. P. Spada, . *Chem.-Eur. J.*, 2009, **15**, 7792.
5. G. Gottarelli, S. Masiero and G. P. Spada, *J. Chem. Soc.-Chem. Commun.*, 1995, 2555.
6. J. T. Davis, S. Tirumala, J. R. Jenssen, E. Radler and D. Fabris, *J. Org. Chem.*, 1995, **60**, 4167.
7. A. L. Marlow, E. Mezzina, G. P. Spada, S. Masiero, J. T. Davis and G. Gottarelli, *J. Org. Chem.*, 1999, **64**, 5116.
8. E. Mezzina, P. Mariani, R. Itri, S. Masiero, S. Pieraccini, G. P. Spada, F. Spinozzi, J. T. Davis and G. Gottarelli, *Chem.-Eur. J.*, 2001, **7**, 388.
9. M. Garcia-Arriaga, G. Hobley and J. M. Rivera, *J. Am. Chem. Soc.*, 2008, **130**, 10492.
10. J. E. Betancourt, M. Martin-Hidalgo, V. Gubala and J. M. Rivera, *J. Am. Chem. Soc.*, 2009, 131, 3186.
11. M. D. Rivera-Sanchez, I. Andujar-de-Sanctis, M. Garcia-Arriaga, V.Gubala, G. Hobley and J. M. Rivera, *J. Am. Chem. Soc.*, 2009, **131**, 10403.
12. D. Gonzalez-Rodriguez, J. L. J. van Dongen, M. Lutz, A. L. Spek, A. Schenning and E. W. Meijer, *Nat. Chem.*, 2009, **1**, 151.
13. N. Sreenivasachary and J. M. Lehn, *Proc. Natl. Acad. Sci. U. S. A.*, 2005, **102**, 5938.
14. G. Gottarelli, S. Masiero, E. Mezzina, G. P. Spada, P. Mariani and M. Recanatini, *Helv. Chim. Acta*, 1998, **81**, 2078.
15. G. Gottarelli, S. Masiero, E. Mezzina, S. Pieraccini, J. P. Rabe, P. Samori and G. P. Spada, *Chem.-Eur. J.*, 2000, **6**, 3242.
16. G. Gottarelli, S. Masiero, E. Mezzina, S. Pieraccini, G. P. Spada and P. Mariani, *Liq. Cryst.*, 1999, **26**, 965.

17  T. Giorgi, F. Grepioni, I. Manet, P. Mariani, S. Masiero, E. Mezzina, S. Pieraccini, L. Saturni, G. P. Spada and G. Gottarelli, *Chem.-Eur. J.*, 2002, **8**, 2143.
18  U. Thewalt, C. E. Bugg and R. E. Marsh, *Acta Crystallogr. Sect. B*, 1970, **26**, 1089.
19  K. Araki, R. Takasawa and I. Yoshikawa, *Chem. Commun.*, 2001, 1826.
20  A. Lesage, M. Bardet, L. Emsley, *J. Am. Chem. Soc.*, 1999, **121**, 10987.
21  F. Fayon, D. Massiot, M. H. Levitt, J. J. Titman, D. H. Gregory, L. Duma, L. Emsley and S. P. Brown, *J. Chem. Phys.*, 2005, **122**, 194313.
22  S. Cadars, J. Sein, L. Duma, A. Lesage, T. N. Pham, J. H. Baltisberger, S. P. Brown and L. Emsley, *J. Magn. Reson.*, 2007, **188**, 24.
23  T. N. Pham, S. Masiero, G. Gottarelli and S. P. Brown, *J. Am. Chem. Soc.*, 2005, **127**, 16018.
24  A. L. Webber, S. Masiero, S. Pieraccini, J. C. Burley, A. S. Tatton, D. Iuga, T. N. Pham, G. P. Spada and S. P. Brown, *J. Am. Chem. Soc.*, 2011, **133**, 19777.
25  T. Giorgi, S. Lena, P. Mariani, M. A. Cremonini, S. Masiero, S. Pieraccini, J. P. Rabe, P. Samori, G. P. Spada and G. Gottarelli, *J. Am. Chem. Soc.*, 2003, **125**, 14741.
26  S. Lena, G. Brancolini, G. Gottarelli, P. Mariani, S. Masiero, A. Venturini, V. Palermo, O. Pandoli, S. Pieraccini, P. Samori and G. P. Spada, *Chem.-Eur. J.*, 2007, **13**, 3757.
27  A. M. S. Kumar, J. D. Fox, L. E. Buerkle, R. E. Marchant and S. J. Rowan, *Langmuir*, 2009, **25**, 653.
28  A. Ciesielski, M. Surin, G. P. Spada, P. Samorì,'Nanopatterning the surface with ordered supramolecular architectures: controlling the self-assembly of guanine-based hydrogen-bonded motifs', in *Guanine quartets – structure and application*, W. Fritzsche, L. Spindler, Eds., RSC, London, UK, 2012, pp. 40-47.
29  V. Sidorov, F. W. Kotch, M. El-Khouedi and J. T. Davis, *Chem. Commun.*, 2000, 2369.
30  C. Zhong, J. Wang, N. Q. Wu, G. Wu, P. Y. Zavalij and X. D. Shi, *Chem. Commun.*, 2007, 3148.
31  C. Arnal-Herault, A. Pasc, M. Michau, D. Cot, E. Petit and M. Barboiu, *Angew. Chem, Int. Ed. Engl.*, 2007, **46**, 8409.
32  A. Likhitsup, S. Yu, Y. H. Ng, C. L. L. Chai and E. K. W. Tam, *Chem. Commun.*, 2009, 4070.
33  S. Mihai, A. Cazacu, C. Arnal-Herault, G. Nasr, A. Meffre, A. van der Lee and M. Barboiu, *New J. Chem.*, 2009, **33**, 2335.
34  M. S. Kaucher, W. A. Harrell and J. T. Davis, *J. Am. Chem. Soc.*, 2006, **128**, 38.
35  L. Ma, M. Melegari, M. Colombini and J. T. Davis, *J. Am. Chem. Soc.*, 2008, **130**, 2938.
36  L. Ma, W. A. Harrell and J. T. Davis, *Org. Lett.*, 2009, **11**, 1599.
37  R. Rinaldi, E. Branca, R. Cingolani, S. Masiero, G. P. Spada and G. Gottarelli, *Appl. Phys. Lett.,* 2001, **78**, 3541.
38  R. Rinaldi, G. Maruccio, A. Biasco, V. Arima, R. Cingolani, T. Giorgi, S. Masiero, G. P. Spada and G. Gottarelli, *Nanotechnology*, 2002, **13**, 398.
39  G. Maruccio, P. Visconti, V. Arima, S. D'Amico, A. Blasco, E. D'Amone, R. Cingolani, R. Rinaldi, S. Masiero, T. Giorgi and G. Gottarelli, *Nano Lett.*, 2003, **3**, 479.
40  A. Neogi, J. Li, P. B. Neogi, A. Sarkar, H. Moroc, *Elec. Lett.,*2004, **40**, 1605.
41  H. Liddar, J. Li, A. Neogi, P. B. Neogi, A. Sarkar, S. Cho and H. Morkoc, *Appl. Phys. Lett.,* 2008, **92**, article number 013309 **DOI:** 10.1063/1.2828405.
42  A. C. Grimsdale and K. Müllen, *Angew. Chem. Int. Ed.*, 2005, **44**, 5592.
43  A. P. H. J.Schenning and E. W. Meijer, *Chem. Commun.*, 2005, 3245.
44  F. J. M. Hoeben, P. Jonkheijm, E. W. Meijer and A. P. H. J. Schenning, *Chem. Rev.*, 2005, **105**, 1491.
45  A. Ajayaghosh, A. and V. K. Praveen, *Acc. Chem. Res.*, 2007, **40**, 644.

46  A. Ajayaghosh, V. K. Praveen and C. Vijayakumar, *Chem. Soc. Rev.*, 2008, **37**, 109.
47  V. Palermo and P. Samorì, *Angew. Chem. Int. Ed.*, 2007, **46**, 4428.
48  P. Samorì, V. Francke, K. Müllen and J. P. Rabe, *Chem. Eur. J.*, 1999, **5**, 2312.
49  E. R. Zubarev, M. U. Pralle, E. D. Sone and S. I. Stupp, *J. Am. Chem. Soc.*, 2001, **123**, 4105.
50  A. P. H. J. Schenning, A. F. M. Kilbinger, F. Biscarini, M. Cavallini, H. J. Cooper, P. J. Derrick, W. J. Feast, R. Lazzaroni, Ph. Leclère, L. A. McDonell, E. W. Meijer and S. C. J. Meskers, *J. Am. Chem. Soc.*, 2002, **124**, 1269.
51  J.-M. Lehn, in *Supramolecular Chemistry, Concepts and Perspectives*, VCH, Weinheim, 1995.
52  J.-M. Lehn and J. P. Sauvage, *J. Am. Chem. Soc.*, 1975, **97**, 6700
53  S. Pieraccini, S. Masiero, O. Pandoli, P. Samorì and G. P. Spada, *Org. Lett.*, 2006, **8**, 3125.
54  S. Masiero, R. Trotta, S. Pieraccini, S. De Tito, R. Perone, A. Randazzo and G. P. Spada, *Org. Biomol. Chem.*, 2010, **8**, 2683.
55  G. Gottarelli, S. Lena, S. Masiero, S. Pieraccini and G. P. Spada, *Chirality*, 2008, **20**, 471.
56  A. I. Karsisiotis, N. Ma'ani Hessari, E. Novellino, G. P. Spada, A. Randazzo and M. Webba da Silva, *Angew. Chem. Int. Ed.*, 2011, **50**, 10645.57  N. Berova andK. K. Nakanishi, in *Circular Dichroism–Principles and Applications*, 2nd Edn., ed.s N. Berova, K. Nakanishi and R. W. Woody, Wiley-VCH, New York, 2000, p. 337.
58  N. Berova, L Di Bari and G. Pescitelli, *Chem. Soc. Rev.*, 2007, **36**, 914.
59  C. Graziano, S. Pieraccini, S. Masiero, M. Lucarini and G. P. Spada, *Org. Lett.*, 2008, **10**, 1739.
60  P. Neviani, E. Mileo, S. Masiero, S. Pieraccini, M. Lucarini and G. P. Spada, *Org. Lett.*, 2009, **11**, 3004.
61  J. E. Wertz and J. R. Bolton, *Electron Spin Resonance*, Chapman and Hall, New York, 1986.
62  A. Ciesielski, S. Lena, S. Masiero, G. P. Spada and P. Samorì, *Angew. Chem. Int. Ed.*, 2010, **49**, 1963.
63  G. P. Spada, S. Lena, S. Masiero, S. Pieraccini, M. Surin and P. Samorì, *Adv. Mater.*, *2008*, **20**, 2433.
64  S. Pieraccini, S. Bonacchi, S. Lena, S. Masiero, M. Montalti, N. Zaccheroni and G. P. Spada, *Org. Biomol. Chem.*, 2010, **8**, 774.
65  S. Lena, P. Neviani, S. Masiero, S. Pieraccini and G. P. Spada, *Angew. Chem. Int. Ed.*, 2010, **49**, 3657.
66  S. Ogasawara and M. Maeda, *Angew. Chem. Int. Ed.*, 2008, **47**, 8839.
67  S. Ogasawara and M. Maeda, *Angew. Chem. Int. Ed.*, 2009, **48**, 6671.

NANOPATTERNING THE SURFACE WITH ORDERED SUPRAMOLECULAR ARCHITECTURES: CONTROLLING THE SELF-ASSEMBLY OF GUANINE-BASED HYDROGEN-BONDED MOTIFS


Artur Ciesielski,[1] Mathieu Surin,[2] Gian Piero Spada,[3] Paolo Samorì[*1]

[1] ISIS &icFRC, Université de Strasbourg & CNRS 7006, 8 allée Gaspard Monge, 67000 Strasbourg (France). E-mail: samori@unistra.fr
[2] Laboratory for Chemistry of Novel Materials, Center for Innovation and Research in Materials and Polymers, University of Mons, 20 Place du Parc B-7000 Mons (Belgium)
[3] Alma Mater Studiorum - Università di Bologna, Dipartimento di Chimica Organica "A. Mangini", Via S. Giacomo 11, 40126 Bologna (Italy)


1 INTRODUCTION

Complex systems and phenomena in nature are dominated by reversible non-covalent interactions. Among weak interactions, H-bonding offers high control over the process of molecular self-assembly[1] because it combines specificity, directionality, reversibility and cooperativity. Such a unique character is the basis of sophisticated programs for self-assembly such as those based on the Watson–Crick base pairing[2] that govern the generation of complex architectures, such as the fascinating DNA double helix. Alongside Watson–Crick base pairing, nucleobases can interact through other types of hydrogen-bonded motifs to form various complex supramolecular architectures,[3] such as guanine (G) quartets and quadruplexes.[4] The G quartet (hereafter $G_4$), an hydrogen-bonded macrocycle typically formed by cation-templated assembly, was first identified in 1962 as the basis for the aggregation of 5'-guanosine monophosphate (5'-GMP),[5] and fits particularly well with contemporary studies in molecular self-assembly and noncovalent synthesis.[6] Among nucleobases, guanine (Scheme 1a) is very versatile:[7] depending on the experimental conditions it can undergo different self-assembly pathways. In the presence of certain

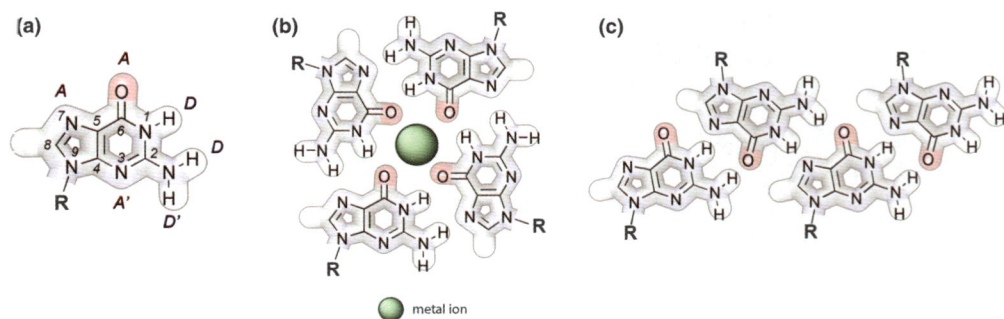

**Scheme 1** a) chemical structure of guanine; examples of H-bonded motifs of guanines: b) quartet based (involving pairing: NH(2)-N(7) and NH(1)-O(6)) and c) ribbon-like (involving pairing: NH(2)-O(6) and NH(1)-N(7))

metal ions, guanines can form G-quartet based architectures (Scheme 1b) such as octamers or columnar polymeric aggregates. In the absence of metal templating centers, guanines can self-assemble, both in solution and in the solid state, into ribbon-like architectures (Scheme 1c).[8]

The need for exploring ordered architectures at the molecular scale has made scanning tunneling microscopy (STM)[9] a widely employed yet extremely powerful tool to investigate supramolecular materials at interfaces with sub-molecular resolution, providing direct insight into the supramolecular world.[10] The working principle of STM is the tunneling of electrons from a sharp scanning tip to a substrate. When adsorbed on graphite, the STM current images show a brighter contrast for conjugated moieties and darker for aliphatic groups. Such a contrast is ruled by the resonant tunneling between the Fermi level of the substrate (in this case graphite) and the frontier orbital of the molecoles. The energy difference between them is inverse proportional to the tunneling probability.[11] The spatial sub-nanometre resolution that can be achieved by STM imaging allows gaining detailed information on molecular interactions. STM is therefore a tool of choice to assist the design of molecular modules that can undergo controlled self-assembly at surfaces under precise conditions (temperature, pressure and concentration) to form the chosen supramolecular structures, and investigate complex functional architectures such as dynamers. Currently, the adsorption of molecules on conductive substrates can be studied by STM under different environmental conditions including ultra-high vacuum (UHV), atmospheric pressure to image dry films or with the tip immersed into a liquid to investigate the solid–liquid interface. The solid–liquid interface provides a particularly interesting environment to carry out the self-assembly experiments and their investigation by STM. Compared to sample preparation and imaging under UHV conditions, the solid–liquid interface has numerous advantages: (1) the experimental approach is straightforward and does not require complicated and expensive infrastructure; (2) the dynamic exchange of molecules adsorbed on the surface and the one in the liquid phase promotes self-healing of defects in the self-assembled layers;[12] (3) the solid–liquid interface provides an excellent environment for *in-situ* chemical modifications of adsorbed molecules. When working under such condition, it is possible to monitor the reversible changes in the monolayers structure, upon addition of external chemical stimuli, e.g. varying pH [13] or by coordination of organic molecules to the metallic centers,[10f, 14] whereas in most examples of molecular re-organization investigated under UHV conditions the changes are irreversible.

Herein, we will discuss the process of supramolecular engineering in the self-assembly of guanine derivatives on conductive solid substrate, i.e. highly oriented pyrolitic graphite (HOPG), as investigated by STM. In the first section (2), we will focus on the exploration of systems engineered to form assemblies through the formation of hydrogen-bonds and van der Waals (vdW) interactions. We then present recent examples in 2D crystal engineering using alkylated guanosine derivatives. Further on, we give a perspective into the strategies employed alkyl chain interdigitation as the main building strategy for nanopatterning surfaces with guanine molecules. In the second section (3) we will describe systems based on metallo–ligand interactions as well as protonation as a way to trigger dynamic processes in guanine-based dynamers. In the final section (4) will discuss the use of guanosine derivatives as scaffolds for the controlled positioning of electrically/optically active groups in 2D.

## 2 SELF-ASSEMBLIES GENERATED *VIA* HYDROGEN-BONDS AND VAN DER WAALS INTERACTIONS

The self-assembly of hydrogen-bonded networks of a lipophilic guanosine derivative can be used to design highly-ordered supramolecular structures, not only in 3D crystal, but also in 2D when adsorbed on solid surfaces and in solution. For instance, micrometer-long nanoribbons with a molecular cross section can be grown on a mica surface from deposits of lipophilic guanosine derivatives bearing two alkyl groups (Fig.1a). This arises from the self-assembly into highly directional A-type ribbons (NH(2)–O(6) and NH(1)–N(7)), ultimately forming two-dimensional polycrystalline structures of parallel ribbons at the solution-graphite interface. (Fig. 1a-c).[8] This latter structure reflects the supramolecular motif that has been detected both in the single crystal[15] and as a metastable state in solution by NMR spectroscopy.

In an effort to modify and enhance electronic properties of guanosine derivatives, we broadened our scope to investigate 8-substituted lipophilic oxoguanosine derivatives.[16] The cooperative effect of hydrogen-bonding and solvophobic interactions induces the 8-oxoguanosines (Fig. 1d) to self-assemble into helical architectures both in the liquid crystalline phase, in solution and at the solid-liquid interface (Fig. 1e,f). These arrangements, which are markedly different from the structures obtained by the spontaneous self-assembly of guanosine derivatives unsubstituted at the C(8) position, are of interest, not only for their optical properties, but also for their ability to rectify currents, making them potential building blocks for the construction of nanoscale bio-electronic devices and circuits.[17]

To achieve an *in-depth* understanding of the self-assembly of guanine at the solid–liquid interface, we performed a sub-molecularly resolved STM study of physisorbed monolayers on graphite of a series of $N^9$-alkylated guanines with linear alkyl side-chains

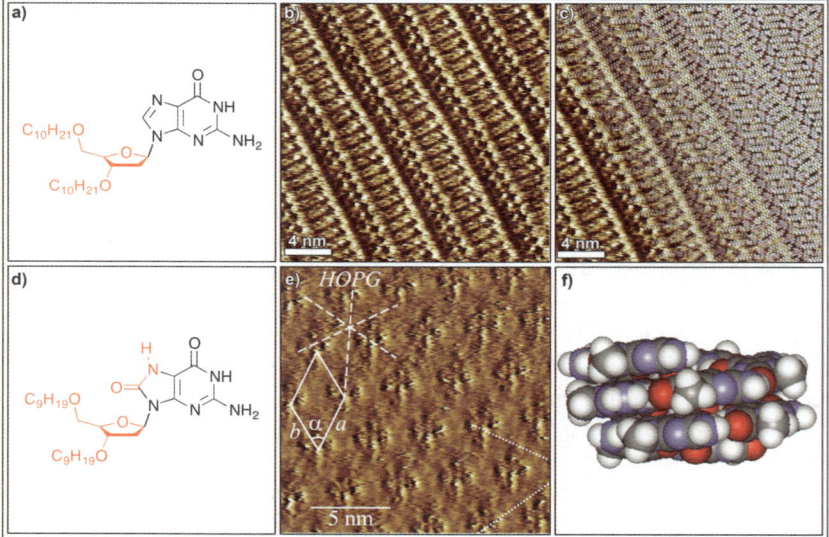

**Figure 1** *(a) chemical structures of studied guanosine derivative, b) STM current images and proposed packing motif (c) of molecular model of an A-type ribbon-like architecture on a graphite surface; (d) chemical representation of investigated oxyguanosine derivative, e) STM images and schematic illustration (f) alkylated oxyguanosine forming helical structures. Images (e,f) adapted from ref.[16], with permission from American Chemical Society.*

from –$C_2H_5$ up to –$C_{18}H_{37}$ (Fig. 2).[18] This comparative study was carried out by applying a drop of a solution of the chosen guanine-based molecule in 1,2,4-trichlorobenzene (TCB) on freshly cleaved HOPG surface. The presence of a long aliphatic side chain and the absence of the sugar with respect to previously studied guanosines were expected to promote the molecular physisorption on HOPG. All guanine derivatives were found to form monomorphic 2D crystals, which are stable on the several tens of minutes timescale and with size of domains exceeding several hundreds of $nm^2$. Subtle changes in the length of the alkyl side-chains dramatically influenced the 2D patterns on graphite. Derivatives with alkyl tails longer or equal to $C_{12}$ (Fig. 2f-i) self-assembled into linear H-bonded ribbons through the NH(2)–O(6) and NH(1)–N(7) pairing with 4 molecules in the unit cell. The same H-bonding pattern was observed for $N^9$-ethylguanine (Fig. 2a), but the packing shows only two molecules per unit cell. For derivatives with tails of intermediate length (from $C_6$ to $C_{10}$), no H-bonded supramolecular polymers were formed at the surface: ordered monolayers of single rows of (non-H-bonded) molecules (Fig. 2b and 2e) or H-bonded dimers (Fig. 2c and 2d) were observed.

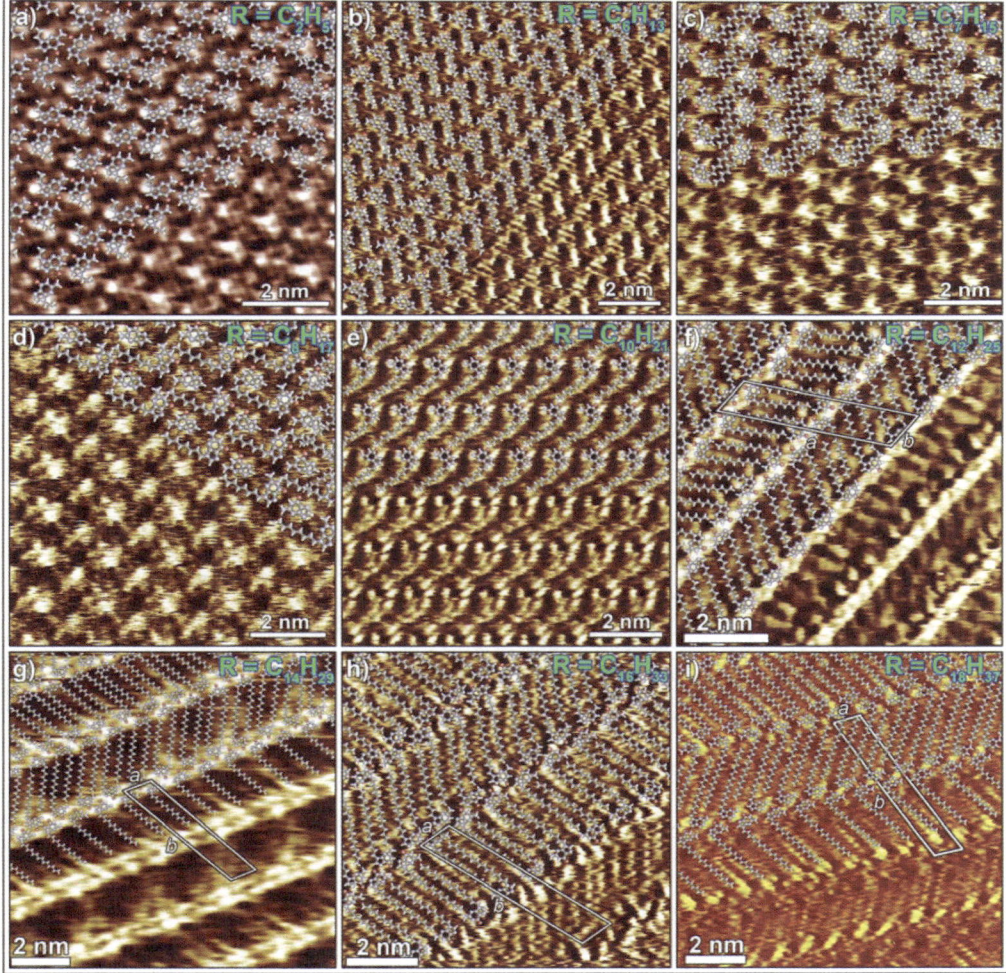

**Figure 2**  *STM images of monolayers of alkylated guanines showing ribbon-like (a, f, g, h and i), crystalline (b, e) and dimeric structures (c, d) formed on HOPG surface. Images (a–i) adapted from ref.[18], with permission from Royal Society of Chemistry.*

## 3 ASSEMBLY/REASSEMBLY PROCESSES ON SURFACES

As a building block for the fabrication of dynamers,[19] we studied lipophilic guanosines,[8, 16, 20] and guanine derivatives.[18, 21] Recently, we have provided direct evidence on the sub-nanometre scale metal-templated reversible assembly/reassembly process of an octadecyl-substituted guanine into highly ordered quartets ($G_4$) and ribbons[21] (Fig. 3). The self-assembly of octadecyl guanines alone on HOPG has been studied, and, upon subsequent addition in stoichiometric ratio of [2.2.2]cryptand, potassium picrate ($K^+$(pic)$^-$), and trifluoromethanesulfonic acid (HTf), the reversible interconversion between two different highly-ordered supramolecular motifs was triggered. In the absence of metal ions, the obtained monolayer shows a crystal-line structure consisting of ribbon-like architectures (Fig. 3b). This self-assembly behavior is in good agreement with previous observations on guanosine derivatives.[8, 16, 20] Upon *in-situ* addition of 10 mM potassium picrate solution in TCB to the initial ribbon-like motif in Figure 3b, the $G_4$ supramolecular motif was obtained (Fig. 3c). Upon subsequent *in-situ* addition of a 10 mM solution of [2.2.2]cryptand in TCB to the $G_4$ supramolecular architectures on HOPG, the guanine reassembled into the original ribbon (Fig. 3d). By adding a 10 mM solution of trifluoromethanesulfonic acid (HTf) in TCB, the potassium ions were released from the cryptate and the $G_4$ assembly was regenerated (Fig. 3e). Upon further addition of a [2.2.2]cryptand solution, the ribbon structure was regenerated (Fig. 3f). This demonstrates the potential of guanine-based structures to behave as a 2D dynamer, whose response to external chemical stimuli can be monitored by STM on the sub-nanometer scale..

## 4 GUANINE SUPRAMOLECULAR ARCHITECTURES AS SCAFFOLDS FOR ORGANIC ELECTRONICS

Given the possibility to functionalize the guanosines in the side-chains, they appear as ideal building blocks for the fabrication of complex architectures with a controlled high

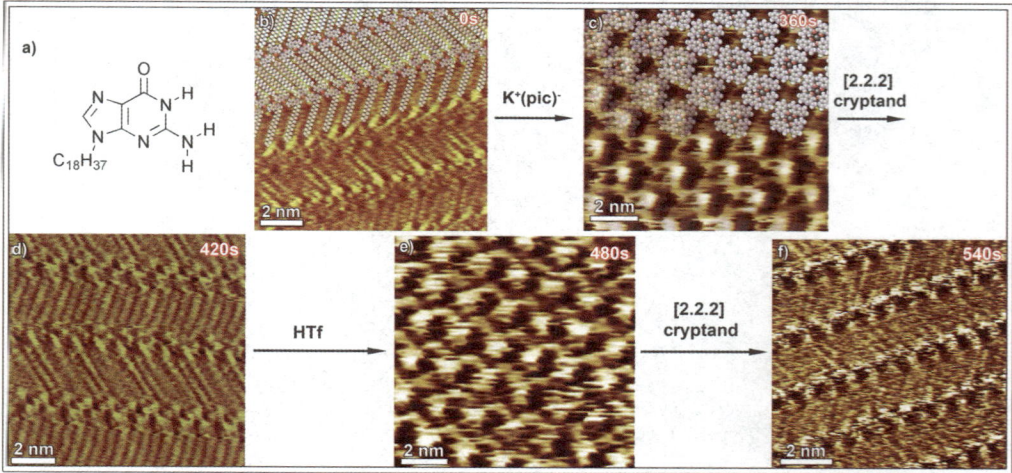

**Figure 3** *(a) Chemical structure of octadecyl guanine; (b–f) consecutive STM images showing the structural evolution of a monolayer of 1 over a 9 minutes time scale (time range displays in the upper right part of the images correspond to the time that was needed to reach the equilibrium after addition of reacting agents). Images (b), (d), and (f) show ribbon-like structure, whereas (c) and (e) exhibit G4-based architectures. Images (b–f) adapted from ref.[21] with permission from Wiley-VCH.*

rigidity, thus paving the way towards their future use for scaffolding, i.e. to locate functional units in pre-programmed positions.[22] Harmonizing the functionalities of single moieties in a supramolecular network represents an adaptable method for generating distinct polymeric architectures with programmed conformations and tailored properties. In this context, we have designed a guanosine derivative bearing a terthiophene moiety linked on the sugar unit (Fig.4a). Indeed, oligo- and poly-thiophenes are part of the most studied structures in organic electronics, because of their interesting optical and electronic properties, with application as active material in field-effect transistors and photovoltaic diodes noticeably. We have shown that this guanosine-terthiophene derivative can form (in solution) different types of H-bonded supramolecular architectures depending on the solution conditions: the reversible inter-conversion fuelled by potassium ion complexation/release allows the switching between ribbons to quartet self-assemblies, thus allowing us to modify the inter-oligothiophene interactions by chemical stimuli. STM and AFM characterization showed that these molecules self-assemble into highly-ordered architectures on surfaces (graphite or mica), see Fig. 4. By combining STM imaging with molecular modelling simulations (Fig. 4), it was shown that the highly-directional structures arises from self-assembly in extended, parallel B-type ribbons (NH(1)-O(6) and NH(2)-N(3), see Fig. 4c,d) on HOPG, which can extend over micrometers as observed with AFM on dry films. This is in contrast with previous results on alkylated guan(os)ine derivatives (Fig.1a-c, 2, and 3), which form A-type ribbons (NH(2)–O(6) and NH(1)–N(7)) on graphite. This difference can be explained by the fact that the guanosine-terthiophene derivative possess only one alkyl group (while guanosine derivatives previously studied were doubly alkylated) and one acetonide group on the sugar unit (pointing perpendicularly to the molecule main plane), both leading to several restrictions that favour the formation of the H-bonding network. Molecular modelling suggest the

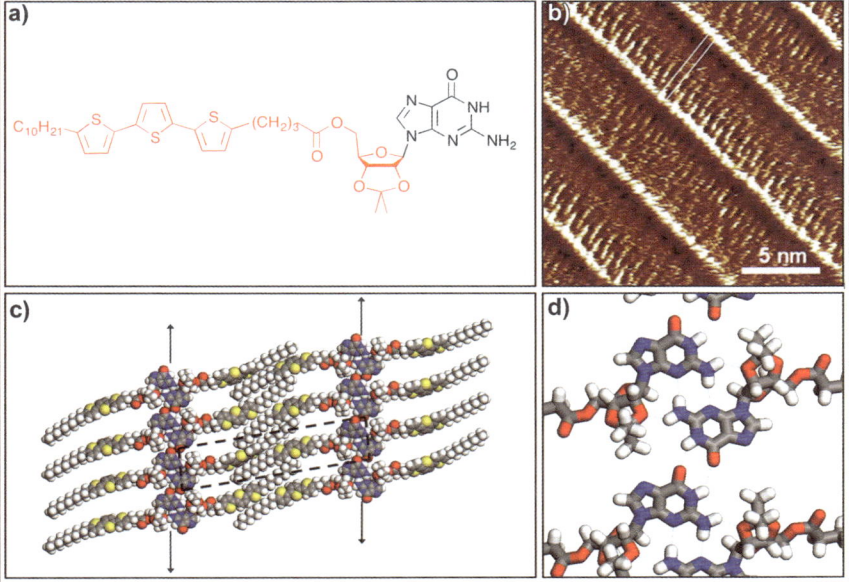

**Figure 4** *(a) Chemical structures of the studied guanosine-terthiophene derivative, b) STM current image of a monolayer on HOPG surface, showing higly-directional parallel structures; c) proposed packing motif into B-type ribbons based on molecular modelling simulations. d) Detailed view of the B-type ribbon packing (NH(1)-O(6) and NH(2)-N(3). Adapted from ref.[22] with permission of Wiley-VCH.*

formation of H-bonds between guanosine NH(2) and the ribose of the adjacent molecule (Fig. 4d), while the spacing between ribbons is dictated by the partial interdigitation of terthiophene-alkyl groups (Fig.4c). Indeed, this self-assembly governed by the formation of H-bonds between guanosines dictates the spatial localization of oligothiophenes, which constitute an elegant strategy to fabricating prototypes of supramolecular nanowires for organic electronics.

## 5 CONCLUSIONS

In summary, we have provided extensive details on how guan(os)ine molecules, decorated with alkyl side chains and functional, are able to self-assemble forming 2D supramolecular crystalline materials. This paves the way towards supramolecular scaffolding, i.e. the use of molecular modules designed to undergo recognition events and incorporating functional units, to form pre-programmed architectures with a sub-ångström resolution, and ultimately functional materials with properties controlled with a high degree of precision.
We have also provided direct evidence on the sub-nanometer scale of a dynamer operating at surfaces. The versatile guanine molecule was reversibly interconverted at the solid–liquid interface between two highly ordered supramolecular motifs, that is, hydrogen-bonded ribbons and $G_4$-based architectures, upon subsequent addition of [2.2.2]cryptand, potassium picrate, and trifluoromethanesulfonic acid. The visualization of such supramolecular interconversion at the solid–liquid interface opens new avenues towards understanding the mechanism of formation and function of complex nucleobase architectures such as DNA or RNA. Furthermore, the *in-situ* reversible assembly and reassembly between two highly ordered supramolecular structures at the surfaces represents the first step towards the generation of nanopatterned responsive architectures.

## Acknowledgements

This work was supported by the COST Network G4-NET (MPNS Action MP0802), the EC Marie Curie ITN-SUPERIOR (PITN-GA-2009-238177) and the International Center for Frontier Research in Chemistry (icFRC, Strasbourg), the University of Bologna, MIUR (Italy) through the National Interest Research Programme (PRIN 2009, grant 2009N5JH4F_002) and the F.R.S.-FNRS (FRFC-BINDER), Belgium. M.S. is research associate of the F.R.S.-FNRS.

## References

1. a) L. Brunsveld, B. J. B. Folmer, E. W. Meijer and R. P. Sijbesma, *Chem. Rev.*, 2001, **101**, 4071-4097; b) J. A. A. W. Elemans, A. E. Rowan and R. J. M. Nolte, *J. Mater. Chem.*, 2003, **13**, 2661-2670; c) T. F. A. Greef and E. W. Meijer, *Nature*, 2008, **453**, 171-173; d) J.-M. Lehn, *Supramolecular chemistry: concepts and perspectives*, VCH New York, 1995; e) J.-M. Lehn, *Polym. Int.*, 2002, **51**, 825-839; f) J. H. van Esch and B. L. Feringa, *Angew. Chem. Int. Ed.*, 2000, **39**, 2263-2266.
2. J. D. Watson and F. H. C. Crick, *Nature*, 1953, **171**, 737-738.
3. S. Sivakova and S. J. Rowan, *Chem. Soc. Rev.*, 2005, **34**, 9-21.
4. J. T. Davis, *Angew. Chem. Int. Ed.*, 2004, **43**, 668-698.
5. M. Gellert, M. N. Lipsett and D. R. Davies, *Proc. Nat. Acad. Sci. U.S.A.*, 1962, **48**, 2013-2018.

6. a) J. S. Lindsey, *New J. Chem.*, 1991, **15**, 153-180; b) D. Philp and J. F. Stoddart, *Angew. Chem. Int. Ed.*, 1996, **35**, 1155-1196; c) D. N. Reinhoudt and M. Crego-Calama, *Science*, 2002, **295**, 2403-2407; d) J. F. Stoddart and H. R. Tseng, *Proc. Nat. Acad. Sci. U.S.A.*, 2002, **99**, 4797-4800; e) G. M. Whitesides, E. E. Simanek, J. P. Mathias, C. T. Seto, D. N. Chin, M. Mammen and D. M. Gordon, *Acc. Chem. Res.*, 1995, **28**, 37-44.
7. J. T. Davis and G. P. Spada, *Chem. Soc. Rev.*, 2007, **36**, 296-313.
8. G. Gottarelli, S. Masiero, E. Mezzina, S. Pieraccini, J. P. Rabe, P. Samorì and G. P. Spada, *Chem. Eur. J.*, 2000, **6**, 3242-3248.
9. a) G. Binnig, H. Rohrer, C. Gerber and E. Weibel, *Phys. Rev. Lett.*, 1982, **49**, 57-61; b) H. Rohrer, *Proc. Nat. Acad. Sci. USA*, 1987, **84**, 4666-4666.
10. a) J. P. Rabe and S. Buchholz, *Science*, 1991, **253**, 424-427; b) J. M. MacLeod, O. Ivasenko, D. F. Perepichka and F. Rosei, *Nanotechnology*, 2007, **18**, 424031-424040; c) A. M. Jackson, J. W. Myerson and F. Stellacci, *Nat. Mater.*, 2004, **3**, 330-336; d) A. Ciesielski, G. Schaeffer, A. Petitjean, J.-M. Lehn and P. Samorì, *Angew. Chem. Int. Ed.*, 2009, **48**, 2039-2043; e) S. De Feyter, A. Gesquiere, M. M. Abdel-Mottaleb, P. C. M. Grim, F. C. De Schryver, C. Meiners, M. Sieffert, S. Valiyaveettil and K. Müllen, *Acc. Chem. Res.*, 2000, **33**, 520-531; f) A. Ciesielski, L. Piot, P. Samorì, A. Jouaiti and M. W. Hosseini, *Adv. Mater.*, 2009, **21**, 1131-1136; g) A. Ciesielski, C.-A. Palma, M. Bonini and P. Samorì, *Adv. Mater.*, 2010, **22**, 3506-3520.
11. R. Lazzaroni, A. Calderone, J. L. Bredas and J. P. Rabe, *J. Chem. Phys.*, 1997, **107**, 99-105.
12. P. Samorì, K. Müllen and H. P. Rabe, *Adv. Mater.*, 2004, **16**, 1761-1765.
13. L. Piot, R. M. Meudtner, T. El Malah, S. Hecht and P. Samorì, *Chem. Eur. J.*, 2009, **15**, 4788-4792.
14. M. Surin, P. Samorì, A. Jouaiti, N. Kyritsakas and M. W. Hosseini, *Angew. Chem. Int. Ed.*, 2007, **46**, 245-249.
15. U. Thewalt, C. E. Bugg and R. Marsh, *Acta Crystallographica Section B: Structural Crystallography and Crystal Chemistry*, 1970, **26**, 1089-1101.
16. T. Giorgi, S. Lena, P. Mariani, M. A. Cremonini, S. Masiero, S. Pieraccini, J. P. Rabe, P. Samorì, G. P. Spada and G. Gottarelli, *J. Am. Chem. Soc.*, 2003, **125**, 14741-14749.
17. G. Maruccio, P. Visconti, V. Arima, S. D'Amico, A. Blasco, E. D'Amone, R. Cingolani, R. Rinaldi, S. Masiero, T. Giorgi and G. Gottarelli, *Nano Lett.*, 2003, **3**, 479-483.
18. A. Ciesielski, R. Perone, S. Pieraccini, G. P. Spada and P. Samorì, *Chem. Commun.*, 2010, **46**, 4493-4495.
19. A. Ciesielski and P. Samorì, *Nanoscale*, 2011, **3**, 1397-1410.
20. a) S. Pieraccini, S. Masiero, O. Pandoli, P. Samorì and G. P. Spada, *Org. Lett.*, 2006, **8**, 3125-3128; b) S. Lena, G. Brancolini, G. Gottarelli, P. Mariani, S. Masiero, A. Venturini, V. Palermo, O. Pandoli, S. Pieraccini, P. Samorì and G. P. Spada, *Chem. Eur. J.*, 2007, **13**, 3757-3764.
21. A. Ciesielski, S. Lena, S. Masiero, G. P. Spada and P. Samorì, *Angew. Chem. Int. Ed.*, 2010, **49**, 1963-1966.
22. G. P. Spada, S. Lena, S. Masiero, S. Pieraccini, M. Surin and P. Samorì, *Adv. Mater.*, 2008, **20**, 2433-2438.

# MORPHOLOGICAL HETEROGENEITY OF SUPRAMOLECULAR G-DNA POLYMERS DERIVED FROM GUANINE RICH OLIGONUCLEOTIDES

T.C. Marsh, Z.M. Henseler and M.A. Klimstra

Department of Chemistry, University of St. Thomas, St. Paul, MN, USA

## 1 INTRODUCTION

Naturally occurring supramolecular assemblies provide us with numerous models for constructing nanoscale materials[1]. Nucleic acids are an excellent example of a class of biomolecules with the capacity to form a variety of higher order structures. The inherent programmability of nucleic acid structure has been exploited to create complex supramolecular assemblies for potential use in nanotechnology[2,3,4]. Elegant two- and three-dimensional assemblies have been created using multiple oligodeoxyribonucleotides that are organized and stabilized by Watson & Crick base paring interactions (i.e. B-DNA based)[5,6]. Supramolecular nucleic acids may also form through self-recognition between guanine-rich oligonucleotides (GROs) that self-assemble into quadruple helical G-DNA structures known as G-wires[7]. These and other supramolecular G-DNAs are of great interest for development of novel nanoscale devices [8,9,10,11] and as potential therapies to treat diseases such as cancer [12,13] and Huntington's Correa[14].

G-DNA is a polymorphic family of nucleic acids that share the characteristic of having four coplanar guanine bases in a cyclically H-bonded structural motif known as a G-quartet or G-tetrad [15,16]. Guanine-rich sequences capable of forming G-DNA found in numerous locations the genome of cells have important biological functions[17,18,19,20,21,22,23]. In addition to base stacking and hydrogen bonding, G-DNA structures are stabilized by specific mono- and divalent metal cations ($K^+$, $Na^+$, $Rb^+$, $Ca^{2+}$, $Sr^{2+}$) or molecular cations ($NH_4^+$) that coordinate within or between stacked G-quartets[24,25,26,27,28]. The O6 carbonyl oxygens of a G-quartet constitute an electrostatic environment in which a single cation may reside. The resulting favourable charge-dipole interaction contributes significantly to the structural stability of a G-DNA [29,30,31]. The degree to which a coordinating cation stabilizes a G-DNA depends partly upon its hydration energy and how well it fits between stacked quartets [30,32]. In addition to conferring a measure of structural stability to the G-DNA, the coordinating cation also plays a role in guiding the overall morphology of the G-DNA[20,28,30]. The sequence of a given guanine-rich nucleic acid and the nature of the cations in its environment collaborate to produce a G-DNA with either a parallel or anti-parallel strand orientation that may be connected through a variety of strand topologies[33,34].

This paper covers the current state of knowledge regarding the formation of supramolecular G-DNA polymers from GROs. Presented here are the key elements that influence the self-assembly of supramolecular G-DNA. These include: the specific GRO sequence composition, cationic conditions that favor self-assembly, the position of covalent modifications to oligonucleotides and temperature of self-assembly. All discussions cover deoxyribonucleotides and the d(N) convention for abbreviating a sequence has been omitted. I am also treating supramolecular G-DNAs derived from GROs as distinct class of polymer from the supramolecular assemblies of guanosine derivatives[35] and the larger monomolecular G4-DNA nanowires[10,36]. I will focus primarily on the GROs that form G-wires and provide a survey of other GRO macromolecular assemblies.

## 1.1 GRO Supramolecular Polymers

There are essentially three major types of supramolecular G-DNAs that have been described to date: G-wires[7], Frayed wires[37] and G-legos[38] (Fig 1). A broader categorization of these supramolecular G-DNA GRO (SGRO) structures divides them into parallel and antiparallel stranded groups with subcategories for the parallel stranded structures that would include G-wires and Frayed-wires. The anti-parallel category has the G-lego as the sole member of this structure class. Aside from the strand orientation difference, supramolecular G-DNAs tend to have the common structural motif of a slipped or out-of-register association of individual strands that form a core G-DNA domain with a bit of G-G duplex at the ends that serve as "sticky" points for additional G-DNA formation. The stoichiometry of this strand addition is revealed by polyacrylamide gel electrophoresis (PAGE) by which a highly regular ladder pattern is observed.

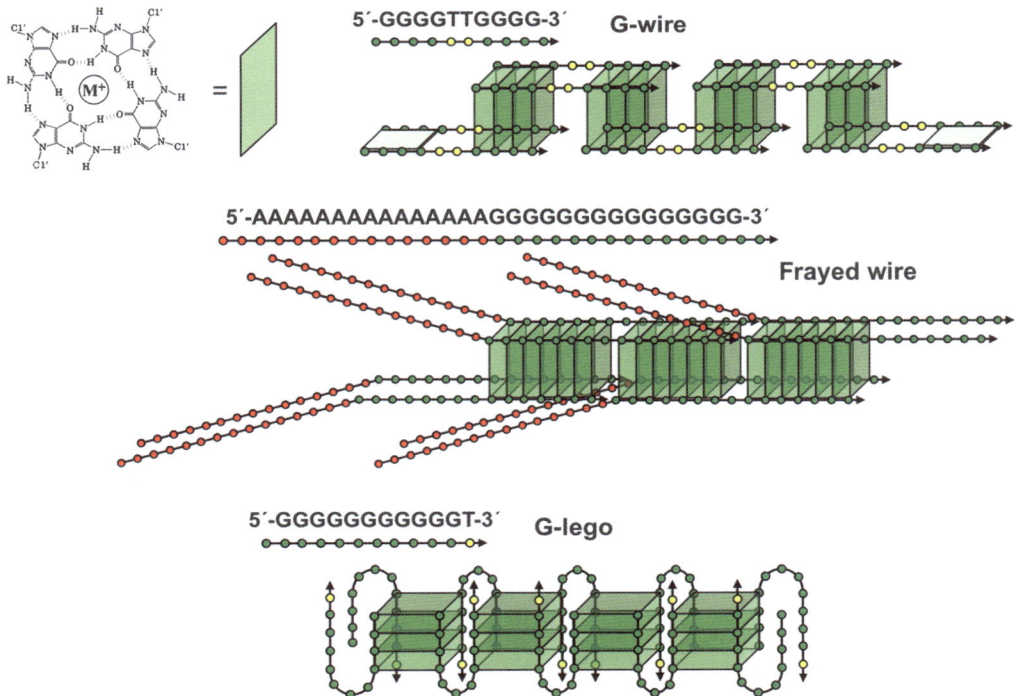

**Figure 1** *Models proposed for G-DNA supramolecular polymers from GROs.*

## 1.2 SGRO Sequences and General Characteristics

A wide variety of sequences have been evaluated for their capacity to form supramolecular G-DNA. The first GROs that were discovered to form ordered multimers were described by Sen and Gilbert[39]. From this initial observation, numerous other studies have explored sequence characteristics and self-assembly conditions from which have emerged several trends that are useful for predicting higher order structure formation. The most significant sequence characteristic required to produce an SGRO is an exposed block of guanine residues at the 5' or 3' terminus of the GRO. A short non-guanine sequence at the 5' end of GRO will produce a sugar cane SGRO and a longer non-guanine sequence will produce a frayed wire. The presence of one, two or three non-guanine residues at the 3' end of a GRO will produce a G-lego. Exposed terminal Gs at the 5'and 3'end of the GRO will produce a G-wire.

**Table 1** *GROs that assemble into multimeric and supramolecular G-DNAs*

| Oligonucleotide Sequence (5'-3') | Cation(s) used | Method(s) of analysis[a] | Strand orientation[b] | Reference |
|---|---|---|---|---|
| *"Sugar cane G-DNA"* | | | | |
| $T_6G_3$ | $K^+$ | PAGE | ∥ slipped htt multimer | 39 |
| $T_{12}G_3$ | | | | |
| $T_nG_4$ | $K^+$ | PAGE, CD | ∥ in-register htt multimer | 40 |
| $(NGG)_4$ N = A,T or C | $K^+$, $Mg^{2+}$ | PAGE, CD | ∥ in-register htt multimer | 41, 42, 43 |
| *"Frayed wire"* | | | | |
| $A_{15}G_{15}$ | $Mg^{2+}$, | PAGE, CD | ∥ slipped | 37, |
| $T_{15}G_{15}$ | $Na^+$, $K^+$, $NH_4^+$ | SPM, Raman, DLS | supramolecular | 44-49 |
| *"G-lego"* | | | | |
| $G_{11}T_{1-3}$ | $Na^+$ | PAGE, CD | anti-∥ slipped supramolecular | 38 |
| $G_{11}C$ | | | | |
| $G_{11}A$ | | | | |
| *"G-wire"* | | | | |
| $G_4T_2G_4$ | $Na^+$, $K^+$, $Mg^{2+}$ | PAGE, AFM | ∥ slipped htt supramolecular | 7, 50, 51 |
| $G_n$ n = 5, 10, 15 or 20 | $Li^+$, $Na^+$, $K^+$, $NH_4^+$ $Mg^{2+}$ $Ca^{2+}$, $Sr^{2+}$ | AFM, CD, PAGE | ∥ slipped supramolecular | 52, 53, 54 and this study |
| $G_{12}$ | $Na^+$ | PAGE | not described | 38 |
|  | $K^+$, $Mg^{2+}$ | CD, AFM | ∥ | 43 |
| $G_{13}$ | $Na^+$ | | | 38 |
| $G_2CGT_4GCG_2$ | $K^+$, $Mg^{2+}$ | PAGE, SPM | ∥ slipped branched | 55, 56 |
| $C_4T_4G_4T_nG_4$ (n = 1-4) | $K^+$, $Mg^{2+}$ | PAGE, CD | ∥ slipped supramolecular | 57 |
| $G_3T_iG_3T_jG_3T_kG_3$ i, j, k = 1-3 | $Na^+$, $K^+$, $NH_4^+$ | PAGE, ESI-MS | ∥ multimers | 58 |
| $GCG_2AG_2CG$ $GCG_2TG_2CG$ $GCG_2TG_4TG_2CG$ $GCG_2(TG_4)_4TG_2CG$ $GCG_2TG_4TG_4TG_2CG$ | $Na^+$ | PAGE, DLS | not described | 59 |

a: CD, Circular dichroism spectropolarimetry; AFM, atomic force microscopy; DLS, Dynamic light scattering; ESI-MS, Electrospray ionization-mass spectrometry
b: ∥ indicates a parallel strand orientation

*1.2.1 Sugar cane G-DNA multimers.* These higher order G-DNA structures are formed by oligonucleotides with a single block of at least three guanines or multiple blocks with at least two guanines and a segment of non-guanine sequence at either the 5' or 3' end of the GRO. Analysis of $T_9G_3$ and $T_{12}G_3$ structures by PAGE (native and/or denaturing) revealed an ordered ladder pattern when incubated in the presence of 1 M KCl. The mobility of the multimers indicated an incremental increase of 4 stranded components following a pattern of 4n+4 (4, 8, 12, 16). No multimers were observed for the GROs that had a T at the 3' terminus. A model was proposed based on methylation protection analysis that included a slipped registration of the G4-DNA components arranged head to tail. The name "sugar cane" was given to these structures to describe the core G-DNA stem with the offset poly-T fronds extending out in a manner reminiscent of the morphology of a sugar cane plant. Though not all GROs in this category have the same sequence pattern, they share a similar strand orientation and incremental size increase of 4N + 4.

Others have investigated the self-assembly potential using GROs with sequences that have a biological context such as telomeric repeat motifs[40, 56] and guanine rich trinucleotide repeat motifs[41,42]. A single repeat of the telomeric G-stand motif from Oxytricha nova, $T_4G_4$, was found to form head to tail G4-DNA multimers in the presence of 200 mM KCl, but only at relatively low temperatures. No multimer formation was observed for $T_4G_4T$.

A greater extent of multimer formation (arguably supramolecular) was observed for "sugar cane" type GROs with tandem telomeric repeat motifs[56]. These supramolecular G-DNAs have much more in common with G-wires than the G-DNA multimers described above. The grow sequences, $C_4T_4G_4T_nG_4$ (n = 1-4), showed highly ordered self-assembly with multimers extending to the origin of the gel. These G-wires were obtained in the presence of 100 mM KCl and 20 mM $MgCl_2$. The additional guanine residues and presence of $Mg^{2+}$ in the self-assembly mixture were key elements in the formation of a supramolecular G-DNA from a sugar cane type GRO.

*1.2.2 Frayed wires.* Frayed wires are formed from oligonucleotides with a track of fifteen guanine residues with a long poly-A or poly-T sequence at the 5'end. These GROs may at first seem similar to the sugar cane multimers; however, they have certain apparent structural differences. Analysis of data from PAGE experiments revealed an increase of 1 strand for each ladder increment as the mobility of the higher order structures decreases (1, 2, 3, 4…) as opposed to the 4N+4 or 2N+2 pattern of strand addition observed with other SGROs. Investigations of strand orientation by CD spectropolarimetry[44] revealed a predominantly parallel stranded structure. Chemical modification and nuclease sensitivity studies suggest there are looped regions and that some anti-parallel character does exist in these SGROs[46,47,48,49]. The formation of Frayed wires show a strong dependence on $Mg^{2+}$ and do not form a regular ladder pattern when incubated with monovalent cations $Na^+$, $K^+$ or $NH_4^+$.

*1.2.3 G-legos.* The most recent supramolecular G-DNA to be reported has many of the properties exhibited by other SGROs. A highly ordered ladder pattern with an incremental increase of two GRO components was observed for oligonucleotides with the sequence $G_{11}T_{1-3}$, $G_{11}A$, $G_{11}C$. Probing the strand orientation of these sequences by CD spectropolarimetry in the presence of 100 mM NaCl, 5 mM $MgCl_2$ buffered at pH 8.5 revealed a hybrid of parallel and anti-parallel strand orientation. The authors also observed that the presence of a 5' phosphoryl group on the G-lego GROs prevented self-assembly. Additionally, GROs with a non-guanine base at both the 5' and 3' termini did not produce a SGRO.

The survey of differ sequences conducted by Biyani and Nishigaki[38] also examined GROs that seemed likely to produce a supramolecular G-DNA but did not yield an ordered supramolecular assembly. For example, $G_6TG_5$ failed to produce an ordered ladder pattern typical of SGRO formation. It must be noted that their study was conducted using only $Na^+$ in the self-assembly buffer for analysis by PAGE; therefore, we predict that some of the sequences the authors evaluated are still viable candidates as SGROs using conditions that are more amenable for self-assembly.

*1.2.4 G-wires.* The first G-wire SGRO was inspired by the "sugar cane" multimers with the idea that if one exposed block of terminal guanines were good, two would be better. The original G-wire GRO, $G_4T_2G_4$, was designed on the basis of the telomeric repeat motif from *Tetrahymena thermophila* and was given the name Tet1.5 (since renamed GRO16). GRO16 exhibited robust self-assembly in the presence of a monovalent cation ($Na^+$ or $K^+$), in combination with $Mg^{2+}$, yielding an ordered ladder pattern that proved to be highly resistant to denaturation. Extensive analysis by PAGE[7,50] supported a parallel stranded slipped G-DNA structure with a ladder increment consistent with the addition of two strands following the pattern of $2N + 2$ (2, 4, 6, 8...stranded multimers). Subsequent studies showed that, in the presence of $K^+/Mg^{2+}$ buffer, $G_4T_2G_4$ yields a CD spectrum consistent with parallel stranded G-DNA[57]. Investigations by Atomic Force Microscopy (AFM) showed that G-wires are linear supramolecular polymers with a length and stability dependent upon the combination of coordinating cation and the presence of $Mg^{2+}$ during self-assembly[50]. The model for assembly of G-wires allows for indefinite addition of material at either end of polymer. Long-term assembly studies done by Vesenka indicate that G-wires can attain remarkably long lengths and looped structures have also been observed[53].

G-wires formed by other GROs include simple tracts of guanines[38,43,52], variations of different telomeric repeat motifs[55,57,58], and sequences that are designed to form not only G-DNA, but also B-DNA duplex interactions[59]. Strategies for self-assembly may include a thermal activation step or dialysis[59]. From these studies it is clear that a wide variety of environmental conditions may be used to form highly stable G-wires. A significant challenge in working with GROs is achieving a measure of control over the size and morphology of the G-wire. The next section will cover specific known factors that influence G-wire self-assembly.

## 2 PARAMETERS FOR G-WIRE SELF-ASSEMBLY

### 2.1 Effect of Cations on G-wire Assembly and Stability.

The vast majority of supramolecular G-DNAs reported to date have been assembled in buffered solutions using 50-100 mM $Na^+$ or $K^+$ with varying use of 1-10 mM concentrations of divalent cations $Mg^{2+}$ or $Ca^{2+}$. There are no clear trends between the different classes of SGROs and within each class vastly different cationic species and concentrations have yielded higher order structure. One trend that is evident is as the number of contiguous Gs in the GRO increases, the concentration of monovalent or divalent cations needed for higher order structure formation decreases. The second trend is that the most extensive higher order structure formation (long/large assemblies) occurs through balancing the cationic species, ratio of cation to GRO, and temperature for a given sequence.

## 2.2 Effect of Temperature on G-wire Formation.

The extent of G-wire formation is greatly influenced by the temperature of self-assembly, which shows a strong thermal activation of higher order structure formation that varies depending upon the species of coordinating cation used to stabilize G-quartets. A comparison of GRO16 G-wires self-assembled in either $Na^+/Mg^{2+}$ and $K^+/Mg^{2+}$ at various temperatures show different efficiencies of higher order structure formation. Peak assembly of GRO16 G-wires in $Na^+/Mg^{2+}$ occurs around 65°C. In contrast, incubation in $K^+/Mg^{2+}$ buffer shows a steady increase through to 75°C. Native gel electrophoresis allows less stable higher order structures to persist and it is not unusual to see a smear of higher order structures as in Figure 2. The smear can be resolved as discrete bands if run under denaturing conditions.[7,50,38] Frayed wires also show a thermal activation[37,47] toward a greater extent of higher order structure formation in similar manner to G-wires.

The resistance to thermal denaturation is also markedly different between $Na^+$ and $K^+$ stabilized G-wires. When $Na^+$ G-wires are placed in a boiling water bath for 10 minutes prior to loading on 10% PAGE, 0.5 X TBE all higher order structure is lost. Subjecting $K^+/Mg^{2+}$ G-wires to 10 minutes in a boiling water bath increases the extent of higher order structure formation and also resolves the polymorphic smear to distinct bands. The temperature at which self-assembly is conducted can have a significant impact on the rate of recruitment of monomeric GRO components to the supramolecular structure and the extent of orderliness of the supramolecular polymer. The self-assembly of frayed wires is also thermally enhanced. A systematic assessment of self-assembly conditions evaluating the effect of temperature and species of coordinating cation shows, again, the preferential stabilization of the G-DNA in the presence of potassium. Numerous studies of thermal

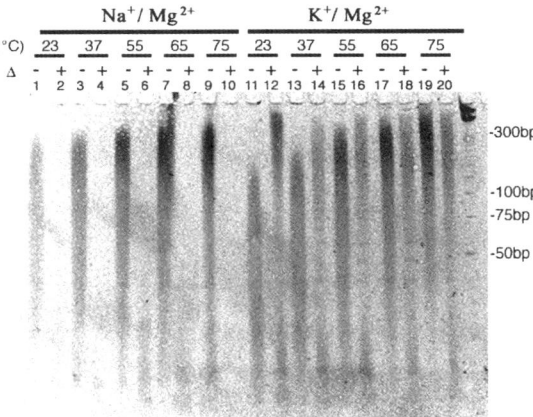

**Figure 2** *Native PAGE analysis of GRO 16 G-wires comparing the effect of temperature on self-assembly and effect of boiling on band resolution. GRO16 G-wires were assembled in either 50 mM NaCl, 10 mM $MgCl_2$ or 50 mM KCl, 10 mM $MgCl_2$ with each buffered in 10 mM Tris-HCl pH7.5. The plus and minus indicate whether a sample was heated for 10 min. in a boiling water bath immediately before loading on a 10% acrylamide (29:1 acrylamide:bis-acrylamide), 0.5 X TBE gel. The electrophoresis was run with a field strength of 25 V/cm. The gel was stained with SYBR green and imaged using a Versadoc 5000 (BioRad, CA) digital image capture device.*

stabilization of G-DNA demonstrate an inability to denature G-DNA structures formed in the presence of potassium. In the case of GRO16 G-wires, boiling samples that were assembled in the presence of potassium has the effect of resolving all higher order structure into a distinct ladder pattern. This pattern of high stability and weeding out of less stable polymorphs is seen with GRO7 ($G_5$) and GRO16 ($G_4T_2G_4$).

## 2.3 Effect of Chemical Modifications to Oligonucleotides on SGRO Formation.

The use of modified oligonucleotides for supramolecular assemblies is an ideal mechanism for adding functionality to a highly ordered robust molecular scaffold. Introducing base analogs and chemical modifications may have adverse and unexpected effects on self-assembly. When the effect of a base analog or modifier are known, they may be used as probes for inferring structural interaction needed for SGRO assembly. A good example of this is methylation analysis using dimethyl sulphate as an alkylating agent to chemically probe for guanines with their N7 exposed to solvent. This technique has been used in a number of studies to confirm G-quartet structure and determine the relative positions of component strands in the structure[37,39,40]. UV crosslinking has also been employed to determine the strand stoichiometry of SGRO repeat units. Thymine residues on different strands that are adjacent to each other in the SGRO become covalently linked when exposed to UV light. Electrophoretic separation of the cross-linked products revealed a number of multimers corresponding to the strand stoichiometry of the G-DNA repeat unit [40,44].

The introduction of modified bases to a GRO often impacts the stability of a G-DNA as well as dictate the morphology of the structure[60,61]. Early investigations of the molecular structure of Frayed wires using methylation analysis indicated that the N7 of guanine was not critical for self-assembly. This would be an unlikely property for a G-DNA structure and an apparent contradiction to the CD data indicated these SGROs are parallel stranded. A recent study of the structure of Frayed wires utilized the isoelectronic guanine analog pyrazolo[3,4,-d]pyrimidine (PPG) to assess the participation of N7 in the assembly of frayed wires[49]. The PPG modified GRO failed to form a Frayed wire and, in concert with other characterization methods, provided strong evidence of G-DNA structure. This study also demonstrated the first G-DNA structure formed from a GRO with the guanine content comprised entirely of 8-bromo-guanine bases.

Non-guanine base analogs may have a subtle to strong impact on the stability of a G-DNA[60] and must be assessed on case-by-case basis to determine this. Modifications at either the 5' or 3' end of the GRO impact the extent of self-assembly depending on the nature of the functional group as well as the conditions used to form the G-wire. It is known that the presence a 5' phosphate group on certain GROs may inhibit assembly. This phenomenon has been observed with G-wire and G-lego GROs. The cause of this inhibition has not been given much attention [7,38]. Never the less, the inhibitory effect provides a clue as to the to the possible strand topology of a G-wire. As indicated in the section of sequence composition, there is a requirement for at least one "naked" guanine base at either end of the GRO to achieve an ordered supramolecular polymer.

## 3 METHODS FOR G-WIRE SELF-ASSEMBLY

The following section provides a general guide for the assembly and preliminary assessment of G-wires. A systematic approach for evaluating the ability of a GRO to form a supramolecular G-DNA varies the: temperature, cationic species, concentration of cations and the combination of cations. The above strategy was applied to the pentamer GRO, $G_5$, named GRO7.

## 3.1 Oligonucleotide Preparation

Guanine rich oligonucleotides sequences were purchased from Integrated DNA Technologies (IDT) [Coralville, IA]. Oligonucleotides were purified using standard purification and desalting methods. Lyophilized samples were resuspended in nanopure water to variable estimated concentrations based on reported mass yields provided by IDT. All final oligonucleotide concentrations were measured with Cary 300 Bio UV-Visible Spectrophotometer [Varian Inc., Palo Alto, CA]. Stock solutions of GRO are stored in small volume aliquots at -25 °C until self-assembly solutions were prepared. A typical self-assembly experiment requires 50 μL of 1 mM GRO. Depending upon the quality of the synthesis, further purification of the GRO may be required by either HPLC or by denaturing PAGE and reverse phase chromatography.

## 3.2 Self-assembly Preparation and Incubation

Solutions of oligonucleotides were thermally denatured at 95°C for 10 minutes then flashed cooled on ice and centrifuged to collect condensate. All self-assembly solutions were prepared in 0.65 mL MicroCentrifuge tubes or similar container (PCR tubes work as well). A total volume of 25 μL is typically used with GRO at a concentration of 100 μM. This is usually a sufficient amount of material to perform 2-3 PAGE analyses, CD measurements and sample preparation for AFM. Self-assembly reactions are run in a10 mM Tris Buffer, pH 7.5, and 10-100 mM XCl (X=$Li^+$, $Na^+$, $K^+$and $NH_4^+$ or $YCl_2$ (Y=$Ca^{2+}$, $Sr^{2+}$). Applicable experimental groups contained 1mM $MgCl_2$. Self-assembly solutions were incubated at variable temperatures, ranging from 37-75°C. Incubation proceeded for 48 hours to allow for appreciable level higher ordered structure formation.

## 3.3 Analysis of Higher Ordered Structure

The extent of supramolecular G-DNA formation and the order of the SGRO are analyzed by native and denaturing PAGE as well as Atomic Force Microscopy (AFM). Ultra-pure AccuGel 19:1, 40% acrylamide: bis-acrylamide [National Diagnostics, Atlanta, GA] was used and polymerized with a final acrylamide concentration of 15% or 20%. Denaturants were used to resolve ordered banding patterns in apparently amorphous self-assembly products. In these cases, formamide [Mallinckrodt Baker Inc., Paris, KY] was combined in equal proportions with self-assembly solutions and optionally heated for 10 minutes at 95°C just prior to loading on the gel prepared with 8 M urea, 15% 19:1 acrylamide:bis-acrylamide. For native gel loading a 20% sucrose in 5x TBE was added to an equal volume of self-assembly sample. Bio-Rad Protean II ix Cell Electrophoresis System [Bio-Rad, Hercules, CA] was used for separation of higher-ordered structure. All gels are stained with SYBR Green I Nucleic Acid Gel Stain [Invitrogen, Eugene, OR] and photographed under UV light using VersaDoc Model 5000 imaging system [Bio-Rad]. Assessment of higher order structure formation by AFM is described in detail in[50,51,65] and else where in this volume[54].

## 3.1 Analysis of $G_5$ by PAGE

A relatively small GRO formed a remarkably stable SGRO in a variety of cationic conditions. The presence of $Mg^{2+}$ significantly enhanced self-assembly in a similar

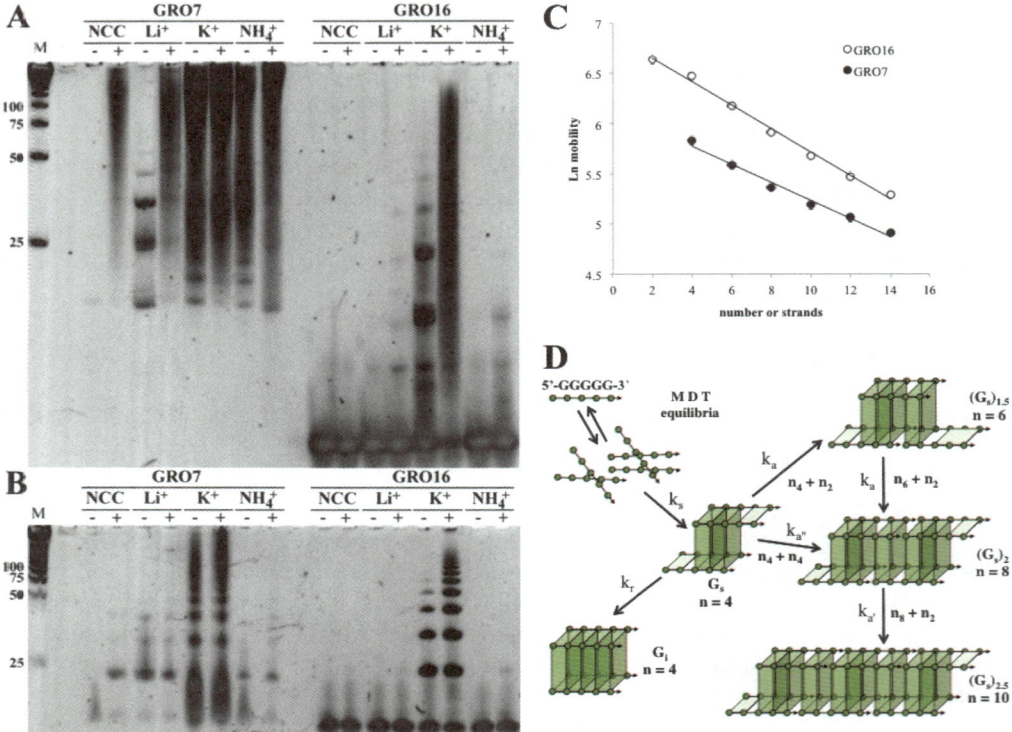

**Figure 3** *A short GRO will self-assemble into a G-wire. A) Native PAGE analysis comparing the self-assembly of two GRO sequences, GRO7 ($dG_5$) and GRO16 ($dG_4T_2G_4$). Samples were incubated for two days at 55°C in the presence of 10 mM concentration of the cations indicated above each sample and each buffered in 10 mM Tris-HCl pH7.5. The plus and minus indicate whether 1 mM $MgCl_2$ was included in the self-assembly buffer. Each sample was loaded on a 15% acrylamide (19:1 acrylamide:bis-acrylamide), 0.5 X TBE gel. The electrophoresis was run with a field strength of 25 V/cm. A 25 bp DNA ladder, M, was run as a relative standard to compare mobility. B) Denaturing PAGE analysis of the same samples in "A". Formamide was combined in equal proportions with self-assembly solutions and heated for 10 minutes at 95°C. Formamide/GRO samples were then loaded directly onto 8 M, 15% (19:1 acrylamide:bis-acrylamide) 0.5 TBE gel run at a constant temperature of 55°C. C) Plot of comparing the log of ladder mobility in native PAGE as a function of strand number between GRO 7 and GRO 16. D) Proposed assembly pathway for GRO7 G-wires. MDT equilibrium represents the monomer, dimer, trimer equilibrium that contribute to the pseudo-fourth order rate constant for the formation of a predicted slipped strand G-DNA intermediate. G-wire formation is favoured if the rates of $k_a$ or $k_{a''} > k_r$.*

manner to GRO16. The stability of GRO7 to mildly denaturing conditions was greater than GRO16 and GRO7 showed structure formation even with LiCl. The plot of mobilities as a function of strand number for each incremental increase in ladder band size fits well with a 2N + 2 pattern (figure 3C). The model shown in figure 3D indicates that the extent of G-wire formation may be due to $G_S$ associating with intermediates from the MDT pool to form additional G-quartets at a greater rate than the Gs intermediate can undergo rearrangement to the in-register structure, $G_I$.

The data collected for GRO7 G-wires support a model for self-assembly that involves slipped strands associating to form a linear supramolecular G-DNA. The proposed assembly path (figure 3D) is consistent with the PAGE analysis and AFM data (figure 4A). The assembly intermediates and details of the proposed pathway are based on predicted intermediates from molecular dynamics simulations[66], ESI-MS[67] data and kinetics of tetramolecular G-DNA assembly[68].

## 4 G-WIRE STRUCTURE AND STRAND TOPOLOGY

Little is know about the structure of G-wires at the molecular level. In the absence of direct physical measurements of G-wire structure, considering the possible pathways for assembly could provide insight into their likely strand topology. Identifying the morphological variations that are possible for G-wires at the molecular level is a key step in developing viable nanomaterials from these SGROs. It is appropriate at this point to revisit the original model for G-wire self-assembly (figure 5) and discuss a problematic aspect of the original proposed structure[7,50] that is, literally and figuratively, the gap in the model. As attractive as the simple indefinite expansion of a G-wire via a slipped registration of four parallel strands happens to be, there is currently no definitive evidence for where the thymine bases reside in a G-wire. Data from PAGE, CD spectropolarimetry, AFM, Mass Spectrometry (MS), and Molecular Dynamics (MD) can be used to evaluate the original model as well as propose a possible alternative: G-wires could maintain Ts in propeller loops.

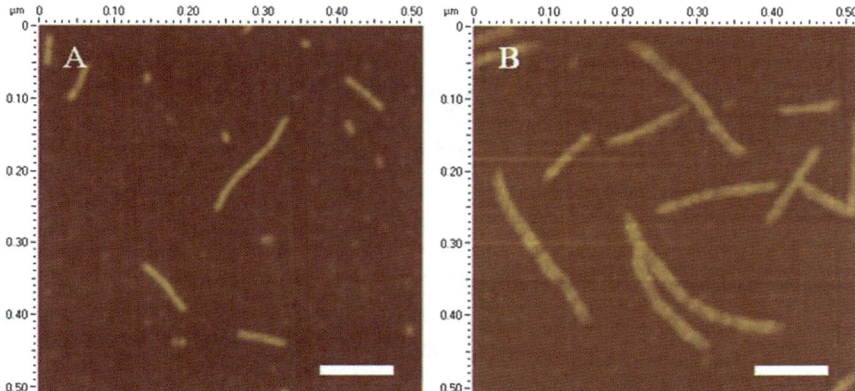

**Figure 4**   *Tapping Mode AFM images of GRO7 A) and GRO16 B) self-assembled in 10 mM KCl, 1 mM MgCl2, 10 mM Tris-HCl as described in figure 3 above. GRO7 G-wires have a mean height of 1.6 nm compared to 2.9 nm for GRO 16. The scale bar is equal to 100 nm. Image A) collected by James Vesenka, Univerity of New England.*

## 4.1 Minding the Gap

In order to conceive of a viable topology for the GRO components of a G-wire, the possible G-DNA strand topologies that a GRO can adopt must be considered. The first candidate to consider that would account for observed G-wire self-assembly is an in-register parallel stranded structure for GRO16 ($G_I$ in figure 5). This structure has a total of eight quartets in two blocks of four separated by a loop of two T bases. Potential multimers could occur through head-to-tail stacking in a manner similar to that reported for $T_4G_4$[40]. The Ts in the loop could either form a tetrad in the gap or bulge out to allow for further stacking of the G-quartets. In either case, the strong argument against this being a viable model for GRO16 and similar GROs is that known head to tail G-DNA multimers, such as $T_4G_4$[40] fall apart near or above room temperature and the more robust $GGGX_iGGGX_jGGGX_kGGG$[58] multimers readily dissociate under denaturing conditions. It is thought that greater interstrand connectivity accounts for some of the high thermal stability of G-wires and may be achieved through a slipped strand orientation. If GRO16 follows a path for G4-DNA assembly that is similar to $TG_5T$ then a slipped strand ($G_S$) intermediate would be a precursor to forming the in-register structure. The slipped orientation of strands is supported by MD simulations[66] of $G_4$ as well as by mass spectrometry of tetramolecular G-DNA intermediates of the GRO $TG_5T$ [68]. As suggested previously for GRO7 G-wires, the rate of rearranging $G_S$ to $G_I$ for GRO16 may be much slower than the formation of additional G-quartet structures at the sticky G-ends. The thymine bases at the 5' and 3' termini of a GRO likely interfere with the annealing of new strands to the slipped intermediate.

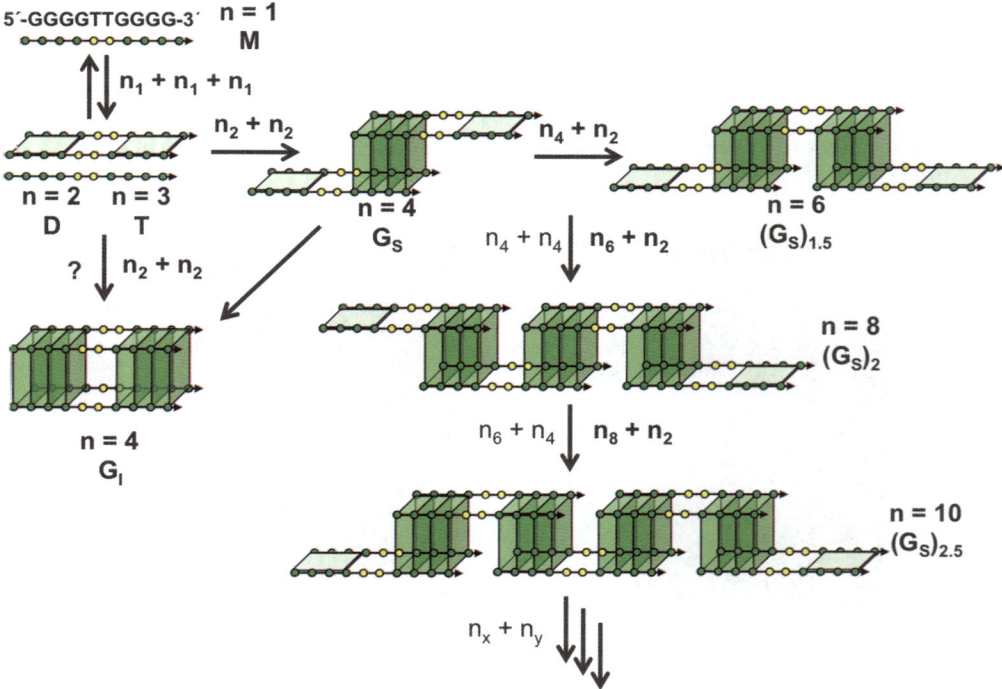

**Figure 5** *Variation of the classical G-wire self-assembly model first proposed to describe formation of highly stable, ordered structures observed by PAGE analysis.*

Even with slipped strands, the same extent of G-quartet formation is eventually achieved as would be expected with head-to-tail multimers of $(G_I)_n$. Despite the benefit of the slipped model of assembly for GRO16 it still places a gap between the adjacent blocks of G-quartets as has also been proposed for $C_4T_2G_4T_{1-4}G_4$[57]. It is possible that the Ts may stack within the structure or bulge out to allow stacking between the two blocks of quartets. Other models[41,42] for SGROs have proposed a bulge mechanism to enable neighbouring quartets to mind the gap and thus achieve greater stability by increasing stacking interactions as well as creating another pocket for binding of cations such as $Na^+$, $NH_4^+$, $K^+$ etc. The positioning of Ts in the slipped registration model of G-wire self-assembly have just two strands in the gap and, therefore, would presumably possess a lower energy barrier to bulging out thus facilitate stacking of neighbouring blocks of G-quartets.

## 4.2 Propeller Loop Topology Model for G-wires

A slipped strand intermediate is not the only conceivable structure compatible with the self-assembly of a G-wire possessing a four strand connectivity with sticky G-ends. The alternative strand topologies for GRO16 that can be explored as candidate G-wire building blocks are shown in Figure 6. It is known that G-DNA structures with distinctly different electophoretic mobilities are observed for GRO16 when incubated in either $Na^+$ or $K^+$ in the absence of magnesium[50]. The specific identity of polymorphs in these experiments are not know, but it is clear from the PAGE data that the $Na^+$ polymorph leads to a dead end in regard to G-wire formation as opposed to $K^+$ which can support self-assembly of G-wires even presence or absence of $Mg^{2+}$. In considering the candidate polymorphs, the edge loop structures may be disqualified as they involve anti-parallel strands. Current CD data indicates that GRO16 and its kin are parallel stranded under conditions that promote self-assembly. The alternative strand topology that fits with CD observations is the double chain reversal or propeller loop.

Propeller loop polymorphs that could lead to self-assembly of as well as account for the observed PAGE and CD analyses are the P2p and P3p conformers (Figure 7). The pathway for G-wire formation through the P4p conformer is may lead to the formation of head to tail multimers in a similar fashion to model proposed for

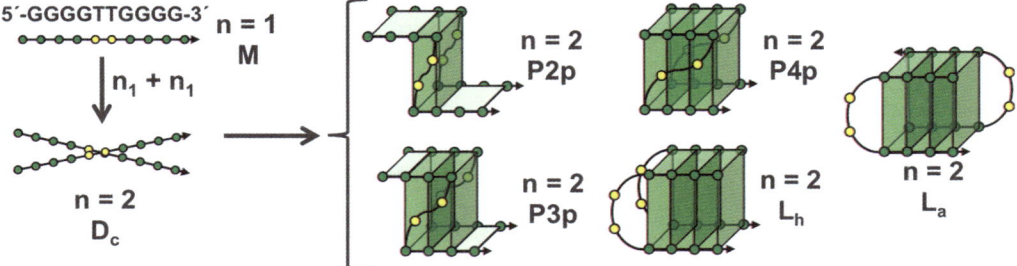

**Figure 6** *Potential looped structural intermediates that GRO16 may adopt. Nomenclature for the structures is as follows: M = monomer, Dc = crossed duplex, P2p = propeller loop spanning two quartets with parallel strands, P3p = propeller loop spanning three quartets with parallel strands, P4p = propeller loop spanning four quartets with parallel strands, Lh = lateral loops on same ends, La = lateral loops on opposite ends.*

GGGX$_i$GGGX$_j$GGGX$_k$GGG[58] though it would not be expected to show the same level of resistance to thermal and chemical denaturation. The formation of a two-stranded parallel G-DNA intermediate fits the 2N+2 pattern of decreasing electrophoretic mobility observed by PAGE analysis. A two strand intermediate bearing two stacked quartets would have greater stability relative to parallel duplex intermediates stabilized by G-G base pairing. Combining two propeller loop intermediates (P2p)$_2$ or (P3p)$_2$ creates a new multimer with 6 and 7 stacked G-quartets respectively. The propeller loop intermediate pathway avoids the gap conundrum while retaining the interstrand connectivity of the original slipped stem model. Recent MD investigations[69] of propeller loop intermediates suggest that loop length is a strong determinant for stability.

## 5 CONCLUSIONS

The new G-wire based on a simple G$_5$ component has remarkable stability in a variety of cationic conditions that may prove useful in specific molecular scaffold applications. GRO7 G-wires also demonstrate the dominant nature of quartet base stacking in stabilizing a G-DNA. When considering the base stacking potential of GRO7 G-wires in the context of GRO16, the "gapped" G-wire may have alternate morphologies that suggest it could function as a dynamic, tunable supramolecular polymer.

In summary, the polymorphic nature of G-DNA is well represented by the variety of SGROs characterized thus far. As knowledge concerning the molecular details of various SGRO structures increases, using GROs to design nanoscale structures and devices will not be merely painting in different shades of grey.

**Acknowledgements**

The authors gratefully acknowledge the assistance of undergraduate researchers Jessica Loehr, Matt Turner, Brandon Goblirsch and Joseph Skaja. I would also like to thank Eric Henderson and Jamie Vesenka for assistance with SPM instrumentation. We acknowledge the financial support from the University of St. Thomas Department of Chemistry and the UST Undergraduate Research and Collaborative Scholarship program.

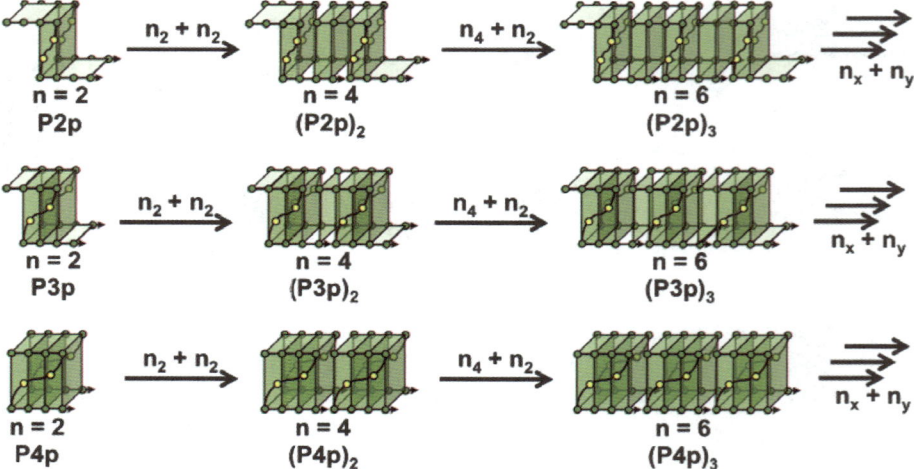

**Figure 7** *Proposed propeller loop intermediate mediated G-wire self-assembly paths.*

# References

1. N.C. Seeman, *Annual Review of Biochemistry*, 2010, **79**, 65.
2. E. Winfree, F. Liu, L.A. Wenzler and N.C. Seeman, *Nature*, 1998, **394**, 539.
3. P.W.K. Rothemund, *Nature*, 2006, **440**, 297.
4. F.A. Aldaye, A. Palmer and H.F. Sleiman, *Science*, 2008, **321**, 1795.
5. B. Wei, M. Dai & P. Y, *Nature*, 2012, **485**, 623.
6. D. Han, S. Pal, J. Nargreave, Z. Deng, Y. Liu and H. Yan, *Science,* 2011, **332**, 342.
7. T.C. Marsh, and E. Henderson. *Biochemistry,* 1994, **33**, 10718.
8. M. Batalia, E. Protozanova, R.B. Macgregor Jr., and D. Erie, *Nano Lett.*, 2002, **2**, 269.
9. R.P. Fahlman, M. Hsing, C. Sporer-Tuhten, and D. Sen, *Nano Letters,* 2003, **3**, 1073.
10. Kotlyar, N. Borovok, T. Molotsky., H. Cohen, E. Shapir, D. Porath, *Adv. Mater.*, 2005, **17**, 1901.
11. C. Leiterer, A. Csaki and W. Fritzsche, *Methods in Molecular Biology,* 2011, **749**, 141.
12. V. Dapic, V. Abdomerovic, R. Marrinton, J. Peberdy, A. Rodger, J. Trent and P. Bates, *Nucleic Acids Res.*, 2003, **31**, 2097.
13. L. Benimetskaya, M. Berton, A. Kolbanovsky, S. Benimetsky and C.A. Stein, *Nucleic Acids Res.*, 1997, **25**, 2648.
14. M. Skogen, J. Roth, S. Yerkes, H. Parekh-Olmedo and E. Kmiec, *BMC Neurosci.*, 2006, **7**, 65.
15. J.R. Williamson, M.K. Raghuraman, and T.R. Cech, *Cell*, 1989, **59**, 871.
16. J.R. Williamson, *Proc. Natl. Acad. Sci. USA.*, 1993, **90**, 3124.
17. D. Sen and W. Gilbert, *Nature*, 1988, **334**, 364.
18. E. Henderson, C.C. Hardin, S.K. Walk, I. Tinoco Jr. and E.H. Blackburn, *Cell*, 1987, **51**, 899.
19. W.I. Sundquist and A. Klug, *Nature,* 1989, **342**, 825.
20. D. Sen and W. Gilbert, *Nature*, 1990, **344**, 410.
21. P. Balagurumoorthy and S.K. Brahmachari, *J. Biol. Chem.*, 1994, **269**, 21858.
22. A.K. Todd, M. Johnston, and S. Neidle, *Nucleic Acids Res.*, 2005, **33**, 2901.
23. J.L Huppert and S. Balasubramanian, *Nucleic Acids Res.*, 2005, **33**, 2908.
24. C. Hardin, E. Henderson, T. Watson, and J.K. Prosser, *Biochemistry*, 1991, **30**, 4460.
25. J.R. Williamson, Annu. Rev. *Biophys. Biomol. Struct.*, 1994, **23**, 703.
26. F.W. Smith, P. Schultze, and J. Feigon, *Structure*, 1995, **3**, 997.
27. G.N. Parkinson, M.P. Lee, and S. Neidle, *Nature*, 2002, **417**, 876.
28. F.M. Chen, *Biochemistry*, 1992, **31**, 3769.
29. J.L. Mergny, A. De Cian, A. Ghelab, B. Sacca, and L. Lacroix, *Nulceic Acids Res.*, 2005, **33**, 81.
30. N. Hud and J. Plavec, *Quadruplex Nucleic Acids*, Royal Society of Chemistry, Cambridge, 2006, Chapter 4, p.100.
31. P. Sket and J. Plavec, *J. Am. Chem. Soc.*, 2010, **132**, 12724.
32. N. Hud, F.W. Smith, F. Anet and Juli Feigon, *Biochemistry*, 1996, **35**, 15383.
33. M.A. Keniry, *Biopolymers*, 2001, **56**, 123.
34. S. Neidle and S. Balasubramanian, *Quadruplex Nucleic Acids*, Royal Society of Chemistry, Cambridge, 2006, Chapter 1, p.15.
35. J.T. Davis, *Angew. Chem. Int. Ed.*, 2004, **43**, 668.
36. N. Borovok, N. Iram, D. Zikich, J.G. Gideon, I. Livshits, D. Porath, and A. Kotlyar, *Nucleic Acids Res.*, 2008, **36**, 5050.
37. E. Protozanova and R.B. Macgregor Jr., *Biochemistry.*, 1996, **35**, 16638.
38. M. Biyani, and K. Nishigaki, *Gene*, 2005, **364**, 130.
39. D. Sen and W. Gilbert, *Biochemistry*, 1992, **31**, 65.

40 M. Lu, Q. Guo and N. Kallenbach, *Biochemistry*, 1992, **31**, 2455.
41 F-M. Chen, *J. Biol. Chem.*, 1995, **270**, 23090.
42 F-M. Chen, *Biophys. J.*, 1997, **73**, 348.
43 F. Sha, R. Mu, D. Henderson and F-M. Chen, *Biophys. J.*, 1999, **77**, 410.
44 E. Protozanova and R.B. Macgregor Jr., *Biophys. J.*, 1998, **75**, 982.
45 K. Poon and R.B. Macgregor Jr., *Biophys. Chem.*, 1999, **79**, 11.
46 K. Poon and R.B. Macgregor Jr., *Biopolymers*, 1998, **45**, 427.
47 E. Protozanova and R.B. Macgregor Jr., *Biophys. Chem.*, 2000, **84**, 137.
48 K. Poon and R.B. Macgregor Jr., *Biophys. Chem.*, 2000, **84**, 205.
49 R. Abu-Ghazalah, J. Irizar, A. Helmy, R.B. Macgregor Jr., *Biophys. Chem.*, 2010, **147**,123.
50 T.C. Marsh, J. Vesenka, and E. Henderson, *Nucleic Acids Res.*, 1995, **23**, 696.
51 J. Vesenka, E. Henderson, & T. Marsh, Ed. W. Fritzsche, *AIP Conference Proceedings*, 2002, **640**, 109.
52 M. Weglarz, J. Vesenka, W. Fritzsche, S. Yerkes and E Kmiec, *AIP Conference Proceedings*, 2008, **1062**, 123.
53 S.Yerkes, J. Vesenka and E. Kmiec, *J. Neurosci. Res.*, 2010, **88**, 335.
54 J. Vesenka, in Chapter 5 of this volume.
55 C. Zhou, C. Wang, Z. Wei, Z. Wang, C. Bai, J. Qin and E, Cao, *J. Biolmol. Struct. Dynamics*, 2001, **18**, 807.
56 X-Y. Zhang, E-H Cao, Y. Zhang, C. Zhou and C. Bai, *J. Biolmol. Struct. Dynamics*, 2003, **18**, 807.
57 T.Y. Dai, S.P. Marotta, and R.D. Sheardy, *Biochemistry.*, 1995, **34**, 3655.
58 N. Smargiasso, F. Rosu, W. Hsia, P. Colson, E.S. Baker, M. Bowers and V. Gabelica, *J. Am. Chem. Soc.*, 2008, **130**, 10208.
59 L. Spindler, M. Rigler, I. Drevensek-Olenik, N. M. Hessari and M. Webba da Silva, *J. Nucl. Acids.*, 2010, **2010**. 43165.
60 D. Miyoshi, A. Nakao, T. Toda and N. Sugimoto, *FEBS Lett.*, 2001, **496**, 128.
61 B. Sacca, L. Lacroix and J-L. Mergny, *Nucl. Acids Res.*, 2005, **33**, 1182.
62 I. Kutyavin, S. Lokhov, I. Afonina, R. Dempcy, A. Gall, V. Gorn, E. Lukhtanov, M. Metcalf, A. Mills, M. Reed, S. Sanders, I. Shishkina, and N.M.J. Vermeulena, *Nucleic Acids Res.* 2002, **30**, 4952.
63 C.C. Hardin, A.G. Perry, and K. White, *Biopolymers*, 2001, **56**, 147.
64 D. Miyoshi, A. Nakao, N. Sugimoto, *Nucleic Acids Res.*, 2003, **31**, 1156.
65 T. Marsh and J. Vesenka, in *Nano and Molecular Electronics Handbook*, Sergy Lyshevski ed., CRC Press, New York, 2007, Chapter 13, pp. 1-15.
66 R. Stefl, T.E. Cheatham 3$^{rd}$, N. Spackova, E. Fadrna, I. Berger, J. Koca, J. Sponer, *Biophys J.*, 2003, **85**, 1787.
67 F. Rosu, V. Gabelica, H. Poncelet, and E. De Pauw, *Nucleic Acids Res.*, 2010, **38**, 5217.
68 J. Gros, F. Rosu, S. Amrane, A. De Cian, V. Gabelica, L. Lacroix and J-L. Mergny, *Nucl. Acids Res.*, 2007, **35**, 3064.
69 X. Cang, J. Sponer and T. Cheatham 3$^{rd}$., *J. Am. Chem. Soc.*, 2011, **133**, 14270.

# THERMODYNAMICS OF G-QUADRUPLEXES

C. Giancola

Department of Chemical Science, University of Naples "Federico II", Via Cintia, Naples, 80126 Italy

## 1 INTRODUCTION

The transient folding and unfolding of G-quadruplex structures at the ends of telomeres may have a specific role in a regulatory mechanism of the cellular life cycle[1-3]. The understanding of the energetics of G-quadruplexes stability is important not only for the stability in itself, but also to identify the target structures for drugs. The thermodynamics of G-quadruplex folding/unfolding has been reported in relatively few papers. Chaires has provided a thermodynamic overview of the stability of G-quadruplexes and commented the different data reported by various laboratories on the stability of G-quadruplexes of similar sequences, in comparable solution conditions, using the following hard words: "the results are unacceptable and discouraging"[4]. Recently, Lane discussed the stability of quadruplex structures in a review, and raised questions about the influence of solvation and steric crowding on stability of intramolecular quadruplexes compared to that of DNA duplexes[5]. G-quadruplex structures are very unstable in comparison to duplex DNA and need monovalent cations as $Na^+$ and $K^+$ for their folding[6]. Further, the exact structure of the folded form depends critically on the cation composition of the solution. Recent reviews summarize all the known structures characterized for human telomeric quadruplexes[7,8]. In any case, the folding of unimolecular telomeric quadruplex sequences is rapid and thermodynamically favoured[4,7,9]. Instead, the formation of tetramolecular G-quadruplexes is very slow, and the thermodynamic parameters should be accurately determined considering the kinetic control in the quadruplex formation. A method to determine the thermodynamic stability of G-quadruplexes from the sequence information alone was implemented by Stegle et al. and applied to quadruplexes in the human genome[10]. However, validation of that method requires experimental data and, in turn, the collected experimental data based on a standardization of measurements should help to improve the prevision of thermodynamic parameters.

## 2 BIOPHYSICAL METHODOLOGIES FOR STUDYING QUADRUPLEX UNFOLDING

A number of biophysical methodologies are available to follow the folding and the unfolding of G-quadruplexes. These methodologies can be used simply for evaluating the melting temperature or to obtain more detailed information about the thermodynamics of G-quadruplex conformational transitions. The main methodologies devoted to gain information on energetics of conformational transitions are based on spectroscopic or calorimetric principles. The spectroscopic techniques such as UV, fluorescence and circular dichroism (CD) are rapid methods and require small amounts of material. A calorimetric technique as differential scanning calorimetry (DSC) allows a deeper insight into the energetics aspects of G-quadruplex stability.

The UV melting of G-quadruplexes can be conveniently monitored by hyperchromic shift due to the unstacking of bases. The experiments measure the change in absorbance at 295 nm, a maximum for G-quadruplex structures[11]. The fluorescence resonance energy transfer (FRET) can be utilized to follow quadruplex melting, but it requires labels which could affect quadruplex stability[12,13]. However, the FRET method offers the advantage that low concentration and small volume of quadruplex solutions can be used. CD melting experiments can be also conveniently performed since this spectroscopic methodology is particularly sensitive to quadruplex conformational transitions[13]. The CD signal of a maximum or a minimum of the quadruplex spectrum can be plotted as function of temperature. $T_m$ and van't Hoff enthalpy can be calculated from the fitting of the obtained sigmoidal curve. In a DSC experiment the change in heat capacity of a macromolecule in a buffer solution is followed as function of temperature.

The most important difference among these methodologies is that DSC directly measures the unfolding enthalpy, whereas in the spectroscopic experiments the enthalpy is calculated by the van't Hoff analysis assuming a two-step melting process, without significantly populated intermediate states. DSC can be utilized to study quadruplexes of any molecularity providing direct information on enthalpy changes due to dissociations, without any model assumption. It is a convenient tool especially to study G-quadruplex which folding/unfolding proceeds without kinetically significant intermediates and thermodynamic parameters are straightforward calculated. In addition, comparison of the calorimetric enthalpies with calculated model-dependent enthalpies provides additional information on the unfolding process as the cooperativity of the melting or the presence of intermediate states[14-17]. A representative DSC curve for the unfolding of a G-quadruplex structure is shown in Figure 1.

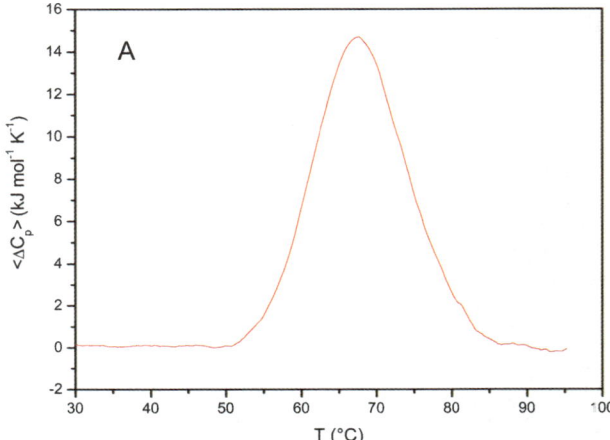

**Figure 1** *Heat capacity change versus temperature profiles for the unimolecular quadruplex of sequence TAGGG(TTAGGG)$_3$.*

In a typical calorimetric experiment the overall enthalpy is obtained directly as the integral of the DSC curve:

$$\Delta H^0 = \int_{T_i}^{T_f} C_P^0 dT \tag{1}$$

Where $C_P^0$ is the heat capacity change referred to a native initial state and $T_i$ and $T_f$ the initial and final melting temperature, respectively.

The overall entropy can be also calculated from DSC curve:

$$\Delta S^0 = \int_{T_i}^{T_f} \frac{C_P^0}{T} dT \tag{2}$$

It is noteworthy that where comparisons are to be made between systems, $\Delta G^0$ must be calculated at a chosen reference temperature, such as 298 K or 310 K. For this purpose, denaturation enthalpies, entropies and Gibbs energies can be calculated as a function of temperature (T). These thermodynamic functions can be obtained according to the well-known Kirkhhoff relationships:

$$\Delta H^0(T) = \Delta H^0(T_m) + \int_{T_m}^{T} \Delta C_P^0 dT \tag{3}$$

$$\Delta S^0(T) = \Delta S^0(T_m) + \int_{T_m}^{T} \frac{\Delta C_P^0}{T} dT \tag{4}$$

Where $T_m$ is the maximum of DSC peak, $\Delta H^0(T_m)$ and $\Delta S^0(T_m)$ are the enthalpy and entropy value calculated from Equations 2 and 3. Finally, the free energy change may be calculated:

$$\Delta G^0(T) = \Delta H^0(T) - T\Delta S^0(T) \tag{5}$$

The heat capacity chance $\Delta C_P^0$ may be obtained from DSC curves, using the differences between pre- and post-transition baselines. In most cases no differences are observed[18] suggesting that $\Delta C_P^0$ is close to zero or that a very small, non-zero heat capacity change is lost in the experimental noise.

In order to apply the equilibrium thermodynamic equations to G-quadruplex systems, it is essential to check the reversibility of the unfolding process. The usual test for reversibility is to perform two DSC scans and check that the second scan gives the same curve observed in the first scan.

## 3 THERMODYNAMICS OF G-QUADRUPLEXES

The "disruption" of a quadruplex structure requires a cost in term of Gibbs energy. Although G-quadruplexes are considered stable structures, their stability when compared with that of other macromolecules is not so high[5]. The unfolding is clearly entropy driven, the enthalpy being an unfavourable contribute to the energetic balance. The gain in entropy is due to the increased freedom degrees of unfolded oligonucleotide chain, and in the case of bi- and tetra-molecular quadruplexes it also due to the contribution of strands dissociation. The unfavourable enthalpy contribution is mainly due to the lost of eight hydrogen bonds per tetrad, the stacking interactions between tetrad planes and the ions release from the quadruplex cavity.

As above mentioned, both calorimetric and spectroscopic signals may be conveniently used to follow G-quadruplex melting. However, determination of thermodynamic parameters from melting curves requires that these profiles correspond to equilibrium curves. To check it, one may record the heating and cooling profiles to follow the unfolding and folding process, respectively. If the unfolding/folding process is fast compared to the temperature gradient the two profiles overlap, and the thermodynamic parameters can be determined from these curves. If they do not overlap a hysteresis phenomenon is present[19]. The hysteresis depends on the temperature gradient. A very low scan rate leads to near-reversible curves, and the extrapolation to zero scan-rate gives a good estimation of the $T_m$ [20]. These long-term experiments are usually not so accurate due to the evaporation and/or degradation of the solutions. In those cases, other methods should be utilized, as described in the paragraph 3.2.

## 3.1 Unimolecular G-quadruplexes

The folding/unfolding of unimolecular G-quadruplexes are in most cases fast processes and melting and annealing curves do not present hysteresis. Therefore, the thermodynamic parameters relative to their stability are easily calculated by both calorimetric and spectroscopic measurements, especially if the thermal unfolding is a two state-transition. However, the unfolding is not always a two-state transition, in that case some experimental tests could help to justify the presence or not of intermediate states. An useful test is the comparison between the calorimetric (model-independent) and the van't Hoff (two-state model) enthalpies: if the $\Delta H^0_{cal}/\Delta H^0_{v.H.}$ ratio is greater than one, intermediate species are present. Another interesting test consists in following the melting at two different wavelengths that should be plotted against one another[21]. A deviation from linear behaviour reveals the presence of intermediate states that are significantly populated. Spectra series collected at different temperatures can be further analysed with the more sophisticated Singular Value Decomposition (SVD) method. SVD analysis allows to evaluate the number of significant spectral species contributing to the observed spectral change in the folding/unfolding process[22-24].

Few papers in literature describe the thermodynamics of unimolecular G-quadruplexes. Marky and co-workers reported a complete thermodynamic analysis of the unfolding of G-quadruplexes containing two G-tetrads with a number of modified loop sequences relative to the thrombin binding aptamer (TBA), whose well known sequence is d(GGTTGGTGTGGTTGG) [25]. The comparison of thermodynamic parameters allowed evaluating the relative contributions of the base modifications in the loops. Changes in the sequence of the loops of G-quadruplexes may, indeed, have a significant effect on their overall stability. In particular, they found that all base substitutions in the loops caused a decreased in the stability. The thermodynamic contributions were discussed in terms of loss of base-base stacking within the loops and their change in hydration. In another paper, Kumar and Maiti compared the experimental thermodynamic values of 36 quadruplex-forming sequences from the promoter regions of various proto-oncogenes compared with estimates of the values for the corresponding duplexes[26]. They found a greater stability for the duplex than the G-quadruplex structure and, interestingly, the increase in loop length increases the difference between duplex-quadruplex stability.

Although the G-quadruplexes deriving from the truncated human telomeric sequence are the most studied among the unimolecular G-quadruplex structures, only few data are present in literature on their thermodynamic stability. In the Table 1 are collected the thermodynamic parameters for similar sequences in $K^+$ containing solution, obtained by spectroscopic or by calorimetric measurements.

The values in Table 1 require some comments:

- $T_m$ values are in the range 60-69 °C, except for the G-quadruplex TGGG(TTAGG)$_3$, whose $T_m$ is higher.
- $\Delta H^0$ and $\Delta G^0$ values show a great variability, although most of $\Delta H^0$ values fall in the range 60-80 kJ mol$^{-1}$ per tetrad.
- Thermodynamic parameters for the G-quadruplex of sequence (TTAGGG)$_4$ determined by CD or DSC measurements are in good agreement.

The main reasons for the discrepancies found for the same sequence studied by different methodologies is probably due to the different data analyses, especially in the choice of the pre- and post-transition baseline[4]. Another discrepancy could arise from the presence of intermediate states in the melting process. Recently, an accurate analysis of the unfolding process of the sequence AGGG(TTAGGG)$_3$ was made analyzing CD and DSC data[27]. The authors suggested that the unfolding may be described as a monomolecular equilibrium three-state process that involves an intermediate state, hypothesised to be a triplex conformation. The overall enthalpy value was 230 kJ mol$^{-1}$ close to that found by Mergny et al.[28] (Table 1) and in the range of the average values for G-tetrad[28-33]. Their findings are suggestive and are indicative of how there are still many unanswered questions concerning G-quadruplex stability and their folding/unfolding mechanism.

**Table 1** *Thermodynamic parameters relative to the unfolding of human telomeric quadruplexes in 100 mM $K^+$ solution.*

| Quadruplex sequence | $T_m$ (°C) | $\Delta H^0$ kJ mol$^{-1}$ | $\Delta S^0$ J mol$^{-1}$ K$^{-1}$ | $\Delta G^0$(310K) kJ mol$^{-1}$ | Ref. |
|---|---|---|---|---|---|
| (TTAGGG)$_4$[a,b] | 63 | 205 | 615 | 14.2 | [34] |
| (TTAGGG)$_4$[c] | 63.1 | 228 | 681 | 16.9 | [35] |
| (TTAGGG)$_4$TT[c] | 59.8 | 213 | 639 | 14.9 | [35] |
| AGGG((TTAGGG)$_3$[d] | 63 | 238 | 707 | 19.2 | [28] |
| AGGG(TTAGGG)$_3$[c] | 66.1 | 144 | 424 | 12.5 | [36] |
| GGG(TTAGGG)$_3$[d] | 65 | 253 | 749 | 20.9 | [28] |
| GGG(TTAGGG)$_3$[b] | 69.3 | 324 | 845 | 61.9 | [37] |
| GGG(TTAGGG)$_3$[b] | 67 | 202 | 592e | 18.5 | [38] |
| TGGG(TTAGGG)$_3$[e] | 81.8 | 278 | 780 | 35.1 | [39] |
| TAGGG(TTAGGG)$_3$[c] | 67.6 | 246 | 720 | 22.8 | u.d. |

[a] Obtained in 70 mM $K^+$ solution. Determined by [b] CD, [c] DSC, [d] UV, [e] fluorescence. u.d. = unpublished data.

## 3.2 Tetramolecular G-quadruplexes

The dissociation of tetramolecular quadruplexes is kinetically controlled. The melting temperatures and the calorimetric profiles are affected by the scan rate, as in the case of G-quadruplex (TGGGGT)$_4$ (Figure 2). This arises when the complexes are not at thermodynamic equilibrium during the temperature scanning, and it is due to the slow rates of dissociation and/or association process. Consequently, the thermodynamic parameters cannot be obtained from DSC melting experiments and the calorimetric $T_m$ values should be considered an upper limit. In these non-equilibrium conditions only the enthalpy change relative to the quadruplex dissociation process may be directly obtained from DSC measurements. The $\Delta H^0$ value obtained for the quadruplex [d(TGGGGT)]$_4$ is 320 kJ mol$^{-1}$, corresponding to a value of 80 kJ mol$^{-1}$ per tetrad, in good agreement with previously reported values for other quadruplexes[28-33].

To achieve a complete thermodynamic analysis for a tetramolecular quadruplex almost two different methods may be utilized. In the first method, the equilibrium melting curve may be obtained collecting CD spectra in an appropriate range of temperatures, between 0 °C and 100 °C, with a step of 5 °C or less. The quadruplex solution should be equilibrated at each temperature for a suitable time necessary to reach the thermodynamic equilibrium before recording the CD signal. The CD spectrum of a native parallel tetramolecular G-quadruplex is well characterized by a negative band at 243 nm and a positive band at 264 nm, which decrease in magnitude and shift to lower wavelengths when the temperature increases (Fig. 3, left) Further, the molar ellipticity at 264 nm is zero in the spectrum of the single strand. Therefore, the change in molar ellipticity of the maximum at 264 nm may be conveniently reported as a function of temperature, and the resulting curve represent the equilibrium melting curve (Fig.3, right). From the computer fitting of this curve, assuming a two-state process, may be easily calculated the melting temperature and the van't Hoff enthalpy and so on all the other thermodynamic parameters.

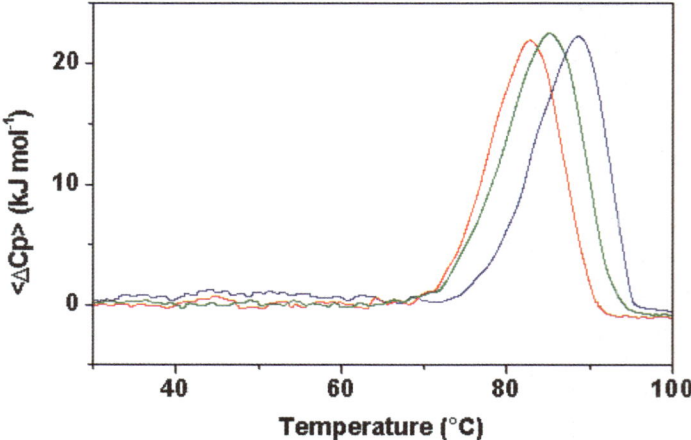

**Figure 2** *Calorimetric heat capacity vs. temperature profiles for the [TGGGGT]4 quadruplex at a scan rate of 1.0 (blue), 0.5 (green) and 0.3 °C min$^{-1}$ (red). Data were collected at 6.6·10$^{-4}$ M single strand concentration. Reprinted with permission from Ref. [18]. Copyright 2005 American Chemical Society.*

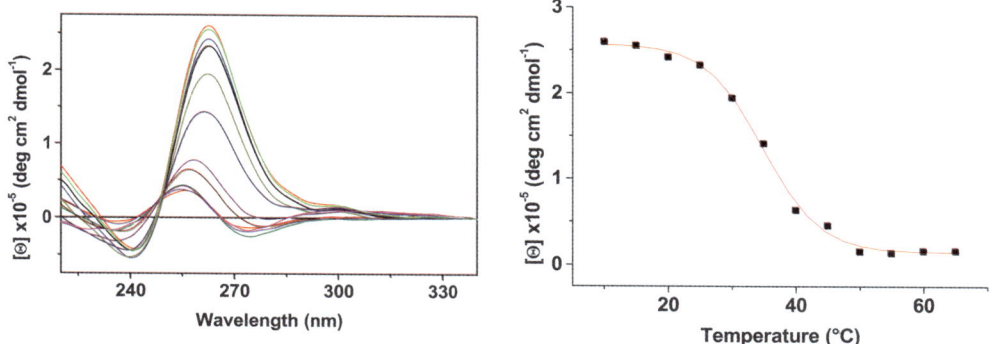

**Figure 3** *CD spectra of (TGGGT)$_4$ collected at different temperatures (on the left) and the corresponding molar ellipticity at 264 nm vs. temperature (on the right). Reprinted with permission from Ref. [18]. Copyright 2005 American Chemical Society.*

In the second method, the thermodynamic values can be extract from kinetic parameters, recording both heating and cooling profiles, i.e. dissociation/association curves[11]. UV absorbance at 295 nm may be conveniently recorded on changing temperature. The heating and cooling curves are not superimposed when a hysteresis phenomenon is present due to the different rates of folding and unfolding processes[19]. In that case, the kinetic constants of association $k_{on}$ and dissociation $k_{off}$ can be extract at each temperature from a heating/cooling cycle. It is then possible to calculate the activation energies, $E_{on}$ and $E_{off}$, from the slopes of Arrhenius plots (logarithm of $k_{on}$ and $k_{off}$ as a function of 1/T). The difference $E_{on} - E_{off}$ represents, $\Delta H^0$ value for the unfolding process.

## 4 THERMODYNAMICS OF LONG TELOMERIC INTRAMOLECULAR G-QUADRUPLEX

The exact structure of the single-strand telomeric overhang is not known. The questions are: How many quadruplexes are formed at the telomeric ends? What is their topology? How stable are the long intramolecular structures? X-ray crystallographic and NMR methods failed in obtaining high-resolution structures. Among the attempts to simulate plausible long intramolecular G-quadruplex structures[38,40-47], the work of Petraccone et al. on the sequence (TTAGGG)$_8$TT gives a convincing evidence for a hybrid 1-hybrid 2 structure, on the basis of computational and experimental studies[48]. The validation of the hybrid 1-hybrid 2 structure was made by sedimentation velocity, circular dichroism and fluorescence measurements, which eliminated other possible structures as the parallel-stranded, the propeller quadruplex or a variety of combinations and arrangements of hybrid and propeller structures. Recently, the formation of a stable structure with three consecutive G-quadruplex was assessed and a mixture of hybrid conformations has been hypothesized, utilizing the above mentioned methodologies[35]. In the same paper, the stability of the structures containing one (monomer), two (dimer) and three (trimer) consecutive G-quadruplexes has been studied by circular dichroism and differential scanning calorimetry. Two different oligonucleotide sequences were utilized: (TTAGGG)$_n$ and (TTAGGG)$_n$TT with n = 4, 8, 12. The reversibility and the lacking of hysteresis in the melting process have allowed to perform a complete thermodynamic analysis for all the G-quadruplexes. The presence of 3'-flanking bases do not affect the thermal stability of the longer DNA sequences as much as they do for the shorter sequences. Interestingly, melting temperatures decrease for higher order structures compared to single quadruplexes: for the (TTAGGG)$_n$TT series $T_m$, determined by DSC, is 59.7°C for the monomer, 57.3°C for the dimer and 57.1°C for the trimer. The enthalpy value for the trimer, 605 kJ mol$^{-1}$, is lower than expected for three consecutive monomers, with the $\Delta H^0$ per monomer of 235 kJ mol$^{-1}$ and 213 kJ mol$^{-1}$ for the sequences (TTAGGG)$_4$ and (TTAGGG)$_4$TT, respectively. Further, the unfolding of the dimer and trimer is not a simple two-state process. A detailed thermodynamic analysis of calorimetric measurements reveals that the unfolding pathway proceeds through two intermediates for the dimer and three for the trimer. In addition, from the $\Delta G^0$ values it was possible to calculate the coupling free energy, $\Delta G_{Coupling}$, for multimeric G-quadruplexes, i.e., the difference between the total free energy of the folding of two or three adjacent quadruplexes and two or three times the folding free energy of a single quadruplex. Surprisingly, these values are positives, 5 kJ mol$^{-1}$ for the dimer and 14 kJ mol$^{-1}$ for the trimer, indicating unfavourable interactions between adjacent quadruplex units, a sort of anti-cooperativity which works on increasing the number of contiguous G-quadruplexes. These results disagree with previous ones on the folding/unfolding of the (TTAGGG)$_n$ series, with n = 4, 8, 12, obtained by UV spectroscopy. The authors found a similar stability for monomer, dimer and trimer structures concluding that the high order structures were formed by monomers of equal conformation[40]. In a previous paper, a number of G-quadruplexes of sequence AGGG(TTAGGG)$_n$, with n = 1-16, have been studied by circular dichroism spectroscopy and PAGE electrophoresis in K$^+$ solutions[38]. The authors found that the increasing number of (TTAGGG) repeats caused thermal destabilization of the relative G-quadruplex structures, a decrease in their $\Delta H^0$ values and a slower migration in the gel electrophoresis. Recently, it has been showed by temperature-gradient gel electrophoresis experiments that the sequences GGG(TTAGGG)$_7$ and GGG(TTAGGG)$_{11}$ do not form the expected maximum number of quadruplex subunits[47]. All these findings, taken together, indicate that structures formed by more than three consecutive G-quadruplexes are thermodynamically unfavoured and their existence *in vivo* is questionable.

## 5 CONCLUSIONS

The folding/unfolding of human telomeric G-quadruplexes may be conveniently followed by a number of spectroscopic methodologies. The thermodynamic parameters can be easily extracted from melting curves of spectroscopic measurements, starting from the basic assumption that the unfolding is a simple two-state process. Unfortunately, in only few cases the folding and unfolding are two-state processes; indeed, for many G-quadruplex structures the unfolding process proceeds along a pathway with intermediate states. In that case, the thermodynamic parameters may be extract from calorimetric measurements, in which the measured enthalpy is model-independent. $\Delta G^0$ values relative to the stability of G-quadruplex structures in physiological conditions are in the tens of kJ mol$^{-1}$, due to opposite enthalpy and entropy contributions. As for other biological macromolecules, the stability is enthalpy driven, whereas the entropy contribution to the stability is unfavourable. In addition, the slow kinetics of the folding/unfolding process could affect the equilibrium thermodynamics, and lead to a wrong evaluation of the thermodynamic parameters. Combined kinetic and thermodynamic measurements should be always performed in order to obtain a complete elucidation of the folding/unfolding mechanism.

Further, the initial folded G-quadruplex structure could be not unique and the thermodynamic values could be affected by the presence of more conformations. Structural information (NMR, X-ray, CD) is required to know the correct initial conformation. Finally, cations, cosolutes or small ligands may induce G-quadruplex unfolding or conformational transitions from one form to another. The knowledge of the stability of different conformations might help to better identify the target structures for drugs.

**Acknowledgements**

This work was supported by the European Commission through the COST Action MP0802, offering a stimulating environment for discussions on the topic, and by the Italian PRIN 2009 funding.

**References**

1. A. M. Zahler, J. R. Williamson, T. R. Cech and D. M. Prescott, *Nature*, 1991, **350**,718.
2. J. Tang, Z. Y. Kan, Y. Yao, Q. Wang, Y. H. Hao and Z. Tan, *Nucleic Acids Res*, 2008, **36**,1200.
3. S. Neidle, *FEBS J*, 2010, **277**,1118.
4. J. B. Chaires, *Febs J.*, 2010, **277**,1098.
5. A. N. Lane, *Biochimie*, 2012, **94**,277.
6. R. Ida and G. Wu, *J. Am. Chem. Soc.*, 2008, **130**,3590.
7. A. N. Lane, J. B. Chaires, R. D. Gray and J. O. Trent, *Nucleic Acids Res*, 2008, **36**,5482.
8. D. Yang and K. Okamoto, *Future Med. Chem.*, **2**,619.
9. C. Antonacci, J. B. Chaires and R. D. Sheardy, *Biochemistry*, 2007, **46**,4654.
10. O. Stegle, L. Payet, J. L. Mergny, D. J. Mackay and J. H. Leon, *Bioinformatics*, 2009, **25**,i374.
11. J. L. Mergny and L. Lacroix, *Oligonucleotides*, 2003, **13**,515.
12. P. A. Rachwal and K. R. Fox, *Methods*, 2007, **43**,291.
13. A. De Cian, L. Guittat, M. Kaiser, B. Sacca, S. Amrane, A. Bourdoncle, P. Alberti, M. P. Teulade-Fichou, L. Lacroix and J. L. Mergny, *Methods*, 2007, **42**,183.
14. P. L. Privalov and S. A. Potekhin, *Methods Enzymol.*, 1986, **131**,4.
15. E. Freire, *Methods Mol. Biol.*, 1995, **40**,191.
16. G. Bruylants, J. Wouters and C. Michaux, *Curr. Med. Chem.*, 2005, **12**,2011.
17. C. Giancola, *J. Therm. Anal. Cal.*, 2008, **91**,79.
18. L. Petraccone, B. Pagano, V. Esposito, A. Randazzo, G. Piccialli, G. Barone, C. A. Mattia and C. Giancola, *J. Am. Chem. Soc.*, 2005, **127**,16215.
19. V. V. Anshelevich, A. V. Vologodskii, A. V. Lukashin and M. D. Frank-Kamenetskii, *Biopolymers*, 1984, **23**,39.
20. L. Petraccone, E. Erra, V. Esposito, A. Randazzo, A. Galeone, G. Barone and C. Giancola, *Biopolymers*, 2005, **77**,75.
21. P. Wallimann, R. J. Kennedy, J. S. Miller, W. Shalongo and D. S. Kemp, *J. Am. Chem. Soc.*, 2003, **125**,1203.
22. R. J. Desa and I. B. Matheson, *Methods Enzymol.*, 2004, **384**,1.
23. I. Haq, B. Z. Chowdhry and J. B. Chaires, *Eur. Biophys. J.*, 1997, **26**,419.
24. R. W. Hendler and R. I. Shrager, *J. Biochem. Biophys. Methods*, 1994, **28**,1.

25  C. M. Olsen, H. T. Lee and L. A. Marky, *J. Phys. Chem. B*, 2009, **113**,2587.
26  N. Kumar, B. Sahoo, K. A. Varun, S. Maiti and S. Maiti, *Nucleic Acids Res.*, 2008, **36**,4433.
27  M. Boncina, J. Lah, I. Prislan and G. Vesnaver, *J. Am. Chem. Soc.*, **134**,9657.
28  J. L. Mergny, A. T. Phan and L. Lacroix, *FEBS Lett.*, 1998, **435**,74.
29  R. Jin, B. L. Gaffney, C. Wang, R. A. Jones and K. J. Breslauer, *Proc. Natl. Acad. Sci. USA*, 1992, **89**,8832.
30  L. Petraccone, E. Erra, L. Nasti, A. Galeone, A. Randazzo, L. Mayol, G. Barone and C. Giancola, *Int. J. Biol. Macromol.*, 2003, **31**,131.
31  L. Petraccone, E. Erra, V. Esposito, A. Randazzo, L. Mayol, L. Nasti, G. Barone and C. Giancola, *Biochemistry*, 2004, **43**,4877.
32  L. Petraccone, E. Erra, A. Randazzo and C. Giancola, *Biopolymers*, 2006, **83**,584.
33  J. L. Mergny, A. De Cian, S. Amrane and M. Webba Da Silva, *Nucleic Acids Res.*, 2006, **34**,2386.
34  P. Balagurumoorthy and S. K. Brahmachari, *J. Biol. Chem.*, 1994, **269**,21858.
35  L. Petraccone, C. Spink, J. O. Trent, N. C. Garbett, C. S. Mekmaysy, C. Giancola and J. B. Chaires, *J. Am. Chem. Soc.*, 2011, **133**,20951.
36  C. M. Olsen, W. H. Gmeiner and L. A. Marky, *J. Phys. Chem. B*, 2006, **110**,6962.
37  W. Li, P. Wu, T. Ohmichi and N. Sugimoto, *FEBS Lett.*, 2002, **526**,77.
38  M. Vorlickova, J. Chladkova, I. Kejnovska, M. Fialova and J. Kypr, *Nucleic Acids Res.*, 2005, **33**,5851.
39  A. Risitano and K. R. Fox, *Biochemistry*, 2003, **42**,6507.
40  H. Q. Yu, D. Miyoshi and N. Sugimoto, *J. Am. Chem. Soc.*, 2006, **128**,15461.
41  I. M. Pedroso, L. F. Duarte, G. Yanez, K. Burkewitz and T. M. Fletcher, *Biopolymers*, 2007, **87**,74.
42  L. P. Bai, M. Hagihara, Z. H. Jiang and K. Nakatani, *Chem.Biochem.*, 2008, **9**,2583.
43  D. Renciuk, I. Kejnovska, P. Skolakova, K. Bednarova, J. Motlova and M. Vorlickova, *Nucleic Acids Res.*, 2009, **37**,6625.
44  Y. Xu, T. Ishizuka, K. Kurabayashi and M. Komiyama, *Angew. Chem. Int. Ed. Engl.*, 2009, **48**,7833.
45  Y. Sannohe, K. Sato, A. Matsugami, K. Shinohara, T. Mashimo, M. Katahira and H. Sugiyama, *Bioorg. Med. Chem.*, 2009, **17**,1870.
46  H. Wang, G. J. Nora, H. Ghodke and P. L. Opresko, *J. Biol. Chem.*, 2011, **286**,7479.
47  L. Bauer, K. Tluckova, P. Tohova and V. Viglasky, *Biochemistry*, 2011, **50**,7484.
48  L. Petraccone, J. O. Trent and J. B. Chaires, *J. Am. Chem. Soc.*, 2008, **130**,16530.

# G-QUADRUPLEX NANOSTRUCTURES PROBED AT THE SINGLE MOLECULAR LEVEL BY FORCE-BASED METHODS

Soma Dhakal[1], Hanbin Mao*[1], Arivazhagan Rajendran[2], Masayuki Endo[3], and Hiroshi Sugiyama*[2,3]

1. Department of Chemistry and Biochemistry, Kent State University, Kent, OH, 44242, USA.
2. Department of Chemistry, Graduate School of Science, Kyoto University, Kitashirakawa-oiwakecho, Sakyo-ku, Kyoto 606-8502, Japan.
3. Institute for Integrated Cell-Material Sciences (iCeMS), Kyoto University, Yoshida-ushinomiyacho, Sakyo-ku, Kyoto 606-8501, Japan.

## 1 INTRODUCTION

Atomic details of G-quadruplex structures have been revealed by NMR and X-ray crystallography for more than a decade. G-quadruplex in human telomeric sequence, 5'-(TTAGGG)$_4$, is the first and the most widely studied DNA tetraplex structure.[1] Albeit its simple repeating sequence, a myriad of conformations have been revealed in the human telomeric region, which include parallel, antiparallel, and hybrid G-quadruplexes.[2-4] The conformation of G-quadruplex is a rather important property for various applications based on this structure. For example, Miyoshi and co-workers have reported the formation of G-wire, a high-order nanostructure assembled by parallel G-quadruplex units.[5] Another example is the use of parallel G-quadruplex to develop $K^+$ ion responsive biomimetic nanochannels,[6] in which the immobilized G-quadruplex serves as a channel with varying pore size due to the conformation change of the G-quadruplex under different $K^+$ concentrations. The conformation-specific catalytic activity of G-quadruplex structures has been demonstrated. The thrombin binding aptamer, which is known to assume an antiparallel structure in a $K^+$ buffer, catalyses an aldol reaction between a ketone and a porphyrin linked aldehyde.[7] Similarly, an intramolecularly folded parallel G-quadruplex in 5'-TGGGTAGGGCGGGTTGGGAAA-3' has shown peroxidase activity when bound with hemin.[8] Thus, the regulation of G-quadruplex conformations can empower novel methodologies to control the property of nanomaterials.

G-quadruplex structures with millimolar (mM) DNA concentrations are determined by ensemble techniques such as NMR and X-ray crystallography. Under such high concentrations, a mixture of conformations may exist, which are rather difficult to resolve for these methods. For applications such as molecular electronics, it is desirable to know specific conformation at the single-molecular level to fully exploit the high sensitivity for folded nanostructures. Again, it is beyond the capability of traditional structural determining techniques. Single-molecule methods, however, are ideal tools to profile conformations under these conditions.

Fluorescence resonance energy transfer (FRET) is the first technique to reveal the dimensions of G-quadruplexes at the single-molecule level. The 6-9 nm measurable distance in the FRET, however, strictly limits its capability to investigate quadruplex structures. Compared to FRET techniques, force based methods can reach Angstrom (Å) resolutions, and therefore, are better equipped for structural characterization at the single-molecule level. In addition, force-based approaches can reveal mechanical stability of a structure and provide thermodynamic information such as change in the free energy of unfolding ($\Delta G_{unfold}$) of the structure. For many materials-related applications that employ G-quadruplexes as essential building blocks, mechanical stability is a rather important indicator to reflect the robustness of a device under strained environment. One example is the transmembrane ion transporter whose structure must have enough mechanical stability to sustain the tension in the phospholipid membrane.[9] Similarly, G-quadruplex nanomotors designated to maneuver nanoparticles, move microtubules, and transport nanodevices,[10] must be strong enough to carried out these mechanical works. Therefore, revealing the mechanical stability of G-quadruplex nanostructures can provide useful insights to design nanomaterial devices for applications in nanotechnology and biotechnology fields. Optical tweezers and atomic force microscopy (AFM) are the two major force-based single-molecule approaches developed since the nineteen nineties.[11] In this review, we discuss the application of optical tweezers and AFM to illustrate G-quadruplex nanostructures. We start with the discussion of one-dimensional structural interpretation of G-quadruplex using laser tweezers, which is followed by 2D quadruplex structure profiling using AFM.

## 2 PROFILING G-QUADRUPLEX NANO-STRUCTURES USING OPTICAL TWEEZERS

### 2.1 Laser tweezers instrument

In a typical dual-beam dual-trap laser tweezers setup,[12,13] two collimated laser beams are focused at separate locations where two nano- or micron-sized particles can be trapped (Figure 1). A single-stranded DNA segment that hosts the G-quadruplex forming sequence is tethered between two dsDNA handles, which prevent the G-quadruplex-hosting DNA from contacting the particle surface. Mechanical unfolding experiments start with moving one of the trapped particles away from the other by adjusting the angle of a mirror that steers one of the laser beams. The increased inter-particle distance as a result of this movement is accompanied with the increment of tension in the DNA tether, which unfolds the secondary structure priorly formed in the sequence.

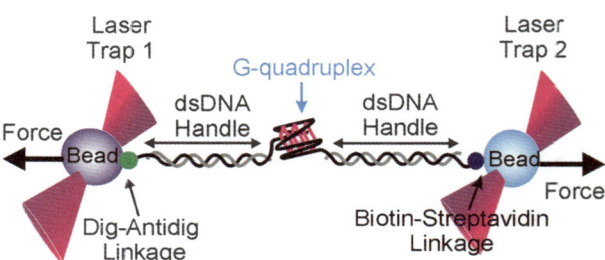

**Figure 1** *Laser tweezers setup for mechanical manipulation of G-quadruplex structures.*

The mechanical stability is measured by the rupture force that unfolds the secondary structure. When force is reduced by bring two trapped particles closer, refolding transition can be observed. In the single-molecule dynamic force spectroscopy, the rupture force for each unfolding is recorded at a specific force ramping rate.[14,15] Such a method allows obtaining the kinetic information of the unfolding or refolding process, which includes rate constant, distance to the transition state, as well as activation energy. In an extreme case where the force ramping rate is rather fast (a few pN in millisecond for example), the rapid change in the force allows the determination of the folding or unfolding rate constant directly as the force jump populates the DNA to either the folded or unfolded state at time zero.[16,17] For slow process such as unfolding or refolding of a G-quadruplex, the nonequilibrium transitions predominate with an experimentally achievable force-ramp rate. Under such a condition, either Jarzynski non-equilibrium method[18] or Crooks approach[19] can be followed to obtain the free energy of unfolding.

The dimension of the folded structure under a specific unfolding geometry, which is defined by the two tethering points in the structure, can be estimated by the change in contour length ($\Delta L$) accompanied with the rupture event. To obtain $\Delta L$, the force-extension ($F$-$X$) curves are fitted separately with the worm-like-chain model[20] (Equation 1) separately, for the region that contains the folded structure and the region that corresponds to the unfolded structures (Figure 2).

$$\frac{x}{L} = 1 - \frac{1}{2}\left(\frac{k_B T}{FP}\right)^{\frac{1}{2}} + \frac{F}{S} \qquad (1)$$

Here, $x$ is the end-to-end distance of the folded G-quadruplex structure, $k_B$ is the Boltzmann constant, $T$ is absolute temperature, $P$ is the persistent length (51.95 nm[20]), $F$ is the force, and $S$ is the elastic stretch modulus[20].

Each fitting allows the determination of the contour length of the corresponding region in the $F$-$X$ curves, which yields $\Delta L$ by taking the difference between the two contour lengths. Nevertheless, such a fitting approach is time consuming and suffers from reduced reproducibility due to the identifiability problem associated with the complex WLC equation.[21] To address these issues, Mao and co-workers[21] first took the difference in the $x$ axis ($\Delta x$) between the extending and returning $F$-$X$ curves at a particular force. Next, they converted $\Delta x$ to $\Delta L$ using the WLC model based on the assumption that the contribution of an unfolded G-quadruplex to the overall contour length is negligible (< 3%) compared to the dsDNA handles. As this approach does not rely on any fitting procedures, the identifiability problem does not exist. Furthermore, since the numerical operations are easily automated, the process is rather convenient to proceed. This new approach has been validated as the $\Delta L$ is comparable with that obtained from the traditional methods described above.[21]

$\Delta L$ is related to the contour length ($L$) and the inter-tether distance ($x$) by the following equation, $\Delta L = L - x$ (Equation 2). As the contour length of DNA can be calculated from the number of nucleotides ($N$) multiplied by the known values of the contour length per nucleotides ($L_{nt}$), $L = N \times L_{nt}$, equation 2 allows the estimation of the inter-tether distance from the $F$-$X$ curves. For a typical DNA construct, two tethering points coincide with two termini of a structure, therefore, this method allows the direct measurement of the end-to-end distance of the structure. Using this approach, the Mao lab[17,21,22] has characterized G-quadruplex nanostructures along one-dimensional trajectories.

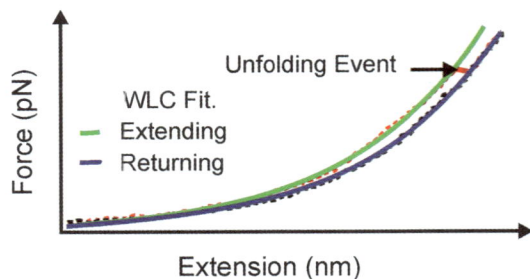

**Figure 2** *A typical force extension (F-X) curve with extending (red) and returning (black) traces fitted with the WLC model. The unfolding of the G-quadruplex has been highlighted by a red solid trace.*

## 2.2 Mechanical unfolding of single g-quadruplex units

In the first example to demonstrate this approach, Yu and co-workers investigated the predominant variant sequence, 5'-(ACAGGGGTGTGGGG)$_2$, in the promoter of the insulin linked polymorphism region (ILPR).[22] Comprising of hundreds of these and similar G-rich tracts, the ILPR is known to enhance insulin production.[23] Understanding the quadruplex nanostructures can therefore help to illuminate the mechanism of this important biological function. In addition, presence of multiple G-rich tracts in the ILPR provides ample opportunity for interactions between quadruplexes (or QQI, quadruplex-quadruplex interaction). Full understanding of these interactions can assist in the design of materials that employ QQI between multiple G-quadruplex units.

Using NMR and gel shift assays, previous investigations revealed an antiparallel G-quadruplex structure in this ILPR sequence.[24] However, the laser-tweezers results are rather surprising. A mixture of two conformations, parallel and antiparallel G-quadruplexes, were observed (Figure 3a).[22] These two structures have different mechanical stabilities. One population has 23 pN in rupture force, while the other has 37 pN. When each rupture force population was analyzed for the corresponding $\Delta L$, two statistically distinct populations were observed (Figure 3b). Compared to the expected $\Delta L$ measured from the known parallel and antiparallel G-quadruplex structures, the 23 pN population has the $\Delta L$ matching with the parallel G-quadruplex whereas the 36.9 pN species has the $\Delta L$ matching with the antiparallel conformation. The observed mechanical stability qualitatively agrees with the finding that antiparallel G-quadruplexes are thermodynamically more stable than parallel quadruplexes. Indeed, using Jarzyski's non-equilibrium theorem,[18] $\Delta G_{unfold}$ has been determined as 14 and 23 kcal/mol for parallel and antiparallel structures, respectively. These values are close to the free energy determined by 290 nm UV melting methods, which validates the single-molecule results.

More recently, other intramolecular DNA nanostructures including telomeric G-quadruplex and hTERT (human telomerase reverse transcriptase) G-quadruplex have been investigated.[17,21] Whereas the former G-quadruplex shows a structure consistent with the basket or the hybrid I conformation depending on buffer conditions, the latter quadruplex shows unexpected complexity, which features a long middle loop that contains a hairpin. Interestingly, all G-quadruplexes demonstrate a mechanical stability (16-40 pN) stronger than the DNA hairpins (~15 pN[25]), suggesting a more rugged nanostructure for the G-quadruplexes. Compared to the hairpins, however, the rupture force of the quadruplex is more spread out, indicating a more dynamic structure of the G-quadruplex with respect to the hairpin.

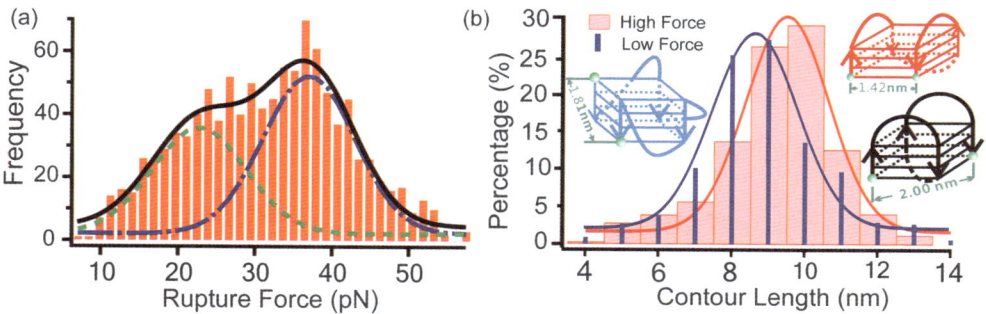

**Figure 3** *(a) Rupture force histogram for the parallel (left population) and antiparallel (right population) G-quadruplex structures in the human ILPR. The overall population is fitted with a two-peak Gaussian (black solid). Individual populations are fitted with single Gaussians. (b) Contour length histograms for the parallel (blue) and antiparallel (red) G-quadruplex in Figure 3a. Insets show the corresponding G-quadruplex structures (blue: parallel; red and black: antiparallel). This Figure is reprinted with permission from J. Am. Chem. Soc. 2009, 131, 1876, © 2009 American Chemical Society.*

## 2.3 Mechanical unfolding of multiple g-quadruplex units

Schonhoft and co-workers investigated the nanostructures that contain multiple G-quadruplexes in the human ILPR.[26] Using a DNA sequence that can form a maximum of two G-quadruplexes at a time, both sequential and cooperative (Figure 4) unfolding of G-quadruplexes were demonstrated. These two types of unfolding events were justified by the appearance of two populations with distinct change in contour length upon unfolding ($\Delta L_{single}$ = 8.6 ± 0.3 nm for the unfolding of individual G-quadruplexes and $\Delta L_{double}$ = 18.1 ± 0.5 nm for the unfolding of the two G-quadruplexes together). The sequential unfolding events suggest the formation of two G-quadruplexes that unfold independently. The occurrence of the cooperative unfolding (33.77%) was higher than that predicted for the two independent G-quadruplexes to unfold simultaneously (10.96%). This suggests the existence of higher order quadruplex-quadruplex interactions (QQI). Recently, Chaires and co-workers also identified similar QQI in human telomeric sequences.[27] Since higher order QQI provides additional interactions between participating G-quadruplexes, it could be useful to design materials with increased mechanical stability.

Higher order interactions also exist between G-quadruplex and other non-B DNA structures. Single-molecule study by Yu and co-workers recently reported the interaction between a G-quadruplex and a hairpin in the hTERT promoter region.[21] The cooperative unfolding of the complex structure in the hTERT provided the evidence of interaction between the G-quadruplex and the hairpin. Such a long range interaction showed an increased $\Delta G_{unfold}$, indicating an increased stability for structures with higher order interactions. The knowledge of the higher order interaction within the promoter core can be crucial for targeting those regions with small molecule drugs in an effort to treat cancer and other age-related diseases.

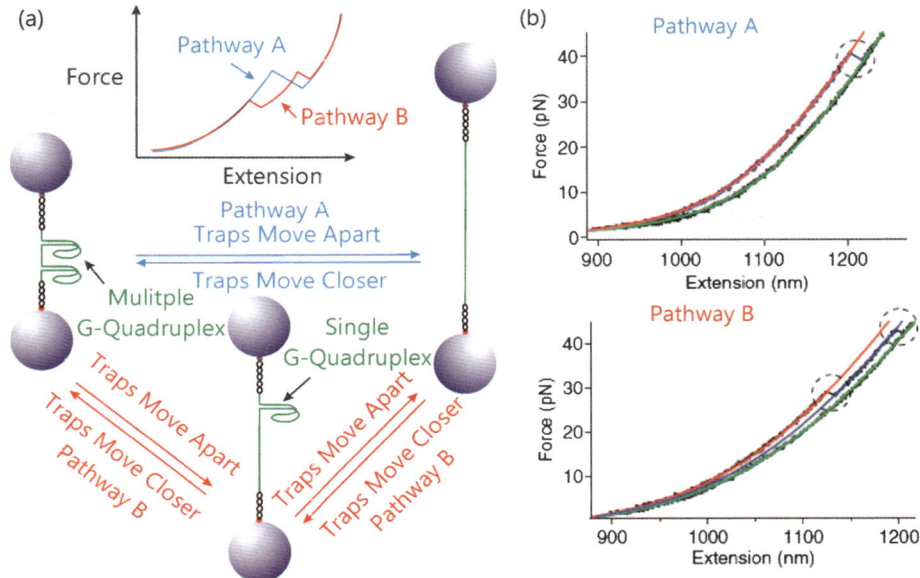

**Figure 4** *(a) Schematic of the unfolding pathways of dual G-quadruplexes in the human ILPR. Pathway A represents the co-operative unfolding while pathway B denotes the sequential unfolding events of two G-quadruplexes. The predicted force-extension (F-X) curves corresponding to either pathway A or B is shown at the top. (b) Experimental F-X curves showing the cooperative (Pathway A, top) and sequential (Pathway B, bottom) unfolding of ILPR G-quadruplex structures. The unfolding events are highlighted by dotted circles and the stretching and relaxing curves are fitted with the WLC model. Figure 4b is reprinted with permission from Nucleic Acids Research, 2009, Vol. 37, No. 10, 3310 © 2009.*

Several G-quadruplex binding small molecules have been explored.[17,28-29] Depending on G-quadruplex conformations, these small molecules have the propensity to bind to different sites of the nanostructures.[30-32] It has been shown that with the binding of a small molecule PDS, the mechanical stability of human telomeric G-quadruplex increases,[17] therefore, more rugged G-quadruplex can be obtained once bound with ligands. This approach can be exploited to manipulate the mechanical property of quadruplex-based nanodevices. When multiple quadruplex units are involved, the binding of small molecules depends on the availability of specific binding sites, which exist not only inside each G-quadruplex, but also in the interface region between G-quadruplexes.[33]

## 2.4 Intermolecular g-quadruplex nanostructures

The diverse G-quadruplex nanostructures are also reflected by the various intermolecular G-quadruplex structures. Using DNA sequences that contain one, two, or four G-rich tracts, monomeric, dimeric, or tetrameric intermolecular G-quadruplexes can be constructed. However, the folding pathways, which provide insights for efficient construction of G-quadruplex based nanomaterials, are not clearly understood. To clarify the diverse structures as well as folding pathways of intermolecular (3+1) G-quadruplex nano-assembly, NMR, AFM, and laser tweezers investigations were performed.[34-36] The

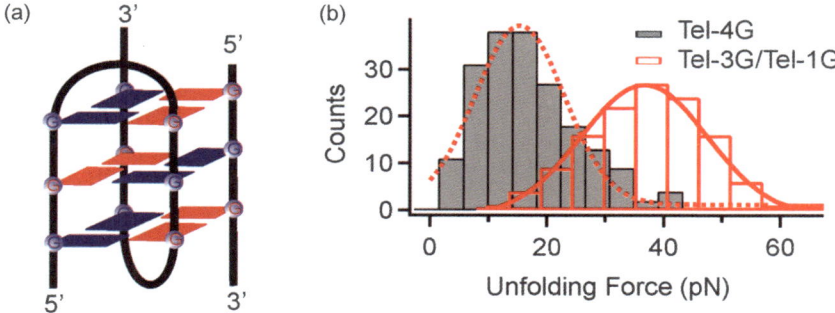

**Figure 5** *(a) Schematic of the intermolecular (3+1) G-quadruplex assembly in a human telomeric fragment. (b) Comparison of the unfolding force histograms for intramolecular 'Tel-4G' (left population), and intermolecular (3+1) G-quadruplex assembly 'Tel-3G/Tel-4G' (right population) in 10 mM Tris buffer with 100 mM Na$^+$. Histograms are fitted with Gaussian functions. This Figure is reprinted from Chem. Commun., 2012, 48, 2006, © The Royal Society of Chemistry 2012.*

(3+1) G-quadruplex structure comprises of three G-tracts from one DNA strand and one G-tract from a separate strand (Figure 5a). Laser tweezers studies by Koirala and co-workers have shown that the (3+1) assembly in a human telomere sequence follows a folding mechanism consistent with the strand-by-strand pathway.[36] In addition, the mechanical stability of the structure demonstrates a strikingly increased value (2.5 fold) ($F_{unfold}$ = 38 pN) compared to that of an intramolecular G-quadruplex ($F_{unfold}$ = 16 pN) (Figure 5b).[36] Such a result implied that intermolecular G-quadruplex nanostructures could be designed for nanomaterials that require high mechanical stabilities.

## 3 TWO-DIMENSIONAL IMAGING OF G-QUADRUPLEX NANO-STRUCTURES

### 3.1 Atomic force microscopy

AFM or scanning force microscopy (SFM) was invented by Binnig, Quate and Gerber in 1986,[37] four years later than the invention of scanning tunneling microscopy (STM). It is a force-based technique and works by scanning the surface of the sample by using the sharp tip of a cantilever, and produces the topographical 2D image of the sample. For imaging, the sample should be immobilized on a flat surface, for instance the widely used mica surface. Images are produced by monitoring the deflection of the cantilever caused by the tip-sample interactions during imaging. AFM is operated in three different modes, namely contact, tapping, and noncontact modes. Initially, the samples were imaged in contact mode. However, this mode is not suitable for the analysis of less robust biological samples, such as proteins and nucleic acids. This is because the continuous contact between the tip and sample leads to the damage of the sample, and also often detaches the sample from the surface. To solve this problem, tapping mode was invented in 1993, in which the cantilever oscillates at a constant frequency and the tip-sample interaction is considerably reduced.[38] Investigations by using the tapping mode demonstrated that this mode is best suited for imaging biological samples. Before the invention of AFM, the direct visualization of the biomolecules was carried out using electron microscopy under vacuum condition. At the early stage of AFM, analyses were performed in air. In 1987, the first report on the imaging under liquid condition was published, where the researchers imaged the surface of a sodium chloride crystal in a paraffin oil environment.[39] Further developments resulted a method suitable for single-molecular analysis of biomolecules in physiologically relevant solutions. A typical instrumental setup of a tapping mode AFM[40] is show in Figure 6.

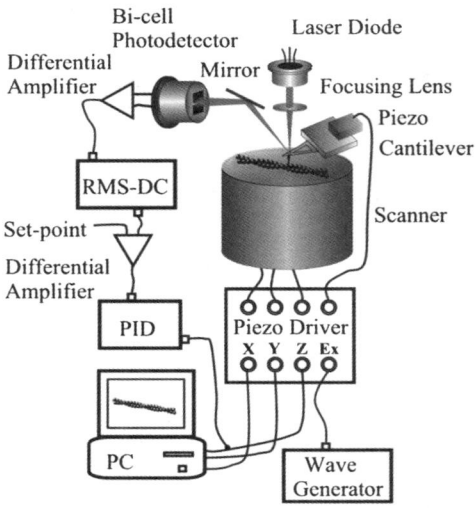

**Figure 6** *Schematic of the instrumentation of AFM for the tapping mode of operation. This Figure is adapted with permission from Advances in Experimental Medicine and Biology, 2003, Vol. 538, 119.*

### 3.2 High-speed atomic force microscopy

Besides the advantages of AFM, the analysis of structural dynamics of biomolecules has been hampered by the slow scan speed of conventional AFM that varies from several seconds to minutes per frame. Thus, the rapidly moving molecules and the processes that take place in subsecond time scale or faster are difficult to study. During last 10 years, several improvements have been carried out to achieve an AFM that is faster and compatible with the study of biological samples. The early stage developments were performed by Hansma et al.,[41-43] and Quate et al.[44,45] The notable attempts by Ando et al. improved the function of several components of AFM, such as small cantilever, improved scanning stage, amplitude-to-DC converter, and dynamic PID controller. These improvements lead to the realization of first-[46,47] and second-generation[48] high-speed AFM (HS-AFM) instruments, with a scan rate of 80 and 50 ms/frame, respectively. The developments were continued and current laboratory-built HS-AFM instrument can record a movie with an imaging rate of up to ~30 ms/frame, suitable for the dynamic analysis of biological samples.[49] Its ability to record almost video imaging is recently demonstrated for the analysis of walking myosin V on actin filament.[49] The details of the instrumental setup and improvements in the components were well reviewed recently.[48,50] The capacity of HS-AFM to analyze various biomolecules, including proteins,[48,51] DNA,[52] and nucleosomes,[53] was also reported recently. Here in this section, we briefly describe the direct observation of the formation and disruption of single G-quadruplex structure using

HS-AFM and DNA origami.[54-56] Static imaging of G-quadruplex structures have been performed by many researchers,[57] however the dynamic imaging is rather new, and to the best of our knowledge we have demonstrated the first example of the dynamics of a (3+1) G-quadruplex structure in real-time.[35]

## 3.3 Real-time monitoring of G-quadruplex folding

We have developed a novel label-free method for the direct and real-time observation of the formation of single G-quadruplex structure under conditions that favor the folding of a four-stranded structure and its disruption under unfavorable conditions.[35] For this analysis we have adopted a frame-shaped DNA origami nanostructure[58,59] as suitable substrate. As shown in Figure 7a, a DNA origami frame with an inner vacant rectangular area was constructed, in which two sets of connection sites (A-B and C-D) were introduced for the attachment of dsDNAs of interest. We have designed two dsDNAs each contained a G-rich

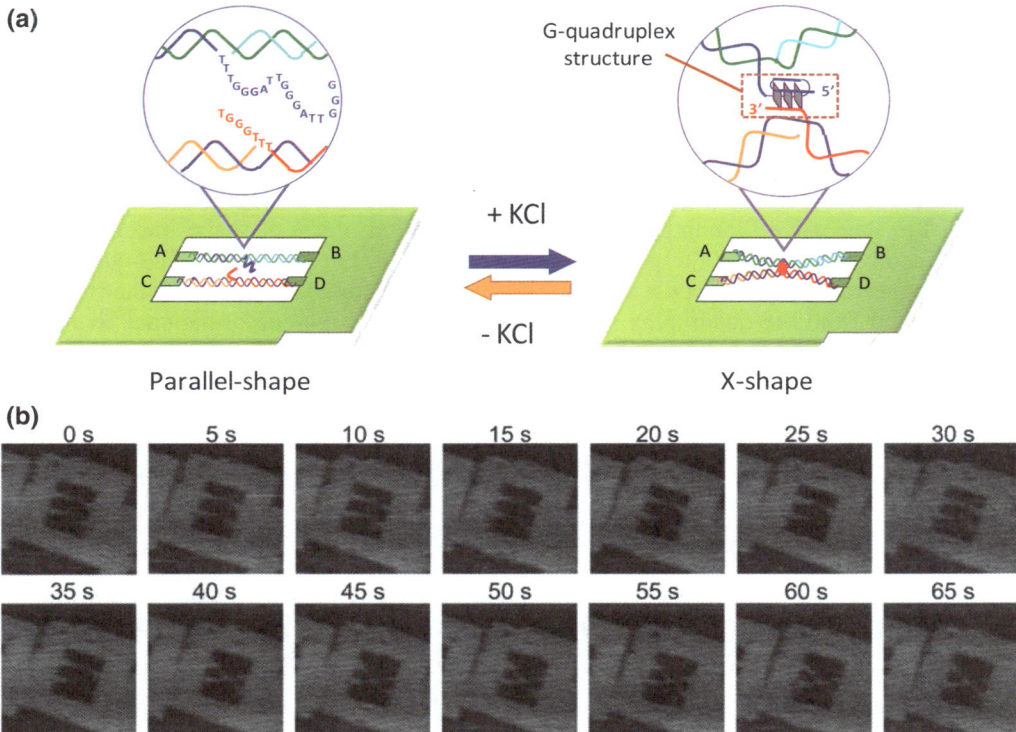

**Figure 7** *(a) Graphical representation of the DNA origami frame with two G-tracts containing dsDNAs attached between A-B and C-D. The strands forming a parallel-state in the absence of KCl (left) and an X-shape in the presence of KCl (right) are also shown. (b) The time lapse HS-AFM images of the formation of a single G-quadruplex structure. Image size: 160 × 160 nm. This Figure is reprinted with permission from J. Am. Chem. Soc., 2010, Vol. 132, 1631.*

single-stranded protrusion: three G-tracts in the upper and one G-tract in the lower strand. In the absence of KCl, G-tracts do not form a quadruplex structure and the duplex strands adopt a parallel-shape. In contrast, addition of KCl induces the protrusions to form (3 + 1) G-quadruplex structure, which brings the dsDNAs in contact at the middle, forming an X-shape which can be visualized in the AFM images. As expected, we were able to monitor the conformational change of the G-tracts from the single-stranded to the G-quadruplex by static imaging. Further, we have performed the real-time analysis of the formation of G-quadruplex structure, as shown in Figure 7b. The formation of DNA origami and the insertion of the G-tracts containing dsDNAs were carried out in KCl-free buffer and the HS-AFM analysis was carried out in a buffer that contained KCl. The DNA strands were initially in the parallel-state (0-35 s), and after some time (at 40 s), converted into an X-shape, which clearly indicated the formation of a G-quadruplex structure. The analysis with G-to-A mutated strands failed to adopt X-shape, suggesting that X-shape is exclusively due to the formation of G-quadruplex. Further, our results showed that the quadruplex folding event completed within 5 s which is in close proximity to the result obtained by stopped-flow experiments.[60]

## 3.4 Unfolding event of single G-quadruplex structure

Likewise, we have attempted to monitor the disruption of a G-quadruplex structure by removing KCl.[35] The origami structure and the G-tracts containing dsDNAs insertion were performed in KCl containing buffer. After immobilizing the origami assembly on mica surface, the images were recorded in a buffer droplet that contained no KCl. The formation of X-shape was initially seen (0 s) in the images which remained for a while (Figure 8). During the successive scanning, the gradual deformation of the X-shape into a parallel-shape was observed. Once the parallel-state was formed, the formation of G-quadruplex and the appearance of X-shape were not observed thereafter. This indicated that the KCl was removed from the quadruplex structure which induced the conformational change to the initial single-stranded structure of the G-tracts. The sequence dependent studies were also investigated using a variety of sequences. We believe that our method can be extended to the analysis of other types of G-quadruplexes[61] and various other conformational changes of nucleic acids.

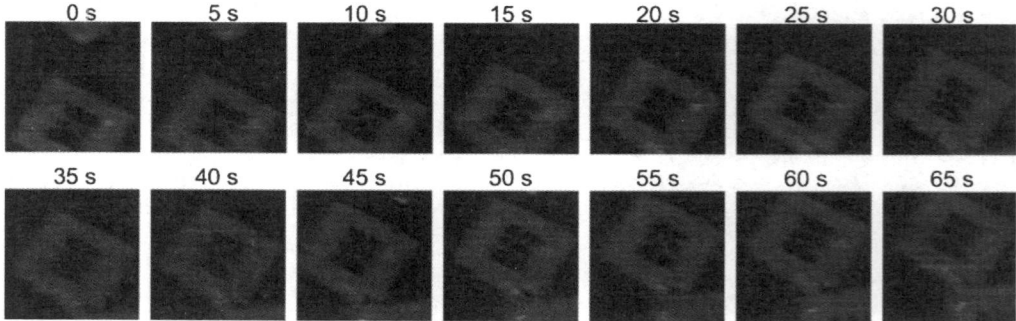

**Figure 8** *Snapshots of the HS-AFM imaging of the disruption of a G-quadruplex structure in the absence of KCl. Image size: 175 × 175 nm. This Figure is reused with permission from J. Am. Chem. Soc., 2010, Vol. 132, 16311.*

## 4 CONCLUSIONS

By using force-based single-molecular approaches, the G-quadruplex nanostructures have been imaged at the 1D or 2D level. The force based profiling enables the evaluation of the mechanical stability of various G-quadruplexes. Whereas the G-quadruplex is mechanically more stable than the DNA hairpin, it is also structurally more dynamic. That intermolecular G-quadruplex is mechanically more stable than the intramolecular quadruplex provides guidelines for the design of materials that employ G-quadruplex as building blocks. Besides profiling the mechanical stability of the nanostructures, the single-molecule approaches discussed here are also useful to identify different populations formed simultaneously in the same preparation. This provides more accurate profiling of nanostructures which often exist as a mixture. Laser tweezers and AFM allow dissecting the single-molecule behaviour and hence could refine the ensemble properties determined by NMR and X-ray crystallography. Moreover, the conformation and kinetic analyses of the G-quadruplex nanostructures at single-molecule level provides practical insights for the use of these materials in nanodevices.

**References**

1   S. Balasubramanian and S. Neidle, *Current Opinion in Chemical Biology*, 2009, **13**, 345.
2   G. N. Parkinson, M. P. Lee and S. Neidle, *Nature*, 2002, **417**, 876.
3   J. Dai, M. Carver and D. Yang, *Biochimie*, 2008, **90**, 1172.
4   J. Li, J. J. Correia, L. Wang, J. O. Trent and J. B. Chaires, *Nucleic Acids Res.*, 2005, **33**, 4649.
5   D. Miyoshi, A. Nakao and N. Sugimoto, *Nucleic Acids Research*, 2003, **31**, 1156.
6   X. Hou, W. Guo, F. Xia, F. Q. Nie, H. Dong, Y. Tian, L. Wen, L. Wang, L. Cao, Y. Yang, J. Xue, Y. Song, Y. Wang, D. Liu and L. Jiang, *J. Am. Chem. Soc.*, 2009, **131**, 7800.
7   Z. Tang, D. P. N. GonÅalves, M. Wieland, A. Marx and J. S. Hartig, *Chem.Bio.Chem.*, 2008, **9**, 1061.
8   D.-M. Kong, L.-L. Cai, J.-H. Guo, J. Wu and H.-X. Shen, *Biopolymers*, 2008, **91**, 331.
9   J.Lee, H.-J. Kim and J. Kim, *J. Am. Chem. Soc.*, 2008, **130**, 5010.
10  J. J. Li and W. Tan, *Nano Lett*, 2002, **2**, 315.
11  W. J. Greenleaf, M. T. Woodside and S. M. Block, *Annu Rev Biophys Biomol Struct*, 2007, **36**, 171.
12  P. Luchette, N. Abiy and H. Mao, *Sensors and Actuators B*, 2007, **128**, 154.
13  H. Mao and P. Luchette, *Sensors and Actuators B*, 2008, **129**, 764.
14  R. Merkel, P. Nassoy, A. Leung, K. Ritchie and E. Evans, *Nature*, 1999, **397**, 50.
15  M. Benoit, D. Gabriel, G. Gerisch and H. E. Gaub, *Nat Cell Biol*, 2000, **2**, 313.
16  P. T. X. Li, C. Bustamante and I. Tinoco, *Proc. Nat. Acad. Sci. USA*, 2006, **103**, 15847.
17  D. Koirala, S. Dhakal, B. Ashbridge, Y. Sannohe, R. Rodriguez, H. Sugiyama, S. Balasubramanian and H. Mao, *Nat. Chem.*, 2011, **3**, 782.
18  C. Jarzynski, *Phys. Rev. Lett.*, 1997, **78**, 2690
19  G. E. Crooks, *Phys. Rev. E*, 1999, **60**, 2721.
20  C. G. Baumann, S. B. Smith, V. A. Bloomfield and C. Bustamante, *Proc. Natl. Acad. Sci. USA*, 1997, **94**, 6185.

21   Z. Yu, V. Gaerig, Y. Cui, H. Kang, V. Gokhale, Y. Zhao, L. H. Hurley and H. Mao, *J. Am. Chem. Soc.*, 2012, **134**, 5157.
22   Z. Yu, J. D. Schonhoft, S. Dhakal, R. Bajracharya, R. Hegde, S. Basu and H. Mao, *J. Am. Chem. Soc.*, 2009, **131**, 1876.
23   G. C. Kennedy, M. S. German and W. J. Rutter, *Nat. Genet.*, 1995, **9**, 293.
24   P. Catasti, X. Chen, R. K. Moyzis, E. M. Bradbury and G. Gupta, *J. Mol. Biol.*, 1996, **264**, 534.
25   M. Rief, H. Clausen-Schaumann and H. E. Gaub, *Nat. Struct. Biol.*, 1999, **6**, 346
26   J. D. Schonhoft, R. Bajracharya, S. Dhakal, Z. Yu, H. Mao and S. Basu, *Nucleic Acids Res.*, 2009, **37**, 3310.
27   L. Petraccone, J. O. Trent and J. B. Chaires, *J. Am. Chem. Soc.*, 2008, **130**, 16530.
28   N. W. Luedtke, *Chimia*, 2009, **63**, 134.
29   M. Bejugam, S. Sewitz, P. S. Shirude, R. Rodriguez, R. Shahid and S. Balasubramanian, *J. Am. Chem. Soc.*, 2007, **31**, 12926.
30   D. J. Patel, A. T. Phan and V. Kuryavyi, *Nucleic Acids Res.*, 2007, **35**, 7429.
31   S. Neidle and G. N. Parkinson, *Biochemie*, 2008, **90**, 1184.
32   D. Monchaud and M.-P. Teulade-Fichou, *Org. Biomol. Chem.*, 2008, **6**, 627.
33   K.-i. Shinohara, Y. Sannohe, S. Kaieda, K.-i. Tanaka, H. Osuga, H. Tahara, Y. Xu, T. Kawase, T. Bando and H. Sugiyama, *J. Am. Chem. Soc.*, 2010, **132**, 3778.
34   N. Zhang, A. T. Phan and D. J. Patel, *J. Am. Chem. Soc.*, 2005, **127**, 17277.
35   Y. Sannohe, M. Endo, Y. Katsuda, K. Hidaka and H. Sugiyama, *J. Am. Chem. Soc.*, 2010, **132**, 16311.
36   D. Koirala, T. Mashimo, Y. Sannohe, Z. Yu, H. Mao and H. Sugiyama, *Chem. Commun.*, 2012, **48**, 2006.
37   G. Binnig, C. F. Quate and C. Gerber, *Phys. Rev. Lett.* 1986, **56**, 930.
38   Q. Zhong, D. Inniss, K. Kjoller and V. B. Elings, *Surface Science Letters* 1993, **290**, L688.
39   O. Marti, B. Drake and P. K. Hansma, *Appl. Phys. Lett.* 1987, **51**, 484.
40   N. Kodera, T. Kinoshita, T. Ito and T. Ando, *Advances in Experimental Medicine and Biology*, 2003, **538**, 119.
41   M. B. Viani, T. E. Schaffer, G. T. Paloczi, L. I. Pietrasanta, B. L. Smith, J. B. Thompson, M. Richter, M. Rief, H. E. Gaub, K. W. Plaxco, A. N. Cleland, H. Hansma and P. K. Hansma, *Rev. Sci. Instrum.*, 1999, **70**, 4300.
42   M. B. Viani, T. E. Schaffer, A. Chand, M. Rief, H. E. Gaub and P. K. Hansma, *J. Appl. Phys.*, 1999, **86**, 2258.
43   M. B. Viani, L. I. Pietrasanta, J. B. Thompson, A. Chand, I. C. Gebeshuber, J. H. Kindt, M. Richter, H. G. Hansma and P. K. Hansma, *Nat. Struct. Biol.*, 2000, **7**, 644.
44   M. Lutwyche, C. Andreoli, G. Binnig, J. Brugger, U. Drechsler, W. Häberle, H. Rohrer, H. Rothuizen, P. Vettiger, G. Yaralioglu and C. F. Quate, *Sensors and Actuators A: Physical*, 1999, **73**, 89.
45   S. R. Manalis, S. C. Minne and C. F. Quate, *Appl. Phys. Lett.* 1996, 68, 871.
46   T. Ando, N. Kodera, E. Takai, D. Maruyama, K. Saito and A. Toda, *Proc. Nat. Acad. Sci. USA*, 2001, **98**, 12468.
47   T. Ando, N. Kodera, D. Maruyama, E. Takai, K. Saito and A. Toda, *Jpn. J. Appl. Phys.*, 2002, **41**, 4851.
48   T. Ando, N. Kodera, T. Uchihashi, A. Miyagi, R. Nakakita, H. Yamashita and K. Matada, *e-J. Surf. Sci. Nanotech.*, 2005, **3**, 384.
49   N. Kodera, D. Yamamoto, R. Ishikawa and T. Ando, *Nature*, 2010, **468**, 72.
50   T. Ando, T. Uchihashi and T. Fukuma, *Progress in Surface Science*, 2008, **83**, 337.

51 A. Rajendran, M. Endo and H. Sugiyama, *Adv. Protein Chem. Struct. Biol.*, 2012, **87**, 5.
52 Y. Suzuki, Y. Yoshikawa, S. H. Yoshimura, K. Yoshikawa and K. Takeyasu, *WIREs Nanomed Nanobiotechnol*, 2011, **3**, 574.
53 A. Miyagi, T. Ando and Y. L. Lyubchenko, *Biochemistry*, 2011, **50**, 7901.
54 P. W. K. Rothemund, *Nature*, 2006, **440**, 297.
55 A. Rajendran, M. Endo and H. Sugiyama, *Curr. Protoc. Nucleic Acid Chem.*, 2012, **48**, 12.9.1.
56 A. Rajendran, M. Endo and H. Sugiyama, *Angew. Chem. Int. Ed.*, 2012, **51**, 874.
57 L.T. Costa, M. Kerkmann, G. Hartmann, S. Endres, P.M. Bisch, W.M. Heckl and S. Thalhammer, *Biochem. Biophys. Res. Commun.*, 2004, **313**, 1065.
58 M. Endo, Y. Katsuda, K. Hidaka and H. Sugiyama, *J. Am. Chem. Soc.*, 2010, **132**, 1592.
59 M. Endo, Y. Katsuda, K. Hidaka and H. Sugiyama, *Angew. Chem. Int. Ed.*, 2010, **49**, 9412.
60 R. D. Gray and J. B. Chaires, *Nucleic Acids Res.*, 2008, **36**, 4191.
61 T. Mashimo, H. Yagi, Y. Sannohe, A. Rajendran and H. Sugiyama, *J. Am. Chem. Soc.*, 2010, **132**, 14910.

# Chapter 2

## Synthesis and Characterization

Chapter Editor: **Gian Piero Spada**

The development of methodology for the chemical synthesis of oligonucleotides is the first step for producing not only the natural G-quadruplex sequences but also a large number of modified G-quadruplexes and other G-based supramolecular motifs; these methodologies will be summarized in this chapter. Natural or chemically modified G-motifs can be characterized by a variety of techniques. This chapter will not give an exhaustive list of methodologies, but simply shows how it is possible to gain structural information on G-quadruplexes by means of relatively uncommon techniques such as: (i) differential pulse voltammetry; (ii) atomic force microscopy; (iii) dynamic light scattering; (iv) solution X-ray scattering; (v) temperature gradient gel-electrophoresis; (vi) Langmuir-Blodgett technique. Many other techniques are more usually adopted in G4 characterization (NMR, Scanning Tunnel Microscopy, Circular Dichroism, X-ray diffraction ...) and examples of such applications can be found in the contributions to the other chapters of this book.

# SYNTHESIS AND PROPERTIES OF OLIGONUCLEOTIDES FORMING G-QUADRUPLEXES

A. Aviñó, and R. Eritja*

Institute for Research in Biomedicine, IQAC-CSIC, CIBER-BBN Networking Centre on Bioengineering, Biomaterials and Nanomedicine. Baldiri Reixac 10, E-08028 Barcelona, Spain. Email: recgma@cid.csic.es

## 1 INTRODUCTION

The oligonucleotides that form G-quadruplexes are essentially G-rich oligonucleotides. The first attempts by synthetic chemists to develop methodology for the synthesis of oligonucleotides were severely encumbered by G-rich sequences. In this regard, several side reactions with the guanine residue were observed during the activation of the monomers and, in some cases, several peaks were observed during purification which together led to a considerable decrease in yield. Later, with the consolidation of the phosphoramidite method in the 80s most of the side reactions on the guanine were avoided and the purification processes were standardized. The emergence of the Polymerase Chain Reaction and especially the use oligonucleotides to inhibit gene expression triggered a large demand for synthetic oligonucleotides, thus generating an unprecedented focus on methodology for the chemical synthesis of nucleic acids. These advances have allowed the production not only the natural G-quadruplex sequences but also a large number of derivatives. Therefore, researchers can now go beyond structural determination and new avenues in the biomedical, technological and material science fields are opening up. This chapter will summarize the developments in the field of the chemical synthesis of modified oligonucleotides with special emphasis on the synthesis of G-quadruplex forming oligonucleotides.

## 2 SOLID-PHASE SYNTHESIS OF OLIGONUCLEOTIDES

The identification of DNA as the genetic material and the elucidation of its structure stimulated interest in attempting the challenging task of synthesizing oligonucleotides of a defined sequence. It took over 30 years to achieve a fast and reliable methodology for the preparation of these compounds. In the following section a small description of this research is described.

## 2.1. Oligonucleotide synthesis

The first approach developed for the production of oligonucleotides of a defined sequence was the "phosphodiester" method.[1] Here, the synthesis of these oligomers involved the condensation of a first nucleoside with a free 3'-hydroxyl group and a second 5'-phosphate nucleoside generating "phosphodiester" intermediates.

This breakthrough was followed by the emergence of the "phosphotriester" methodology. Here, a protected 3'-O-phosphodiester was reacted with the 5'-hydroxyl group. Aryl groups, especially the 2-chlorophenyl group, were developed for the protection of phosphate groups.[2] Other phosphate-protecting groups, such as 2-cyanoethyl and methyl, were also designed at this time and they were of extreme relevance during the implementation of the phosphite-triester method.

Another milestone during the development of the "phosphotriester" method was the use of solid supports for oligonucleotides synthesis. The first successful synthesis was reported in 1965,[3] shortly after Merrifield's description of the solid-phase synthesis of the first peptide. In this approach linkers and supports were developed for solid-phase oligonucleotide synthesis. In the mid-1970s automation of the synthetic cycles started, and HPLC began to be used for the purification of oligonucleotides. DNA synthesizers appeared and oligonucleotides started to become available for biochemical and molecular biology studies.

At that time, recognition of the potential of phosphorous III derivatives, as a result of their exceptional reactivity[4], by Letsinger's group led to the arrival of "phosphite-triester" method. Shortly afterwards, Beaucage and Caruthers developed nucleoside phosphoramidites[5]. These derivatives could be synthesized, stored, and activated by tetrazole to yield phosphite-triester intermediates which were later transformed into phosphate-triester by oxidation (Figure 1). Nowadays, phosphoramidite derivatives are the most widely used monomers for oligonucleotide synthesis as they allow the preparation of oligonucleotides with natural phosphodiester bonds and also with modified phosphates such as phosphorothiotes. Many breakthroughs in nucleic acid chemistry achieved using the phosphoramidite approach are reported in a number of excellent reviews.[6,7]

Soon after the appearance of phosphoramidite monomers, the H-phosphonate method came about.[8,9] In this method, protected 3'-O-hydrogen phosphonate nucleosides were activated with acid chlorides, such as pivaloyl chloride or preferably, adamantoyl chloride, to yield H-phosphonate diester intermediates. Oxidation of these compounds to achive phosphodiesters was performed at the end of the synthesis in a single step.

## 2.2. Solid-phase synthesis

Solution-phase methods were used for the synthesis of oligonucleotides for the first 20 years.[1,2] At the end of the 70s, solid-phase methods were developed and became the most used approaches for the preparation of oligonucleotides. In these methods the 3'-end of the first nucleoside is attached covalently to a polymeric solid support and the excess of the chemicals used during the synthesis are washed out, thus simplifying the synthesis. Also, solid-phase techniques allow the automation of the entire synthetic process and oligonucleotide synthesizers became available to most laboratories.

Most the machine-assisted synthesis of oligonucleotides is performed on silica-derived supports[10], such as controlled-pore glass (CPG). Some supports such highly cross-linked polystyrene is suitable for very small scales (10-50 nmol) as a result of its hydrophobic properties and the lack of surface reactivity.[11] For large scales several alternative supports have been described.[12,13]

Synthesis and Characterization

The synthesis of an oligonucleotide on solid-phase is accomplished by a cycle of reactions that are repeated until the desired oligonucleotide has been assembled on the support. The oligonucleotide is then cleaved from the support for subsequent purification. The addition of one nucleotide using phosphoramidite chemistry requires four reactions: detritylation, coupling, capping and oxidation (see Figure 1). Exocyclic amino groups of phosphoramidite monomers are usually protected with a benzoyl group (A and C) and isobutyryl group (G). The capping step not only blocks the unreacted hydroxyl groups but also reverses phosphitylation in the guanine residues.[14] The complete cycle for the addition of one nucleotide unit takes around 3-5 minutes on an oligonucleotide synthesizer.

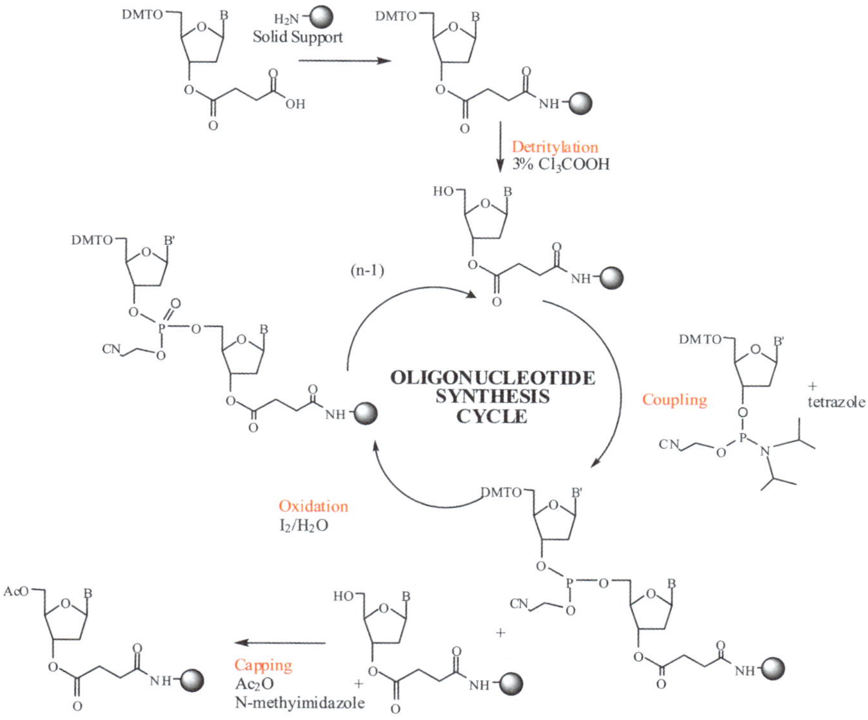

**Figure 1** *Scheme of the reactions used to assemble oligodeoxynucleotides by the solid-phase phosphoramidite methodology.*

## 2.3. Purification and characterization of oligonucleotides

For some applications, synthetic oligonucleotides can be used directly without purification or after a small clean-up using gel filtration chromatography or ethanol precipitation. Nevertheless, others call for a careful separation of shorter sequences generated during the synthesis. In these cases, synthetic oligonucleotides are usually purified either by polyacrylamide gel electrophoresis (PAGE) or HPLC techniques. Gel electrophoresis is performed on a preparative 15-20% polyacrylamide gel in denaturing conditions (7M urea). In general, satisfactory resolution of the n-1 sequence is obtained for long synthetic oligonucleotides but the recoveries of the product are often low. HPLC purification can be divided into two basic methods: ion-exchange and reversed-phase. The former also provides an acceptable resolution of full-length product from failed sequences for oligonucleotides shorter than 30-40 bases. HPLC on reversed-phase columns is also used for this purpose. In this case, the DMT group of the last phosphoramidite addition is left on the oligonucleotide during ammonia treatment (DMT on protocol). The DMT group is highly hydrophobic and will delay the elution of the full-length product, thus increasing the separation from shorter sequences. A similar result can be obtained using reversed-phase cartridges (oligonucleotide purification cartridges) which contain a polymeric support with reversed-phase material that is stable to ammonia solution. Thus, using this approach, no special equipment is required, the deprotection solution can be used directly and purification is faster.

Oligonucleotide characterization is a difficult task due to the small amounts prepared and the intrinsic complexity of the molecule. Most the synthetic oligonucleotides are controlled by monitoring the progress of the synthesis by the absorbance of DMT and by analysing the purity of the final product by HPLC or PAGE.

Mass spectrometry (MS) and capillary electrophoresis (CE) are the methods used to characterize oligonucleotides. In MS, namely in MALDI-(matrix-Assisted Laser Desorption Ionization) or ES-(Electrospray Ionization) based methods, the oligonucleotide is ionized and characterized by mass/charge ratio. These methods are useful for the determination of the molecular mass of synthetic oligonucleotides especially for non-degradable analogs. In CE, the product which is subjected to a constant electrical field in a capillary is separated into component sizes in a similar manner to PAGE.

Further quality characterization can be achieved by enzyme degradation followed by HPLC analysis of the nucleosides. To degrade DNA, snake venom or spleen phosphodiesterase (PDE) is used obtaining a mixture of 5' (snake venom PDE) or 3'- (spleen PDE) nucleoside monophosphates. These are dephosphorylated by alkaline phosphatase to form nucleosides which are then analyzed by reversed-phase HPLC. Finally, NMR techniques can be applied to characterize oligonucleotides, although the most common use of NMR is the conformational analysis and structural determination of oligonucleotides.

## 3  SYNTHESIS OF MODIFIED OLIGONUCLEOTIDES

The phosphoramidite approach allows the preparation of oligonucleotides that contain not only natural bases but also a large number of non-natural nucleosides, as well as non-natural backbones. Considerable research effort has been devoted to elucidating the role of individual residues in G-quadruplex in terms of structure and function. In particular, several modified nucleotides have been synthesized to study the stability, binding affinity and biological activity of G-quadruplex-forming oligonucleotides. In the following section we will described the most relevant developments in the preparation of modified G-quadruplex-forming oligonucleotides.

### 3.1. Synthesis of oligonucleotides carrying guanine derivatives

One of the critical questions in G-quadruplex research is whether guanine can be substituted by other non-natural bases. Figure 2 shows guanine derivatives that have been introduced in G-quadruplex structures. Most of these modifications influence the G-tetrad and contribute to the overall stability and biological activity of the G-quadruplex structures. The thrombin binding aptamer (TBA) is a unimolecular antiparallel quadruplex that has been extensively used as model compound to test single position modifications with the aim to improve quadruplex stability, and the binding or inhibitory properties of the aptamer. In addition to the TBA, several biologically relevant quadruplexes, such as telomerase sequences and aptamers associated with HIV or cancer have also been modified with several derivatives.

Hypoxanthine and 7-deazaguanine derivatives were first introduced in the G-tetrads positions to determine the tertiary structure of the TBA by NMR. These guanine derivatives are unable to form the H-bond required for the formation of the G-tetrad. Consequently, they produced the disruption of the chair-like structure.[15] In addition, these derivatives were studied in the parallel tetramolecular G-quadruplex [TG$_4$T]$_4$.[16] These authors analyzed the thermodynamic and kinetic properties of sequence variants of parallel quadruplexes in which each guanine of the central block was systematically substituted with a different nucleobase. Hypoxanthine and 7-deazaguanine were deleterious to tetramolecular quadruplex structure.

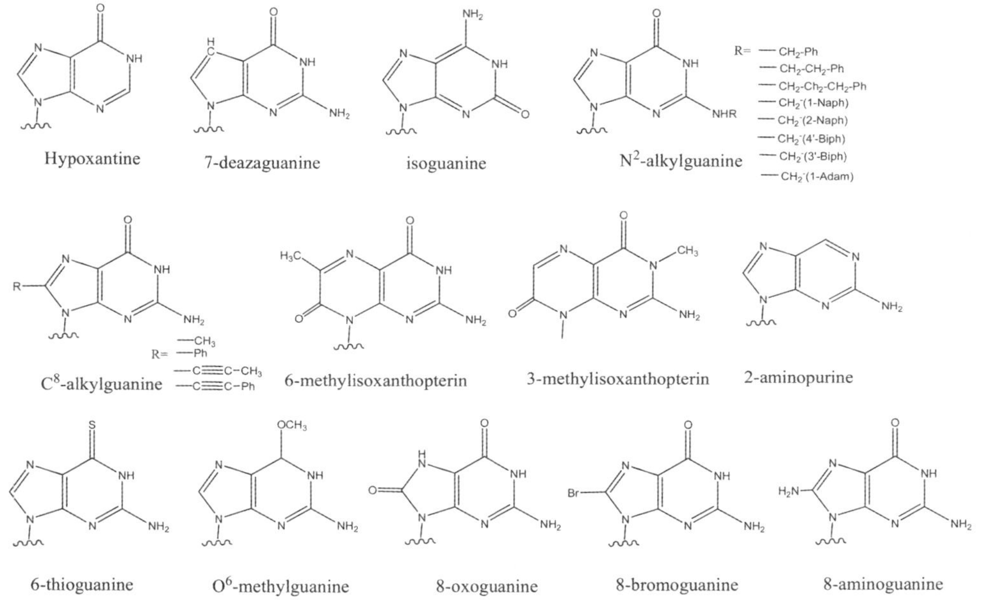

**Figure 2** *Structures of the bases introduced into oligonucleotides to study non-natural G-tetrads.*

$N^2$ and $C^8$-alkylguanine were also introduced into the TBA G-tetrads.[17] These substitutions caused relatively small perturbation of the quadruplex structure as these positions do not form the H-bonding of the tetrads. Substituents in the $C^8$ positions favor the *syn* conformation of the guanine, therefore the introduction of these modifications at positions that show *syn* conformations stabilized antiparallel G-quadruplex structure.

A combinatorial library of TBAs containing isoguanine was synthesized by split and mix synthesis. These modified TBAs showed improved binding affinity to thrombin.[18]

3-Methylisoxanthopterin and 6-methylisoxanthopterin were also studied in tetramolecular quadruplex structures.[16] The former did not enhance structural stability while the latter accelerated quadruplex formation when present at the 5' end. This nucleobase was also used as a reporter fluorescent molecule in human Telomeric Repeat DNA for studying protein binding and the destabilization of G-tetrad structures.[19]

The fluorescent base 2-aminopurine was incorporated into the human telomerase sequence to monitor the duplex to quadruplex conformational change. In addition, this base has the capacity to recognize two G-quadruplex structures of human telomerase sequence and to monitor specific ligand complexes of G-quadruplex.[20]

Several 8-substituted guanines were introduced into G-quadruplex structures. Although the presence of the 8-amino group in guanine did not affect hydrogen bonding or purine-ion interaction, it clearly reduced the strength of stacking interactions of TBA tetrads producing minor destabilization of the quadruplex.[21]

Human telomeric sequences may adopt parallel or antiparallel folds depending on several conditions: however, programmed folding can be accomplished by hydrogen bonding patterns within tetrads composed of nucleobases other than guanine. Harting and co-workers synthesized human telomeric sequences containing two 8-oxoguanine and two xantosine modifications instead of the natural guanine tetrad.[22] In this case, the number of hydrogen bonds was the same as in the G-tetrad but these 8-oxoguanosine/xantosine tetrads allowed the programming of antiparallel and parallel topologies. However, the incorporation of this unnatural tetrad slightly destabilizes the quadruplex structure compared to a conventional tetrad formed by guanines. 8-Oxoguanine was also studied in tetramolecular parallel structures. Since this nucleoside favors the *syn*-conformation, parallel structures are disadvantaged because this type of quadruplex derivatives adopts the *anti* configuration.

In contrast, 8-bromoguanine which also adopts a *syn* conformation, produced altered quadruplex structures. This base was used to study intramolecular human telomere sequences by NMR as this modification forces nucleoside conformation.[23] NMR, molecular dynamics and mechanics calculations, and CD spectroscopy were used to study the parallel quadruplex structure of $[TG_3T]_4$ containing 8-bromoguanine tetrads in various positions.[24] The thermodynamic and kinetic stability of the same quadruplex structures were also addressed.[16] In this case, 8-bromoguanine accelerated quadruplex formation when located at the 5'-end. Similar properties were found in other 8-substituted guanine derivatives, such as 8-aminoguanine.[21]

6-Methylguanine was also examined in this tetramolecular parallel system. This substitution was detrimental to quadruplex stability as it led to a decrease in the association rate and in thermal stability.[25]

The replacement of guanines by 6-thioguanine in G-quadruplexes inhibited G-tetrad formation. The reduced stability of G-quadruplex is due to the increased radius and decreased electronegativity of the sulphur atom, which destabilizes hydrogen bonding. In addition, the decreased electronegativity, caused by the presence of the thio group, may also lead to decrease cation interaction.[26] 6-Thioguanine has also been introduced into G-quadruplexes for optical, photocrosslinking and metal ion binding studies.[27]

## 3.2. Synthesis of oligonucleotides carrying modified backbones

In addition to the guanine derivatives located in tetrads, a large number of modified backbones, including the sugar and internucleotide phosphate modifications, have been introduced into quadruplex structures. Figure 3 shows the most relevant modifications.

Most of the studies addressing the G-quadruplex are based on the DNA quadruplex: however, there is increasing interest in the contribution of the RNA quadruplex. To date, all the RNA quadruplexes observed have been parallel. This finding is attributed to the fact that ribonucleotides with a 3'-endo sugar pucker adopt the *anti* orientation of the glycosidic bond. Several studies showed the incorporation of ribonucleotides in G-quadruplex structures. Shafer and co-workers studied the incorporation of riboG in the G-tetrads of TBA. These authors observed a change in topology and molecularity of the TBA due to the strong preference of riboG for the *anti* conformation.[28] Similarly, RNA/DNA hybrids of *Oxytricha* Telomere G$_4$T$_4$G$_4$ fragment were examined by Vorlíčková's group. They showed that the substitution of a single dG by rG also changed quadruplex topology.[29] In addition, the topologies and stabilities of DNA and RNA quadruplexes have recently been compared.[30]

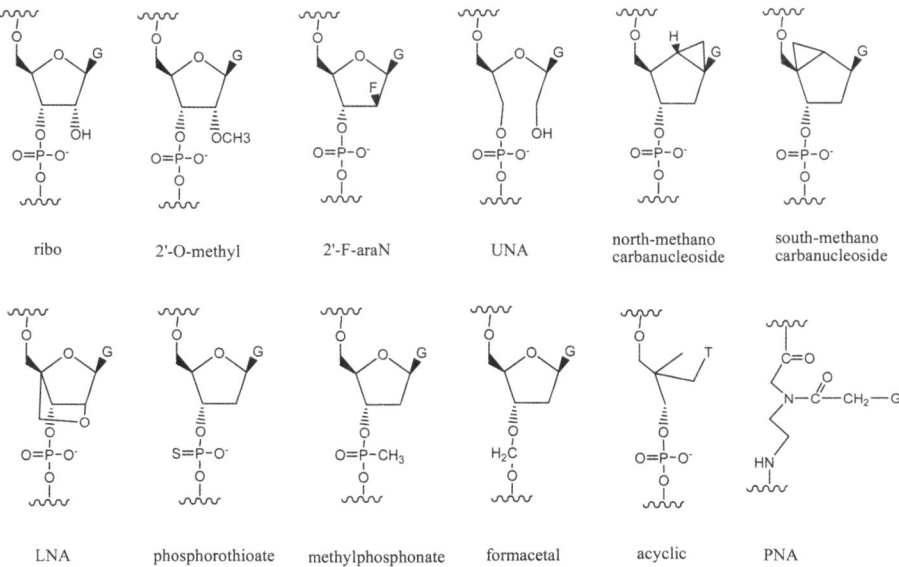

**Figure 3** *Some of the modified backbones introduced into G-quadruplex-forming oligonucleotides.*

GRO29A is a G-quadruplex growth-inhibitory aptamer. Analogs of this oligonucleotide, including 2'-O-methyl RNA, mixed DNA/2'-O-methyl RNA and phosphorothioate, were synthesized to study the antiproliferative properties, protein binding and quadruplex formation of GRO29A.[31] In this context, modifications of the backbone charge and atom size and the effect of modification at the 2'-position of several G-quadruplexes were analyzed by UV spectroscopy.[32] The lack of negative charge at the phosphate internucleotide (methylphosphonate) yielded a general destabilization of the G-quadruplex structure. The formacetal group was incorporated in the TBA. This modification caused a decrease in tissue uptake and an increase in the *in vivo* half-life of the aptamer. Substitution of the oxygen atoms of the phosphate backbone with sulphur atoms (phosphorothioate) appeared to affect the stability of the G-quadruplex structures in a molecularity-dependent way.[32] In addition, a remarkable role was observed in the 2'-position of the sugar moiety (2'-OH or 2'-OMe). The same 2'-sugar modification had a different structural effect depending on the molecularity of the quadruplex. In particular, loss of hydrogen bond capacity affected the conformation of the G-quadruplex.[32]

The impact of 2'-deoxy-2-fluoroarabinonucleoside residues (2'-F-araN) on several G-quadruplex structures was studied.[33] 2'-Deoxy-2'-fluoro-D-arabinonucleic acids conferred a DNA-like (*South/East*) conformation to the oligonucleotides while rendering them more nuclease-resistant. Generally, replacement of *anti*-Gs with 2'-F-araN stabilized the G-tetrads that require *anti*-Gs and maintained the quadruplex conformation. In contrast, replacement of *syn*-Gs with 2'-F-araN was not favoured and resulted in complete conformational change of the G-quadruplexes.

Conformationally rigid nucleosides LNA (Locked Nucleic Acid) and methanocarbanucleosides have been inserted in several G-quadruplex structures. LNA are 2'-O-4'-C-methylene-linked ribonucleotide nucleic acid analogs with a C3'-*endo* puckering and *anti* glycosidic bond configuration. Bonifacio *et al.* studied the effect of single LNA substitutions on several positions of the TBA.[34] The LNA substitutions had either a moderate stabilizing or destabilizing effect on the folded structure, depending on the position of the LNA in the TBA. Tetrameric quadruplexes of [TG$_4$T]$_4$ were also modified with LNA to study local structural implication of incorporation of these derivatives.[35]

Carbacyclic bicyclo [3.1.0] hexane locked nucleoside analogs (*North* and *South*) were prepared by shifting the position of the fused cyclopropane ring. The introduction of methanocarba nucleosides with locked-*N* (*anti*) and locked-*S* (*syn*) conformations fixed the conformational state of these nucleosides and helped to unveil the impact of conformational restrictions on the TBA. These studies revealed that the glycosidic conformation is more restrictive for TBA stability than sugar puckering.[36]

Furthermore, acyclic nucleoside derivatives with no restrictive conformation have been introduced into the G-quadruplex. Specifically, Wengel's group examined the influence of unlocked nucleic acids (UNA) on the thermodynamic stability, binding affinity and biological activity of the TBA. Modification of any position within the G-tetrads was unfavourable for quadruplex formation.[37]

Several modifications previously described in G-tetrads were also used in the loops of G-quadruplex structures. Loop sequences that connect G-tetrads are crucial for the folding, stability and biological properties of the G-quadruplex. Most of the modifications were based on changes in the composition or in the length of the loops.[38,39]

Since the discovery of the G-quadruplex, several applications of this structure have been reported, ranging from biomedical to nanotechnology purposes. For this reason, several modifications have been introduced to adapt the G-quadruplex to fulfil the increasing requirements of their implementations.

Exploration of G-quadruplex stabilization by the binding of diverse types of molecules revealed a role of this structure as therapeutic targets in drug design.[40] Similarly, several molecules, such as acridines, have been conjugated to stabilize G-quadruplex structures.[41]

Given their well-defined structure and self-recognition properties, G-quadruplexes exhibit great potential in applications that involve nanomolar devices. For example, the conformational change of the TBA during folding/unfolding was exploited for building sensors for metal ions and for the detection of thrombin. In this case, quadruplexes were conjugated to several molecules, such as amino or thiol moieties, to functionalize surfaces. In addition, several quadruplex-based FRET probes with diverse fluorophore and quencher molecules have been developed as biosensors.[42]

### 3.3. Synthesis of four-stranded oligonucleotides

In addition to single-stranded oligonucleotides, it is possible to prepare multi-stranded or branched oligonucleotides. Recently, branched oligonucleotides carrying four G-rich strands linked through their 5' or 3'-ends have been described to result in the formation of highly stable quadruplexes.[43-48] These compounds can be synthetized using a combination of symmetric and asymmetric doubler phosphoramidites.[43-47] Moreover, the synthesis of quadruplex structures formed by four strands, of which three are identical and one contains single modifications, is also possible. In this case, a single strand is prepared first and then the use of the trebler phosphoramidite allows the synthesis of the remaining three strands to complete the quadruplex structure.[48] These tetra-end-linked oligonucleotides are characterized by fast association and slow dissociation rates and, more importantly, they have higher thermal stability than the corresponding tetramolecular quadruplexes.[43-48] A preference for the formation of parallel quadruplexes is observed even when the strands are synthesized in antiparallel direction.[42,48] Finally, the formation of dimeric G-quadruplex structures is also observed in addition to the desired monomolecular quadruplex structure.[48]

## 4 CONCLUSIONS

In spite of the relatively large number of modified bases studied, guanine has been proven to be the best base for quadruplex formation. 8-Substituted guanine derivatives can be used to increase association rates and to freeze the *syn/anti* topology of guanine. Several studies have addressed modified backbones and demonstrated that some of them can provide efficient nuclease resistance. Tetra-end linked structures have higher thermal stability and higher association rates than the corresponding tetramolecular quadruplexes. All together, these findings are an excellent starting-point for the generation of new quadruplex assemblies for the development of novel biomedical and technological applications.

**Acknowledgements**

Financial support from the Spanish Ministry of Education (CTQ2010-20541), the EU COST project (MP0802) and the *Generalitat de Catalunya* (2009/SGR/208) is gratefully acknowledged. CIBER-BBN is an initiative funded by the VI National R&D&i Plan 2008-2011, *Iniciativa Ingenio* 2010, Consolider Program, CIBER Actions and financed by the *Instituto de Salud Carlos III* with assistance from the European Regional Development Fund).

**References**

1. H.G. Khorana, *Science*, 1979, **203**, 614.
2. C.B. Reese, *Tetrahedron*, 1978, **34**, 3143.
3. R.L. Letsinger, and V. Mahadevan, *J. Am. Chem. Soc.*, 1965, **87**, 3526.
4. R.L. Letsinger, and W.B. Lunsford, *J. Am. Chem. Soc.*, 1976, **98**, 3655.
5. S.L. Beaucage, and M.H. Caruthers, *Tetrahedron Lett.*, 1981, **22**, 1859.
6. S.L. Beaucage and R.P. Iyer, *Tetrahedron*, 1992, **48**, 2223.
7. S.L. Beaucage and R.P. Iyer, *Tetrahedron*, 1993, **49**, 6123.
8. P.J. Garegg, I. Lindh, T. Regberg, J. Stawinski, R. Stromberg, and C. Henrichson, *Tetrahedron Lett.*, 1986, **27**, 4051.
9. B.C. Froehler, P.G. Ng, and M.D. Matteucci, *Tetrahedron Lett.*, 1986, **27**, 5399.
10. M.H. Caruthers, A.D. Barone, S.L. Beaucage, D.R. Dodds, E.F. Fisher, L.J. McBride, M. Matteucci, Z. Stabinsky, and J.Y. Tang, *Methods in Enzymol.* 1987, **154**, 287.
11. C. McColum, and A. Andrus, *Tetrahedron Lett.*, 1991, **32**, 4069.
12. F.X. Montserrat, A. Grandas, R. Eritja, and E. Pedroso, *Tetrahedron*, 1994, **50**, 2617.
13. P. Wright, D. Lloyd, W. Rapp, and A. Andrus, *Tetrahedron Lett.*, 1993, **34**, 3373.
14. J.S. Eadie, and D.S. Davidson, *Nucleic Acids Res.*, 1987, **15**, 8333.
15. K.Y. Wang, S.H. Krawczyk, N. Bischofberger, S. Swaminathan, and P.H. Bolton, *Biochemistry*, 1993, **32**, 11285.
16. J. Gros, F. Rosu, S. Amrane, A. De Cian, V. Gabelica, L. Lacroix, and J.L. Mergny, *Nucleic Acids Res.*, 2007, **35**, 3064.

17 G.X. He, S.H. Krawczyk, S. Swaminathan, R.G. Shea, J.P. Dougherty, T. Terhorst, V.S. Law, L.C. Griffin, S. Coutre, and N. Bischofberger, *J. Med. Chem.*, 1998, **41**, 2234.
18 R.S. Nallagatla, B. Heuberger, A. Haque, and C. Switzer, *J. Comb. Chem.*, 2009, **11**, 364.
19 J.C. Myers, S.A. Moore, and Y. Shamoo, *J. Biol. Chem.*, 1993, **43**, 42300.
20 T. Kimura, K. Kawai, M. Fujitsuka, and T. Majima, *Tetrahedron*, 2007, **63**, 3585.
21 J.P. de la Osa, C. González, R. Gargallo, M. Rueda, E. Cubero, M. Orozco, A. Aviñó, and R. Eritja, *ChemBioChem*, 2006, **7**, 46.
22 A. Benz, and J.S. Harting, *Chem. Commun.*, 2008, 4010.
23 A.T. Phan, V. Kuryavyi, K.N. Luu, and D. J. Patel, *Nucleic Acids Res.*, 2007, **35**, 6517.
24 V. Esposito, A. Randazzo, G. Piccialli, L. Petraccone, C. Giancola, and L. Mayol, *Org. Biomol. Chem.*, 2004, **2**, 313.
25 M. Tintoré, A. Aviñó, F.M. Ruiz, R. Eritja, and C.Fàbrega, *J. Nucleic Acids*, 2010, 9 pages, article ID 632041, doi:10.4061/2010/632041.
26 V.M. Marathias, M.J. Sawicki, and P.H. Bolton, *Nucleic Acids Res.*, 1999, **27**, 2860.
27 T.S. Rao, R.H. Durland, D.M Seth, M.A. Myrick, V.Bodepudi, and G.R. Revankar, *Biochemistry*, 1995, **34**, 765.
28 F. Tang , and R.H. Shafer, *J. Am. Chem. Soc.*, 2006, **128**, 5966.
29 Vondrušková, J. Kypr, I. Kejnovská, M. Filanová, and M. Vorlíčková, *Biopolymers*, 2008, **89**, 797.
30 A. Joachimi, A. Benz, and J.S. Harting, *Biorg. Med. Chem.*, 2009, **17**, 6811.
31 V. Dapić, P.J. Bates, J.O. Trent, A. Rodger, S.D. Thomas, and D.M. Miller, *Biochemistry*, 2002, **41**, 3676.
32 B. Saccà, L. Lacroix, and J.L. Mergny, *Nucleic Acids Res.*, 2005, **33**, 1182.
33 C.G. Peng, and M.J. Damha, *Nucleic Acids Res.*, 2007, **35**, 4977.
34 L. Bonifacio, F.C.Church, and M.B. Jarstfe, *Int. J. Mol. Sci.* , 2008, **9**, 422.
35 J.T. Nielsen, K. Arar, and M. Pedersen, *Nucleic Acids Res.*, 2006, **34**, 2006.
36 H. Saneyoshi, S. Mazzini, A. Aviñó, G. Portella, C. González, M. Orozco, V.E. Marquez, and R. Eritja, *Nucleic Acids Res.* 2009, **37**, 5589.
37 A. Pasternak, F.J. Hernández, L.M. Rasmussen, B. Vester, and J. Wengel, *Nucleic Acids Research*, 2010, **39**, 1155.
38 I. Smirnov, and R.H. Shafer, *Biochemistry*, 2000, **39**, 1462.
39 P. Hazel, J. Huppert, S. Balasubramanian, and S. Neidle, *J. Am. Chem. Soc.*, 2004, **126**, 16405.
40 See contribution on G-quadruplex forming oligonucleotides with tailor-made modifications as effective aptamers for potential therapeutic applications by D. Musumeci and D. Montesarchio, this book.
41 A. Aviñó, S. Mazzini, R. Ferreira, and R. Eritja, *Biorg. Med. Chem.*, 2010, **18**, 7348.
42 B. Juskowiak, *Anal. Chim. Acta*, 2006, **568**, 171.
43 G. Oliveiro, J. Amato, N. Borbone, A. Galeone, L. Petraccone, M. Varra, G. Piccialli, L. Mayol, *Bioconjugate Chem.*, 2006, **17**, 889.
44 G. Oliveiro, J. Amato, N. Borbone, S. D'Errico, A. Galeone, L. Mayol; S. Haider; O. Olubiyi, B. Hoorelbeke; J. Balzarini, G. Piccialli, *Chem. Comm.*, 2010, **46**, 8971.
45 L. Petraccone, L. Martino, I. Duro, G. Oliveiro, N. Borbone, G. Piccialli, C. Giancola, *Int. J. Biol. Macromol.*, 2007, **40**, 242.
46 G. Oliveiro, N. Borbone, J. Amato, S. D'Errico, A. Galeone, G. Piccialli, M. Varra, L. Mayol, *Biopolymers*, 2009, **91**, 466.

47 P. Murat, R. Bonnet, A. Van der Heyden, N. Spinelli, P. Labbé, D. Monchaud, M.-P. Teulade-Fichou, P. Dumy, E. Defrancq, *Chem. Eur. J.* 2010, **16**, 6106.
48 R. Ferreira, M. Alvira, A. Aviñó, I. Gómez-Pinto, C. González, V. Gabelica, R. Eritja, *ChemistryOpen*, 2012, **1**, 106.

# ELECTROCHEMICAL CHARACTERIZATION OF GUANINE QUADRUPLEXES

A.–M. Chiorcea–Paquim[1], P. Santos[1], V.C. Diculescu[1], R. Eritja[2] and A.M. Oliveira-Brett[1*]

[1] Departamento de Química, Faculdade de Ciências e Tecnologia, Universidade de Coimbra, Portugal. Email: brett@ci.uc.pt
[2] Institute for Research in Biomedicine, IQAC-CSIC, CIBER-BBN Networking Centre on Bioengineering, Biomaterials and Nanomedicine, Barcelona, Spain

## 1 INTRODUCTION

G-rich nucleic acid sequences that contain guanine repeats can form four-stranded nucleic acid secondary structures named G-quadruplexes.[1–5] The building-blocks of G-quadruplexes are structures known as G-quartets, comprise by four guanine bases held together by Hoogsteen hydrogen bonds in a square planar arrangement (Figure 1A).

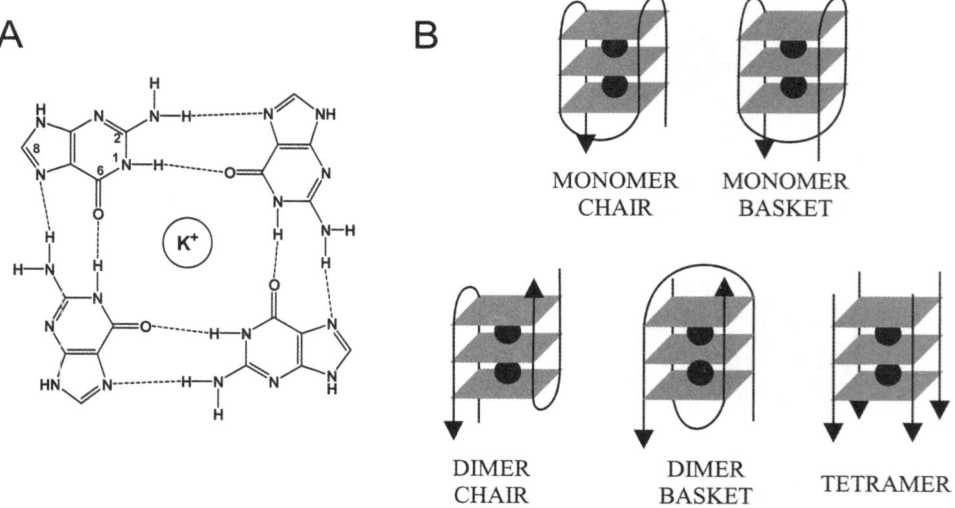

**Figure 1** *Schematic representation of the: (A) G-quartet and (B) molecularity and loop orientation of G-quadruplexes composed of three base stacked G-quartets. The K+ cations that stabilise the G-quadruplex structures are shown as balls.*

The G-quartets are stack on each other in a helical fashion via π–π interactions, being stabilized by monovalent cations such as $K^+$ and $Na^+$, and the order of preference is $K^+ > Na^+$. The ions are positioned in the interior channel that is formed at the centre of each G-quartet, coordinating the O6 atoms of the guanines. A variety of G-quadruplex structures exist (Figure 1B) and can be classified in terms of their molecularity, monomer, dimer or tetramer, and strand orientation, chair or basket.

The formation of G-quadruplexes is a current research interest because they are found in telomeric regions of chromosomes, oncogene promoter sequences, and other biologically relevant regions of the genome, such as immunoglobulin switch regions, ribosomal DNA and RNA.[1,2] In addition, G-rich oligodeoxynucleotides (ODNs) are able to self-organize in two-dimensional reticulated networks and long nanowires, relevant for nanotechnology applications.[6,7]

Atomic force microscopy (AFM) is a directly probing technique that can resolve with extraordinary resolution and accuracy the surface morphological characteristics of nucleic acids nanostructures immobilized on a solid substrate.[8,9] Differential pulse (DP) voltammetry is a technique that presents a high sensitivity and selectivity and can be successfully employed for the rapid detection of small perturbations of the nucleic acid secondary structure.

This chapter describes a systematic study performed by AFM on a highly oriented pyrolytic graphite (HOPG) surface and DP voltammetry on a glassy carbon (GC) electrode, to elucidate the adsorption mechanism and the redox behaviour of five G-rich ODN sequences, $d(G)_{10}$, $d(TG_9)$, $d(TG_8T)$, thrombin binding aptamer (TBA, 5'-GGTTGGTGTGGTTGG-3') and extended TBA (eTBA, 5'-GGGTTGGGTGTGGGTTGGG-3'), with respect to their ability to form G-quadruplex nanostructures.

## 2 METHOD AND RESULTS

The morphological characteristics of the G-rich ODNs were investigated by AFM on HOPG, because HOPG is atomically flat, with less than 0.06 nm of root-mean-square (*r.m.s.*) roughness for a 1000 x 1000 $nm^2$ surface area, while the GC electrode used for the voltammetric characterisation is much rougher, with 2.10 nm *r.m.s.* roughness for the same surface area, therefore unsuitable for AFM surface characterisation. The electrochemical experiments show similar behaviour using GC and HOPG electrodes.

The adsorption mechanism of the nucleic acids onto HOPG and GC is mainly driven by hydrophobic interactions[10], being strongly influenced by the hydrophobicity of the constituent nucleotides, the ODNs molecular mass and the formation of secondary structures. The guanine residues of $d(G)_{10}$, $d(TG_9)$, $d(TG_8T)$, TBA and eTBA influence the ODN hydrophobicity directly, through the intrinsic hydrophobic character of the aromatic rings, and indirectly, by allowing these ODNs to establish G-quadruplex conformations. The interaction of the G-quadruplex ODNs with the HOPG or GC surface is more difficult because the G-quadruplex structure have the bases protected by the sugar-phosphate backbones, when compared with the single-stranded ODNs that have the bases exposed to the solution and free to undergo hydrophobic interactions with the surface.

The formation and stabilisation of G-quadruplex configurations is influenced by the presence of $K^+$ cations, and the ODNs adsorption morphology and voltammetric behaviour after incubation with $K^+$, during different periods of time, was investigated.

## 2.1 AFM and electrochemical characterisation of d(G)$_{10}$

AFM topographical images in air of d(G)$_{10}$ molecules adsorbed spontaneously onto HOPG (Figure 2A) showed the formation of tilted polymeric structures of 0.8 nm height, corresponding to the immobilisation of single-stranded d(G)$_{10}$ molecules.

In the presence of K$^+$ cations that stabilize the G-quadruplex configurations (Figure 2B), AFM images showed the formation of both tilted polymeric structures of 1.0 nm height, associated with the adsorption of single-stranded d(G)$_{10}$ molecules, as well as larger 1.5–3.5 nm height spherical and rod-like shape aggregates, due to the adsorption of quadruplex d(G)$_{10}$ molecules, as was also observed for telomeric G-rich sequences on mica.[8] This is due to the fact that, at pH = 7.0, d(G)$_{10}$ exhibit complex and heterogeneous configurations, consisting on a mixture of single-, double- or quadruple helical morphologies. Increasing the K$^+$ ions concentration and/or incubation time, the number of G-quadruplexes observed in the images increased and the aggregates grew to form small nanowires, with lengths up to 100 nm.

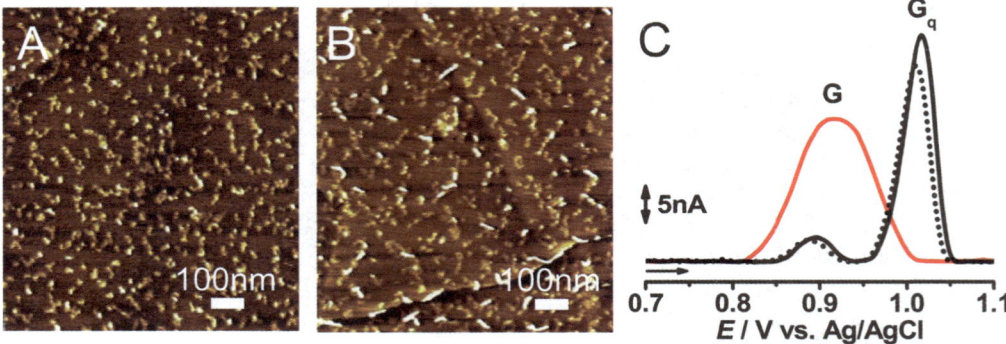

**Figure 2** *(A, B) AFM topographical images in air of d(G)$_{10}$ modified HOPG, prepared by spontaneous adsorption during 3 min, from a solution of 0.3 µM d(G)$_{10}$ in 0.1 M phosphate buffer pH = 7.0: (A) in the absence and (B) in the presence of 0.1 M K$^+$, 24 h of incubation. (C) DP voltammograms baseline corrected obtained with the GC electrode in a solution of 3 µM d(G)$_{10}$ in 0.1 M phosphate buffer pH = 7.0, (—) in the absence and in the presence of 0.1 M K$^+$, (—) 0 h and (•••) 24 h of incubation.*

The voltammetric study showed that d(G)$_{10}$ molecules are oxidized at GC electrode (Figure 2C), and the only anodic peak G, at $E_{pa}$ = + 0.92 V, corresponds to the oxidation of guanine residues. The addition of K$^+$ ions stabilizes the G-quadruplex molecules and this process was detected by the decrease of the guanine oxidation peak G in a time-dependent manner and the occurrence of a new peak G$_q$ at a higher potential, due to the oxidation of the G-quartets.

Upon the formation of the G-quartets, the guanine electroactive centres are hidden inside the rigid quadruplex being unable to reach the GC electrode surface. The decrease of the guanine oxidation peak is therefore related to a decrease of the concentration of guanine residues not involved in the G-quartet hydrogen bonding and still available for oxidation. The difference in the oxidation potential of guanine and G-quartets is due to the difficulty of the transition of the electrons from the inside of the rigid quadruplex to the GC electrode surface than from the more flexible single-stranded d(G)$_{10}$ molecules, with guanine residues that can easily reach the surface.

Moreover, the G$_q$ oxidation peak decreased with increasing the incubation time, which is associated with the formation of long G-wires composed of multiple d(G)$_{10}$ sequences that self-assemble together via inter-molecular G-quartets. The transition of electrons from the inside of the very rigid and stable G-wires to the GC electrode surface is even more difficult. The formation of G-wires was confirmed by AFM, observed as different length small nanowires (Figure 2B).

## 2.2 AFM and electrochemical characterisation of d(TG$_9$) and d(TG$_8$T)

AFM topographical images in air of d(TG$_9$) molecules adsorbed spontaneously onto HOPG (Figure 3A) showed the formation of tilted polymeric structures of 0.8 nm height, as also observed for d(G)$_{10}$ molecules (Figure 2A).

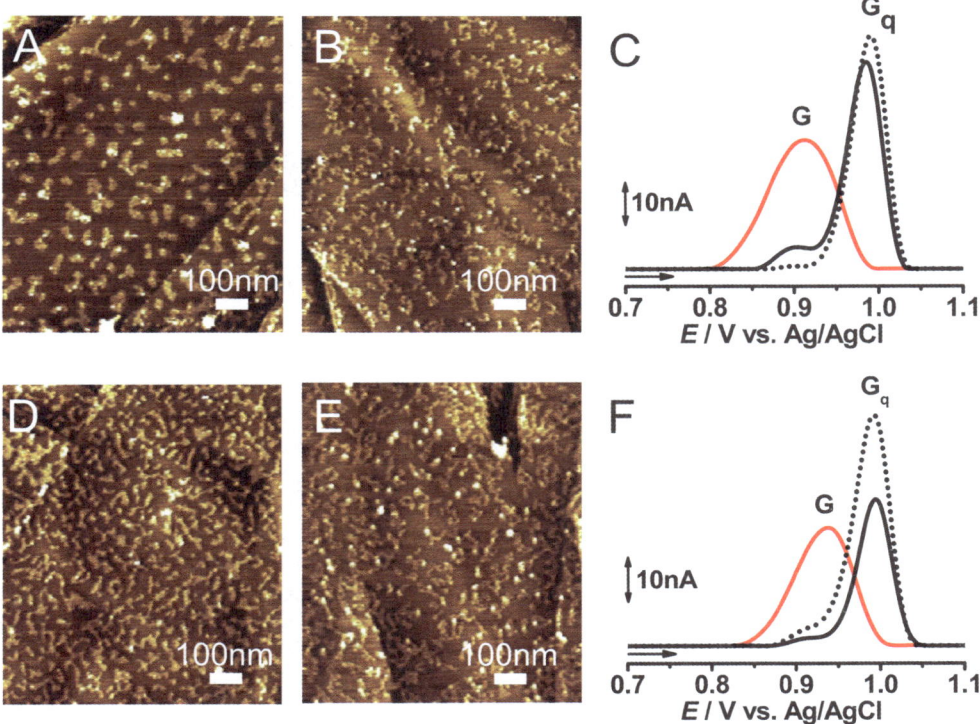

**Figure 3** *(A, B, D, E) AFM topographical images in air: (A, B) d(TG$_9$) and (D, E) d(TG$_8$T) modified HOPG, prepared by spontaneous adsorption during 3 min from solutions of 0.3 µM ODNs in 0.1 M phosphate buffer pH = 7.0: (A, D) in the absence and (B, E) in the presence of 0.1 M K$^+$, 24 h of incubation. (C, F) DP voltammograms baseline corrected obtained with the GC electrode in a solution of 3 µM (C) d(TG$_9$) and (F) d(TG$_8$T) in 0.1 M phosphate buffer pH = 7.0, (—) in the absence and in the presence of 0.1 M K$^+$, (—) 0 h and (•••) 24 h of incubation.*

Similar results were obtained for d(TG$_8$T) molecules, which present a substitution of one guanine by a thymine in the base sequence (Figure 3C). In both cases the adsorption pattern corresponds to the immobilisation of single-stranded d(TG$_9$) or d(TG$_8$T).

In the presence of K$^+$ cations, d(TG$_9$) (Figure 3B) and d(TG$_8$T) (Figure 3D), both lead to the formation of tilted polymeric structures of 1.0 nm height, associated with the adsorption of single-stranded d(TG$_9$) or d(TG$_8$T) molecules, and small 2.0 nm height spherical aggregates, corresponding to the adsorption of d(TG$_9$) or d(TG$_8$T) in G-quadruplex configurations. With increasing the K$^+$ ions concentration and/or incubation time, the number of G-quadruplexes observed in the images also increased.

DP voltammograms obtained with the GC electrode in solutions of d(TG$_9$) (Figure 3C) and d(TG$_8$T) (Figure 3F) showed a main oxidation peak G, at $E_{pa}$ = + 0.92 V, due to the oxidation of guanine residues, similarly to d(G)$_{10}$. The oxidation of the thymine residues occurs at much higher positive potential, near the potential of oxygen evolution and therefore it is more difficult to detect.[11]

Upon the formation of stable d(TG$_9$) and d(TG$_8$T) quadruplex configurations in the presence of K$^+$ ions, DP voltammograms showed, for both ODNs, a smaller oxidation peak G and a new peak G$_q$ at more positive potential, specific of the G-quartets oxidation.

The G$_q$ oxidation peak of both d(TG$_9$) and d(TG$_8$T) molecules increased with increasing the incubation time, while for the d(G)$_{10}$ molecules a decrease was observed. Indeed, in the AFM images the formation of molecular nanowires was never observed for any concentration of K$^+$ and/or incubation times. This is due to the presence of thymine residues at the 5' end, in the case of d(TG$_9$), and at both 5' and 3' ends, in the case of d(TG$_8$T), that inhibits the ability of the adjacent ODNs to self-assemble into long G-wires.

## 2.3 AFM and electrochemical characterisation of thrombin binding aptamers and their interaction with thrombin

AFM topographical images in air of TBA (5'-GGTTGGTGTGGTTGG-3') (Figures 4A and 4B) and eTBA (5'-GGGTTGGGTGTGGGTTGGG-3') (Figures 4D and 4E) modified HOPG showed that, in the presence of Na$^+$ ions (0.1 M phosphate buffer solutions pH =7.0), both TBA and eTBA adsorb spontaneously onto HOPG. Two different morphologies of the adsorbed structures were observed: a thin network film of 0.8 nm height, associated with the adsorption of the single-strands, and aggregates with spherical and rod-like shapes and 1.5 – 1.9 nm of height, due to the adsorption of G-quadruplexes. The adsorption onto HOPG of eTBA was greater when compared with TBA, due to an increased hydrophobic character given by the larger number of guanine bases at the molecule extremities.

In the presence of K$^+$, TBA (Figure 4C) and eTBA (Figure 4F) adsorb less onto HOPG, when compared to the adsorption in the presence of only Na$^+$ ions. The adsorption pattern corresponds only to the spontaneous adsorption of a small number of single-stranded molecules, while no quadruplex TBA or eTBA were observed. This is due to the fact that, in the presence of K$^+$ cations, TBA and eTBA form monomeric intra-molecular quadruplexes arranged in a chair-like structure and consisting of two G-quartets connected by two TT loops and a single TGT loop (Figure 5).[12] These unimolecular quadruplex structures are very stable, compact and rigid, preventing the interaction with the HOPG of the hydrophobic bases.[13]

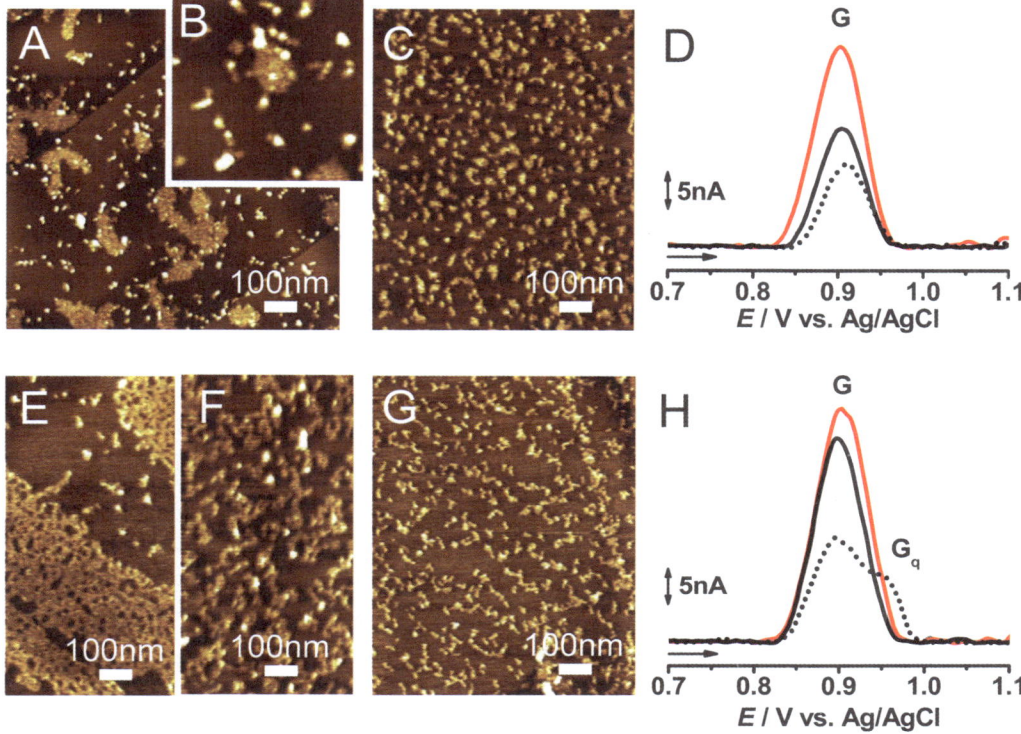

**Figure 4** *(A–C, E–G) AFM topographical images in air: (A–C) TBA and (E–G) eTBA modified HOPG, prepared by spontaneous adsorption during 3 min, from solutions of 1 µg mL$^{-1}$ ODNs in 0.1 M phosphate buffer pH = 7.0: (A, B, D, E) in the absence and (C, F) in the presence of 0.1 M K$^+$, 24 h of incubation. (D, H) DP voltammograms baseline corrected obtained with the GC electrode in a solution of 1 µg mL$^{-1}$ (D) TBA and (H) eTBA in 0.1 M phosphate buffer pH = 7.0, (—) in the absence and in the presence of 0.1 M K$^+$, (—) 0 h and (•••) 24 h of incubation.*

On the contrary, the presence of only Na$^+$ ions leads to the formation of less stable quadruplex TBA or eTBA morphologies that are locally destabilised by the HOPG hydrophobic surface, inducing their adsorption. In comparison, d(G)$_{10}$, d(TG$_9$) and d(TG$_8$T) establish manly inter-molecular G-quadruplexes in the presence of K$^+$ ions, that are less compact and stable, being easily destabilised and adsorbed onto HOPG (Figures 2B, 3B and 3A).

DP voltammograms obtained in the solutions of TBA (Figure 4D) and eTBA (Figure 4H) showed in both cases only the anodic peak G, at $E_{pa}$ = + 0.91 V, due to the oxidation of guanine. Similarly to d(TG$_9$) and d(TG$_8$T), the oxidation of thymine residues occurs at very high positive potential and therefore is not detected.

Upon addition of K$^+$ ions, both TBA and eTBA form compact and stable monomeric intra-molecular quadruplexes (Figure 5), observed by the decrease of the guanine oxidation peak G, and, in the case of eTBA (Figure 4H), the occurrence of the G$_q$ oxidation peak.

The $G_q$ oxidation peak did not appeared in the experiments with TBA, due to the very little concentration of G-quartets, below DP voltammetry detection limit. Whereas, a higher number of guanine residues was available with eTBA and thus a higher number of G-quartets occurred, which lead to a larger $G_q$ oxidation peak.

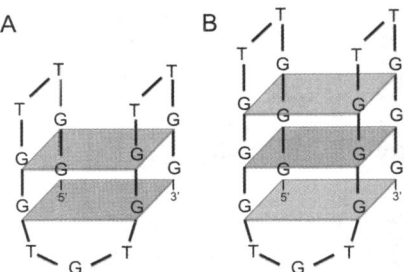

**Figure 5** *Schematic representation of unimolecular antiparallel G-quadruplex structures: (A) TBA and (B) eTBA.*

Thrombin is a serine protease and a coagulation protein in the blood stream that has effects on the coagulation mechanism.[14-16] The activation of thrombin is a crucial process in the physiological and pathological coagulation of various rare diseases. The thrombin binding aptamers TBA and eTBA specifically bind to thrombin, inhibiting its activity[17-19], therefore, the characterization of the thrombin-TBA complex is essential for the understanding of the factors involved in the thrombin-TBA complex binding process.[20] A study of the interaction between TBA and eTBA sequences with thrombin was carried out using AFM and voltammetry and the mechanism of interaction was established.

AFM images obtained after incubation of thrombin with single-stranded TBA (Figures 6A and 6B) or single-stranded eTBA (Figure 6E) showed the formation of a network film with embedded aggregates that cover the HOPG surface almost completely. The films were formed due to the co-adsorption of single-stranded TBA–thrombin or eTBA–thrombin complexes, free single-stranded TBA or TBA and free thrombin molecules.

The DP voltammograms recorded immediately after the addition of TBA or eTBA to the solution of thrombin showed the two peaks of thrombin, peak $T_1$, at $E_{pa}^1 = + 0.76$ V, and peak $T_2$, at $E_{pa}^2 = + 0.90$ V, and the oxidation peak G of single-stranded TBA or eTBA.

After single-stranded TBA–thrombin or eTBA–thrombin complex formation, obtained by increasing the incubation time, both thrombin oxidation peaks $T_1$ and $T_2$ occurred at more positive potentials. Moreover, peak G is not observed and, due to the overlapping of TBA or eTBA oxidation peak G with thrombin oxidation peak $T_2$, an increase of the peak at $E_p \sim + 0.90$ V was observed. This behaviour is explained considering coiling of the single stranded aptamer molecules around the thrombin structure, leading to the formation of a more robust complex that maintains the conformation of the thrombin molecules.

The influence of the $K^+$ cations on the TBA or eTBA interaction with thrombin was investigated in two different experiments. In one, the incubation of the TBA or eTBA with thrombin was performed in the presence of $K^+$, in such a way that the formation of the quadruplex TBA or eTBA occurred in the presence of thrombin. In the other, TBA or eTBA were incubated with $K^+$, in order to form the quadruplex TBA or eTBA, and only afterwards incubated with thrombin. Similar results were obtained in both experiments.

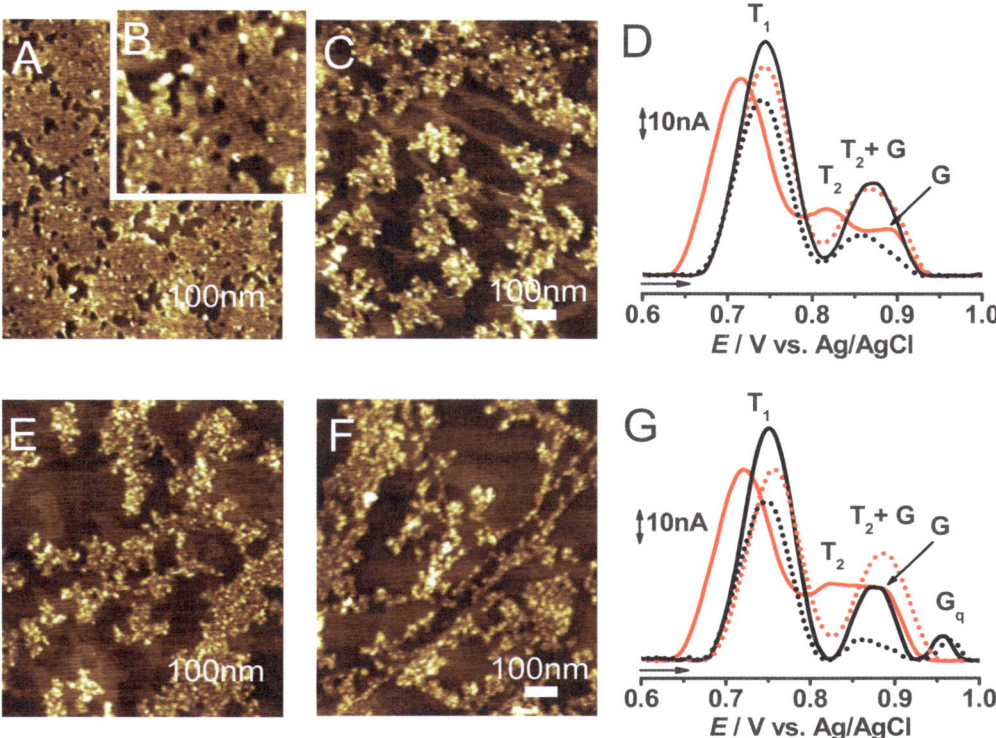

**Figure 6** *(A–C, E, F) AFM topographical images in air: (A–C) TBA–thrombin and (E, F) eTBA–thrombin modified HOPG, obtained by spontaneous adsorption during 5 min from incubated solutions of 1 µg mL$^{-1}$ TBA or eTBA with 10 µg mL$^{-1}$ thrombin in 0.1 M phosphate buffer pH = 7.0, (A, B, E) in the absence and (C, F) in the presence of 0.1 M K$^+$, 24 h of incubation. (D, G) DP voltammograms baseline corrected obtained with the GC electrode in a solution of 1 µg mL$^{-1}$ thrombin incubated with 1 µg mL$^{-1}$ (D) TBA and (G) eTBA in 0.1 M phosphate buffer pH = 7.0, in the absence (—) 0 h and (•••) 48 h of incubation, and in the presence of 0.1 M K$^+$, (—) 0 h and (•••) 48 h of incubation.*

After the incubation of thrombin with quadruplex TBA or eTBA, the AFM images showed patches of a thick layer with a bulky, knotty appearance due to the existence of a large number of embedded aggregates into its structure (Figures 6C and 6F). The adsorption decreased when compared with the adsorption obtained after the interaction between single-stranded TBA or eTBA with thrombin, since a reduced number of single-stranded aptamers adsorbed onto HOPG. Considering that quadruplex TBA or eTBA can interact with either both or any of the thrombin exocites, this layer is due to the adsorption of at least three types of aggregates corresponding to the interaction with either one or both thrombin exocites.

DP voltammograms of quadruplex TBA or eTBA incubated with thrombin showed the thrombin peak $T_1$, and the peak at $E_p \sim +0.90$ V corresponding to the overlapping peaks $T_2$ and G, and in the case of eTBA the occurrence of a new peak $G_q$ due to the oxidation of the G-quartets. With increasing the incubation time, a gradual decrease of the oxidation peaks $T_1$ and $T_2$ + G was observed, while peak $G_q$ remained constant. Therefore, upon the interaction with thrombin, the quadruplexes were always in contact with the electrode surface, whereas the thrombin molecules lied on top of the aptamer quadruplex molecules: the reduced contact between thrombin molecules and the GC electrode surface lead to the occurrence of the lower oxidation peak currents observed for both $T_1$ and $T_2$.

CONCLUSION

The redox behaviour and the adsorption morphology of five different G-rich ODN sequences, $d(G)_{10}$, $d(TG_9)$, $d(TG_8T)$, TBA and eTBA, was studied using DP voltammetry at GC electrode and AFM on HOPG, concerning their ability to form G-quadruplex configurations. The different adsorption patterns and degree of surface coverage were correlated with the ODN sequence base composition, presence or absence of $K^+$ and voltammetric behaviour.

Single-stranded ODNs, formed mainly in $Na^+$ containing solutions and short incubation times, were oxidized at the GC electrode and the only anodic peak observed corresponds to guanine oxidation. The AFM study showed the formation of thin network films onto HOPG, since the single-strands presented the bases more exposed and free to interact with hydrophobic HOPG.

The formation of stable G-quadruplex configurations stabilized by $K^+$ cations was observed in DP voltammetry by the decrease of the guanine oxidation peak and the occurrence of a new peak at higher potential due to the oxidation of G-quartets. The higher oxidation potential of G-quartets is due to the greater difficulty of electron transfer from the inside of the quadruplex to the electrode surface compared with the more flexible single-strands.

In the presence of $K^+$ ions, $d(G)_{10}$, $d(TG_9)$ and $d(TG_8T)$ formed tetrameric inter-molecular G-quadruplexes that adsorbed onto HOPG as spherical and rod-like shape aggregates, while TBA and eTBA formed monomeric intra-molecular G-quadruplexes that presented almost no adsorption onto HOPG. $d(G)_{10}$ was the only ODN sequence that showed the ability to form long G-wires, observed as small nanowires in the AFM images.

Acknowledgements

Financial support from Fundação para a Ciência e Tecnologia (FCT), projects PTDC/SAU-BEB/104643/2008, PTDC/QUI-QUI/098562/2008 and PTDC/SAU-BMA/118531/2010, POPH (co-financed by the European Community Funds FSE and FEDER/COMPETE), and CEMUC-R (Research Unit 285), is gratefully acknowledged.

References

1. *Quadruplex Nucleic Acids*, eds. S. Neidl, S. Balasubramanian, The Royal Society of Chemistry, Cambridge, UK, 2006.
2. C. Punchihewa, D. Yang, 'Therapeutic Targets and Drugs II: G-Quadruplex and G-Quadruplex Inhibitors' in *Telomeres and Telomerase in Cancer*, ed. K. Hiyama, Humana Press, Springer, 2009, Chapter 11, pp. 251–280.
3. J.L. Huppert, *Biochimie*, 2008, **90**, 1140.
4. J.T. Davis, G.P. Spada, *Chem. Soc. Rev.*, 2007, **36**, 296.

5   V. Dapic, V. Abdomerovic, R. Marrington, J. Peberdy, A. Rodger, J.O. Trent, P.J. Bates, *Nucl. Acids Res.*, 2003, **31**, 2097.
6   N. Borovok, N. Iram, D. Zikich, J. Ghabboun, G.I. Livshits, D. Porath, A.B. Kotlyar, *Nucl. Acids Res.*, 2008, **36**, 5050.
7   E. Shapir, L. Sagiv, N. Borovok, T. Molotski, A.B. Kotlyar, D. Porath, *J. Phys. Chem. B,* 2008, **112**, 9267.
8   T.C. Marsh, J. Vesenka, E. Henderson, *Nucl. Acids Res.*, 1995, **23**, 696.
9   L.T. Costa, M. Kerkmann, G. Hartmann, S. Endres, P.M. Bisch, W.M. Heckl, S. Thalhammer, *Biochem Biophys Res Commun.*, 2004, **313**, 1065.
10  A.-M. Chiorcea Paquim, T.S. Oretskaya, A.M. Oliveira Brett, *Biophys. Chem.*, 2006, **121**, 131.
11  A.M. Oliveira Brett, V.C. Diculescu, A.M. Chiorcea-Paquim, S.H.P. Serrano, 'DNA-electrochemical biosensors for investigating DNA damage' in *Comprehensive Analytical Chemistry*, 2007, Volume 49, Chapter 20, 413–437.
12  R.F. Macaya, P. Schultze, F.W. Smith, J.A. Roe, J. Feigon, *Proc. Natl. Acad. Sci. USA*, 1993, **90**, 3745.
13  V.C. Diculescu, A.-M. Chiorcea-Paquim, R. Eritja, A.M. Oliveira-Brett, *Journal of Nucleic Acids*, 2010, article ID 841932, 8 pages, doi:10.4061/2010/8419321-8.
14  M.T. Stubbs, W. Bode, *Trends Biochem. Sci.*, 1995, **20**, 23.
15  W. Bode, *Blood Cell. Mol. Dis.*, 2006, **36**, 122.
16  W. Bode, I. Mayr, U. Baumann, R. Huber, S.R.Stone1, J. Hofsteenge, *EMBO J.*, 1989, **8**, 3467.
17  S. Chandra, B. Gopinath, *Thromb. Res.*, 2008, **122**, 838.
18  J. A. Kelly, J. Feigon, T. O. Yeates, *J. Mol. Biol.*, 1996, **256**, 417.
19  B. Pagano, L. Martino, A. Randazzo, C. Giancola, *Biophys. J.*, 2008, **94**, 562.
20  V.C. Diculescu, A.-M. Chiorcea-Paquim, R. Eritja, A.M. Oliveira-Brett, *J. Electroanal. Chem.*, 2011, **656**, 159.

# AFM OF GUANINE RICH OLIGONUCLEOTIDE SURFACE STRUCTURES

James Vesenka
University of New England, 11 Hills Beach Road, Biddeford, ME 04005. Email: jvesenka@une.edu

## 1 INTRODUCTION

Atomic force microscopy (AFM) of self-assembled guanine-rich oligonucleotide (GRO) surface structures is described. The sequences used to build "G-wire" DNA in this work are from the $G_4T_2G_4$ polymer, or from pure guanine sequences ranging from 5 to 25 nucleotides in length. Samples are prepared by incubating GROs and appropriate cations in solution for times ranging from seconds to months. At an initial concentration of 1nM the average lengths of these molecules are around 100 nm long after 24 hours of growth. After many months longer G-wire DNA molecules (many micrometers) have been found to assemble in rafts, in crystalline forms, and always with some hint of orientating with the underlying substrate (freshly cleaved muscovite). The oriented G-wire structures are interesting candidates for molecular templates.

G-wires belong to a polymorphic family of quadruple helical nucleic acids collectively known as G-DNA. The common structural motif of G-DNA is the guanine tetrad (G-quartet)[1,2] in which four guanine bases are H-bonded in a cyclic planar array (Fig. 1a). The formation and stability of G-DNA is greatly enhanced by the coordination of monovalent and/or divalent metal cations[3,4,5,6,7]. G-DNA is known to occur in a variety of functional settings within the genomes of living cells[8,9,10,11,12,13,14]. The important biological role of guanine self-recognition may also be viewed as a useful materials property for the development of self-assembling supramolecular structures. Guanine-rich oligonucleotides (GRO's) spontaneously form G-DNA under appropriate conditions [12,15,16]. A number of GRO have lead to the development of self-assembling supramolecular G-DNA structures [17,18,19,20] from a single sequence component. One supramolecular G-DNA in particular, the G-wire, has been extensively characterized by atomic force microscopy (AFM) [21,22]. In this paper we examine the growth kinetics and surface structures of two types of four-stranded "G-wire" DNA characterized by AFM: $dG_4T_2G_4$ (GRO16) and $dG_n$ oligo where n = 5 to 25 guanine bases.

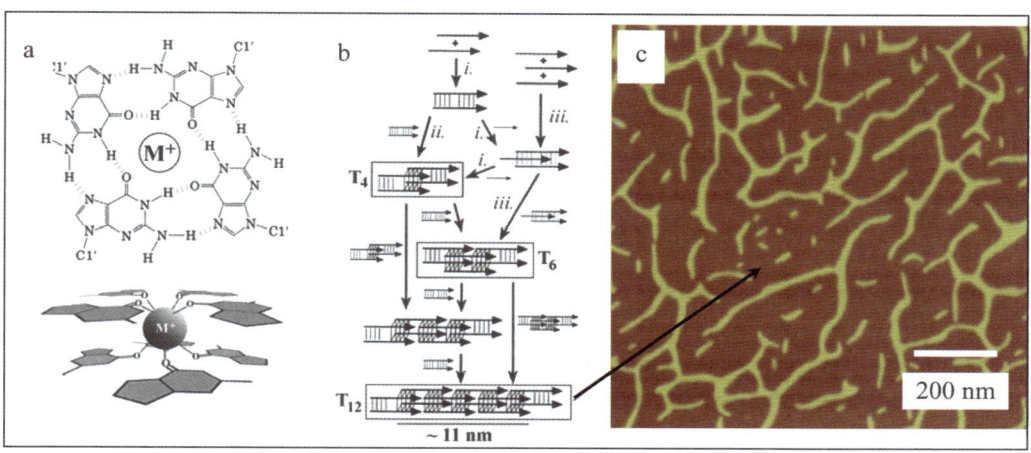

**Figure 1**  a) Structure of a G-quartet showing cation coordination. b) Hypothetical pathway of G-wire self-assembly. The thin arrows represent the sequence $d(G_4T_2G_4)$ and parallelograms represent G. This pathway is expected to be more staggered for pure guanine sequences. Initial formation of a G-DNA nucleating structure may occur through multiple paths: The italic i, ii, and iii represent sequential stepwise assembly, dimererization, and triplex disproportionation respectively as potential models for the initial G-DNA structure formation [23]. c) AFM image of typical G-wire sample in a $Na^+/Mg^{2+}$ buffer and deposited on a mica substrate. Cation species such as potassium, sodium or magnesium are thought to help stabilize the G-wires in the base-stacked core of the structure as seen in (a). Given enough time, 60 days in (c) superstructures emerge with greater flexibility than the short strands seen at the head of the arrow. From [24].

A schematic of the 10mer telomere-like sequence d(GGGGTTGGGG) designated "GRO16" is shown in Figure 1b. Self-assembly occurs in the presence of group I and group II metal cations over temperatures ranging from above freezing to below the DNA melting point (around 95 °C) to produce a supramolecular G-DNA that appears as rigid linear fibers when imaged by AFM (Fig. 1c). Whereas duplex DNA tends to collapse on the surface of mica, G-wire DNA appears to hold its cylindrical shape (Figs. 2) [25]. Intermittent Contact (a.k.a. "Tapping Mode") Atomic Force Microscopy [26] is a high resolution, near-field, three dimensional dynamic imaging technique that was used to characterize G-wires prepared in different growth media and imaged under controlled humidity. Humidity affects the measured height of the structures atop the substrate for these experiments, freshly cleaved and atomically smooth Muscovite (common mica). Mica is hydrophilic and will rapidly form a hydration layer [27] by adsorbing water from the surrounding air in order to hydrate residual ions left from the sample preparation process.

Monovalent cations ($Na^+$ and $K^+$) are assumed to stabilize a G-wire internally by residing in the coordination pocket provided by the four O6 groups of the G-quartet (Fig. 1a). However, self-assembly into relatively long supramolecular structures (Fig. 1c) requires divalent cations such as magnesium, zinc, or strontium [21]. In particular $Mg^{2+}$ is known to stabilize a nucleic acid structure to a 100 fold or more effectively primarily functioning as a counter ion for the negatively charged phosphates in the backbone. The manner in which divalent cations interact with G-DNA to stabilize a quadruplex is more complex. Studies of divalent cation stabilization of various G-DNA's have shown a general trend for divalent metal cations to stabilize G-DNA structures at lower concentrations ($M^{2+}$ below 10 mM) and destabilize G-DNA at higher $M^{2+}$ concentrations [28]. Oligoucleotides such as d(GGGGTTGGGG) and d(GGGGTTTTGGGG) are an apparent exception to this general trend. In the case of the hairpin dimer forming $dG_4T_4G_4$, increasing $Ca^{2+}$ concentration induces a switch in the conformation of a $d(G_4T_4G_4)$ G-DNA from an anti-parallel hairpin dimer to a parallel tetramer G-DNA leading to a supramolecular assembly similar to the GRO16 oligonucleotide[29]. The model presented in Fig. 1b suggests a slipped or out-of register strand(s) promote the further growth of G-wires once a stable G-DNA is formed. In this scheme a combination monovalent and divalent cation (G-DNA stabilizing and destabilizing) effectors promote longer G-wires by reducing the number of G-DNA intermediate species that are exposed to solvent. Initial assembly of a slipped strand G-DNA may occur through several paths as indicated in Fig. 1b. Molecular dynamic simulations of precursor G-DNA structures predict stable slipped strand structures stabilized by $Na^+$ or $K^+$ occur in possible assembly pathways [30,31] and lend support to this model. The precise role divalent cations play in G-DNA assembly and by extension in G-wire self-assembly is under intense investigation [32].

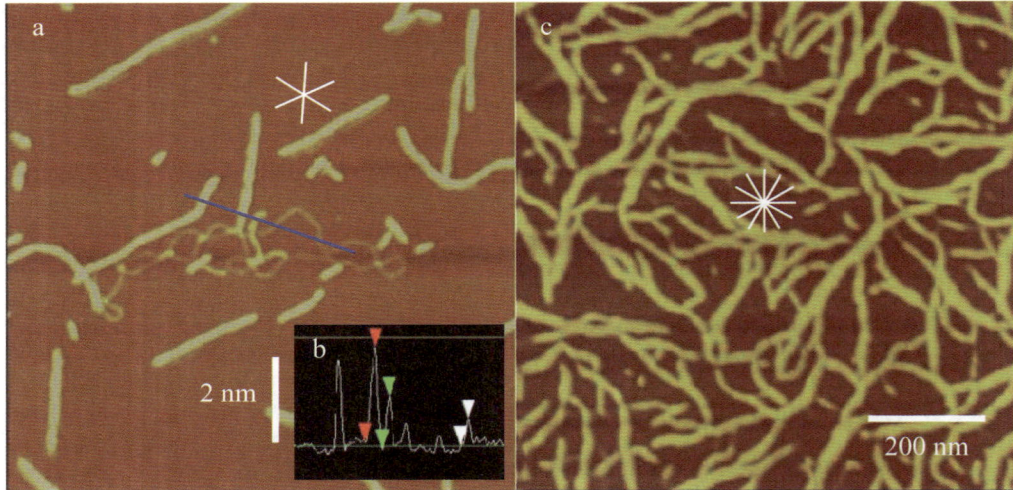

**Figure 2** *Topographic comparison between duplex and quadruplex "G-wire" DNA co-adsorbed on the same substrate (a). The duplex DNA collapses on the surface of mica to a height of 0.5 nm above the surface as seen in the cross section of (b). Even the supercoiled segments of the double stranded DNA measure only about 1.0 nm above the mica substrate. This is about half the diameter expected from Watson-Crick duplex DNA in solution. However, the quadruplex DNA is uniformly about 2.2 nm in diameter, very close to the NMR and x-ray spectra of G-quartet DNA (2.4 nm in diameter). G-wire DNA tends to "auto-orient" along specific directions (arms of the paddle wheel). c) Other forms of mica (biotite) also yield auto-orientation, though longer strands or G-wire superstructures yield greater deviation in the hexagonal ordering, possibly due to competition with next-next nearest neighbour potassium vacancies. Details below.*

## 2 METHODS

### 2.1 Sample Preparation

*2.1.1 Materials.*

Quadruplex G-wire DNA was prepared according to the procedure of Marsh et al. [21]. Briefly, lyophilized GRO16 d(GGGGTTGGGG) was purchased from Biosynthesis (Lewisville TX) and dissolved in a growth cocktail consisting of 50 mM NaCl, 10 mM $MgCl_2$, 10 mM Tris-HCl pH 7.5 at a concentration of about 1mM. The samples were heated to 95°C for ten minutes to promote the melting of fortuitous G-4 structures, to ensure a monomeric concentration of G-wires. The dissolved GRO16 was then incubated at 37°C in the growth cocktail for different amounts of time ranging from minutes to months. Samples of concentrated G-wires were diluted by a factor of 10 to 100 in either in the growth cocktail or an imaging buffer consisting of 10 mM Tris (pH 7.6), and 1 mM $MCl_x$, where M = copper, magnesium, potassium, silver, sodium, strontium or zinc and "x" is the appropriate number of chlorine counterions. Pure guanine oligonucleotide sequences of $dG_5$, $dG_{10}$, $dG_{15}$, and $dG_{20}$ were directly diluted either into the growth cocktail or an imaging buffer.

*2.1.2 Imaging*

10µL samples were then adsorbed onto freshly cleaved muscovite mica and accepted only if the sample spread uniformly over the entire surface. The samples were then incubated on the mica between minutes and months by keeping control of the humidity to inhibit drying. Samples selected for AFM imaging were first rinsed with 1 mL deionized water to remove buffer salts, dried in a stream of dry nitrogen and allowed to stabilize in a 37°C oven before imaging. Contact imaging employed 125 µm x 20 µm etched silicon nitride probes (k≈1nN/nm) using a Nanoscope E controller (Digital Instruments, Santa Barbara CA) in nitrogen-purged dry air. A Nanoscope IIIa controller (Digital Instruments, Santa Barbara CA) and Multimode AFM employing intermittent contact was used to image freshly prepared and desiccated samples in nitrogen-purged dry air (silicon probe resonance frequencies between 75 and 330Khz).

## 3 RESULTS

### 3.1 Auto-orientation

Auto-orientation of G-wire DNA (Fig. 3) has been identified as being a result of a match between the mica lattice and the quadruplex DNA (Fig. 4). This auto-orientation is always present to some degree in AFM images of both dried and aqueous G-wire DNA constructs atop freshly cleaved mica [33]. Auto-orientation appears to be due to a lattice match between the G-wire phosphate backbone and nearest neighbour potassium vacancies of mica [34]. These vacancies are important for enabling the presence of "tethering" cations (principally magnesium) that allow DNA of all types to be bound to mica for the purposes of AFM imaging [25].

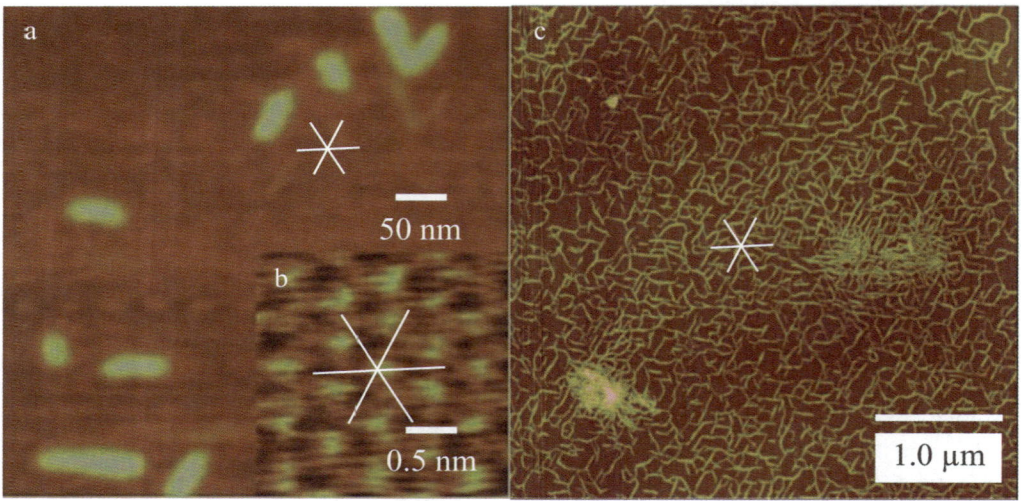

**Figure 3**  a) Oriented 1μM $G_4T_2G_4$ G-wire DNA deposited 24 hours after incubation at 37°C in growth cocktail onto freshly cleaved mica. Note the G-wires tend to automatically orient in one of three directions indicated by the embedded paddle wheel. b) Insert: A high resolution image of the same mica substrate as (a) of G-wire orientation a indicates alignment preferences at three angles oriented 60° from each, suggesting an auto-orientation the next nearest neighbour potassium vacancies (bright atoms). c) 10μM $G_4T_2G_4$ G-wire DNA after 24 hour incubation on mica. These orientation effects are more easily observed in shorter segments of G-wires and can take place over vast regions of the substrate. Vertical scale 10nm dark to light.

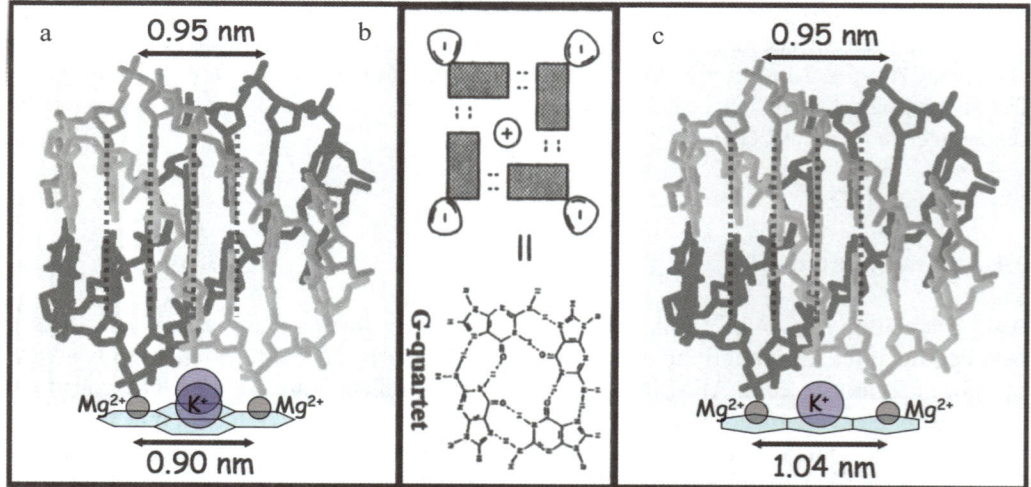

**Figure 4**  Mica has a hexagonal surface structure parallel to the electronegative aluminium-silcate cleavage planes. The potassium vacancies (blue spheres in a and c) are thought to be responsible for tethering cation species such as magnesium between the DNA and mica [25]. There are two arrangements that might lead to orientation of the packed G-quartets (b), either due to a

*phosphate backbone lattice match with the next nearest neighbour (a) or the next-next nearest neighbour (c). High resolution measurements of underlying mica substrate indicate that (a) is more common for short segments of G-wires, there being only a 0.05nm mismatch between the guanine backbone and the next nearest neighbour potassium vacancies. Longer segments tend to favour combinations of (a) and (c) (e.g. Figure 2c).*

## 3.2 G-wire Rafts

The density of the G-wire "rafts" resulting from our sample preparation procedure appears to depend primarily on initial GRO concentration. The G-wires in these rafts appear flexible and are associated with $K^+$ and $Mg^{2+}$ found in the growth cocktail. Figure 3a corresponds to concentration of 100μM GRO16 monomer, which appears to be include super-bundles of G-wires. At longer incubation times (Figure 5b; 10μM, 48hr deposition) rafting was observed in GRO16. Organization of a different form takes place quickly with GRO20 monomers (Figure 5c; 100μM, 1 minute deposition)

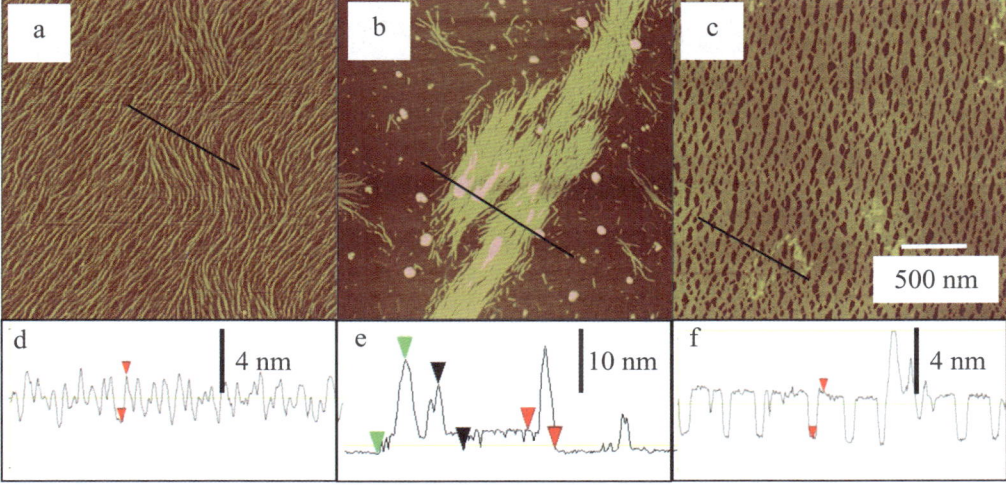

**Figure 5** *a) 100μM GRO16, 1 minute deposition at 37°C in growth cocktail. b) 10μM GRO16, 48 hour deposition at 37°C in growth cocktail. c) 100μM GRO20, 1 minute deposition in growth cocktail. At high concentrations GRO20 forms interconnected rafts, consistent with its ability to create greater numbers of random G-quartets. Again, notice the obvious tendencies to orient along axis at 60° to each other. Individual G-wires strands average 2.2nm in height above the mica substrate, but not for concentrated samples. d) Cross section of (a) with G-wire superstructures of about 3.0 nm in diameter. e) Cross section of (b) with average GRO16 raft thickness ranging from 3.2 nm (red arrows) to 15nm (green arrows) from 5 stacks of wires. f) Cross section of (c) with average GRO20 thickness of about 3.2nm. The additional thickness of the wires may be due to a significant contribution of dried buffer cations on the densely packed samples. Indeed, substitution of multivalent cations for magnesium results in a variety of interesting G-wire morphologies. All images have same horizontal scale and vertical scale is 10 nm from dark to light.*

## 3.2 G-wire Loops

Occasionally, as seen in the remaining figures, G-wire superstructures are observed. Figure 6 includes three examples of flexible G-wire loops, similar in shape and size to supercoiled DNA or double stranded plasmid DNA. These structures retain a minimum diameter of 2.2 nm when imaged in dry air in the brightest parts of Figures 6a-c. Double stranded DNA collapses on the surface of mica due to interactions between the substrate and the DNA. Four-stranded G-wire DNA always maintains structural integrity on the surface of mica in dry air. The G-wire loops might be kinetically stable because, by self-assembling into a closed structure, disassembly is discouraged. There is no "end" on a closed loop for the ladder building blocks to vacate (in the absence of cross-linkers). The conditions under which such long structures develop, when the surrounding solution contains only smaller 100 nm length G-wire DNA, appears to depend on high concentrations and long incubation times.

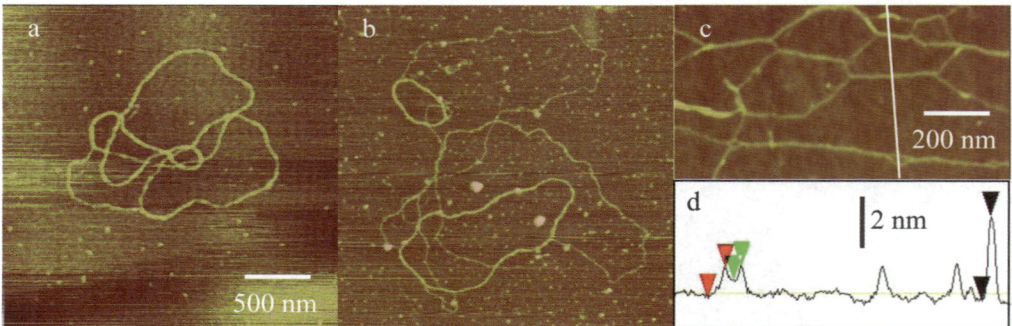

**Figure 6** *a) Interesting GRO16 structures sometimes emerge after long incubation times including these examples of loops of single and multiple stranded G-wires. b) These results are found with numerous smaller G-wires surrounding the immediate vicinity. The looped G-wire structures are stable against disassembly after dilution and rinsing, perhaps because of adsorption on mica prior to rinsing. Scale marker same as in (a). The contour lengths shown here are much larger than the sub-micrometer average length of individual G-wires. Multiple-stranding of G-wire DNA is common as seen in (b) and (c), a result that is surprising in view of the strong electrostatic repulsion that is expected from the four strands of phosphate backbones on each G-wire. d) Is a cross section of (c) with details that suggests G-wires may not consists of G-quartets and consist of only double or even single strands, with vertical heights of about 1.1 nm (red arrows) to 0.5nm (green arrows). These may be artifacts of sample preparation and excessive rinsing. The black arrows record the height of a supercoiled strand of G-wire (4 nm). Vertical height range is 10 nm from dark to light.*

## 3.2 G-wire Crystals

G-wire crystals present themselves in two forms. Figure 7a is an image of "angel hair" like strands of G-wires of various thicknesses. These structures were very long, relatively inflexible, as seen by kinks, and tended not to auto-orient with the mica substrate. The image taken in Figure 7 was incubated for several months at 37°C in a growth cocktail in which zinc chloride was substituted for magnesium chloride. The dimensions of these G-wire fibers (b and c) were typically much larger than 2.2nm in diameter, suggesting either G-wire superstructures or metallization of the G-wire scaffold by buffer ions.

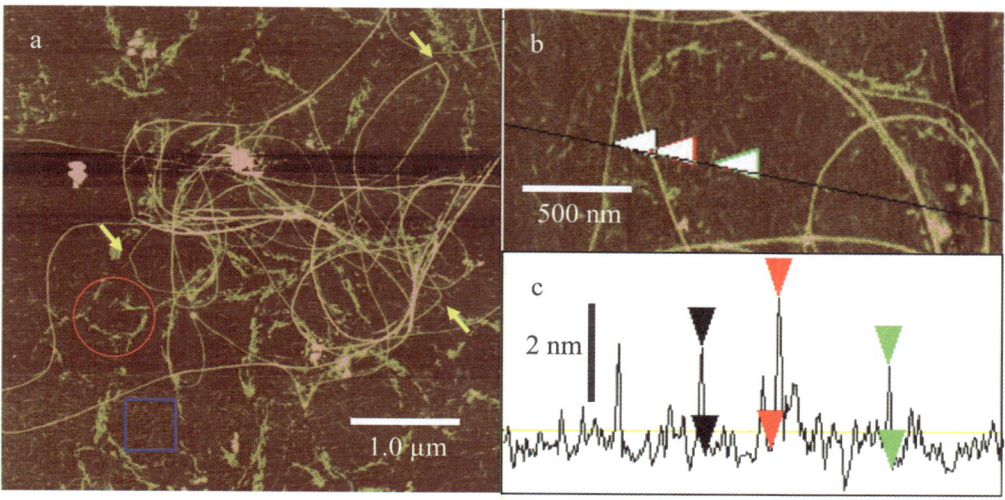

**Figure 7** *a) Two successive images from the same sample can be seen with three types of GRO16 G-wires. Typical flexible 2 nm diameter G-wires (within blue square), rafts (inside the red circle) and thick wires consisting of extremely long (hundreds of micrometers), spaghetti-like structures. These unusual presentations are the result of G-wire self assembly in the presence of zinc substituted for magnesium in the growth cocktail. Note the yellow arrows in the image identify kinked regions of the spaghetti-like structures suggesting these superstructures are not as flexible. b) The diameter of these long structures (as measured by the height above the surface) are 3nm or greater, indicating more complex structure than the simple G-wires found in the blue box of (a). c) is a cross section of (b) which presents a range of G-wire diameters from about 2nm to 3nm in diameter.*

Zinc substitution for magnesium generated a range of interesting structures as summarized in Figure 7. The question we asked was if G-wires might integrate other types of multivalent ion species. Figure 8 includes two examples of GRO-20 G-wires assembled in the presence of silver (in the form of $AgNO_3$) and copper ($CuCl_2$) without buffering or any other form of counterions. The structures that formed in the presence of the cations tended to clump together in larger aggregates, presumably due to the strong electrostatic interactions that $Ag^{3+}$ and $Cu^{2+}$ with other G-wire scaffolds. These clumps could be broken apart by sonication (Figure 7a). In the clumped form the G-wires were sufficiently large to be easily detected by the optical navigator.

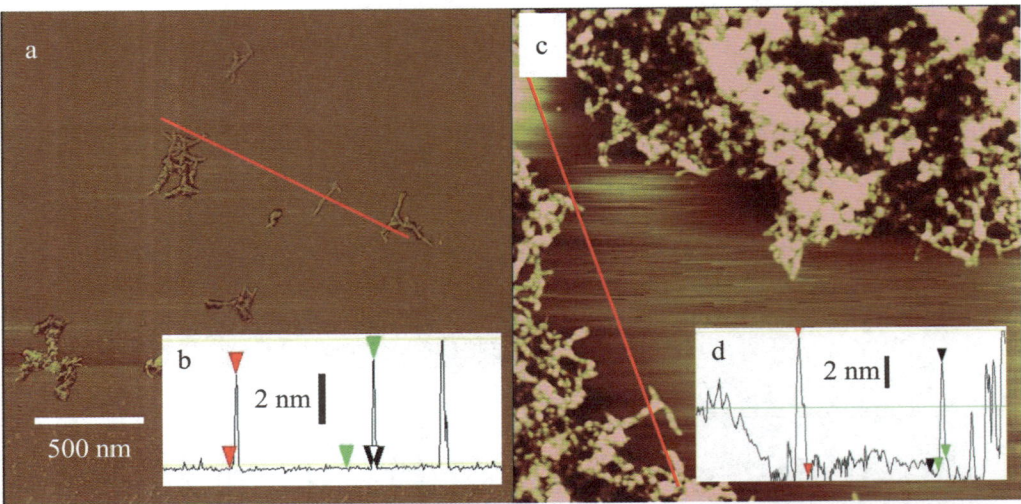

**Figure 8** *a) 1μM GRO20 oligonucleotide in a solution of 1mM $AgNO_3$. In the absence of the GRO not even crystal of silver nitrate were found on the surface. The GRO20 act as a scaffold for the formation of complex metalized G-wires. Phase image with a range of ±30°. b) Inset is taken from a height image, indicating these metalized GRO20 DNA were 4nm in diameter or larger. c) 1μM GRO20 oligonucleotide in a solution of 1mM copper chloride. Large clumps were common, but could be broken apart by sonication. The rest of the surface was devoid of G-wires. d) Inset of (c) indicating G-wire diameters of between 6 and 8 nm in the presence of copper ions.*

Evidence for more rigid structures is seen in Fig. 9a and 9c. Alignment with the potassium vacancies in the mica substrate is strongly evident in these pictures. The hexagonal arrangement of the potassium vacancies demands orientation at 60° intervals, as seen in these images. Unlike Fig. 8, the structures in Fig. 9 are rigid. Furthermore they appear to be multi-layered (Fig. 9b), as seen in the cross section image of Fig. 9a. The layers are integrals of half diameter of G-wires (about 1.1 nm). These structures might be examples of ladder G-DNA ($L_1$) crystals as seen in Fig. 5. Since these structures are much wider than the artificial broadening generated by finite tip geometry, the greater width could be explained by parallel packing of the ladder structures. A lattice match with the ladders and the potassium vacancy sites of the mica substrate (1.04 nm) would create a surface that the ladder DNA could also adsorb to, i.e. create multiple layers of ladder G-DNA. Note in Fig. 9c that flexible strands of G-wire DNA are also seen aligning in the same direction as the crystalline structures.

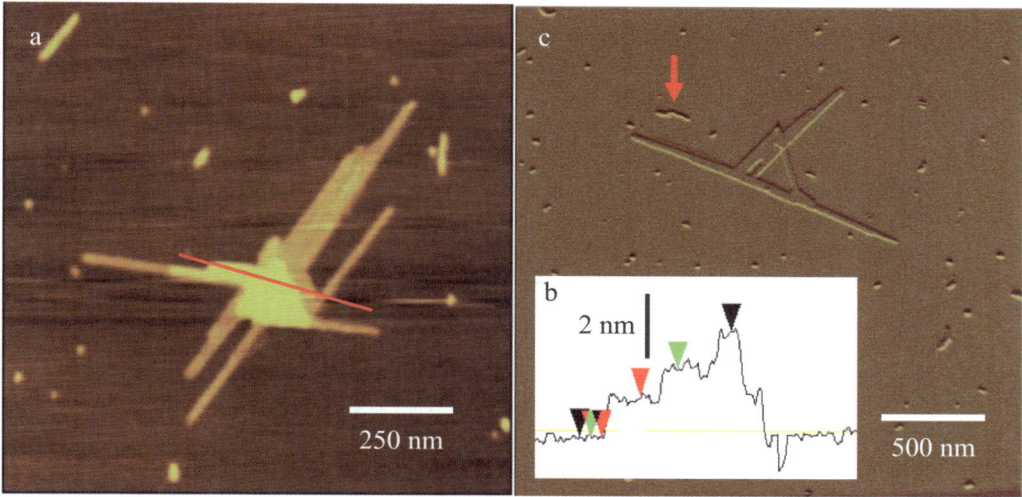

**Figure 9** *(a) One-dimensional GRO16 crystals (initial concentration of 1µM) observed at 60° intervals, orienting themselves with the underlying potassium vacancies in the surface of mica. Contact AFM image taken in dry air with a vertical height scale of 10 nm from dark to light. The red line through the crystal is a cross section described in (b). The cross section indicates the different heights of the crystals, which form three distinct layers of heights 1.4 nm, 2.2 nm and 3.6 nm tall (each ±0.2nm), exactly incremental steps of one half the diameter of G-wire DNA. (c) Deflection image example of the purported crystalline of G-wires, indicating the micrometer lengths that these structures can attain. Note the flexible G-wire DNA segments in (c – red arrow) and their alignment with the crystals, evidence of the lattice match with the underlying potassium vacancies of mica, and indirect evidence of the similar nature of the crystalline and flexible structures on the same surface.*

## Acknowledgment

The author gratefully acknowledges the undergraduate research assistance of Tamieka Armstrong, David Bagg, Geoffrey Champagne, Nicholas Demers, Kristin Eccleston, Matthew Fletcher, Brandon Goblirsch, Mellissa Holden, Peter Hulsey, Marci Luhrs, Bethany Rioux, Joe Skaja and graduate students Robert Baron, Robert Kretschmer, Patrick Spinney, and Matthias Urban. We acknowledge the financial support from the University of New England, Research Corporation Cottrel College Science Award. This work was supported in part by NSF Major Research Instrumentation award DMR-0116398.

## References

1. J.R. Williamson, M.K. Raghuraman, and T.R. Cech, *Cell.* **59**:871-880 (1989).
2. J.R. Williamson, *Proc. Natl. Acad. Sci. USA.* **90**, 3124-3124 (1993).
3. D. Sen and W. Gilbert, *Nature* **344**, 410–414 (1990).
4. C. Hardin, E. Henderson, T. Watson, and J.K. Prosser, *Biochemistry*, **30**, 4460–4472(1991).
5. P. Balagurumoorthy and S.K. Brahmachari, *J. Biol. Chem.* **269**, 21858–21869 (1994).
6. M.A. Keniry, *Biopolymers,* **56**, 123–146 (2001).

7   G.N. Parkinson, M.P. Lee, and S. Neidle, *Nature* **417**, 876–880 (2002).
8   E. Henderson, C.C. Hardin, S.K. Walk, I. Tinoco Jr. and E.H. Blackburn, *Cell* **51**, 899–908 (1987).
9   D. Sen and W. Gilbert, *Nature* **334**, 364–366 (1988).
10  J.R. Williamson, *Annu. Rev. Biophys. Biomol. Struct.*, **23**, 703–730 (1994).
11  A. Siddiqui-Jain, C.L. Grand, D.J. Bearss, L.H. Hurley, *Proc. Natl. Acad. Sci.* **99**, 11593–11598 (2002).
12  J.L Huppert and S. Balasubramanian, *Nucleic Acids Res.* **33**, 2908–2916 (2005).
13  A.K. Todd, M. Johnston, and S. Neidle, *Nucleic Acids Res.,* **33**, 2901-2907 (2005).
14  P. Rawal, V. Bhadra R. Kummarasetti, J. Ravindran, N. Kumar, K. Halder, R. Sharma, M. Mukerji, S. Kumar Das and S. Chowdhury, *Genome Research,* **16**, 644-655 (2006).
15  D. Sen and W. Gilbert, *Nature* **344**, 410–414 (1990).
16  F.W. Smith, P. Schultze, and J. Feigon, *Structure*, **3**, 997–1008 (1995).
17  T.C. Marsh, and E. Henderson. *Biochem.* **33**,10718-1072 (1994).
18  D. Sen and W. Gilbert, *Biochem.* **31**, 65-70 (1992).
19  T.Y. Dai, S.P. Marotta, and R.D. Sheardy, *Biochem.* **34**, 3655-3662 (1995).
20  E. Protozanova and R.B. Macgregor Jr., *Biochem.* **35**,16638-16645 (1996).
21  T.C. Marsh, J. Vesenka, and E. Henderson, *Nucleic Acids Res.* **23**, 696-700 (1995).
22  J. Vesenka, E. Henderson, & T. Marsh, Ed. W. Fritzsche, *AIP Conference Proceedings* **640**, 109-122 (2002).
23  C.C. Hardin, M. Corregan, B.A. Brown, L.N. Frederick, *Biochemistry* **32**, 5870-5880 (1993).
24  T. Marsh and J. Vesenka, in *Nano and Molecular Electronics Handbook*, Sergy Lyshevski ed., CRC Press, New York, 13, pp. 1-15 (2007).
25  T. Muir, E. Morales, J. Root, I. Kumar, B. Garcia, C. Vellandi, T. Marsh, E. Henderson, and J. Vesenka, *J. Vac. Sci. Technol. A.* **16**, 1172-1177 (1998).
26  Zhong, Q; Inniss, D; Kjoller, K; Elings, V. *Surface Science Letters* **290**: L688 (1993).
27  C. Park, P.A. Fenter, K.L. Nagy and N.C. Sturchio, Physical Review Letters, **97**, 016101(1-4) (2006).
28  C.C. Hardin, A.G. Perry, and K. White, *Biopolymers* **56**, 147-194 (2001).
29  D. Miyoshi, A. Nakao, N. Sugimoto, *Nucleic Acids Res.* **31**, 1156-63 (2003).
30  N. Spackova, I. Berger, and J. Sponer, *J. Am. Chem. Soc.* **121**:5519-553 (1999).
31  R. Stefl, T.E. Cheatham 3rd, N. Spackova, E. Fadrna, I. Berger, J. Koca, J. Sponer, *Biophys J.* **85**, 1787-1804 (2003).
32  Thomas Marsh, this book.
33  K Kunstelj, F Federiconi, L Spindler, & I Drevensek-Olenik . *Colloids and Surfaces B: Biointerfaces.* **59**, pp.120-127 (2007).
34  J. Vesenka, D. Bagg, A. Wolff, A. Reichert, & W. Fritzsche, *Colloids and Surfaces B: Biointerfaces*, **58**, pp. 256-263 (2007).

# SOLUTION DYNAMICS AND STRUCTURE OF G-QUADRUPLEXES STUDIED BY DYNAMIC LIGHT SCATTERING

Lea Spindler

Faculty of Mechanical Engineering, University of Maribor, Smetanova 17, SI-2000 Maribor *also with* J. Stefan Institute, Jamova 39, SI-1000 Ljubljana, Slovenia. E-mail: lea.spindler@uni-mb.si

## 1 INTRODUCTION

Different experimental techniques can be used to elucidate biomolecular structures in solution and among them dynamic light scattering (DLS) is a valuable non-invasive tool which gives information on the dynamics of macromolecular systems. DLS monitors the translational and rotational motion of particles in solution and is sensitive to the size and shape of the molecules as well as to their intermolecular, especially electrostatic, interactions. With DLS the conformational changes of molecules can be investigated and the response of the system to environmental parameters such as temperature, pH or added salt. The interpretation of experimental DLS results can be additionally refined in combination with hydrodynamic model calculations, which provide prediction for the translational and rotational diffusion coefficients based on detailed model structures.

In this work we explore the use of DLS as a successful tool for investigating the formation and conformations of G-quadruplex structures formed either from guanine-rich DNA sequences or from self-assembled stacks of guanosine mononucleotides. We first give the theoretical basis for solution dynamics described by the diffusion equation in dilute solutions and for solutions dominated by electrostatic forces. In a light scattering experiment the correlation functions of the intensity of scattered light are calculated and their relation to the diffusion coefficients is given. Finally, the use of DLS on self-assembled guanosine molecules and G-rich oligonucleotides is demonstrated.

## 2 SOLUTION DYNAMICS

Molecules in solution randomly change their position due to thermal agitation and solution interactions. This process is characterised by the diffusion coefficient which, in turn, is related to the structure and dimensions of the diffusing objects.

### 2.1 Diffusion of noninteracting particles in dilute solutions

The most simple solution dynamics is Brownian motion: small particles concentrated at a certain point will spread out in time because of thermal fluctuations. The process is phenomenologically described by the diffusion equation[1]

$$\frac{\partial c(\vec{r},t)}{\partial t} = D\nabla^2 c(\vec{r},t) \tag{1}$$

where $c$ is the molecular concentration at position $\vec{r}$ and time $t$, $D$ is the diffusion coefficient, and $\nabla^2$ the Laplacian operator. Molecular motion induces concentration fluctuations, which can be described as

$$c(\vec{r},t) = \langle c \rangle + \delta c(\vec{r},t) \tag{2}$$

where $\langle c \rangle$ is the average solution concentration and $\delta c(\vec{r},t)$ the fluctuation term. It is often convenient to express the fluctuations in the wavevector $\vec{q}$ space. Introducing

$$\delta c(\vec{r},t) = \left(\frac{1}{2\pi}\right)^{3/2} \int_{-\infty}^{\infty} \delta c(\vec{q},t) e^{-i\vec{q}\cdot\vec{r}} d^3\vec{q} \tag{3}$$

into Equation (1) and solving it gives

$$\delta c(\vec{q},t) = \delta c(\vec{q},0) e^{-Dq^2 t}. \tag{4}$$

The translational diffusion coefficient $D_T$ of noninteracting spherical particles is given by the Einstein relation[2]

$$D_T = \frac{k_B T}{6\pi \eta_0 R} \tag{5}$$

where $k_B$ is the Boltzmann constant, $T$ is the temperature, $R$ the particle radius and $\eta_0$ the viscosity of the solvent. If the exact particle dimensions are not known then from the diffusion coefficient its effective dimensions are obtained assuming spherical shape with an effective hydrodynamic radius $r_H$.

The shape of many biomolecules, including DNA, can be approximated by a cylindrical rod. The corresponding hydrodynamic theory was developed by Tirado and Garcia de la Torre[3,4] who modelled circular cylinders as stacks of rings composed of spherical subunits, extrapolated to zero bead size. They made polynomial approximations of their numerical calculations, which provide an analytical form for the transport coefficients of cylindrically shaped molecules with a relatively small length to diameter ratio, $2 \leq p = L/d \leq 30$. The translational diffusion coefficient for a rod with length $L$ and diameter $d$ is given by

$$D_T = \frac{k_B T}{3\pi \eta_0 L}(\ln p + \nu) \tag{6}$$

where $\nu = 0.312 + 0.565/p - 0.100/p^2$ is the end-effect correction term. Rodlike molecules in solution experience not only random translational motion but also frequent reorientations. The corresponding rotational diffusion coefficient for rotation of the long axis is given by

$$\theta = \frac{3k_B T}{\pi \eta_0 L^3}(\ln p + \delta_\perp) \tag{7}$$

with $\delta_\perp = -0.662 + 0.917/p - 0.050/p^2$.

Tirado and Garcia de la Torre proposed a simple method to obtain the length and the diameter of the molecule from the experimental data for $D_T$ and $\theta$[5]. Combining Equations (6) and (7) gives a function of the axial ratio, $h(p)$, expressed as

$$h(p) = \frac{(9\pi\eta_0/k_B T)^{2/3} D}{\sqrt[3]{\theta}} = \frac{\ln p + \nu}{\sqrt[3]{\ln p + \delta_\perp}}. \tag{8}$$

The value of the axial ratio thus depends only on the experimental data and no previous knowledge on the $L$ and $D$ values is needed to calculate it. From $p$ the length and the diameter are obtained with less error than as if they were directly calculated from Equations (6) or (7).

For molecules with small length to diameter ratios the Tirado and Garcia de la Torre theory cannot be applied. The theoretical values of the translational diffusion coefficients are in this case much better calculated by the bead model where the molecular structure is modelled by an assembly of spherical subunits[6,7]. For DNA nucleotides the molecule is divided into two groups of atoms: the first containing the nitrogen base and the second one the sugar and phosphate residue. Each of these groups is then replaced by a bead of 0.5 nm radius and for the resulting system of beads the diffusion coefficients are numerically calculated[8].

## 2.2 Effect of electrostatic forces on particle diffusion

Polyelectrolytes like DNA have many ionisable groups that in solution dissociate into polyvalent macroions (polyions) and a large number of small ions of opposite charge (counterions). Both polyions and counterions are charged species creating a common electrostatic field, which fluctuates due to their motion and reversibly influences their dynamics. The motion of the polyions thus becomes coupled to the dynamics of the small and much faster counterions.

The polyion diffusion can be qualitatively described by the coupled mode theory[2,9] based on the linearized Poisson-Boltzmann equation

$$\nabla^2 \phi(r,t) = -\frac{4\pi e_0}{\varepsilon \varepsilon_0} \left[ Z_p \delta n_p(r,t) + Z_a \delta n_a(r,t) + Z_c \delta n_c(r,t) \right] \tag{9}$$

where $\phi(r,t)$ is the electrostatic potential at position $r$ and time $t$, $e_0$ is the absolute value of the electron charge, $\varepsilon\varepsilon_0$ is the dielectric constant of the solution, $Z_\alpha$ is the charge and $\delta n_\alpha(r,t)$ the number concentration fluctuation of species $\alpha$ ($\alpha$ = "p": polyion, "a": anion, "c": cation). Equation (9) is valid only for low potentials, $Z_\alpha e_0 \phi(r) \ll k_B T$, when the local number concentration $n_\alpha(r,t)$ of ion species $\alpha$ may be approximated as

$$n_\alpha(r,t) = <n_\alpha> + \delta n_\alpha(r,t) \tag{10}$$

The value of $n_\alpha(r,t)$ fluctuates around the average value $<n_\alpha>$ by concentration fluctuations $\delta n_\alpha(r,t)$, and according to the diffusion Equation (1) it fluctuates with an additional term taking into account the electrostatic interactions

$$\frac{\partial \delta n_\alpha(r,t)}{\partial t} = D_\alpha \nabla^2 \delta n_\alpha(r,t) + D_\alpha \frac{Z_\alpha e_0 <n_\alpha>}{k_B T} \nabla^2 \phi(r) \tag{11}$$

where $D_\alpha$ is the diffusion coefficient of species $\alpha$. Using the spatial Fourier transformation $\nabla^2 \delta n_\alpha(r,t) \to -q^2 \delta n_\alpha(q,t)$ of combined Equations (9) and (11), where $q$ is the wavevector of the fluctuation, gives three coupled equations describing the ion dynamics in the solution[9]. The three solutions of this system represent the decay rates of the normal modes. The first solution $\lambda_1$ represents the decay rate of the charge fluctuations, which is $q$-independent. The other two solutions are for $q \to 0$ proportional to $q^2$: $\lambda_2$ is associated with the apparent polyion diffusion coefficient defined as

$$D_{app}(q) = \frac{\lambda_2}{q^2}, \tag{12}$$

while $\lambda_3$ is related to small ion diffusion.

The simplest analytical expression of the apparent polyion diffusion coefficient is given upon several assumptions: the added salt is a symmetric 1-1 electrolyte, it has common counterions with the polyion, and there is the same charge and the same diffusion coefficient of the counterion and the co-ion, respectively

$$D_{app}(q \to 0) = \frac{1}{2}\left[D_p(1-\Omega) + D_s(1+\Omega)\right] \quad (13)$$

with

$$\Omega = \frac{D_p Z_p - D_s[1+(2c_s/Z_p c_p)]}{D_p Z_p + D_s[1+(2c_s/Z_p c_p)]} \quad (14)$$

where $c_p$ and $c_s$ are the molar concentrations of the polyion and the small ions, respectively, and $D_s$ is the small ion diffusion coefficient that includes the contributions of added ions as well as the counterions (Figure 1). Interesting are the asymptotic limits of Equation (13):

$$D_{app}(q \to 0) = D_p \quad (2c_s/c_p \to \infty) \quad (15a)$$

$$D_{app}(q \to 0) = \frac{D_p D_s (2+Z_p)}{D_p Z_p + D_s} \quad (2c_s/c_p \to Z_p). \quad (15b)$$

At high added salt conditions (Equation (15a) and Figure 1(b)) there is a strong screening of the polyion charge and the polyions diffuse like uncharged particles with no coupling to the dynamics of the small ions. On the other hand, highly charged polyions ($D_p Z_p \gg D_s$) under very strong ionic-strength conditions ($2c_s/c_p \to Z_p$) diffuse at the same rate as their counterions and $D_{app}(q \to 0) \sim D_s$ (Equation (15b) and Figure 1(a)). The diffusion dynamics in unscreened polyelectrolyte systems thus becomes extremely fast and the obtained apparent diffusion coefficient cannot be considered as a measure of the dimensions of diffusing molecules.

## 3 DYNAMIC LIGHT SCATTERING

An appropriate non-invasive technique for characterizing the dynamics and structure of macromolecules in the solution is dynamic light scattering also called photon correlation

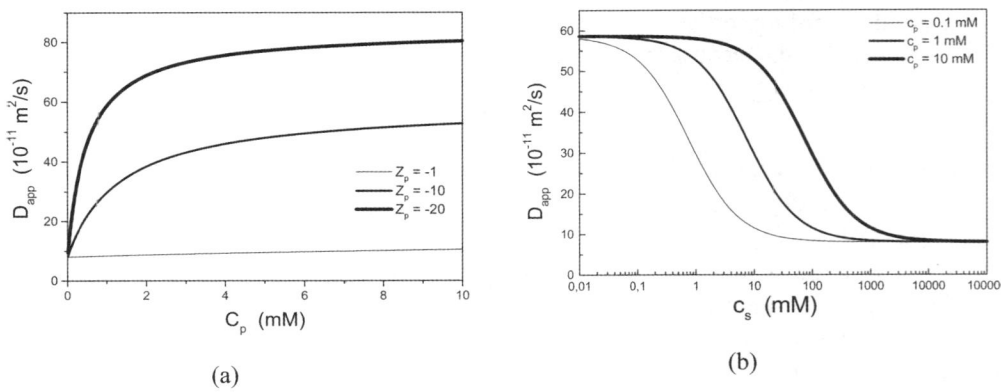

**Figure 1** *Predictions of the coupled-mode theory: The dependence of the apparent polyion diffusion coefficient on the polyion concentration at fixed $c_s = 10$ mM (a) and on the added salt concentration at fixed $Z_p = -10$ (b). The data are given for rodlike particles with the values $L = 10$ nm, $d = 2.5$ nm, $D_p = 8.0 \cdot 10^{-11}$ m²/s and $D_s = 160 \cdot 10^{-11}$ m²/s.*

spectroscopy. Depolarized DLS provides the rotational diffusion coefficient of the molecules, polarized DLS is used to obtain information on the translational motion, and corresponding hydrodynamic theories relate both transport coefficients to the molecular dimensions. Therefore, a combination of both light scattering methods provides means to determine the size and shape of different molecular structures in the solution.

## 3.1 Light scattering theory

In the classical theory of light scattering a light beam of given polarization impinges on a molecule, inducing a dipole moment that subsequently radiates. As the molecules in the medium are constantly moving, the instantaneous dielectric constant, which depends on the positions and orientations of the molecules, fluctuates in time. The local dielectric tensor can be expressed as

$$\underline{\varepsilon}(\vec{r},t) = \langle \varepsilon \rangle \underline{I} + \underline{\delta\varepsilon}(\vec{r},t) \tag{16}$$

where $\langle \varepsilon \rangle$ is the average value, $\underline{\delta\varepsilon}(\vec{r},t)$ is the fluctuation at position $\vec{r}$ and time $t$ and $\underline{I}$ is the second rank unit tensor. Let the incident electric field be a plane wave of the form

$$\vec{E}_i(\vec{r},t) = \vec{i} E_0 \exp(i\vec{k}_i \cdot \vec{r} - i\omega_i t) \tag{17}$$

The wave propagates in the direction of the wave vector $\vec{k}_i$ with the angular frequency $\omega_i$ and amplitude $E_0$. The unit vector $\vec{i}$ denotes the polarization of the incident wave. The position of the detector determines the scattering angle $\vartheta$ between the incident and the scattered light (Figure 2). The polarization $\vec{f}$ of the scattered light, propagating in the direction of the wave vector $\vec{k}_f$ with the angular frequency $\omega_f$, can be selected with an analyzer.

The total radiated field is a superposition of the fields radiated from all molecules in the scattering volume $V$. The scattered electric field at a large distance ($R \gg \lambda$) from the scattering volume is[2]

$$E_s(R,t) = \frac{E_0 e^{ik_i R}}{4\pi\varepsilon_0 R} \int_V \exp(i\vec{q} \cdot \vec{r} - i\omega_i t) [\vec{f} \cdot [\vec{k}_f \times (\vec{k} \times (\underline{\delta\varepsilon}(\vec{r},t) \cdot \vec{i}))]] d^3\vec{r} \tag{18}$$

where the subscript $V$ indicates that the integral is taken over the whole scattering volume. The scattering vector $\vec{q}$ is defined in terms of the scattering geometry as $\vec{q} = \vec{k}_i - \vec{k}_f$.

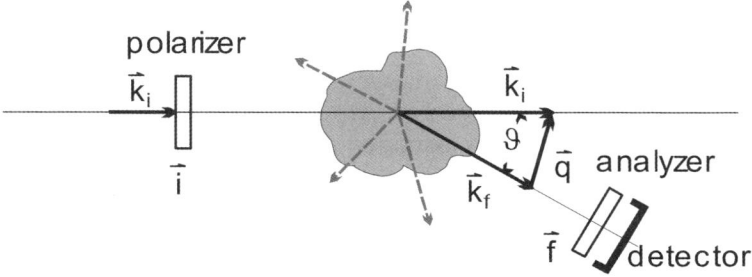

**Figure 2** *Illustration of the light scattering process: the incident light beam with wave vector $\vec{k}_i$ is scattered in all directions. Only scattered light with wave vector $\vec{k}_f$ arrives to the detector. The scattering vector $\vec{q}$ is selected by the position of the detector.*

When molecular dynamics that induces the inhomogeneties of $\underline{\varepsilon}$ is slow compared to the optical oscillation time it can be assumed that $\omega_i \approx \omega_f$ and $k_i \approx k_f$. The magnitude of the scattering vector $\vec{q}$ thus depends only on the scattering angle:

$$q = \frac{4\pi n}{\lambda} \sin\frac{\vartheta}{2} \quad (19)$$

where $n = \sqrt{\langle\varepsilon\rangle}$ is the refractive index of the scattering medium and $\lambda$ is the wavelength of light in vacuo. By varying the scattering angle, fluctuations with different scattering vectors are monitored. Equation (18) can be expressed in terms of the spatial Fourier transform of the dielectric tensor fluctuation as

$$E_s(R,t) = \frac{-k_f^2 E_0}{4\pi\varepsilon_0 R} \exp(ik_f R - i\omega_i t)\delta\varepsilon_{if}(\vec{q},t) \quad (20)$$

where the component of the dielectric tensor fluctuation along the initial and the final polarization directions is given as

$$\delta\varepsilon_{if}(\vec{q},t) = \vec{f} \cdot \underline{\delta\varepsilon}(\vec{q},t) \cdot \vec{i}. \quad (21)$$

The dielectric tensor is in general a function of the solution concentration, which fluctuates around an average value as given by Equation (2). Since the concentration fluctuations are expected to be quite small, the dielectric tensor can be expanded in a power series of $\delta c(\vec{r},t)$ and the relation $\delta\varepsilon_{ij}(\vec{r},t) \propto \delta c(\vec{r},t)$ is obtained. The dielectric constant fluctuations are in general proportional to the concentration fluctuations described by the extended diffusion Equation (11) taking into account the fluctuations due to coupled polyion – small ion dynamics.

## 3.2 Detection and analysis of scattered light

The rate, at which the dielectric constant fluctuations decay to the equilibrium value, is directly dependent upon the dynamics of the molecules and is contained in the autocorrelation function

$$G_1(t) = \langle E_s^*(0)E_s(t)\rangle. \quad (22)$$

According to equation (20) the electric field time-correlation function is

$$G_1(t) = \frac{E_0^2 k_f^4}{16\pi^2 \varepsilon_0^2 R^2} \exp(-i\omega_i t)\langle\delta\varepsilon_{if}(\vec{q},0)\delta\varepsilon_{if}(\vec{q},t)\rangle. \quad (23)$$

In order to calculate the correlation function of the scattered electric field one must have a model for the mechanism by which dielectric fluctuations decay. Since dielectric fluctuations are proportional to the concentration fluctuations, the electric field time-correlation function $G_1(t)$ is proportional to $\exp(-Dq^2 t)$ as follows from Equation (4). For the rotational diffusion the relation[2] is expressed as $G_1(t) \propto \exp(-6\theta t)$.

In a light-scattering experiment the scattered light impinges on the detector, usually a photomultiplier tube. The instantaneous current output of the detector is proportional to the number of photons hitting the tube and therefore to the intensity of the impinging light $I(t) = |E(t)|^2$. The detector output is passed into a digital autocorrelator which calculates the intensity time-correlation function

$$G_2(t) = <I(0)I(t)> = \langle|E(0)|^2|E(t)|^2\rangle. \quad (24)$$

There are two different detection methods that lead to different interpretations of the measured data. In the homodyne regime only the light scattered from the sample impinges on the detector and Equation (24) reduces to

$$G_2(t) = \left\langle \left|E_s^*(0)\right|^2 \left|E_s(t)\right|^2 \right\rangle = \left|G_1(0)\right|^2 + \left|G_1(t)\right|^2 \qquad (25)$$

where $E_s(t)$ is the electric field of the scattered beam. In heterodyne detection regime the unscattered laser light is mixed with the scattered light on the photomultiplier cathode. If $E_O(t)$ represents the unscattered field, then the electric field at the detector is a superposition of $E_O(t)$ and $E_s(t)$ and thus the intensity correlation function becomes

$$G_2(t) = \left\langle \left|E_O(0) + E_s(0)\right|^2 \left|E_O(t) + E_s(t)\right|^2 \right\rangle. \qquad (26)$$

The above expression leads to

$$G_2(t) = I_O^2 + 2I_O \, \mathrm{Re}\left\langle E_s^*(0) E_s(t) \right\rangle \qquad (27)$$

where $\langle E_s^*(0) E_s(t)\rangle = G_1(t)$. In the heterodyne regime the measured intensity correlation function $G_2(t)$ is simply proportional to the scattered field correlation function $G_1(t)$.

The translational diffusion of a molecule in solution is a first-order decay process and the corresponding normalized scattered electric field correlation function is given as

$$g_1(t) = \frac{G_1(t)}{G_1(0)} = \exp(-t/\tau) \qquad (28)$$

where $\tau$ is the relaxation time of the molecule. In general, several different relaxation modes characterize the dynamics of the solution and the field correlation function is expressed as

$$g_1(t) = \sum_i A_i \exp(-t/\tau_i) \qquad (29)$$

where $A_i$ is the amplitude and $\tau_i$ the relaxation time of the $i$-th diffusive mode. A polydisperse distribution of the sizes of the scattering objects results in a distribution of the corresponding relaxation times. For a broad distribution of relaxation times the field correlation function can be approximated by a stretch exponential or Kohlrausch – Williams – Watts (KWW) function[1]

$$g_1(t) = \sum_i A_i \exp(-t/b_i)^{\beta_i} \qquad (30)$$

with the constraint $0 \leq \beta_i \leq 1$. The average relaxation time of the $i$-th diffusive mode is given by

$$\tau_i = \frac{b_i}{\beta_i} \Gamma(1/\beta_i) \qquad (31)$$

where $\Gamma(1/\beta_i)$ is the gamma function. The parameter $\beta_i$ is a measure of the width of the distribution: the value $\beta_i = 1$ corresponds to the exponential relaxation, and hence a very narrow distribution with $\tau_i = b_i$, while smaller $\beta_i$ correspond to broader distributions.

The characteristic relaxation times in homodyne detection are obtained by fitting the measured normalized time-correlation functions to

$$g_2(t) = \frac{G_2(t)}{G_2(0)} = B + \left(\sum_{i=1}^n A_i \exp(-t/b_i)^{\beta_i}\right)^2 \qquad (32)$$

where $B$ represents the baseline (Figure 3(a)). For heterodyne regime the relaxation times are obtained from the normalised time-correlation function given as[10]

$$g_2(t) = 1 + 2(1 - j_d) j_d g_1(t) + j_d^2 g_1^2(t) \qquad (33)$$

with $j_d$ being the ratio between the intensity of the light that is scattered inelastically and the total scattered intensity. With the calculated relaxation time of the diffusive modes the translational diffusion coefficient is obtained from polarized scattering ($\vec{i} \parallel \vec{f}$) as

$$D_{T(i)} = \frac{1}{\tau_i q^2} \tag{34}$$

where $q$ is the length of the scattering vector defined by Equation (19) and the relation between $\tau_i$ and $b_i$ is given by Equation (31). The corresponding dispersion curves ($1/\tau_i$ versus $q^2$) are presented in Figure 3(b). From depolarized scattering ($\vec{i} \perp \vec{f}$) the rotational diffusion coefficient is obtained as

$$\theta = \frac{1}{6\tau_{depol}}. \tag{35}$$

Translational diffusion coefficients can also be obtained by using the standard data analysis program CONTIN[11]. Correlation functions as given by Equation (23) are Laplace transforms of the spectrum of relaxation times

$$g_1(t) = \int_0^\infty A(\tau) \exp(-t/\tau) d\tau. \tag{36}$$

The spectrum of relaxation times obtained from correlation functions by inverse Laplace transform is frequently a multimodal distribution where separate peaks can be ascribed to different diffusive modes. The position of the peak on the time axis corresponds to the relaxation time $\tau_i$ of the particular mode and the peak area corresponds to the portion of scattering intensity due to this diffusive mode.

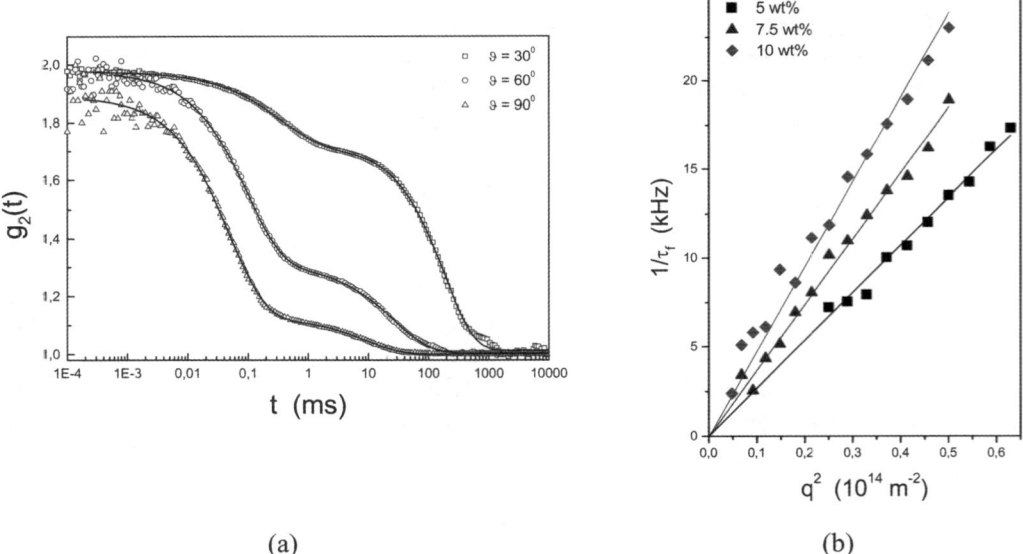

(a)  (b)

**Figure 3** *Typical time-correlation functions taken at different scattering angles show the presence of two diffusive modes. Solid lines are fits to Equation (32). Data are given for 7.5 wt% $(NH_4)_2(5'\text{-}dGMP)$ solution (a). The inverse relaxation time of the faster diffusive mode versus $q^2$ in $(NH_4)_2(5'\text{-}dGMP)$ solutions. Solid lines are fits to Equation (34) (b).*

## 4 SELF-ASSEMBLED GUANOSINE STACKS

The unique ability of guanosine 5'-monophospate (5'-GMP) and its derivatives to self-assemble into G-quadruplex structures has been known for decades[12]. The basic building unit is a planar G-quartet, a cyclic hydrogen-bonded array of four guanosine molecules (Figure 4). At high solution concentration G-quartets stack on each other and a highly stable quadruplex structure is formed even without a linking sugar-phosphate backbone. The extend of stacking and hence the length of self-assembled stacks is highly sensitive to solution parameters such as GMP concentration, temperature, pH value and added ions[13].

### 4.1 Guanosine 5'-monophosphate

Eimer and Dorfmüller demonstrated already in 1992 that DLS is a convenient tool for investigating the dimensions of self-assembled G-stacks[14]. They studied $Na_2$(5'-GMP) in aqueous solutions at different NaCl concentrations by combining depolarised and polarised scattering. To analyse the experimental data they used the hydrodynamic theory of Tirado and Garcia de la Torre for cylinder-symmetric molecules (Equations (6-8)). Rather surprisingly, they found that depending on concentration and temperature 5'-GMP forms both, vertically stacked monomers and stacked planar G-quartets (Figure 5). At a molar concentration of about 0.1 mol/L the average stack size of G-quartets was 9 units and for the monomer stacks about 4 units. Aggregates grew with increasing concentration and at 1.0 mol/L reached the length of 18 stacked G-quartets and 10 stacked monomers. These results, however, were in serious contradiction with previous NMR studies that postulated that primarily octamer, dodecamer, and hexadecamer structures exist in solution[15-17].

As the theory of Tirado and Garcia de la Torre is applicable only for length to diameter ratios of $p \geq 2$ it cannot be used for stacks composed of a few G-quartets or a few GMP monomers. Jurga-Nowak et al. therefore used bead modelling of the hydrodynamic parameters to distinguish between different structures formed in aqueous solutions of 5'-GMP acid[18]. At a 5'-GMP concentration of 55 mM in water solutions with pH 2 they identified the presence of only unstacked GMP monomers in the whole temperature range studied (2 – 60 °C). Their experimentally determined diffusion coefficient at 20 °C was $D_T$ = (3.93 ± 0.24) · $10^{-10}$ m$^2$/s, which was in good agreement with the value calculated using the bead model $D_T$ = (3.98 ± 0.27) · $10^{-10}$ m$^2$/s. From the measurements the hydrodynamic

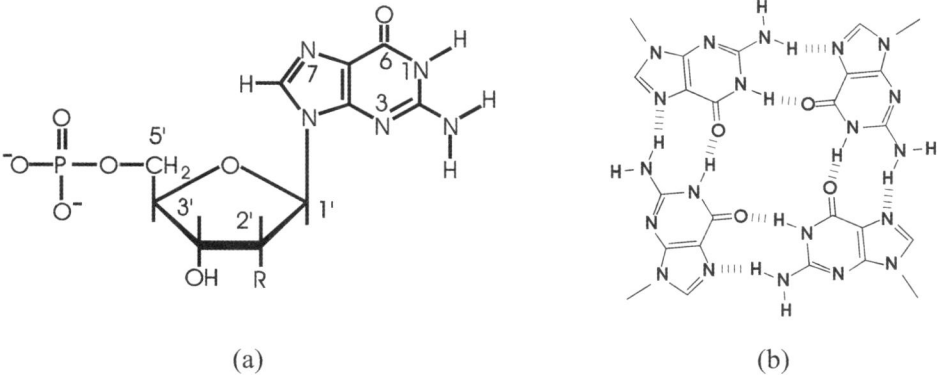

(a)            (b)

**Figure 4** *The structure of guanosine 5'-monophosphate (R = OH) and 2'-deoxyguanosine 5'-monophosphate (R = H) (a). The G-quartet (the sugar-phosphate groups are omitted for clarity) (b).*

radius of GMP molecules was estimated to be $r_H = 0.54$ nm. This corresponds to a GMP monomer hydrated with 12 water molecules and to a hydration layer of 0.135 nm. In highly concentrated water solutions 5'-GMP formed self-assembled structures composed of 32 or more stacked G-quartets and with a length of more than 10 nm. Their size depends on the concentration of GMP and on temperature. With increasing temperature the G-quartet stacks decompose and continuously transform to structures formed from single GMP monomers. In solutions with acidic pH values G-quadruplexes were found to be larger than those formed in solutions with basic pH. This is consistent with early studies on GMP which reported that GMP in acidic conditions forms well-ordered gels[12,19].

The effect of pH was further investigated by Wong et al.[20] in a combined diffusion NMR and DLS study of $Na_2$(5'-GMP) aggregates at pH 8. The results from these two different techniques unambiguously established that $Na_2$(5'-GMP) aggregates are on the nanometer scale and are thus significantly larger than what has been believed for years. The length of G-quartet stacks was 8 – 30 nm for GMP concentrations between 18 and 33 wt%. For stacked monomers aggregate length was 3 – 11 nm for the same GMP concentrations. While the length of monomer aggregates increased slightly with temperature, the length of G-quartet aggregates was essentially independent of temperature. Both, monomer and G-quartet stacks, formed molecular cylinders with the same axial ratio. This observation suggests that the same stacking mechanism is in action for both types of aggregates. A small but notable solvent isotope effect was also detected for G-quartet aggregates, which are about 20 % longer in $D_2O$ than in $H_2O$. Similar effect was reported previously[15,21] and has to be taken into account when interpreting diffusion measurements that are generally performed for samples dissolved in $D_2O$. Finally, the effect of added NaCl was studied and revealed that while NaCl increases the population of G-quartets, the length of these aggregates seems to be insensitive to the presence of NaCl in solution. The combined NMR and DLS study showed the advantage of diffusion NMR over DLS in cases where aggregates of similar sizes are monitored. While different aggregates can be measured simultaneously by NMR the DLS cannot distinguish between them and gives and averaged value of the diffusion coefficient.

## 4.2 Deoxyguanosine 5'-monophosphate

Although Wong et al. detected G-quartet stacks with lengths up to 30 nm, even larger aggregates, with a length of 45 nm, were observed by DLS in aqueous solutions of

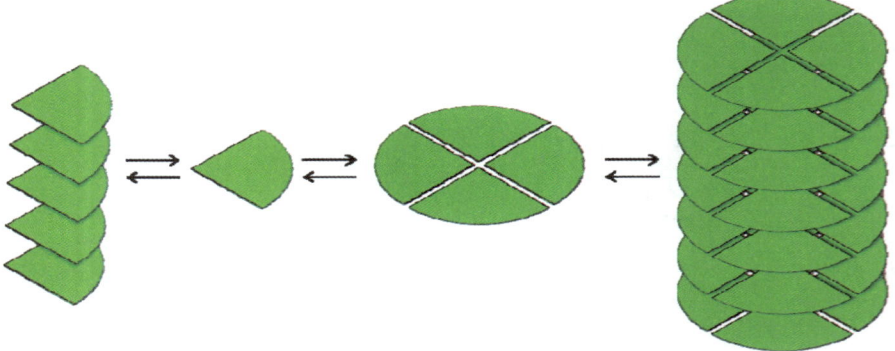

**Figure 5** *In solutions of $Na_2$(5'-GMP) vertical stacks from monomers and G-quartets are evidenced by dynamic light scattering.*

diammonium 2'-deoxyguanosine 5'-monophosphate[22]. At dGMP concentrations $c \geq 4$ wt% a sharp crossover between a dilute regime of weak aggregation and a regime of extensive aggregate growth takes place. This is reflected in the appearance of a fast diffusive mode in the DLS response, which is connected to the translational diffusion of the aggregates. The appearance of G-quartet stacks was additionally confirmed by the appearance and growth of an additional peak in $^{31}$P NMR spectra. With increasing temperature the aggregates abruptly disassemble and the fast diffusive DLS mode dissappears[23]. The sharp transition from monomers to large stacked aggregates could be assigned to the polyelectrolyte nature of the aggregates species. The diffusion coefficient resulting from the translational motion of G-quartet stacks increased with increasing dGMP concentration. Such behaviour is typical for polyelectrolyte systems and predicted by the couple mode theory (see Figure 1(a)). DLS measurements revealed also the presence of a slow diffusive mode in the whole concentration range investigated. This slow mode was assigned to the translational motion of large unspecific aggregates with hydrodynamic radii of $r_H \sim 200$ nm. The existence of large globular aggregates was confirmed by freeze fracture electron microscopy. The slow diffusive mode has been reported for numerous polyelectrolyte solutions, including the DNA[24-28] but the mechanism responsible for the formation of the globules is still not clearly resolved[29,30].

The addition of KCl to $(NH_4)_2$(5'-dGMP) resulted in a complex solution dynamics of aggregates that originates from an interplay of $K^+$-induced aggregate growth and electrostatic screening[31]. At a fixed dGMP concentration of 4 wt% already small amounts of added KCl result in an increase of the value of the apparent polyion diffusion coefficient from $D_{app} = (0.148 \pm 0.003) \cdot 10^{-10}$ m$^2$/s (no added KCl) to $D_{app} = (0.495 \pm 0.017) \cdot 10^{-10}$ m$^2$/s in 0.1 M KCl. After that, a gradual decrease is observed until the same value $D_{app} = (0.148 \pm 0.008) \cdot 10^{-10}$ m$^2$/s is reached at 1.0 M KCl. This result strongly suggests that the aggregate size of 45 nm is independent of added $K^+$ amount and that just the aggregate population increases as evidenced by $^{31}$P NMR. This observation is in good agreement with the added NaCl studies on $Na_2$(5'-GMP)[20]. Depolarised scattering on $(NH_4)_2$(5'-dGMP) revealed rotational dynamics of G-quartet stacks, which exhibited pronounced slowing down of orientational fluctuations due to the gelation of the solution when approaching the isotropic – liquid crystalline transition.

## 5 QUADRUPLEXES FROM G-RICH DNA SEQUENCES

The translational and rotational diffusion coefficients are highly sensitive to the size and shape of the molecules and therefore to their conformational changes. Bolten et al.[6] were the first to use polarised and depolarised DLS to distinguish between alternative intra- and intermolecular quadruplex structures. They investigated three G-rich oligonucleotides where conformation was known from previous X-ray crystallography or from NMR spectroscopy: the 15-mer d(GGTTGGTGTGGTTGG) known as the thrombin binding aptamer, the 14-mer d(TTGGGGTTGGGGTT), and the 8-mer d(TTGGGGTT), see Figure 6. To interpret the experimental data hydrodynamic model calculations using the bead model were applied. The 15-mer exhibited in the presence of KCl and unusually short relaxation time as well as a high translational diffusion coefficient $D_T = 1.8 \cdot 10^{-10}$ m$^2$/s at 20 °C. Model calculations showed that these transport coefficients could be assigned to compact intramolecular quadruplex conformations with two G-quartet planes. The 14-mer proved to be a bimolecular quadruplex formed through association of two "hairpin" molecules with $D_T = 1.48 \cdot 10^{-10}$ m²/s. And for the 8-mer with a single guanine-rich region a parallel intermolecular quadruplex could be confirmed with $D_T = 1.34 \cdot 10^{-10}$ m²/s.

A DLS study of the *Tetrahymena thermophila* telomeric sequence d(TGGGGT)$_4$ showed the presence of two quadruplex structures: an intramolecular quadruplex and an intermolecular or tetramolecular quadruplex[32]. Experimental values for the translation diffusion coefficients at 20 °C were $D_T = 1.4 \cdot 10^{-10}$ m²/s for the monomer and $D_T = 0.8 \cdot 10^{-10}$ m²/s for the tetramolecular structure. The relative weighted concentrations of the two structures were investigated as a function of DNA concentration and depending on the presence of NaCl, KCl and SrCl$_2$. The results obtained in the presence of monovalent ions could be presented in a coherent plot in which the concentration of salt was expressed by the number of ions per DNA molecule[33]. A large number of ions per DNA molecule favoured tetrameric quadruplex formation while a small number favoured the monomeric form. $K^+$ ions exhibited greater effectiveness in formations of the tetraplex as $Na^+$. In the presence of $Sr^{2+}$ ions the formation of extremely large aggregates with $D_T = 0.3 \cdot 10^{-10}$ m²/s was reported.

The same authors[34] also investigated the effect of different ions on the formation and behaviour of quadruplex structures of the human telomere sequence d(TTAGGG)$_4$. In the presence of $Na^+$ and $K^+$ ions only one collective diffusion mode with $D_T = 1.4 \cdot 10^{-10}$ m²/s was observed. Comparison of those values with calculated results based on the bead model allowed identification of the molecules corresponding to the diffusion mode as the monomeric quadruplex but it was not possible to differentiate between the parallel or antiparallel monomer. As DNA molecules in neutral pH are highly charged polyions their solution dynamics is dominated by strong electrostatic intermolecular interactions. Therefore their solution behaviour was analysed via the coupled mode theory (Equations (11-14)) and the effective charge of the monomeric quadruplex could be estimated to $z_{eff} = -5 \pm 1$ in the presence of NaCl and $z_{eff} = -4 \pm 1$ in the presence of KCl. The effect of different ions was additionally studied by circular dichroism and proved that the effect of $Sr^{2+}$ ions on the formation of monomeric quadruplexes is the most pronounced. DNA solutions containing $Sr^{2+}$ ions showed the presence of two diffusive modes: one related to the translational motion of the monomeric species and the other that was assigned to the tetrameric quadruplex. With decreasing DNA concentration a conformational tetramer-monomer transition was observed. The effect of $Sr^{2+}$ ions on the quadruplex structures

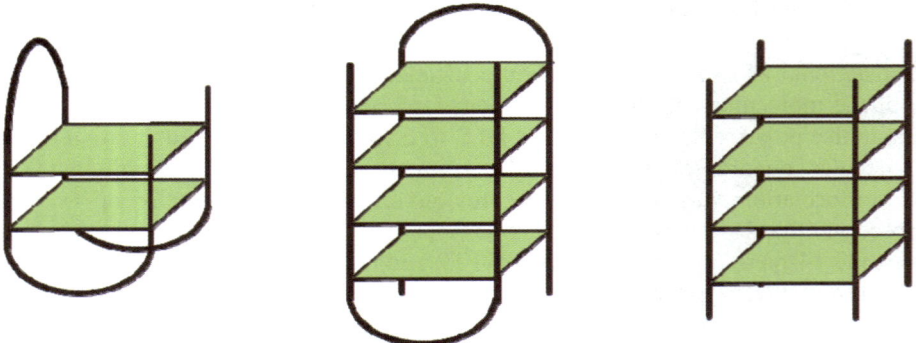

**Figure 6** *Schematic representation of different quadruplex structures: intramolecular quadruplex from d(GGTTGGTGTGGTTGG) (left), dimer through association of two "hairpin" strands of d(TTGGGGTTGGGGTT) (middle) and tetrameric intermolecular quadruplex from sequence d(TTGGGGTT) (right).*

formed by the human telomere sequence was found to be much stronger and of different character than that of the ions $K^+$ and $Na^+$. This observation matches the effect of $Sr^{2+}$ ions on the *Tetrahymena thermophila* sequence[33].

Another study used DLS to investigate the formation of stable multistranded complexes from $d(A_{15}G_{15})$ called frayed wires[35]. Guanine-guanine interactions stabilise the quadruplex stem while non-guanine portions are disposed away from the stem and form single stranded arms. DLS measurements showed a polydisperse distribution of the aggregate length arising from the stacking of the quadruplex stems. The aggregation of frayed wires was promoted by the presence of $Mg^{2+}$ ions and incubation at high temperature. DLS autocorrelation functions showed also the presence of the slow mode that is typical for polyelectrolyte solutions. The addition of NaCl decreased the intensity of the slow mode but did not affect the diffusion coefficients of the two diffusive modes.

The slow diffusive mode was also reported by the most recent DLS studies[36,37] thus underlining the polyelectrolyte nature of the G-quadruplex structures. Zimbone *et al.*[36] investigated the appearance of the slow mode in *Micrococcus luteus* DNA with high GC content and the effect of guanine sequences on changes of DNA physical state and conformational transitions. The authors concluded that an unconventional structural transition due to the presence of long guanine stretches could occur and is revealed by the slow mode. The other study by Spindler *et al.*[37] investigated short G-rich DNA sequences with GC termini but focused on the fast mode assigned to the translational diffusion of the quadruplex structures. DLS measurements revealed the formation of multimer structures arising from quadruplex associations via G:C:G:C tetrads. The existence of such multimeric structures was confirmed by polyacrylamide gel electrophoresis.

## 6 CONCLUSION

Several studies on guanosine nucleotides and G-rich oligonucleotides proved that dynamic light scattering is a convenient non-invasive tool for studying macromolecular systems. Although it does not approach the high resolution of X-ray crystallography or NMR spectroscopy it provides valuable complementary information about translational and rotational motion of particles in solution. By using an appropriate hydrodynamic theory the size and shape of G-quadruplexes under different environmental conditions in solution can be investigated.

**Acknowledgements**

COST Action MP0802 *"Self-assembled guanosine structures for molecular electronic devices"* is gratefully acknowledged for financial support in scope of Short Term Scientific Missions and for providing a stimulative environment for the exchange of ideas and scientific knowledge.

**References**

1  K.S. Schmitz, *An Introduction to Dynamic Light Scattering by Macromolecules*, Academic Press, San Diego, 1990, Chapter 2, p. 11.
2  B.J. Berne and R. Pecora, *Dynamic Light Scattering*, Wiley, New York, 1976.
3  M.M. Tirado and J. Garcia de la Torre, *J. Chem. Phys.*, 1979, **71**, 2581.
4  M.M. Tirado and J. Garcia de la Torre, *J. Chem. Phys.*, 1980, **73**, 1986.

5. M.M. Tirado, M.C. Lopez Martinez and J. Garcia de la Torre, *Biopolymers*, 1984, **23**, 611.
6. M. Bolten, M. Niermann and W. Eimer, *Biochemistry*, 1999, **38**, 12416.
7. E. Banachowicz, J. Gapinski and A. Patkowski, *Biophys. J.*, 2000, **78**, 70.
8. J. Rotne and S. Prager, *J. Chem. Phys.*, 1969, **50**, 4831.
9. K.S. Schmitz, *An Introduction to Dynamic Light Scattering by Macromolecules*, Academic Press, San Diego, 1990, Chapter 7, p. 205.
10. A. Mertelj, L. Cmok, and M. Čopič, *Phys. Rev. E*, 2009, **79**, 041402.
11. S.W. Provencher, *Comput. Phys. Commun.*, 1982, **27**, 213.
12. M. Gellert, M.N. Lipsett and D.R. Davies, *Proc. Natl. Acad. Sci. USA*, 1962, **48**, 2013.
13. J.T. Davis, *Angew. Chem. Int. Ed.*, 2004, **43**, 668.
14. W. Eimer and Th. Dorfmüller, *J. Phys. Chem.*, 1992, **96**, 6790.
15. M. Borzo, C. Detellier, P. Laszlo and A. Paris, *J. Am. Chem. Soc.*, 1980, **102**, 1124.
16. Ch.L. Fisk, E.D. Becker, H.T. Miles and T.J. Pinnavaia, *J. Am. Chem. Soc.*, 1982, **104**, 3307.
17. J. A. Walmsley, R. G. Barr, E. Bouhoutsos-Brown and T. J. Pinnavaia, *J. Phys. Chem.*, 1984, **88**, 2599.
18. H. Jurga-Nowak, E. Banachowicz, A. Dobek and A. Patkowski, *J. Phys. Chem. B*, 2004, **108**, 2744.
19. I. Bang, *Biochem. Z.*, 1910, **26**, 293.
20. A. Wong, R. Ida, L. Spindler and G. Wu, *J. Am Chem. Soc.*, 2005, **127**, 6990.
21. F. Carsughi, M. Ceretti and P. Mariani, *Eur. Biophys. J.*, 1992, **21**, 155.
22. L. Spindler, I. Drevenšek-Olenik, M. Čopič, R. Romih, J. Cerar, J. Škerjanc and P. Mariani, *Eur. Phys. J. E*, 2002, **7**, 95. Note that there is an error in this article. The length of the cylinder should be $L = 45$ nm corresponding to a stack of 133 G-quartets.
23. L. Spindler, F. Federiconi, P. Mariani, I. Drevenšek Olenik, M. Čopič, M. Tomšič and A. Jamnik, *Mol. Cryst. Liq. Cryst.*, 2005, **435**, 1/[661].
24. H.Tj. Goinga and R. Pecora, *Macromolecules*, 1991, **24**, 6128.
25. P. Wissenburg, T. Odijk, P. Cirkel and M. Mandel, *Macromolecules*, 1995, **28**, 2315.
26. M. Sedlak, *Langmuir*, 1999 **15**, 4045.
27. L. Skibinska, J. Gapinski, H. Liu, A. Patkowski, E.W. Fischer and R. Pecora, *J. Chem. Phys.*, 1999, **110**, 1794.
28. E. Buhler and M. Rinaudo, *Macromolecules*, 2000, **33**, 2098.
29. T. Odijk, *Macromolecules*, 1994, **27**, 4998.
30. J. Ray and G.S. Manning, *Macromolecules*, 2000, **33**, 2901.
31. L. Spindler, I. Drevenšek Olenik, M. Čopič, J. Cerar, J. Škerjanc, R. Romih and P. Mariani, *Eur. Phys. J. E*, 2004, **13**, 27.
32. A. Włodarczyk, J. Gapiński, A. Patkowski and A. Dobek, *Acta Biochimica Polonica*, 1999, **46**, 609.
33. A. Włodarczyk, A. Patkowski, P. Grzybowski and A. Dobek, *Acta Biochimica Polonica*, 2004, **51**, 971.
34. A. Włodarczyk, P. Grzybowski, A. Patkowski and A. Dobek, *J. Phys. Chem. B*, 2005, **109**, 3594.
35. E. Protozanova and R.B. Macgregor Jr., *Biophysical Chemistry*, 2000, **84**, 137.
36. M. Zimbone, G. Bonaventura, P. Baeri and M.L. Barcellona, *Eur. Biophys. J.*, 2012, **41**, 425.
37. L. Spindler, M. Rigler, I. Drevenšek-Olenik, N. Ma'ani Hessari and M. Webba da Silva, *Journal of Nucleic Acids*, 2010, **2010**, 431651.

# GMP-QUADRUPLEX STRUCTURES IN DILUTE SOLUTIONS AND IN CONDENSED PHASES: AN X-RAY SCATTERING ANALYSIS

E. Jr. Baldassarri, M. G. Ortore, A. Gonnelli, M. Marcinekova, S. Mazzoni, M. L. Travaglini and P. Mariani*

Università Politecnica delle Marche, Dept. of Life and Environmental Sciences – Biophysical Research Group, Via Brecce Bianche, Ancona, 60131, Italy. Email: mariani@univpm.it

## 1 INTRODUCTION

Among the four DNA bases, guanosine 5'-monophosphate (GMP) is specifically interesting because of the self-assembling ability in aqueous solutions to form four-stranded helical structures (G-quadruplex). G-quadruplexes are the supramolecular organization of basic units, called G-quartets, made by planar rings constituted by four guanosines linked by Hoogsten hydrogen bonds[1] (in this configuration each base behaves as hydrogen–bond donor and hydrogen-bond acceptor[2]). In aqueous solution, tetrads stack one on the top of another, the stacking being determined in the absence of a sugar-phosphate backbone by attractive π-π interactions and by coordination of eight oxygen atoms in the central cavity of two adjacent tetramers by the proper cation[3]. Concerning the cation, it has been observed that it must be appropriately sized to fit the central cavity (e.g. $Na^+$) or the space between two tetramers[4] (e.g. $K^+$)[1]. The final G-quadruplex structure presents characteristic structural parameters like a stacking distance of 3.4 Å and an external diameter of 26 Å[5,6]. Moreover, due to the repulsive electrostatic interactions among the phosphate, each tetramer is rotated of approximately 30° relative to the previous one (Figure 1).

The GMP-quadruplex length depends on the experimental conditions. Indeed, in dilute solutions G-quadruplexes coexist with GMP monomers, G-quartets, G-octamers, G-dodecamers and so on, and the balance of the different species varies according to several factors, such as kind and amount of counter-ions, concentration and temperature[7,8]. At higher concentrations, GMP shows a liquid crystalline polymorphism[3,9,10] (Figure 1e), the G-quadruplexes being organized in a cholesteric fashion (low concentration, low ionic strength) or in an hexagonal lattice (high concentration, high counter-ion content). The self-assembling process of GMP has already been studied[3,7-10], although the mechanisms leading to different assembly paths in dilute conditions are not still completely clarified. Moreover, the structural characteristics induced by different counter-ions are not fully established.

Considering the special case of GMP dipotassium salt in aqueous solutions, we analyse in this paper by different X-ray scattering methods the structural properties and the size changes of GMP-quadruplexes in different experimental conditions, from dilute solutions to the formation of condensed phases[8,10]. In the last case, high-pressure was used to derive information on GMP-quadruplex hydration properties.

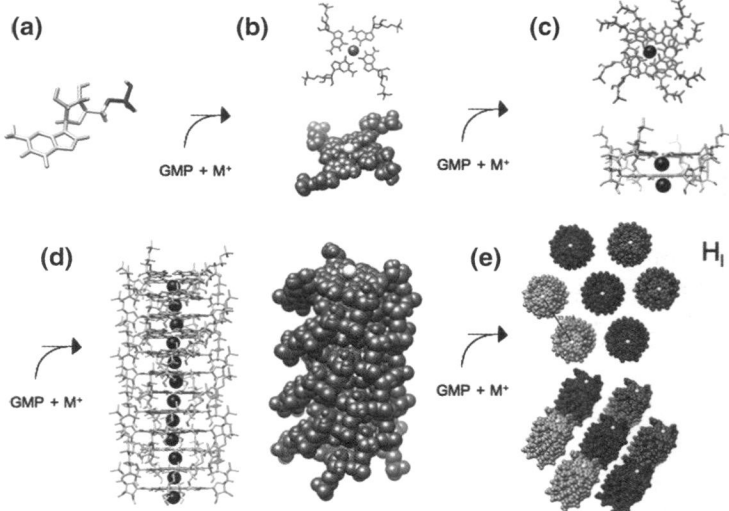

**Figure 1** *GMP self-assembling in dilute solutions, from monomers (a), tetramers (b), octamers (c), dodecamers (not shown), G-quadruplex structures (d) to the formation of an hexagonal condensed phase (e). Note the absence of the axial phosphate backbone.*

## 2 METHODS AND RESULTS

### 2.1 Sample Preparation

GMP dipotassium salt was obtained from GMP free acid (MP Biomedicals) by acid-base titration with potassium hydroxide solution (Sigma). GMP free acid solution was diluted by bidistilled water and the resulting solution was heated at temperature of 40°C to promote the solubility of the GMP free acid. In order to obtain the GMP/$K_2$ salt, 1M potassium hydroxide was added to GMP solution, monitoring the pH until its value was 9. Ethanol was added to the solution to induce GMP/$K_2$ precipitation. The solution was centrifuged at 8000 rpm and the supernatant removed. The pellet was washed twice with ethanol and centrifuged again at the same rate. Supernatant was removed and the pellet lyophilized by ThermoSavant SPD 111V SpeedVac. Finally, a lyophilized powder of GMP/$K_2$ was obtained.

Samples for Small-Angle X-ray Scattering experiments were prepared dissolving the GMP/$K_2$ powder at $c_{GMP}$ concentrations of 5%$_{w/w}$ and 8%$_{w/w}$ in bidistilled water and in 20 mM and 60 mM KCl solutions. Samples for X-ray diffraction analysis were prepared dissolving the powder in bidistilled water at $c_{GMP}$ concentrations ranging from 50 to 70%$_{w/w}$. All samples were analysed after an equilibration time of at least 24 hours at room temperature.

Sample water volume concentrations, $c_{v,w}$, were calculated using $v_{wat}$ =1.0 cm$^3$/g and $v_{GMP}$ =0.651 cm$^3$/g as specific volumes for water and GMP[3], respectively, by:

$$c_{v,w} = v_{wat}(100-c_{GMP})/(v_{wat}(100-c_{GMP})+v_{GMP}c_{GMP}) \qquad (1)$$

**Table 1**  *List of the investigated samples*

| SAXS | | High-Pressure X-Ray Diffraction | | |
| --- | --- | --- | --- | --- |
| 5%$_{w/w}$ | 8%$_{w/w}$ | 50%$_{w/w}$ | 60%$_{w/w}$ | 70%$_{w/w}$ |
| H$_2$O | H$_2$O | H$_2$O | H$_2$O | H$_2$O |
| 20 mM KCl | 20 mM KCl | - | - | - |
| 60 mM KCl | 60 mM KCl | - | - | - |

## 2.2 Methods

*2.2.1 SAXS – Small Angle X-Ray Scattering.* SAXS experiments were performed at the Desy Synchrotron radiation source in Hamburg, Germany, on the A2 beamline. The wavelength λ of the incident beam was 1.5 Å and the investigated $Q$-range (being $Q = 4\pi \cdot \sin\theta / \lambda$, where $2\theta$ is the scattering angle, $Q$ the modulus of the scattering vector) was from 0.015 to 0.33 Å$^{-1}$. The sample was enclosed between two kapton films and the temperature was changed from 24°C to 50°C. The scattering intensity was recorded on a bidimensional CCD camera of 1024x1024 pixels, radially averaged and corrected for the dark, detector efficiency, buffer contribution and sample transmission.

*2.2.2 X-Ray Diffraction.* Data at high-pressures were collected on the SAXS beamline at the Elettra Synchrotron (Trieste, Italy). The wavelength λ of the incident beam was 1.54Å and the investigated $Q$-range was from 0.05 to 7 Å$^{-1}$. The sample was wrapped into a Teflon cylinder and inserted in a pressure cell closed by two diamond windows[12]. The high pressure, ranging from 1 to 2000 bar, was provided by a hydraulic system as described by Pressl *et al.*[12]. Sample temperature was maintained at 25°C by a thermostat. The measurements were performed in steps of 100 bar. Particular attention was devoted to check for equilibrium conditions and to monitor and avoid radiation damage. The scattering intensity was corrected for background, buffer and detector inhomogeneities.

## 2.3 Results – Small Angle X-Ray Scattering

GMP/K$_2$ dilute solutions were analysed by SAXS. It should be clear that SAXS data provide an estimation of G-quadruplex size, hence let us to determine how monovalent ions (potassium ion, K$^+$, in this case), as well as GMP/K$_2$ concentration and temperature, affect the G-quadruplex stability[8,10]. The experiment has been performed considering two different points of view. First, temperature effects on G-quadruplex structures in the absence of external added counter-ions have been considered; in this condition, potassium ions are then only those due to phosphate dissociation. Second, the stabilization of G-quadruplexes due to excess counter-ions has been investigated.

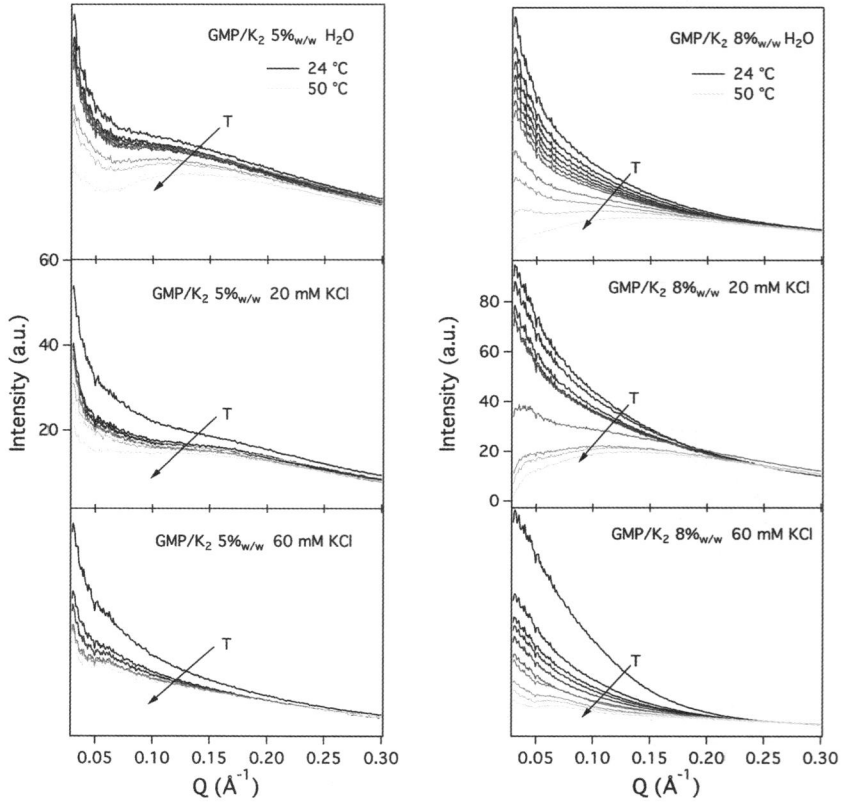

**Figure 2** *SAXS spectra observed at different GMP/$K_2$ concentrations (5%$_{w/w}$ left, 8%$_{w/w}$ right) in absence of KCl and in 20-60 mM KCl solutions, as reported in the legends. Each panel provides the ensemble of SAXS spectra recorded at increasing temperatures, from 24 to 50°C. Arrows indicate the direction of increasing temperatures.*

SAXS curves measured at different sample compositions and temperatures are reported in Fig. 2. Data clearly evidence that the scattering intensity increases when the concentration of excess counter-ions increases, confirming that potassium ion promotes the assembly of quartets, producing a larger amount of longer G-quadruplexes. On the other side, the scattering intensity decreases on heating. Hence, it could be derived that temperature effect is opposite to counter-ion effect: a temperature increase destabilizes the aggregate structure, and the consequence of quadruplex breaking is a decrease of the scattering intensity[8,13].

In a simple approach, Guinier's approximation[14] can be used to derive from a SAXS curve the particle gyration radius ($Rg$), which is related to the average size of the quadruplexes in solution, and the cross-sectional radius of gyration ($Rc$), which represents the radius of gyration of the particle section and is then related to the dimension of the the G-quadruplex cross-section. In the present case, the derived $Rc$ values resulted to be quite constant, i.e. insensitive to GMP/$K_2$ and salt concentrations, and centred around 8.5 Å (±10%), confirming that the scattering particles have a cylindrical shape[14]. Instead, the measured $Rg$ values showed a clear dependence on the experimental conditions, as shown in Fig. 3.

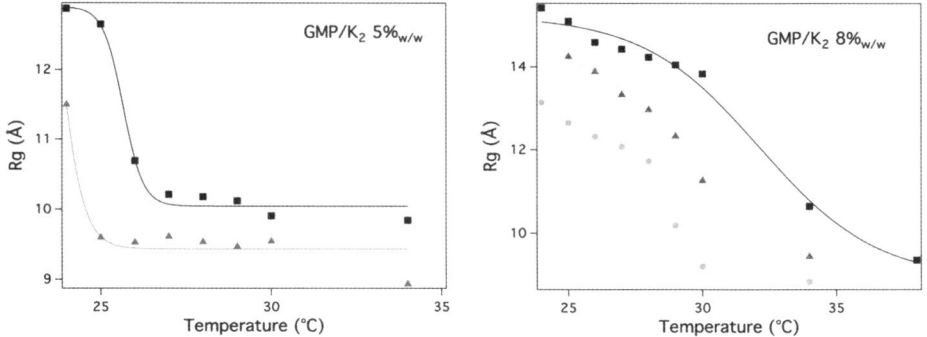

**Figure 3** *Temperature dependence of the radii of gyration of GMP/K$_2$ particles in solution (left 5%$_{w/w}$ and right 8%$_{w/w}$ samples). The radii of gyration of 5%$_{w/w}$ sample in water are absent, due the very low aggregation. Continuous lines are sigmoidal fit to the data* (● H$_2$O; ▲ 20 mM KCl; ■ 60 mM KCl)

According to the qualitative SAXS curve analysis, $Rg$ data shows that quadruplex dimensions reduce on heating. Considering that the GMP-octamer has a $Rg$ of about 9 Å, a temperature induced particle break-up can be easily suggested. However, fragmentation occurs at a critical temperature (fragmentation temperature, $Tf$), which clearly depends on GMP/K$_2$ concentration and on the presence of excess potassium. In particular, the fragmentation temperature increases by approaching the isotrope-to-liquid crystalline phase transition (expected at $c_{GMP} \approx 20\%_{w/w}$ at 25°C[3,11,15]) and by increasing the KCl concentration, demonstrating that counter-ions provide a strong stabilizing effect. $Tf$ values derived by sigmoid fit of the $Rg$ curves are reported in Table 2.

**Table 2** *Fragmentation temperatures (in °C) resulting from sigmoid fitting of Rg data reported in Fig.3.*

|  | $H_2O$ | 20 mM KCl | 60 mM KCl |
| --- | --- | --- | --- |
| GMP/K$_2$ 5%$_{w/w}$ | - | 23.7 ± 2.3 | 25.6 ± 0.1 |
| GMP/K$_2$ 8%$_{w/w}$ | 28.6 ± 0.3 | 29.4 ± 0.3 | 32.1 ± 0.6 |

The stabilizing effect is particularly relevant in the more concentrated samples, where quadruplex fragmentation appears to occur in a rather large temperature range.
From $Rg$ and $Rc$ data, the quadruplex length ($L$) can be calculated by using[14]:

$$L^2 = 12\,(Rg^2 - Rc^2) \qquad (2)$$

The estimated average lengths of G-quadruplexes are reported in Table 3. Temperature and counter-ion effects can be clearly accounted for. Indeed, quadruplex elongation is induced by increasing GMP/K$_2$ and excess K$^+$ concentrations, and the effect is visible at all the investigated temperatures. Moreover, when the temperature raises, the average length of the quadruplex decreases according to different efficiencies linked to the counter-ion amount in solution. In particular, larger is the counter-ion concentration, higher is the transition temperature: this behaviour clearly indicates a stabilizing effect of excess K$^+$ on the quadruplex structure, which will be further confirmed by high-pressure X-ray diffraction experiments described in the next paragraphs.

**Table 3** *GMP-quadruplex length (Å). Errors are in the order of 10%.*

| Temperature (°C) | GMP/K$_2$ 5%$_{w/w}$ 20 mM KCl | GMP/K$_2$ 5%$_{w/w}$ 60 mM KCl | GMP/K$_2$ 8%$_{w/w}$ H$_2$O | GMP/K$_2$ 8%$_{w/w}$ 20 mM KCl | GMP/K$_2$ 8%$_{w/w}$ 60 mM KCl |
|---|---|---|---|---|---|
| 24 | 23.9 | 31.3 | 32.5 | 37.7 | 42.9 |
| 25 | 9.5 | 30.1 | 30.1 | 36.0 | 41.5 |
| 26 | 8.6 | 18.9 | 28.4 | 33.4 | 39.3 |
| 27 | 9.7 | 15.4 | 27.1 | 31.7 | 38.5 |
| 28 | 8.7 | 15.1 | 25.2 | 28.5 | 37.6 |
| 29 | 7.9 | 14.7 | 15.2 | 22.5 | 36.8 |
| 30 | 8.9 | 12.8 | - | 7.6 | 35.8 |
| 34 | - | 12.2 | - | - | 18.6 |
| 38 | - | - | - | - | 6.2 |

### 2.4 Results – X-Ray Diffraction

GMP/K$_2$ condensed phases were analysed by X-ray Diffraction technique at several concentrations and as a function of the hydrostatic pressure, from ambient until 1.5 kbar. A few representative X-ray diffraction profiles obtained at ambient pressure are reported in Fig. 4. Indeed, at high concentration, GMP-quadruplexes are typically forming a hexagonal *H* columnar phase, while in diluted intermediate conditions a columnar cholesteric *Ch* phase usually forms[3,15]. In the hexagonal phase, quadruplexes are parallel and packed in a 2D hexagonal lattice; in the cholesteric phase, adjacent quadruplexes are slightly rotated, due to the helicoidal distribution of charges on their surface. As a result, a long-range chiral order occurs (see Fig.5). In both phases, the lateral distance between the columns is mainly determined by the sample composition, but preferential hydration effects (so that lateral or axial quadruplex distances differently increase on swelling) were detected due to the strong lateral repulsive interactions[16,17].

*2.4.1 Concentration Effects.* To evaluate the concentration effect, GMP/K$_2$ samples prepared in bidistilled water were first investigated at a temperature of 25°C and ambient pressure. Low-angle X-ray diffraction spectra reported in Fig. 4 show a series of well defined peaks, whose spacing ratios in the order 1:√3:√4... confirm the hexagonal lattice at all the investigated composition.

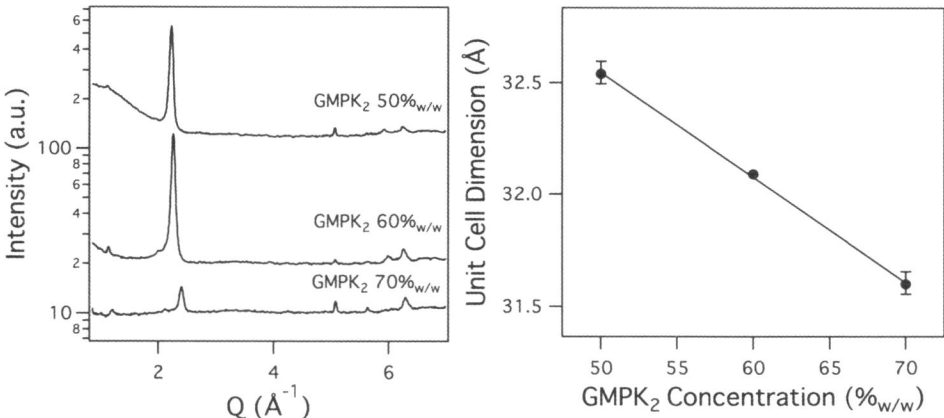

**Figure 4** *Left Panel: X-ray diffraction profiles obtained from GMP/K$_2$ samples at different concentrations (as reported in the legend). Curves are scaled for clarity. Right Panel: Hexagonal unit cell dependence on GMP/K$_2$ concentration.*

The positions of the peaks corresponding to the *H* phase provide the unit cell dimension *a* (see Fig. 5), according to the equation:

$$a = (4\pi/\sqrt{3}) \cdot (h^2 + k^2 + hk)^{0.5} / Q_{hk} \qquad (3)$$

where *h* and *k* are the Miller indices which define the Bragg peak centred at $Q_{hk}$. Data reported in Fig. 4 (right panel) show that by increasing GMP/K$_2$ concentration, the unit cell dimension decreases. This trend is in agreement with literature[3,15] and the mechanisms at the basis of this phenomenon can be easily traced by considering a classical swelling behaviour. In fact, for increasing concentration, the amount of water is reducing and quadruplexes should laterally approach one to another, in contrast with the electrostatic repulsive forces due to the charged phosphate groups on the quadruplex surfaces[16].

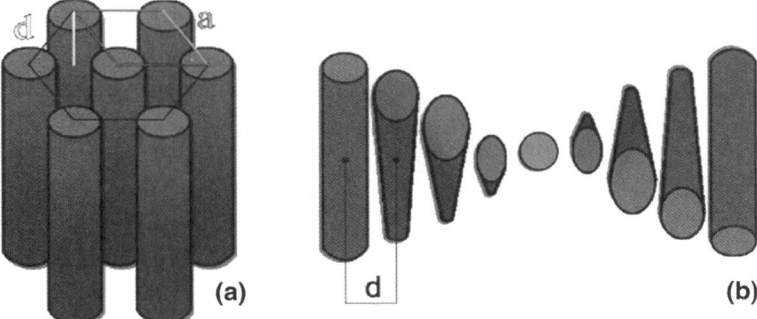

**Figure 5** *Structural models for the (a) Hexagonal (H) and (b) Cholesteric (Ch) phases. Quadruplexes are represented as cylinders.*

According to swelling models described for short and rigid or long and flexible particles[16,18,19], the observed behaviour suggests that GMP-quadruplexes follow a preferential hydration scheme along the *H* domain. Indeed, assuming that the *H* phase consists of cylindrical particles of constant radius *R* (about 12 Å for quadruplexes) and average length $L$, the relationship between the unit cell dimension $a$ and the fraction of water volume in the sample $c_{v,w}$, is given by[18]:

$$L\pi R^2 = C(\sqrt{3}/2)a^2(1-c_{v,w}) \quad (4)$$

where $C$ is the average inter-axial distance between the centres of the cylinders, orthogonal to the plane defined by the hexagonal 2D cell (see Fig. 6). Rearranging Eq. (4), the $L/C$ ratio, which essentially provides an indication of the fraction of water located in the direction of $C$ (the interaxial region), can be derived:

$$L/C = (\sqrt{3}/2) \cdot a^2 \cdot (1-c_{v,w})/(\pi R^2) \quad (5)$$

Note that for cylindrical elements of infinite length, $L/C$ can be approximated with the unit value. The $L/C$ values obtained in the present case are reported in Table 4.

**Table 4** *Structural parameters obtained from the analysis of X-ray Diffraction experiments on $GMP/K_2$ hexagonal phase.*

| Samples (%w/w) | $c_{v,w}$ | $L/C$ ratio | $a$ parameter |
|---|---|---|---|
| 50 | 0.606 | 0.68 ± 0.02 | 32.5 ± 0.1 |
| 60 | 0.506 | 0.83 ± 0.03 | 32.1 ± 0.1 |
| 70 | 0.397 | 0.98 ± 0.01 | 31.6 ± 0.1 |

In Fig. 7, $L/C$ values and unit cell parameters are reported as a function of the volume concentration of water. Is can be observed that $L/C$ ratio tends to unit at low hydration, but decreases rapidly with increasing water concentration. Moreover, the hexagonal cell parameter $a$ is found to vary with the water volume concentration as $(1-c_{v,w})^{-0.07}$ which is

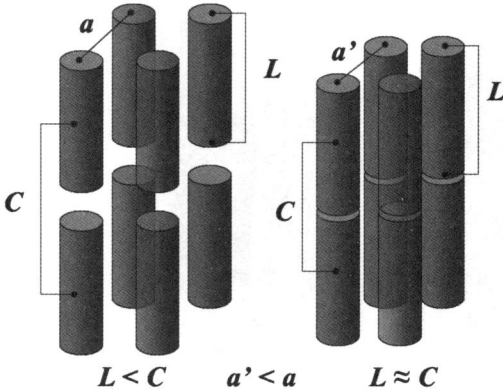

**Figure 6** *Structural models for the hexagonal phase. Left side, "finite rigid" rods, The a, L and C parameters are indicated.*

rather distant from both 1/3 and 1/2 exponents which are theoretically expected for "finite rigid" and "infinite flexible" rods, respectively[18,19]. In the first case, the hydration is essentially 3D (the lateral and interaxial quadruplex distances increase of the same extent during swelling), while in the second one the hydration is 2D (only the lateral quadruplex distance increases during swelling). In the present case, the observed behaviour is a clear evidence that preferential hydration occurs along the *H* domain[18]: due to the very different lateral and axial interactions, the volume around the finite, rather rigid GMP-quadruplexes decreases anisotropically as a function of concentration.

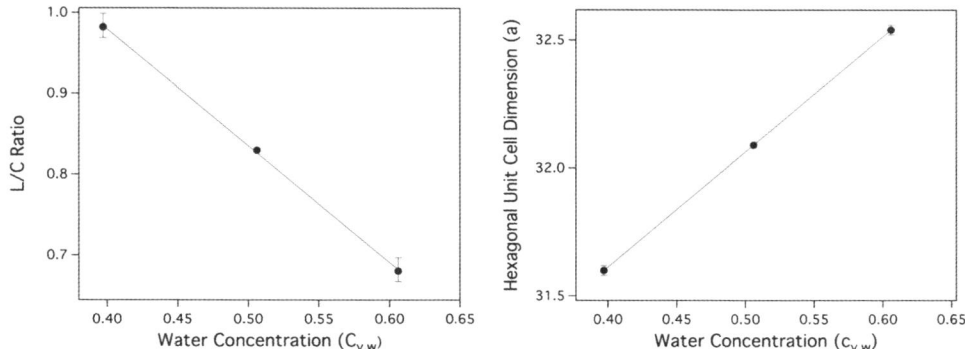

**Figure 7**  *L/C ratio (Left) and Hexagonal unit cell dimension (Right) as a function of water concentration. The parameters of the line fitting the unit cell vs. $c_{v,w}$ are $a = 30.6 \, (1 - c_{v,w})^{-0.07}$.*

*2.4.2 Pressure Effects.* High-pressure can be conveniently used to monitor structural changes and to evaluate interactions on macromolecular structures[17,20]. In the case of G-quadruplexes, relevant information on self-assemblig, stability and preferential hydration mechanism have been obtained[11,17,21]. In particular, pressure has been detected to induce the *H-Ch* phase transition in GMP/Na$_2$ hydrated samples (where tetramer-tetramer interactions are weak), but the *Ch-H* phase transition in GMP/(NH$_4$)$_2$ samples (in which stacking interactions are strong)[11]. The different counter-ion stabilization, which control quadruplex length, was also suggested to drive the pressure-induced water rearrangement inside the *H* cell[11,17,21] (see for example Fig. 8).

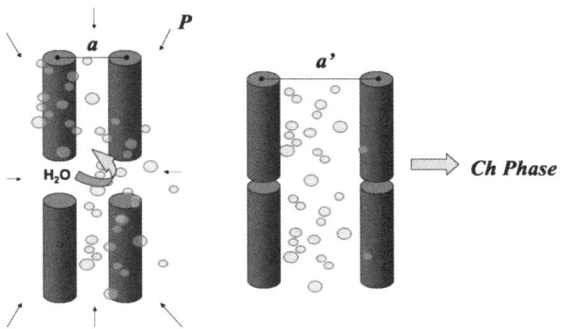

**Figure 8**  *Schematic view of high-pressure effects on G-quadruplexes: in the case of weak stacking interactions, the system has a high inter-axial compressibility but a low lateral compressibility[15].*

In the present case, X-ray diffraction experiments were performed as a function of pressure, from ambient to 1.5 kbar. High-pressure X-ray diffraction profiles are shown in Fig. 9 (top row). At all the investigated conditions, a series of well-defined peaks, which can be indexed considering a 2D hexagonal lattice, was detected; however, by increasing

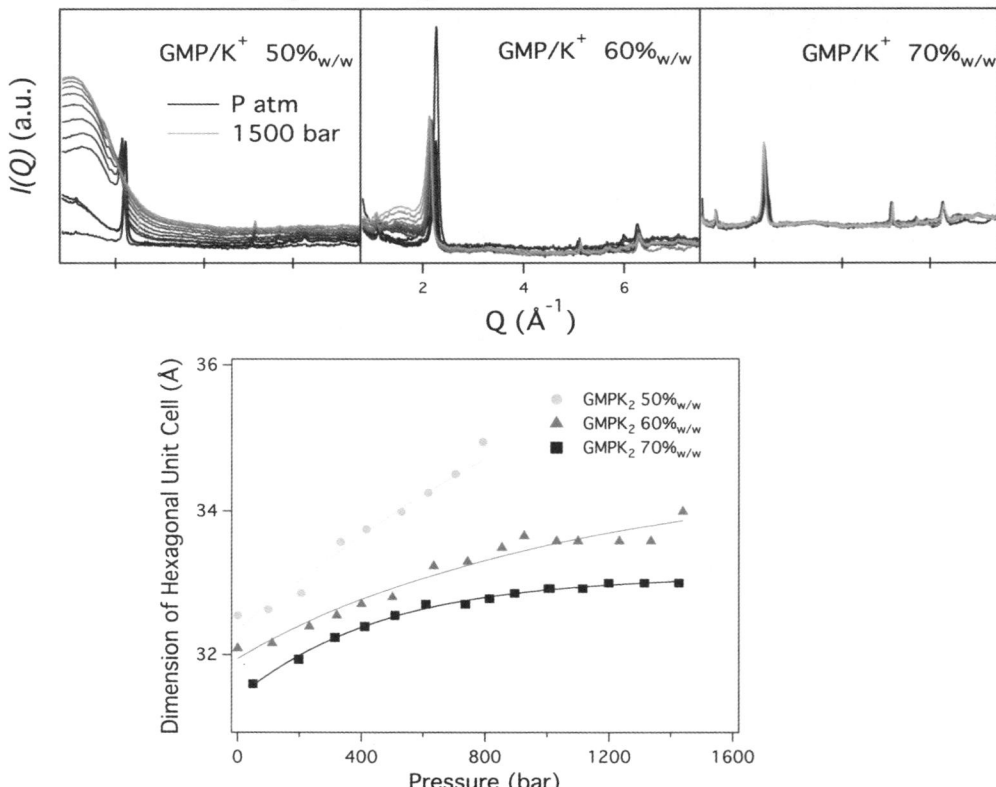

**Figure 9** *X-Ray diffraction profiles obtained as a function of pressure at three different concentrations, as indicated. The bottom graph shows the pressure dependence of the hexagonal unit cell.*

the pressure, a *H-Ch* phase transition was detected in the two less concentrated samples. The transition is complete in the GMP/$K_2$ 50%$_{w/w}$ sample, as shown by the disappearance of the well-defined peak at about 2.2 Å$^{-1}$ and the only presence at pressures higher than 0.8 kbar of the low-angle broad band characteristic of the cholesteric phase. Instead, the phase transition is not complete in the GMP/$K_2$ 60%$_{w/w}$ sample: according to the contemporaneous presence of the broad band and narrow peaks, the system appears to persist in the two-phase region even at 1.5 kbar.

From the peak positions, the unit cell parameter of the *H* phase has been derived (Eq.3), and its pressure dependence is shown in the lower panel of Fig.9. Data are very interesting, because it can be observed that the *H* cell size *increases* under compression (2D lateral compressibility[17] $\beta_{2D} = -1/\sigma \, (d\sigma/dP)$, where $\sigma = a^2(\sqrt{3}/2)$ is the 2D unit cell area, is *negative*). Such an unexpected result has a simple explanation, based on the presence of finite GMP-quadruplexes and on a preferential hydration mechanism (see Fig. 8). Indeed, under compression, axial water moves to the lateral position, because of the strong quadruplex lateral repulsive forces and the net sample volume reduction due to a larger hydration of charged groups and the consequent release of counter-ions. Therefore, the *a* parameter increases with external pressure. Clearly, water rearrangement depends on sample composition and quadruplex length (even when the three GMP/$K_2$ samples at different concentrations present the same unit cell parameter, e.g. $a$=32.5Å observed at ambient pressure, ≈300 bar and ≈500 bar for 50%$_{w/w}$, 60%$_{w/w,}$ and 70%$_{w/w}$ samples, respectively, an increase of the same pressure value does not modify the unit cell parameters in a comparable way). As much obviously, when the $H$ unit cell approaches a critical value, the cholesteric phase forms. According to $L/C$ data previously discussed (Fig. 7), at GMP/$K_2$ 70%$_{w/w}$ concentration the amount of axial water is very low: a very few water redistribution occurs and no transitions are detected in the investigated pressure range.

## 3 CONCLUSION

X-ray scattering experiments have been performed on GMP/$K_2$ samples prepared in water at different compositions and in the presence of increasing amounts of excess KCl. The whole results show that quadruplexes of finite size forms. In diluted solutions, the length of GMP-quadruplexes has been determined, and they appear to be formed by about 10-15 stacked tetramers at GMP/$K_2$ 8%$_{w/w}$. In any case, addition of excess $K^+$ increases the quadruplex stability. The presence of quadruplexes of finite size is also deduced from structural data obtained by X-ray diffraction experiments performed on high concentrated samples. Indeed, structural data for columnar hexagonal and cholesteric phases suggest preferential hydration and pressure-induced water redistribution around finite, and stable, GMP-quadruplexes.

According to previous results[3,11,16,17,21], it can be concluded that ions play a major role in determining the final structure of quadruplexes, and then the properties of the $H$ phase. Stacking stabilization produced by excess ions is the main reason for the larger self-assembling observed at very low GMP/$K_2$ concentration, as well as stacking stabilization drives the preferential hydration mechanism observed in condensed phases. It is evident that these effects might be very relevant in biological systems, where associated cations are strongly implied in the determination of the type of quadruplex eventually formed and on telomeric quadruplex stability[22].

## Acknowledgments

We gratefully acknowledge funding from the Italian Ministry of Education, University and Research (MIUR) through the PRIN project "Self-assembling and structural properties of guanosine derivatives in aqueous solutions" (code 2008F3734A). We also thank M. Steinhart and H. Amenitsch at Elettra and S. Funari at Desy for experimental assistance.

## References

1. M. Gellert, M. N. Lipsett, D. R. Davis, Helix formation by Guanylic Acid, *Proc. Natl. Acad. Sci., USA,* 1962, **48**, 2013.
2. S. Neidle, *Quadruplex Nucleic Acids* (Royal Society of Chemistry, 2006)
3. S. Bonazzi, M. Capobianco, M. M. De Morais, A. Garbesi, G. Gottarelli, P. Mariani, M. G. Ponzi Bossi, G. P. Spada, L. Tondelli, *J. Am. Chem. Soc.,* 1991, **113**, 5809.

4   G. N. Parkinson, *Quadruplex Nucleic Acids* (Royal Society of Chemistry, 2006, pp.1-27).
5   S. Arnott, R. Chandrasekaran, C. M. Marttila, *Biochem. J.* 1974, **141**, 537.
6   S. B. Zimmerman, G.H. Cohen, D. R. Davies, *J. Mol. Biol.*, **1975**, 92, 181.
7   L. Spindler, I. Drevensek-Olenik, M. Copic, J. Cerar, J. Skerjanc, P. Mariani, *Eur. Phys. J. E,* 2004, **13**, 27.
8   P. Mariani, F. Spinozzi, F. Federiconi, H. Amenitsch, L. Spindler, and I. Drevensek-Olenik, *J. Phys. Chem. B,* 2009, **113**, 7934
9   P. Mariani, C. Mazabard, A. Garbesi, G. P. Spada, *J. Am. Chem.Soc.* 1989, **111**, 6369.
10  G. Gottarelli, G. P. Spada, P. Mariani, In *Crystallography of Supramolecular Compounds*; Kluwer Academic: Dordrecht, The Netherlands, 1996.
11  E. J. Baldassarri, M. G. Ortore C. Ferrero, S. Finet, F. Spinozzi, P. Mariani, *International Review of Biophysical Chemistry (I.RE.BI.C.),* 2011, **2**, N.4.
12  K. Pressl, M. Kriechbaum, M. Steinhart, P. Laggner, *Rev. Sci. Instrum.* 1997, **68**, 4588.
13  P. Mariani, F. Spinozzi, F. Federiconi, M. G. Ortore, H. Amenitsch, L. Spindler, I. Drevensek-Olenik, *Journal of Nucleic Acids*, 2010, **2010**, Article ID 472478.
14  A. Guinier, G. Fournet, Small Angle Scattering of X-Ray, John Wiley & Sons, New York, NY, USA, 1955.
15  H. Franz, F. Ciuchi, G. Di Nicola, M. M. De Morais, P. Mariani, *Phys. Rev.,* 1994, **50**, 395.
16  P. Mariani, L. Saturni, *Biophysics J.* 1996, **70**, 2867.
17  P. Ausili, M. Pisani, S. Finet, H. Amenitsch, C. Ferrero, P. Mariani, *J.Phys. Chem.,* 2004, **108**, 1783.
18  L. Q. Amaral, A. Gulik, R. Itri, P. Mariani, *Physical Review A,* 1992, **46**, 3548.
19  R. Hentschke, M. P. Taylor, J. Herzfeld, *Phys. Rev. A* 1989, **40**, 1678; *Phys. Rev. Lett.* 1989, **62**, 800.
20  M. G. Ortore, F. Spinozzi, P. Mariani, A. Paciaroni, L. R. S. Barbosa, H. Amenitsch, M. Steinhart, J. Ollivier, D. Russo, *J. R. Soc. Interface*, 2009, **6**, S61.
21  F. Federiconi, M. Mattioni, E.J. Baldassarri, M. G. Ortore, P. Mariani, *European Biophysics Journal*, 2011, **40**, 1225.
22  A. N. Lane, J. Brad Chaires, R. D. Gray, J. O. Trent, *Nucleic Acids Res.*, 2008, **36**, 5482.

# TEMPERATURE-GRADIENT GEL ELECTROPHORESIS: UNFOLDING OF G-QUADRUPLEXES

Viktor Víglaský, Katarína Tlučková, Petra Tóthová and Ľuboš Bauer

P.J.Šafárik University, Faculty of Sciences, Institute of Chemistry, Department of Biochemistry, Moyzesova 11, 04011 Košice, Slovakia

## 1 INRODUCTION

Many recent works have shown that G-quadruplex is a highly polymorphic molecule.[1] Several experimental techniques have been employed to study G-quadruplexes, e.g. spectroscopy, microcalorimetry, viscometry and electrophoresis, although relevant structural data can only be determined through X-ray crystallography and NMR spectroscopy. However, the crystal structure of a G-quadruplex may not reflect the real arrangement of atoms occurring in cellular condition.[2,3] NMR usually offers valuable structural information under specific conditions. In most studies, the coexistence of multiple conformations has often resulted in spectral overlap, rendering the structural analysis at high resolution very complicated, if not impossible.[4] Nevertheless, CD spectroscopy also gives preliminary information regarding the strands orientation in G-quadruplexes. Parallel and antiparallel arrangements of G-quadruplexes show some characteristic spectral features in the UV range which are used as benchmarks for G-quadruplex folding.[5] Although CD spectroscopy is a valuable and sensitive technique, it is unable to offer complete structural information.[6,7] Again, the spectral overlap of multiple conformers is the main reason for data misinterpretation.
Another routine method used in many laboratories is electrophoresis which provides information about the electrophoretic mobility of macromolecules. This parameter is strongly dependent on the molecular mass of nucleic acid, conformation and molecularity.[8] Therefore, a combination of CD spectroscopy and electrophoresis is often valuable in providing a low-cost and "rough" evaluation of G-quadruplex structure and thermodynamic properties.
The mechanism of the G- quadruplex melting process can be deduced based on the shape of the melting curves obtained by spectral methods. However, this method shares the same problems as the techniques discussed above, namely that it is practically impossible to bypass the overlapping signals of the coexisting structures. Additionally, microcalorimetry is often unable to distinguish between two thermodynamically similar states of one molecule. In some cases a special type of electrophoresis, temperature gradient gel electrophoresis (TGGE), may be able to give useful information towards overcoming this problem.[9,10]

In this study, we focus on a description of the general aspects of G-quadruplexes and data obtained with CD spectroscopy; we then proceed to a description of electrophoretic separation using TGGE in which at least two different conformers coexist.

## 2 METHOD AND RESULTS

### 2.1 Polyacrylamide gel Electrophoresis (PAGE)

DNA oligomers were separated in 15% gel; the electrophoretic mobility of folded G-quadruplex is typically different to that of unfolded DNA sequences. Electrophoresis was run in a temperature-controlled vertical electrophoretic apparatus. ~2 µg of DNA were loaded into the well. Electrophoreses were run at 10°C for 4 hours at ~8 V.cm$^{-1}$. DNA oligomers were visualized with Stains-all following electrophoresis, and the electrophoretic record was photographed with a digital camera. More than a hundred different sequences forming G-quadruplexes were evaluated using the same methods, but in this study four oligomers derived from natural sequences are used as representative samples;
i) human promoter oncogenic sequence of *ret* (GGGG C GGGG C GGGG C GGG),[11,12]
ii) *kras* (AGGG CGGTGT GGG AAGA GGG AAGA GGGGG A GG),[13]
iii) *hif-1α* (GCG A GGG C GGGGG AGA GGGG A GGGG C GCGC GGG),[14] and
iv) thrombin binding aptamer, *TBA* (GG TT GG TGT GG TT GG ).[8]
The fact that oncogenic sequences form at least two variously folded and coexisting conformers under a given condition was confirmed by PAGE (Figure 1a). In addition, MS spectrometry confirms the coexistence of intramolecular G-quadruplexes.[15] *TBA* forms antiparallel G-quadruplex consisting of two-quartets; it unfolds through a two-state mechanism in the presence of K$^+$.[8,16] Other oligomers represent a wide scale of sequences which can form coexisting conformers, and/or intermediates.

### 2.2 Circular dichroism (CD)

Both short and long molecules in a wide range of flexible experimental conditions can be studied with this method. The formation of G-quadruplexes was verified by CD spectrometry under the same conditions as were used in PAGE and TGGE.
The positive signal at 265 and 295 nm corresponds to parallel and antiparallel folds respectively.[9,16] CD spectra of oligonucleotides were recorded on a JASCO J-810 spectropolarimeter equipped with a PTC-423L temperature controller using a quartz cell of 1 mm optical path length in a reaction volume of 300 µl and an instrument scanning speed of 100 nm/min, 1 nm pitch and 1 nm bandwidth, with a response time of 2 s, over a wavelength range of 220–320 nm. All the CD spectra are baseline-corrected for signal contributions caused by the buffer. Three averaged CD scans were taken at a temperature range of (20–95°C). The experimental conditions in CD and PAGE were as follows: strand concentration, 10-20 µM; temperature, 0-95°C; 2.5 or 50 mM KCl; 25 mM modified Britton-Robinson buffer, 25 mM phosphoric acid, 25 mM boric acid, 25 mM acetic acid, and supplemented by appropriate concentration of KCl; pH was adjusted by TRIS to a final value of 7.0.[9,10] In all experimental measurements evaluating a thermal stability, 2.5 mM KCl was used for *hif-1α* and *ret* sequences due to the very high thermal stability of G-quadruplexes formed from these sequences. At higher concentration of K$^+$ the melting

transitions are beyond the range of TGGE, >70 °C. The promoter sequence *kras* and *TBA* allow the use of 50 mM KCl.

## 2.3 TGGE principle

Temperature Gradient Gel Electrophoresis (TGGE) is a form of electrophoresis in which the temperature gradient is used to unfold molecules as they move through the gel matrix. TGGE can be applied in the analysis of DNA and RNA G-quadruplexes. In contrast to conventional PAGE, the separation of folded DNA molecules during TGGE relies on temperature-dependent melting of the G-quadruplex into single stranded DNA. TGGE can be performed in either a parallel or a perpendicular mode depending on the orientation of electric intensity and temperature gradient. The former is used to analyze multiple samples in the same gel, whereas the later provides the melting profile analysis of a single sample.
A description of the TGGE equipment used has been published previously.[17] The main limitation of TGGE is the range of temperature gradient and the use of low-ionic buffers because of the electric conductivity. The linear temperature gradient is uniform within the range of 20 to 75°C throughout the gel.
A temperature increase may lead to melting of G-quadruplexes which would either slow down or speed up their migration through the gel. When a molecule increases its volume without any change of surface charge then its mobility decreases, and vice versa. For example, intermolecular G-quadruplex, dimer, trimer, tetrad, etc., move more slowly than intramolecular monomeric conformers after denaturation induced by a temperature increase, and vice versa. In contrast, the mobility of unimolecular G-quadruplex is higher at lower temperatures and decreases during the unfolding process as was observed in studies of human telomeric repeats $d(G_3T_2A)_3G_3$.[10,11] However, unfolding is often accompanied by a change in the surface charge of a molecule, which in turn significantly influences migration speed.
Herein, we focus on the evaluation of G-quadruplexes forming DNA sequences using polyacrylamide TGGE in the perpendicular mode, and interpret the results in relation to CD measurements.

## 2.4 CD and PAGE results

*Hif-1α*, and *kras* form at least 2 different structures under the given conditions. Oncogenic promoter sequences show a parallel fold; the positive peak is detected at 265 nm. The faster band of *ret* sequence moves a little more slowly than the molecular standard $(d(AC)_9)$, therefore this band does not correspond to the unfolded form of DNA. This suggestion is also supported by the TGGE results (Figure 2). An increase of $K^+$ concentration increases the signal at 265 nm in CD measurements and causes this band to diminish at PAGE (marked by "x"). The most intensive band represents the main population of intramolecular G-quadruplexes. Sequences *hif-1α* and *kras* form at least two different conformers. An increase in $K^+$ concentration results in a disappearance of some bands; their positions are marked by symbol "x". On that account, it is suggested that these sequences may fold into both intra- and intermolecular G-quadruplexes. *TBA* was used as a contrary benchmark; only one band representing antiparallel conformer was observed and its mobility does not correspond to that of an unfolded oligomer.

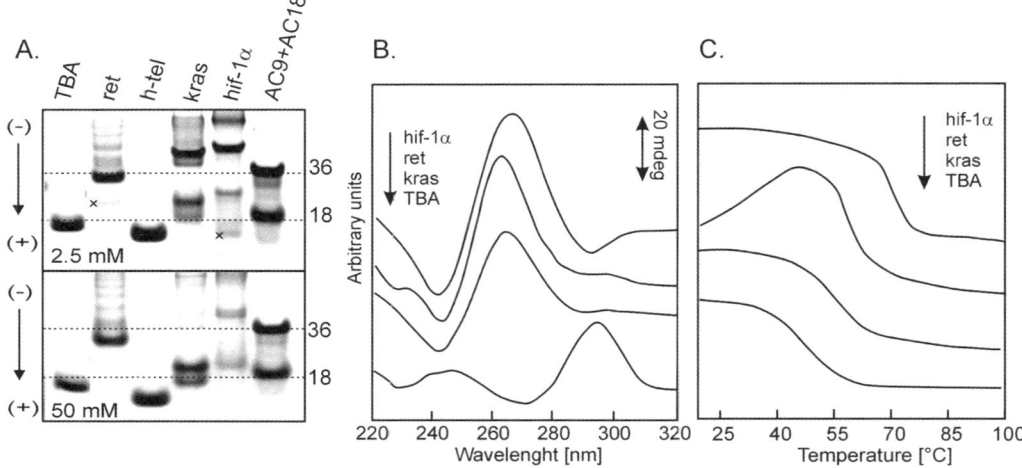

**Figure 1** (A) PAGE of TBA, ret, h-tel, kras and hif-1α sequences in modified Britton-Robinson buffer in presence of 2.5 mM KCl (up) and 50mM KCl (down). The h-tel is human telomeric repeat used as control: d(G$_3$T$_2$A)$_3$G$_3$.[9,10] Molecular standards, d(AC)9 and d(AC)18, were loaded to the right. (B) CD spectra: Hif-1α and ret were dissolved in Britton-Robinson buffer containing 2.5 mM KCl. The solution of TBA and kras contains 50 mM KCl. Measurements were performed at 15°C. Only TBA shows an antiparallel fold; the positive peak is detected at 295 nm. Other sequences form parallel G-quadruplexes. (C) CD melting curves were collected at 265 nm for all oncogenic sequences and 295 nm for TBA at the same condition as in B. The bands marked by "x" disappear at 50 mM KCl at room temperature.

## 2.5 CD and TGGE melting curves

The CD melting profiles were collected at 265 and 293 nm. The thermal stability of different quadruplexes was also measured by recording the UV absorbance at 293 nm (not shown) and the CD ellipticity at 265/293 nm as a function of temperature, using a method similar to that previously published.[10] The temperature ranged from 5 to 100 °C, the heating rate was 0.25°C per minute. The melting temperature ($T_m$) was defined as the temperature of the mid-transition point. $T_m$ was estimated from the peak value of the first derivative of the fitted curve. With the exception of the melting curve of *TBA* sequence, the other three sequences are not governed by two-state mechanism, and the shape of melting curves is significantly different from that of a sigmoid curve (Figure 1c). Sometimes a linear slope in UV and CD melting spectra in the pre- and post-melting regimes may arise from an intrinsic physical phenomenon, such as the intrinsic temperature dependence of absorbance changes resulting from solvent expansion.[1] However, the temperature-dependent inter-conversion of one conformer/intermediate to another, which are collected at one fixed wavelength, has a more significant effect on the slopes of melting curves.[10] The profiles of CD melting curves of two coexisting, topologically different conformers are different, e.g. parallel and antiparallel conformers. This means that the melting curves obtained at 265 nm and 295 nm do not reflect the melting transition of antiparallel and parallel conformers, respectively. In cases in which topologically different conformers coexist only in parallel configurations, the melting curve is unable to give detailed information about the coexistence of more conformers. The non linear slope of *ret*, an increase of CD signal at 265 nm, prior to unfolding is caused by the inter-conversion of partially unfolded or antiparallel conformers to the parallel ones. The same property during unfolding of $(G_4T_2)_3G_4$ oligomer was observed by CD.[10] This effect was not observed at $K^+$ concentrations in excess of 40 mM.

TGGE is unable to provide any information about the topology of G-quadruplexes, but the electrophoretic record obtained through TGGE offers unambiguous spectral data interpretation (Figure 2).

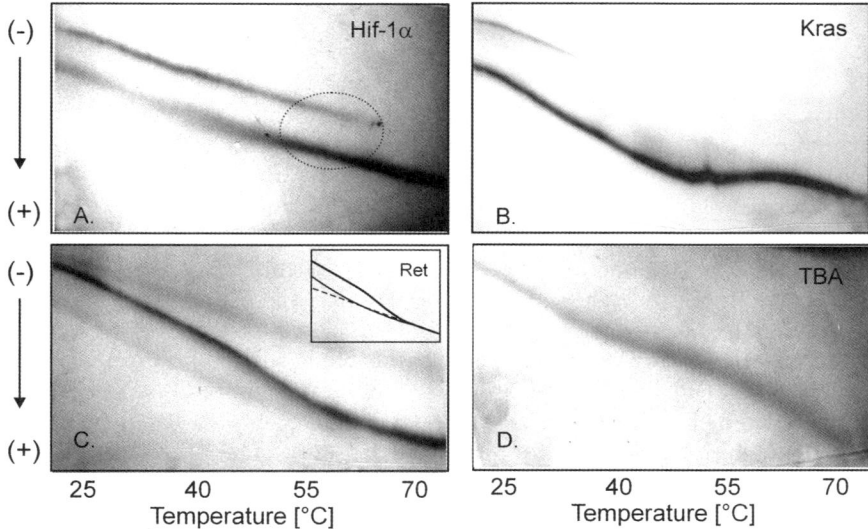

**Figure 2** *TGGE records of DNA oligomers in Britton-Robinsson buffer in the presence of KCl: hif-1α and ret at 2.5 mM KCl (A and C); kras and TBA at 50 mM KCl, (B and D). The circle in (A) highlights a discontinuous transition. Kras sequence also shows a similar effect, but this transition is observed at lower temperatures. Inset in panel (C) represents mobility of the main fraction and the fastest fraction of the ret oligomer. Here, we demonstrate that the position of the post-melting linear prolongation of unfolded oligomer disagrees with the fastest electrophoretic curve. In addition, there is an explanation why the mobility of the fastest band of the ret oligomer in Figure 1a does not correspond to the intersection with y-axis at 20°C which reaches molecular standard $d(AC)_9$.*

TGGE of *TBA* confirms the fact that this oligomer only forms conformers, the melting process of which is governed by a two-state mechanism. However, TGGE of *hif-1α* at 2.5 mM KCl shows two discontinuous bands. The faster band becomes more intensive and, on the contrary, the slower band clearly disappears with increasing temperature. When the kinetic of transition from one state to another is very slow in comparison to the time of electrophoretic migration, a discontinuous band can be observed.[17] The location of the structural transition in the electrophoretic record is marked with a circle. The same effect was also observed for the *kras* sequence, but another continuous transition is observed at higher temperatures, ~58 °C. Generally, the decrease of mobility during the unfolding process usually determines the unfolding of intramolecular G-quadruplexes, whereas the discontinuous transition represents the melting of intermolecular G-quadruplexes; e.g. *hif-1α* and *kras*. The most intensive fraction of *ret* oligomer shows a non linear and continuous increase of mobility with an increase in temperature. The TGGE melting profile confirms that *ret* occurs in intermolecular parallel dimeric form. The fastest band, which is very fine in appearance, represents a small population of intermolecular antiparallel conformer which could be the source of the slight CD peak at 293 nm because this fine band and CD peak at 293 nm (Figure 1b) totally disappears in presence of 50 mM KCl. The slope in the pre-melting trend of CD melting curve increases (Figure 1c) and the peak at 265 nm increases with increasing temperature (Figure 1b), and we therefore suggest that misfolded oligomers convert to a correctly folded parallel G-quadruplex.

## 3 CONCLUSION

The presence of several repeats containing guanines, G-runs, in DNA molecules can favor the formation of topologically various scales of G-quadruplex structures. These structures can coexist under given conditions. The propensity of G-rich sequences to form coexisting structures leads to many occasions when there is coexistence of several inter-converting structures in solution.[4] The conformational plasticity of G-quadruplexes depends on the environment (ions, pH, temperature, etc.) and on the sequence of the particular DNA molecule. For example, a higher concentration of $K^+$ and $Na^+$ ions can result in one preference among many occurring conformers. Here, on selected oligomers, we demonstrate why the spectral evaluation may be unable to offer full information regarding the types of topologically different conformers formed in solution. It is important to note again that spectral measurements of any type, including microcalorimetry, do not always allow us to distinguish between conformers occurring in solution. The separation power of electrophoresis can make a significant contribution towards the correct interpretation of other data and gives a more accurate view of the melting mechanism of a mixture of conformers presented in solution. Both the type of G-quadruplexes and their mutual population among coexisting conformers is dependent on many factors. Small changes in molar ratio between thermodynamically different conformers can lead to inaccurate determinations of melting transitions and other thermodynamic parameters when other methods of measurement are used. TGGE is therefore a great improvement on existing methods, as it can be used to examine both separation and the mobility of molecules in large numbers at any temperature. TGGE is a powerful method which helps to solve the problem of quadruplex polymorphism, because electrophoresis allows us to distinguish between the different conformers and to evaluate the most abundant conformers. This means that for a mixed population of conformers in any sequence, TGGE allows an objective detection of melting profiles over a temperature gradient.[10] The TGGE results clearly demonstrate that CD spectroscopy results are a convolution of several conformers of G-quadruplexes. Interestingly, intramolecular quadruplexes after denaturation move more slowly due to strand unfolding, but intermolecular quadruplexes dissociate to single molecules which usually move faster than the folded structures (Figure 2).

The selected results clearly demonstrate that sequence and ambient conditions are some of the determining factors in the formation of coexisting G-quadruplexes. The TGGE results elucidate and confirm many of the crucial suggestions described above.

**Acknowledgements**

This study was supported by grants from the Slovak Grant Agency (1/0153/09 and 1/0504/12), Slovak Research and Development Agency (APVV-0280-11) and European Cooperation in Science and Technology (COST MP802).

## References

1. A.N. Lane, J.B. Chaires, R.D. Gray and J.O. Trent, *Nucleic Acids Res.*, 2008, **36**, 5482.
2. G.N. Parkinson, M.P. Lee and S. Neidle, *Nature*, 2002, ***417***, 876.
3. J. Li, J.J. Correia, L. Wang, J.O. Trent and J.B. Chaires, *Nucleic Acids Res.*, 2005, ***33***, 4649.
4. A.T. Phan, Y.S. Modi and D.J. Patel, *J. Mol. Biol.*, 2004, ***338***, *93*.
5. J.Kypr, I. Kejnovská, D. Renciuk and M. Vorlícková, *Nucleic Acids Res.*, 2009, **37**, 1713.
6. S. Masiero, R. Trotta, S. Pieraccini, S. De Tito, R. Perone, A. Randazzo and G.P. Spada, *Org. Biomol. Chem.*, 2010, **8**, 2683.
7. A. Virgilio, V. Esposito, G. Citarella, A. Pepe, L. Mayol and A. Galeone, *Nucleic Acids Res.*, 2012, **40**, 461.
8. M. Fialová, J. Kypr and M. Vorlícková, *Biochem. Biophys. Res. Commun.*, 2006, **344**, 50.
9. L. Bauer, K. Tlučková, P. Tóhová and V. Viglaský, *Biochemistry*, 2011, **50**, 7484.
10. V. Víglaský, L. Bauer and K. Tlucková, *Biochemistry*, 2010, **49**, 2110.
11. K. Guo, A. Pourpak, K. Beetz-Rogers, V. Gokhale, D. Sun and L.H. Hurley, *J. Am. Chem. Soc.*, 2007, **129**, 10220.
12. X. Tong, W. Lan, X. Zhang, H. Wu, M. Liu and C. Cao, *Nucleic Acids Res.*, 2011, **39**, 6753.
13. S. Cogoi, M. Paramasivam, B. Spolaore and L.E. Xodo, *Nucleic Acids Res.*, 2008, **36**, 3765.
14. R. De Armond, S. Wood, D. Sun, L.H. Hurley and S.W. Ebbinghaus SW, *Biochemistry*, 2005, **44**, 16341.
15. V. Viglasky, K. Tluckova and V. Gabelica, unpublished results.
16. C.M. Olsen, H.T. Lee and L.A. Marky, *J. Phys. Chem B.*, 2009, **113**, 2587.
17. V. Víglaský, M. Antalík, J. Bagel'ová, Z. Tomori and D. Podhradský, 2000, *Electrophoresis*, **21**, 850.

# SPECIFIC BEHAVIOUR OF GUANOSINE IN LIPONUCLEOSIDE THIN FILMS

L. Čoga[1], M. Devetak[2,3], S. Masiero[4], G.P. Spada[4], I. Drevenšek-Olenik*[1,2]

[1] Faculty of Mathematics and Physics, University of Ljubljana, Jadranska 19, 1000 Ljubljana, Slovenia. Email: irena.drevensek@ijs.si
[2] Department of Complex Matter, J. Stefan Institute, Jamova 39, 1000 Ljubljana, Slovenia
[3] Center of Excellence for Polymer Materials and Technologies, Tehnološki park 24, SI 1000 Ljubljana, Slovenia
[4] Dipartimento di Chimica Organica "A. Mangini", Alma Mater Studiorum - Università di Bologna, via San Giacomo 11, 40126 Bologna, Italy

## 1 INTRODUCTION

Practically all kinds of nanoscale systems vitally depend on surface and interface properties of materials. Therefore, development of new methods for controllable construction of surface architectures is a challenging goal lying in the heart of the nanotechnology progress. Highly ordered thin layers of organic molecules on solid substrates can be fabricated using Langmuir-Blodgett (LB) technique,[1] which is receiving increased attention as a procedure for fabrication of molecular electronic devices.[2] In this chapter we describe investigations on the suitability of the LB technique for fabrication of liponucleoside thin films. We show that films fabricated from guanosine (G) derivatives exhibit very peculiar properties with respect to films fabricated from derivatives of other nucleosides.

The compounds investigated in our study were lipophilic derivatives of nucleosides (or deoxynucleosides) with one, two or three hydrocarbon tails. G-derivatives of this type exhibit a tendency to assemble into supramolecular nanoribbons. This formation has been observed in organic solutions, in the solid state, and also in thin surface films.[3,4] Surface films with ribbon structure possess interesting charge transfer properties, such as optical rectification and photoconductivity, which makes them very promising for applications in nano-electronic devices.[5-8] First thin films of lipophilic G-derivatives were fabricated by drop-casting procedure and consequently their two-dimensional structure was not well controlled. To improve the control of surface organization, we switched to films fabricated by the LB technique.[9]

Despite the fact that investigations of Langmuir and LB films of nucleolipids have significantly expanded during the last two decades,[10,11] surprisingly few studies of molecules containing guanine in their head group were reported.[12,13] The reason for "avoiding" guanine derivatives comes from complications related to their synthesis, as they are known as materials with "difficult" behavior.[12]

## 2 MATERIALS AND METHODS

### 2.1 Synthesis of liponucleosides

The compounds investigated are derivatives of deoxynucleosides with two 10-carbon long alkanoyl tails and derivatives of nucleosides with one or three 10-carbon long alkanoyl tails (Figure 1). All alcoholic functions have been derivatised. In the case of the derivatives with one tail, both secondary alcoholic functions in 2' and 3' positions are blocked with isopropilidene protecting group. The synthesis of guanosine and deoxyguanosine derivatives has been described elsewhere.[9,14] The synthesis of other liponucleosides has been carried out in a similar fashion, by direct esterification of the (deoxy)ribose alcoholic functions with decanoic anhydride in MeCN (Figure 2).

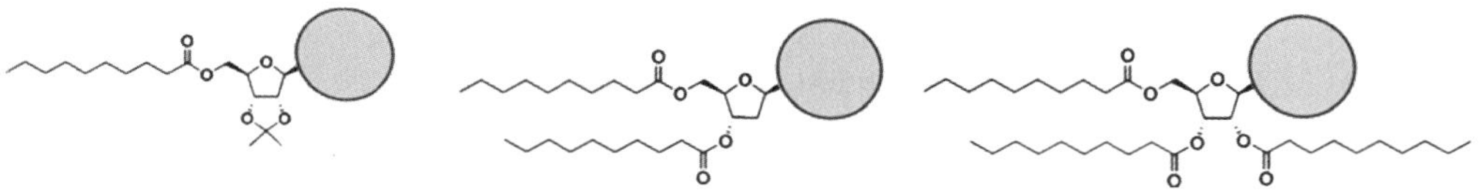

**Figure 1** *Schematic structure of the derivatives used in our study. Circles denote nucleobases.*

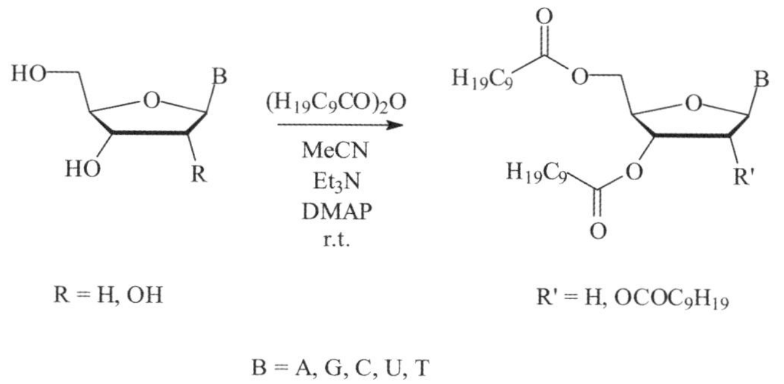

**Figure 2** *The synthetic route for lipophilic nucleoside diesters/triesters.*

### 2.2 Preparation and characterization of Langmuir and LB films

Monolayer surface films (Langmuir films) of nucleoside derivatives deposited at air-water interface were prepared from chloroform solutions with concentration of $10^{-3}$ M. The surface area of the film was continuously varied by moving two teflon barriers of the LB trough (KSV NIMA). Surface pressure $\pi = \gamma_0 - \gamma$, where $\gamma_0$ is surface tension of pure water and $\gamma$ is surface tension of water covered with adsorbed layer, was measured in the centre of the trough by the Wilhelmy plate-method.[1] Schematic illustration of surface pressure versus molecular area isotherms $\pi(A)$ for typical surface active molecules, such as fatty acids, is shown in Figure 3.[15] The region of large surface areas corresponds to the liquid expanded (LE) phase, while the region of small surface areas corresponds to the condensed phase (C). A plateau of two-phase coexistence can be noticed at some intermediate temperatures. The regions of the gas phase and its coexistence with the LE phase, which appear at very large surface areas and small surface pressures, are not shown because they were not investigated in our experiments. All our measurements started in the coexistence region of the gas and the LE phases.

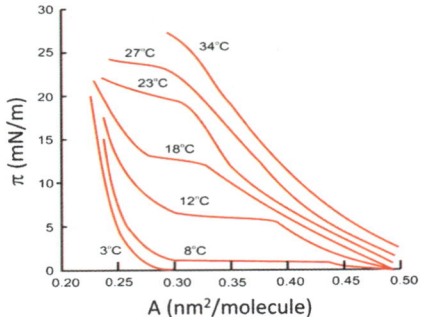

**Figure 3**  *Schematic illustration of typical $\pi(A)$ isotherms at different temperatures.*

In our investigations freshly prepared UV sterilized and distilled water (Millipore) with resistivity of 18 MΩ and temperature of 23°C was used as a subphase. The desired amount of chloroform solution was deposited onto the water subphase and let to relax for 30 min in non-compressed state to allow solvent evaporation and establishment of an internal equilibrium. Then $\pi(A)$ isotherms were recorded by compressing a monolayer with a barrier speed of 5 mm/min. After compression, the film was expanded to the initial surface area with the same speed of the barriers as during compression. The compression/expansion cycles were repeated several times.

Brewster Angle Microscopy (BAM) images of Langmuir films were acquired with a CCD camera mounted on an imaging ellipsometer (Accurion, nanofilm_ep3se) adapted to the Langmuir trough. The images were taken in "real-time" during the compression process at different points of the $\pi(A)$ isotherm.

For the preparation of monolayer LB films, freshly prepared water was placed in the bath of the LB trough. Then a freshly cleaved plate of muscovite mica was dipped into the water by a motorized vertical dipper. After this the solution of a selected liponucleoside was spread on the surface and the non-compressed surface film was allowed to relax for 30 minutes for all chloroform to evaporate. Then the film was compressed to the target pressure and again allowed to relax for another 30 minutes to reach equilibrium. After this the mica plate was lifted from the bath at the rate of 3 mm/min (Figure 4). The adsorbed Z-type films were then dried in air for several hours.

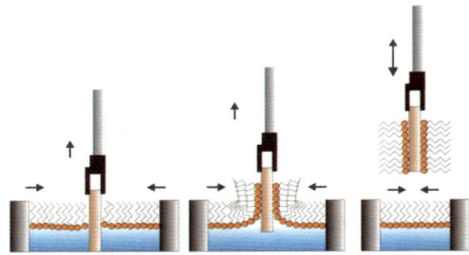

**Figure 4**  *Deposition of Z-type monolayers from the air-water interface to a mica plate.*

Different surface regions of deposited LB films were scanned using Atomic Force Microscope (AFM) (Nanoscope Dimension DI3100 with Nanoscope IV controller from Veeco Instruments). Areas of 10 μm x 10 μm were imaged using soft tapping scan mode. The measurements were carried out with silicon cantilevers having spring constants of 12-100 N/m, thickness of 4.6 μm, and maximum tip radius of 7-12 nm. Vertical set point was set at 7 μm. The vertical precision was 0.1 nm. The scan rate was 1 line/s. The captured images were analysed with Digital Instruments Nanoscope software. Height diagrams and vertical cross-sections were evaluated.

## 3 MOLECULAR LAYERS AT THE AIR-WATER INTERFACE

### 3.1 Two-dimensional phases and phase transitions

Figure 5 shows the $\pi(A)$ isotherms of the G-derivatives measured during the first three expansion-compression cycles. All derivatives display a first order phase transition from the liquid expanded to the condensed phase.[16] The plateau region of phase coexistence starts with a noticeable kink. A subsequent expansion of the films reveals a considerable hysteresis of the $\pi(A)$ isotherms with a prompt drop of surface pressure in the beginning of the expansion process. This indicates formation of a collapsed structure with slow and irreversible dissociation mechanism.[10,17] The subsequent compression-expansion cycles the $\pi(A)$ isotherms exhibits significant shifts towards smaller surface areas. This phenomenon is most profound for the derivative with one tail and signifies formation of loose structures, which are diffusing into the subphase.[10]

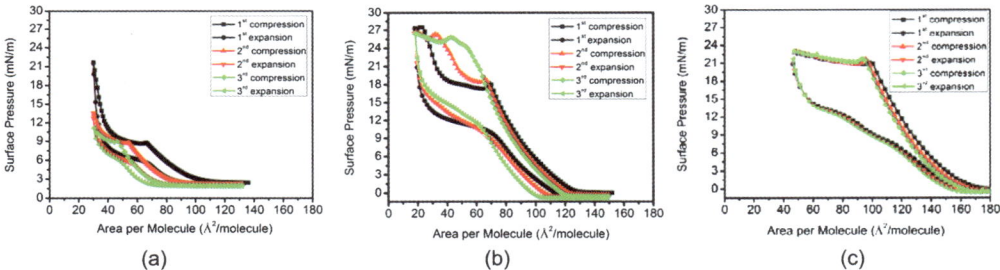

**Figure 5** *Surface pressure versus area isotherms of guanosine derivatives with (a) one, (b) two, and (c) three hydrocarbon tails.*

Figure 6 displays BAM images of the surface layer of G-derivative with two tails captured during the first compression. Surface area of about 400 μm x 300 μm in size is shown. Figures 6(a) and 6(b) were taken in the region of the liquid expanded phase and Figure 6(c) in the region of two phase coexistence. The islands of condensed phase are clearly visible in Figure 6(c). With further compression the island become more and more dense and merge together. The observed behaviour is characteristic for first order phase transitions.[18]

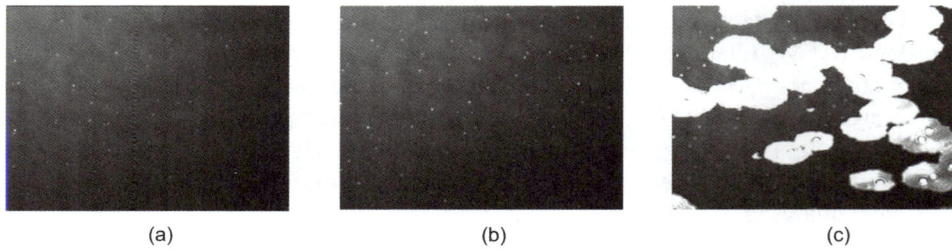

**Figure 6** *BAM images of Langmuir film of G-derivative with two tails taken at: (a) $\pi=10$ mN/m and $A=85$ Å$^2$, (b) $\pi=15$ mN/m and $A=75$ Å$^2$, (c) $\pi=17$ mN/m, $A=65$ Å$^2$.*

Figure 7 shows $\pi(A)$ isotherms of adenosine derivatives measured during the first three expansion-compression cycles. The surface pressure is significantly higher than for G-derivatives and monotonously increases with decreasing area per molecule. The plateau region emerges smoothly and continuously. The observed behaviour is very similar to the behaviour reported for the dioleoylphosphatidyl-adenosine exhibiting considerably longer hydrocarbon tails.[19]

In contrast to the G-derivatives that show very different forms of the $\pi(A)$ curves during compression and expansion processes, the adenosine derivatives behave practically the same during compression and expansion. However, during expansion the $\pi(A)$ isotherms are significantly shifted towards lower areas per molecule. The shifts appear also in consecutive compression-expansion cycles.

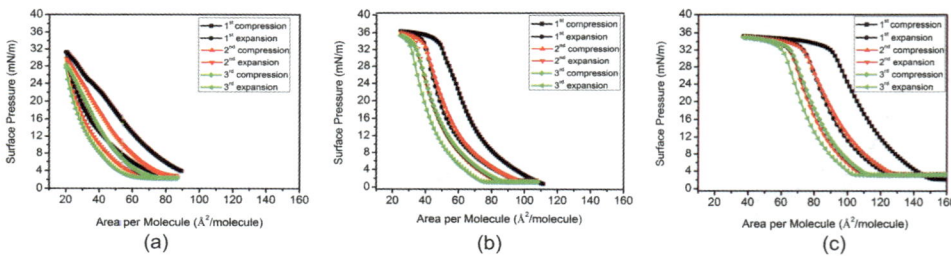

**Figure 7** *Surface pressure versus area isotherms of adenosine derivatives with (a) one, (b) two, and (c) three hydrocarbon tails.*

The BAM images of the film formed by the adenosine derivative with two tails are shown in Figure 8. The film exhibits a similar homogeneous morphology at all points of the $\pi(A)$ isotherm, also in the plateau region. This signifies a continuous 2$^{nd}$ order phase transition.

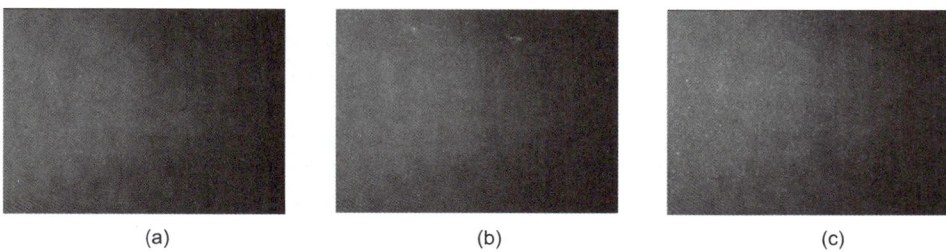

**Figure 8**  BAM images of Langmuir film of adenosine derivative with two tails at: (a) $\pi=10$ mN/m and $A=70$ Å$^2$, (b) $\pi=15.5$ mN/m and $A=60$ Å$^2$, (c) $\pi =36$ mN/m, $A=40$ Å$^2$.

## 3.2 Comparison between different nucleobases

Figure 9 shows $\pi(A)$ isotherms for a complete series of liponucleosides with 2 and 3 tails measured during the first compression-expansion cycle. One can immediately notice that the G-derivatives exhibit lower surface pressure and different shape of isotherms than analogous derivatives with adenosine (A), cytidine (C) and thymidine (T) or uracil (U). The behaviour of G-derivatives apparently corresponds to lower temperature than the behaviour of other nucleosides (see Figure 3), which indicates better stability of their surface assembly.

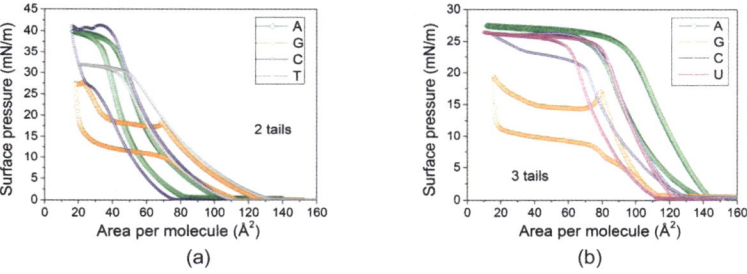

**Figure 9**  Surface pressure versus area isotherms of nucleoside derivatives with (a) two and (b) three hydrocarbon tails. A, G, C, T and U denote nucleobases.

The question that naturally arises is, whether the observed differences originate from Hoogsteen base pairing typical for guanosine assemblies in nanoribbon structures, or whether they are related to some other specific property of the guanine base. The surface areas of flat lying purine and pyrimidine nucleobases are around 45 Å$^2$ and 35 Å$^2$, respectively [13, 20], while the starting of the plateau region of the first order phase transition in surface layers of G-derivatives is observed at much larger areas per molecule (65 Å$^2$, 80 Å$^2$). Consequently, it cannot be attributed to the Hoogsteen interaction between the headgroups, but to the assembling properties of the hydrocarbon tails.

For phospholipid materials it is known that derivatives with hydrocarbon tails that are only 10-carbon atoms long are on the lower limit of their capability to form stable Langmuir films at room temperature,[21] so a relatively good stability of the films investigated in our study should originate from the hydrophilic/hydrophobic properties of the nucleobase headgroups. The hydrophilic strength of a headgroup depends on its water solubility and polarity. Adenine, thymine and uracil possess a multipolar-type of charge distribution with molecular dipole moment of about 3 D, while guanine and cytosine exhibit a profoundly dipolar-type of charge arrangement with molecular dipole moment of about 6 D.[22] These differences can be the reason for the liquid-like nature of Langmuir films of adenosine, thymidine and uridine derivatives and a relatively better stability of the films of cytidine and especially of guanosine derivatives. In addition to this, guanine and its nucleosides are special because their solubility in water is more than one order of magnitude lower than the solubility of other nucleobases and their nucleosides.[23,24] The low solubility of guanine head group provides a good balance to the low hydrophobicity of the decanoyl tails, which is to our opinion the main reason why guanosine derivatives are capable of forming stable surface layers, while other derivatives exhibit a tendency to diffuse into the subphase. The unique behaviour of guanosine derivatives was reported also by Rädler et al. who investigated Lagmuir films of a full series of monoalkylated nucleotides.[12]

## 4 MOLECULAR LAYERS TRASFERED ON MICA SURFACE

Figure 10 shows AFM images of the LB monolayers of G-derivatives transferred onto freshly cleaved mica substrates. The images were recorded at least 24 h after the LB deposition. Upper parts of the images show height profiles of surface areas of 10 µm x 10 µm. Lower parts show vertical cross-sections along the selected lines denoted in the upper parts. Vertical differences between three sets of markers are also given.

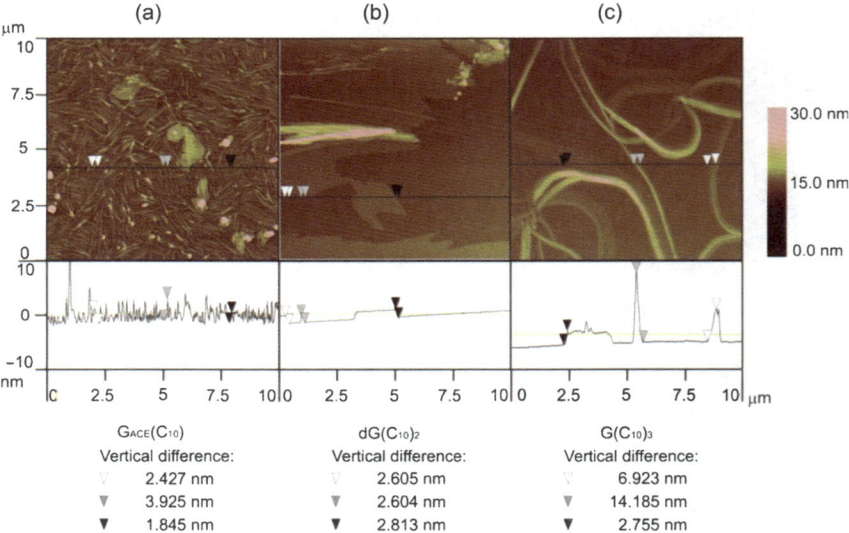

**Figure 10** *AFM images of LB films of G-derivatives with (a) one (b) two and (c) three hydrocarbon tails deposited on mica surface.*

Profound differences between surface structures of different derivatives can be immediately noticed. The derivative with one alkanoyl tail shows a ribbon-like structure. The derivative with two tails shows flat terraces of micrometer size with a height of around 2.7 µm. This is much larger than the length of the decanoic acid ($L_t \sim 1.3$ nm) and also significantly larger than the maximum dimension of the entire liponucleoside molecule ($L \sim 2.2$ nm),[9, 20] so, very probably molecular bilayer is formed. The derivative with three tails shows irregular thread-like assemblies with heights of several micrometers. We attribute large differences between the surface morphologies of the transferred films to surface interaction between the hydrophilic mica surface and the deposited monolayer.[25] Due to this interaction similar behaviour of the $\pi(A)$ isotherms (Figure 5) does not necessarily mean also similar quality of the LB films. These findings are in agreement with the results of recent work of Jin at al., who studied lipid derivatives of acyclovir with one and two 18-carbon long hydrocarbon tails and reported that double-chained derivative formed rigid surface layers, while single-chained derivative formed loose structures.[26,27]

From the point of view of possible applications in nanotechnology, the G-derivative with two tails seems to be most promising. It forms relatively homogeneous LB films with well-defined thickness, that can possibly be tuned to a desired value via the multilayer LB deposition process. The ribbon-like structures formed by the derivative with one tail are on the other hand interesting for applications requiring wire-like surface architectures, such as various interconnects and orientation-sensitive elements.

The LB films of adenosine derivatives, which AFM images are shown in Figure 11, exhibit a droplet-like morphology with droplets heights in the range of tens of micrometers. In between the droplets some patches of a very thin surface film can be noticed. This type of surface structures is not very suitable for applications.

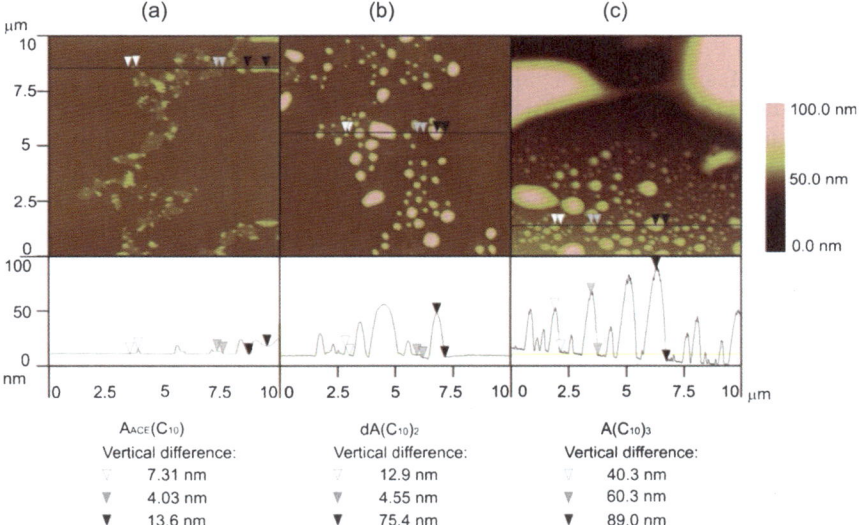

**Figure 11** *AFM images of LB films of adenosine derivatives with (a) one (b) two and (c) three hydrocarbon tails deposited on mica surface.*

The LB films of cytidine, thymidine and uridine derivatives still need to be systematically investigated. However, because of their similarities with the adenosine films revealed at the air-water interface, we assume that LB films of these liponucleosides will exhibit a behaviour very similar to the LB films of the adenosine derivatives. The main efforts to construct liponucleoside surface architectures with the use of LB technique should hence focus on guanosine derivatives.

## 5 CONCLUSIONS

A comparative study of surface assembly of liponucleoside derivatives containing different nucleobases was performed. Thin surface films at air-water interface and LB films deposited on mica substrate were investigated. The results show that liponucleosides possessing guanine in their head group exhibit considerably better surface ordering than liponucleosides containing other nuclobases. This behavior is attributed to the optimaly balanced hydrophilic/lipophilic character of the molecules with the guanosine headgroup, which is associated with the low water solubility of guanosine derivatives.

## References

1.  A. Ulman, *An Introduction to Ultrathin Organic Films: From Langmuir-Blodgett to self-Assembly*, Academic Press, San Diego, 1991.
2.  J.R. Health, *Annu. Rev. Mater. Res.*, 2209, **39**, 1.
3.  J.T. Davis, G.P. Spada, *Chem. Soc. Rev.*, 2007, **36**, 296.
4.  S. Lena, S. Masiero, S. Pieraccini, G. P. Spada, *Chem. Eur. J.*, 2007, **13**, 3757.
5.  R. Rinaldi, G. Marrucio, A. Biasco, V. Arima, R. Cingolani, T. Giorgi, S. Masiero, G.P. Spada, G. Gottarelli, *Nanotechnology*, 2002, **13**, 398.
6.  G. Maruccio, P. Visconti, V. Arima, S. D'Amico, A. Biasco, E. D'Amone, R. Cingolani, R. Rinaldi, S. Masiero, T. Giorgi, G. Gottarelli, *Nano Lett.*, 2003, **3**, 479.
7.  R. Rinaldi, E. Branca, R. Cingolani, S. Masiero, G.P. Spada, G. Gottarelli, *Appl. Phys. Lett.*, 2001, **78**, 3541.
8.  H. Liddar, J. Li, A. Neogi, P. B. Neogi, A. Sarkar, S. Cho, H. Morkoc, *Appl. Phys. Lett.*, 2008, **92**, 0133091.
9.  M. Devetak, S. Masiero, S. Pieraccini, G. P. Spada, M. Čopič, I. Drevenšek-Olenik, *Appl. Surf. Sci.*, 2010, **256**, 2038.
10. P. Dynarowicz-Latka, A. Dhanabalan, O. Oliveira Jr., *Adv. Colloid Interface Sci.*, 2001, **91**, 221.
11. H. Rosemeyer, Chem. Biodiversity, 2005, **2**, 977.
12. U. Rädler, C. Heinz, P.L. Luisi, R. Tampe, *Langmuir*, 1998, **14**, 6620.
13. Y.C. Wang, X. Z. Du, W.G. Miao, Y.Q. Liang, *J. Phys. Chem. B*, 2006, **110**, 4914.
14. T. Giorgi, F. Grepioni, I. Manet, P. Mariani, S. Masiero, E. Mezzina, S. Pieraccini, L. Saturni, G.P. Spada, G. Gottarelli, *Chem. Eur. J.*, 2002, **15**, 2143.
15. G. T. Barnes, I. R. Gentle, *Interfacial Science - an Introduction*, Oxford University Press, New York, 2005.
16. P. Berndt, K. Kurihara, T. Kunitake, *Langmuir*, 1995, **11**, 3083.
17. L. Gambut, J. P. Chauvet, C. Garcia, B. Berge, A. Renault, S. Riviere, J. Meunier, A. Collet, *Langmuir*, 1996, **12**, 5407.
18. C.A. de Vries, J.J. Haycraft, Q. Han, F. Noor-e-Ain, J. Raible, P.H. Dussault, C.J. Eckhardt, *Thin Solid Films*, 2011, **519**, 2430.
19. D. Berti, L. Franchi, P. Baglioni, *Langmuir*, 1997, **13**, 3438.
20. O. Pandoli, 2010, modelling by using utility Chem3D, Cambridge Soft Corp., Cambridge, MA, USA, private communication.
21. H. Yun, Y.W. Choi, N.J. Kim, D. Sohn, *Bull. Korean Chem. Soc.*, 2003, **24**, 377.

22 V.A. Bloomfield, D.M. Crothers, I.T. Tinoco, Jr., *Nucleic Acids: Structures, Properties, and Functions*, University Science Books, Sausalito, CA, 2000.
23 H. Devoe, S.P. Wasik, *J. Solution Chem.*, 1984, **13**, 51.
24 T.T. Herskovits, J.P. Harrington, *Biochemistry*, 1972, **11**, 4800.
25 P. Samori, S. Pieraccini, S. Masiero, G. P. Spada, G. Gottarelli, J.P. Rabe, *Colloids Surf. B: Biointerf.*, 2002, **23**, 283.
26 Y. Yin, Y. Qiao, M. Li, P. Ai, X. Hou, *Colloids Surf. B: Biointerf.*, 2005, **42**, 45.
27 Y. Yin, Y. Qiao, X. Hou, *Appl. Surf. Sci.*, 2006, **252**, 7926.

# Chapter 3

## Theoretical Modelling, Analysis and Prediction

Chapter Editor: **Shozeb Haider**

An important part in studying Quadruplexes is via theoretical modeling, analysis and prediction. This primarily provides a theoretical background of basic self-assembly processes through which structural and electronic properties of example molecular devices can be calculated. Combined with experimental results, such studies help in predicting new properties and design novel of novel materials.

The folding of G-quadruplexes is directed by controllable factors. The structural basis underlying the precise relationship between DNA sequences and folding into quadruplexes has been investigated towards prediction of structural folds. The theoretical modeling and analysis includes a variety of methods that span from *ab initio* quantum chemical techniques to classical molecular dynamics. Different computer-based methodologies have been used to study structural properties of quadruplexes with main emphasis on biological applications and drug discovery. Development of multi-scale approaches to treat different theory levels can aid in studying different parts of complex devices (molecule, surface, solution, electrodes etc.).

# FUNDMENTALS AND APPLICATIONS OF THE GEOMETRIC FORMALISM OF QUADRUPLEX FOLDING

Mateus Webba da Silva and Andreas Ioannis Karsisiotis

School of Biomedical Sciences, University of Ulster, Coleraine, Co. Londonderry, Northern Ireland, BT52 1SA, United Kingdom.
(*m.webba-da-silva@ulster.ac.uk*)

## 1 INTRODUCTION

Double-stranded DNA is defined by its Watson-Crick base-pairing that defines a major, and a minor grooves in the molecule. It may adopt a small number of topologies. Its normal, most abundant state is as B-DNA, but it also adopts a A-DNA topology, or Z-DNA. Notably, as B-DNA its sugar puckers adopt a C2'-endo conformation. In A-DNA the sugar puckers adopt a C3'-endo conformation leading to a somewhat larger diameter for the double-stranded stem. DNA regions rich in sequential adenines that number above four tend to adopt this topology. Z-DNA is characterized by regions rich in sequential CpG steps. This results in the guanosines adopting a *syn* glycosidic bond angle (GBA); Figure 1. For both B-DNA and A-DNA the GBA is generally *anti*. The backbone angles of Z-DNA are more strained and are at the basis of the biological functions attributed to this DNA topology.

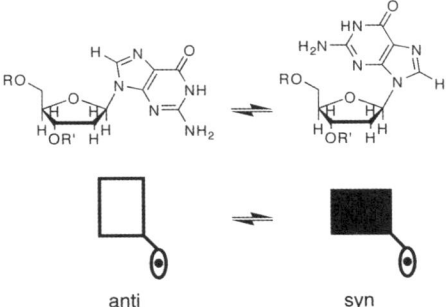

Figure 1. *Chemical structures of anti and syn glycosidic bond angles and their schematic representation.*

In a four-stranded DNA stem both sugar puckers as well as the relationship between base and sugar pucker can be modulated. In their structural versatility DNA quadruplexes vary these two parameters. Specifically, the GBA can be utilized to define the quadruplex topology.

We have previously proposed a formalism of quadruplex folding based on the geometric disposition defined by the glycosidic bond angle.(1) In this Chapter we attempt to cover some of the premises and consequences, and explore the applications, it has had thus far.

## 2 THE NEED FOR A FORMALISM OF QUADRUPLEX FOLDING

Have you tried to compare quadruplex structures? In order to identify the sequence how many times did you start by stating who did determine the structure, how many loops it has? Or, did you call it "basket type"? Or, by calling it "a structure like hybrid-2"? The description may continue by mentioning the "top" or the "bottom" of the quadruplex stem; or, by some transient definition of the groove. These descriptions are usually a poor starting point to describe the topological characteristics unique to the architecture. DNA quadruplexes adopt a variety of architectures just as proteins do. There are many protein families as there can be a variety of quadruplex topologies. In contrast to proteins, quadruplexes and double-stranded DNA, can be defined by an axis that runs perpendicular to the planes defined by its base-pairs. For double-stranded DNA the definition has been instinctive, since the stem is defined by two anti-parallel oriented strands. For a four-stranded stem it is necessary to establish a referencing strand.

## 3 THE QUADRUPLEX REFERENCING SYSTEM

Four hydrogen-bond aligned guanosines, each forming hydrogen bonds through their minor groove edge and their Watson-Crick edge to two other bases, define the pseudo-planar tetrad. On a first instance a quadruplex is characterized by at least two stacking tetrads. By choosing the strand closest to the 5'-end of the oligonucleotide sequence as the first of the four in the stem, a level of definition customary for double-stranded DNA is established. Now, it is feasible to consider the four grooves present in a four-stranded stem. These are defined by the orientation between the base and its sugar pucker; the GBA. Here a simplification is warranted. Instead of a range of values for the GBA, we define a two-state: *syn* or *anti* GBA.

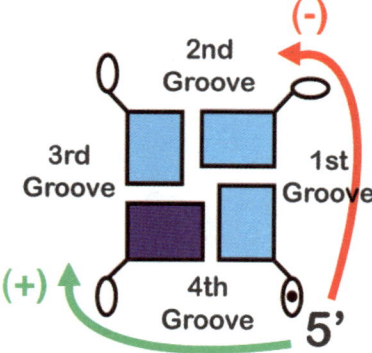

*Figure 2. The frame of reference for describing quadruplex topologies.(1)*

We have thus derived a descriptor for quadruplexes (Figure 2). In this descriptor a tetrad is represented with the orientation of the base relative to the sugar pucker. In this manner the relative disposition of the sugar puckers defines the groove width at the pseudo-plane. Various such stacked pseudo-planes define the four groove widths of the four-stranded stem.

## 4 CONSEQUENCES OF THE QUADRUPLEX DESCRIPTOR

The definition of this reference frame for quadruplexes has various consequences that *a priori* appear to be oversimplifications. A fundamental assertion relates to the definition of quadruplexes. We define quadruplexes as formed by at least two *tetrads with the same groove-width combination*. This leaves out any other hydrogen bond alignments that stabilize the topology; or that are stabilized by the topology. These may be mismatches, triads, or even tetrads with other groove-width combinations as defined by the descriptor.

A notable feature of the descriptor is that the first groove clockwise from the origin (fourth groove) cannot be a wide-groove. Furthermore, the first groove clockwise from the origin (first groove) cannot be a narrow groove. These groove features are fundamental to extending the utility of the descriptor to multimeric assemblies, as we will demonstrate later.

Double-stranded DNA topologies differ in the dimensions of their major and minor grooves. As stated previously, in quadruplex DNA the relative orientations of the sugar pucker with respect to the base define the groove width within the plane of the tetrad. As a consequence the width of the grooves may be comprised of guanosines with the same GBA. In this case a "medium groove" is defined. If the two guanosines have different GBA, the groove may be defined as a "narrow groove", or a "wide groove". In the descriptor a narrow groove is described by the groove with the shortest length between sugar puckers.

What groove widths are possible? In Figure 3 we describe all the possible groove widths. In quadruplexes there are eight possible groove widths. Each groove width is defined by two possible GBA combinations of four guanosines. This limits the number of possible GBA combinations for quadruplexes to sixteen.

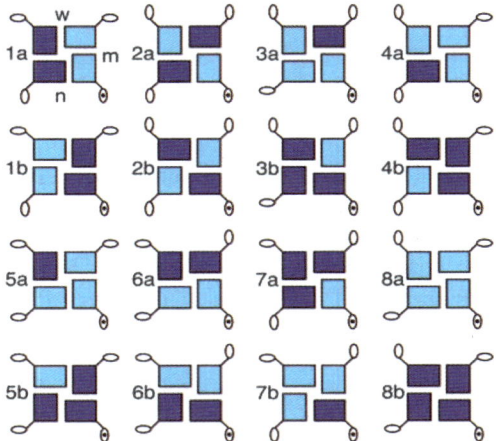

*Figure 3. The possible GBA combinations for tetrads. Pairs a and b define a groove witdh combination.(1)*

The descriptor allows for defining the orientation of the progression of the bases in the sequence from the 5'- to the 3'-end along the four strands of the stem. However, denoting 'four-strands' to the quadruplex stem is a misnomer. A strand may be interrupted. Connecting the strands are single stranded regions, abasic backbone, (2,3) or simply oligonucleotides participating in other hydrogen-bond alignments. These are generally known as loops. Thus defining unimolecular quadruplexes as oligonucleotides with a general formula d($G_xL_yG_xL_yG_xL_yG_x$; where $x$ is the number of G residues, and $y$ is the number of bases in a loop) is of no consequence for our purposes. This is due to the fact that more than four G sequential segments, or even combinations with a single G, may be part of a quadruplex stem.

## 5 LOOPS IN QUADRUPLEXES

Loops simply bridge grooves, or diagonally opposed ends of a tetrad. In canonical unimolecular quadruplexes they link guanosine segments at either end. They may also link segments of Gs that are not sequential in reference to the DNA sequence, but are sequential in the stem.(4,5) Thus, a unimolecular quadruplex may have more than three loops bridging its stem. However, by simplifying topological referencing to three loops without stem interruptions it is feasible to understand the structural interdependencies that define quadruplexes. This approach still allows for interpreting the same characteristics in unimolecular quadruplexes with more than three loops. Furthermore, it remains applicable to multimolecular self-assemblies.

What loop combinations are possible for unimolecular topologies of three loops? Firstly, the question simplifies the universe of quadruplex folds to the general formula d($G_xL_yG_xL_yG_xL_yG_x$; where $x$ is the number of G residues, and $y$ is any number of bases in a loop). This is certainly not the complete picture for unimolecular folds. However, it allows for inferences to be made in relation to the positions of GBA before and after a loop that are valid for unimolecular topologies with more than three loops; as we will see later.

There are 26 feasible unimolecular topologies of three loops- see Figure 4, adapted from (1). Essentially, the first loop can bridge medium grooves in an either clockwise or anti-clockwise manner. Furthermore, it can also bridge a tetrad diagonally. But, it can only bridge a wide grove in an anti-clockwise step, and a narrow groove in a clockwise step. For each loop combination description in which the first loop goes anti-clockwise, there is a loop combination for which the first loop progresses in a clockwise manner. Thus the letters *a* and *b* have been utilized to describe either progression. For the cases that involve a first diagonal loop, the next loop is utilized to describe this progression. The sequence of Arabic numerals was chosen without any further considerations. The set of topologies described in the figure contains three different loop types. A *propeller loop* (p) bridges a medium groove by linking guanosines in different tetrads. A *diagonal loop* (d) bridges a tetrad diagonally across. It starts and ends with guanosines of different GBA. The same change in GBA applies to *lateral loops* (l). These bridge two adjacent guanosines of the same tetrad, thus they bridge either narrow or wide grooves. By knowing the looping progression it is thus possible to select the combination of GBA that defines the tetrad, and consequently its groove width composition. This is done by starting at the top tetrad of the topology and selecting an antiG in the 5'-end origin. Indeed, we assume that the last base of the first strand will adopt an *anti* GBA. If for example the first loop is an anti-clockwise propeller the corresponding base of the tetrad will also adopt an *anti* GBA to form a medium groove.

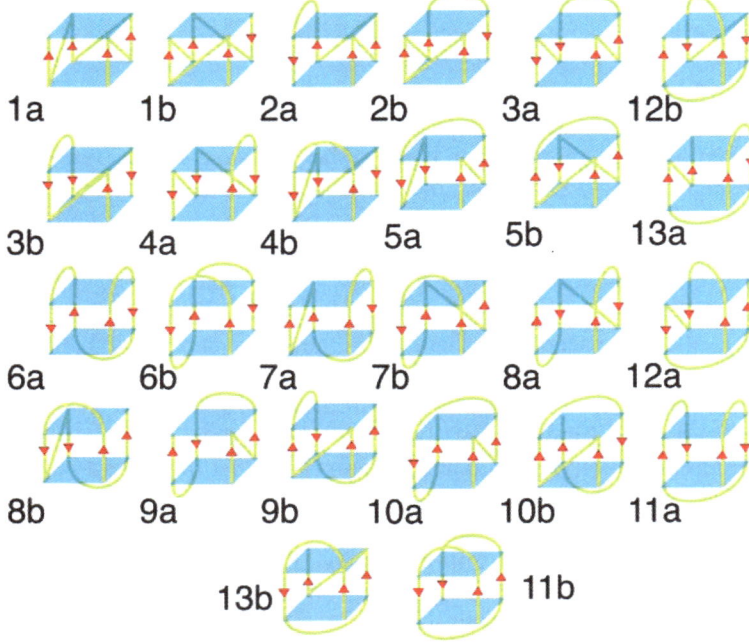

*Figure 4. The quadruplex looping topologies for unimolecular quadruplexes of three loops.(1)*

A partial description of a quadruplex topology can make use of its loop types with their orientation. For example 10b of Figure 4, describes a topology in which the first loop is a propeller loop progressing in a clockwise manner, followed by a diagonal loop, and finally a lateral loop progressing in an anti-clockwise manner. This can be simply described as (+pd-l). For this topology the groove widths can be described in the following manner. We place the 5'-end origin of the quadruplex descriptor on the first guanosine of the first loop- the top tetrad. Thus after +p the guanosine is also anti and the groove this loop bridges is a medium groove. Next, the diagonal loop (d) leads to an inversion of the GBA. Thus the guanosine diagonally across adopts a *syn* GBA. The last loop, -l, will also invert the GBA, and the last guanosine of the tetrad adopts an *anti* GBA. Thus, the anti-clockwise sequence of grooves is wide – narrow – medium – medium. This is defined by anti – syn – anti – anti GBA in the tetrad (also in an anti-clockwise sequence). The description of the topology can be complete by mentioning the number of stacked tetrads. Most topologies may be comprised of two or more stacked tetrads of the same combination of groove widths. The set of topologies shown is the starting point for describing the feasible unimolecular topologies of four or more loops.

## 6  THE SEQUENCE OF GBA IN THE QUADRUPLEX STEM

After establishing the GBA combinations for tetrads in a specific topology it is necessary to know the identity of the sequence of tetrads within the stem of the quadruplex. All the GBA in the stem of (-p-p-p) are *anti*. Thus, all its grooves are medium. There is a single tetrad GBA combination for the topology since there is no inversion of GBA configuration. The same would apply if all the strands of a topology would be *syn*. Other topologies could also be described by stacking of a single tetrad GBA combination only if other hydrogen bond alignments would stabilize the architecture. However, at this point we cannot predict the structural motifs necessary to achieve this stabilization.

More generally, stable quadruplex topologies will have specific stacking sequence of GBA tetrad combinations.(6) This is the case for all topologies with exception of (-p-p-p) and (+p+p+p). We have termed these, Type I quadruplexes. For quadruplexes that have at least two sequential GBA of the same type we have termed them Type II, and for those with no sequential GBA in the quadruplex stem we have termed them Type III. Type II quadruplexes will have at least one propeller loop bridging a medium groove. Whilst Type III quadruplexes may have a medium groove they do not have propeller loops. There are four Type III quadruplexes: (-l-l-l), (+l+l+l), (-ld+l) and (+ld-l). The remaining 20 topologies are all Type II.

But, how to place the sequence of tetrads in the stem? One can construct the sequence of GBA from the sequence of bases in the DNA strand and keeping with the groove widths available for the topology. For Type III quadruplexes it is simple: the GBA alternate in the sequence within the quadruplex stem. It is thus straightforward to place the descriptor GBA tetrad combination on the top tetrad. This implies that the 5'-end of the descriptor is *anti*. Let us examine the example of a three stacked (-ld+l) topology. First, we place the correct descriptor following the sequence of loops. The top tetrad will thus have the following angles described in an anti-clockwise manner: (*anti:syn:syn:anti*). Since the sequence alternates the middle tetrad will be (*syn:anti:anti:syn*), and the lower tetrad (*anti:syn:syn:anti*). This complies with the rules for placing loops. The sequence of GBA of the first strand of the quadruplex stem is *anti-syn-anti*. Since an anti GBA is at the beginning of a lateral loop (-l) the next guanosine of the quadruplex stem will adopt a *syn* GBA. This strand of the quadruplex stem will adopt the sequence *syn-anti-syn*. The second loop is a diagonal loop and thus an inversion of configuration ensues. The first guanosine of the quadruplex stem after the diagonal loop adopts an *anti* GBA. This strand of the quadruplex stem will adopt the sequence *anti-syn-anti*. After placing the GBAs after the third loop the sequence of GBA tetrad combinations will fit the appropriate groove width combination for this topology.

These procedures can be applied to describe quadruplexes with any number of tetrads. For Type II quadruplexes the topology is also simply described by the placing the descriptor on the top tetrad with the 5'origin adopting an *anti* GBA. However, the sequence of tetrads of different GBA dispositions but same groove width combinations becomes less obvious. The first consideration relates to the balance of GBA of guanosines of the quadruplex stem. Due to their relatively perturbed backbone angles, i.e. greater deviation from norm, quadruplexes defined by a greater number of *anti* GBA should be expected to be more accessible for folding. If this is not possible an equal number of *syn* and *anti* GBA should be expected. To comply with this and inversion of GBA configurations after diagonal or lateral loops, the first strand must start with a syn GBA. These rules are sufficient to define two-stacked tetrads since they define both the top and the bottom tetrads. However, for quadruplexes with more tetrads one must define the middle tetrads.

For three stacked tetrads the first strand of the quadruplex stem is predominantly *anti*. Thus the sequence is *syn-anti-anti*. For 18 of the topologies the balance of *syn* vs *anti* GBAs, either favors *anti*, or does not favor either GBA. The exceptions are topologies (-l-p-p) and (+l+p+p). For these a first strand with sequence syn-syn-anti results in more anti guanosines in the quadruplex stem. For four-stacked quadruplexes of Type II the rules are yet to be established.

## 7 EXAMPLES OF PREDICTION

The Geometric Formalism of Quadruplex Folding can be utilized to predict the structural characteristics of quadruplex topologies yet to be discovered. Indeed, there are at least three such examples since the formalism was first published.(7-9) For the topologies that are known, the formalism can describe the GBA tetrad combinations for different number of tetrads within this topology.

Consider the topology that describes the Thrombin binding aptamer; a two tetrad (+l+l+l). For this topology the first loop bridges a narrow groove, the second a wide groove, and the third loop bridges another narrow groove. Since publishing the geometric formalism, Phan's group has determined the structure of Bombyx Mori telomeric sequence.(9) It is a two-stacked tetrad of topology (-l-l-l). For this topology the formalism predicts that the first lateral loop progressing in a lateral anticlockwise manner bridges a lateral wide groove. This defines the subsequent loops as bridging first a narrow, and then a wide groove. Furthermore, the formalism places the tetrad that defines the topology, the one with an antiG in the 5'-end, as the top tetrad. These two rules, together with the rule for GBA assignment of loops, were sufficient for the correct description of groove widths and GBAs within the topology.

A three-tetrad (-pd+l) structure was determined by the Plavec's group.(10) Notably, this topology has all three types of loop. The topology described by the structure perfectly fits the predicted groove width combinations, as well as sequence of GBA tetrad combinations.(1,6)

## 8 IDENTIFYING GBA STEPS IN SOLUTION NMR STUDIES

G-Quadruplex sequence specific assignments are based on sequential NOE connectivities of the type H8/H6(i)-H1'(i)-H8/H6(i+1) (Figure 5). For reasons explained below, tracing these connectivities is heavily dependent on the sequence of GBAs in the quadruplex stem. There are interesting patterns in characterizing quadruplex topologies. Even without full structural assignment it is possible to establish single GBA steps. Here we will describe possible patterns that can be found in a high mixing time ($t_m$>200 ms) NOESY spectrum of a sample in $^2H_2O$.

The *syn-anti* GBA step is the easiest to assign using sequential NOEs. By considering i a specific guanosine in the quadruplex stem, both H8(i)-H1'(i+1) and H8(i+1)-H1'(i) crosspeaks are observed creating a readily indentifiable rectangular pattern. The best observed crosspeaks are of the type H8(i+1)-H2'/2''(i) since distances bettween the aromatic H8 proton of residue i+1 and any of the H2' or H2'' protons of residue (i) are within 3.5 Å (Table 1, Figure 6).

The anti-syn GBA step is impossible to observe using sequential connectivities involving the aromatic H8 proton. This is due to the large distances resulting from the geometry of the step. Both forward (H8 of (i) - H1', H2'/H2'', H3', H4', H5'/H5'' of (i+1)) and reverse (H8 of (i+1) - H1', H2'/H2'', H3', H4', H5'/H5'' of (i)) connectivities are equally hard to observe (Tables 1 and 2, Figures 6 and 7).

The syn-syn GBA step is almost as hard to observe using sequential connectivities involving the aromatic H8 proton as the anti-syn step. However, some of the forward (H8 of (i) - H1', H2'/H2'', H3', H4', H5'/H5'' of (i+1)) connectivities may be within the limit of detection (Table 1, Figure 6) and any of these should be assigned if observed. Reverse (H8 of (i+1) - H1', H2'/H2'', H3', H4', H5'/H5'' of (i)) connectivities, aside from H8(i+1)-H8(i) crosspeaks are not usually observed.

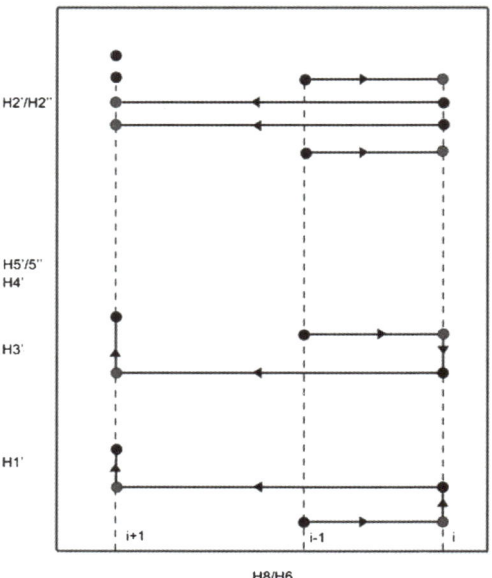

*Figure 5. Scheme for sequential assignments in G-Quadruplexes utilizing sequential NOE connectivities between the aromatic H8 proton and the H1', H3' and H2'/2'' protons of the sugar ring in a NOESY spectrum at high mixing time.*

The anti-anti GBA step can be usually identified and does not interrupt sequential connectivities in the G-Quadruplex stem. It is however much more likely to be assigned using reverse (H8 of (i+1) - H1', H2'/H2'', H3', H4', H5'/H5'' of (i)) connectivities than forward (Tables 1 and 2, Figures 6 and 7). Distances observed for reverse connectivities are much shorter (Table 2) than the equivalent forward ones and this fact should be exploited when making sequential assignments.

*Table 1. Sequential distances between H8 of residue (i+1) and H8 plus sugar ring protons of residue (i). Distances are extracted from PDB entries 201D (G1-G2, syn-anti and G2-G3, anti-syn) and 2GKU (G4-G5, anti-anti and G15-G16. syn-syn).*

|  | Syn-Anti H8(i+1) | Anti-Syn H8(i+1) | Anti-Anti H8(i+1) | Syn-Syn H8(i+1) |
|---|---|---|---|---|
| H8(i) | 5.0 | 9.0 | 5.3 | 5.8 |
| H1'(i) | 4.8 | 6.5 | 3.6 | 7.9 |
| H2'(i) | 3.4 | 8.7 | 3.6 | 6.2 |
| H2''(i) | 2.7 | 7.3 | 2.3 | 7.7 |
| H3'(i) | 4.9 | 9.6 | 5.0 | 7.9 |
| H4'(i) | 6.7 | 9.1 | 5.9 | 10.1 |
| H5'(i) | 7.3 | 10.8 | 7.0 | 10.2 |
| H5''(i) | 7.1 | 11.2 | 7.5 | 9.6 |

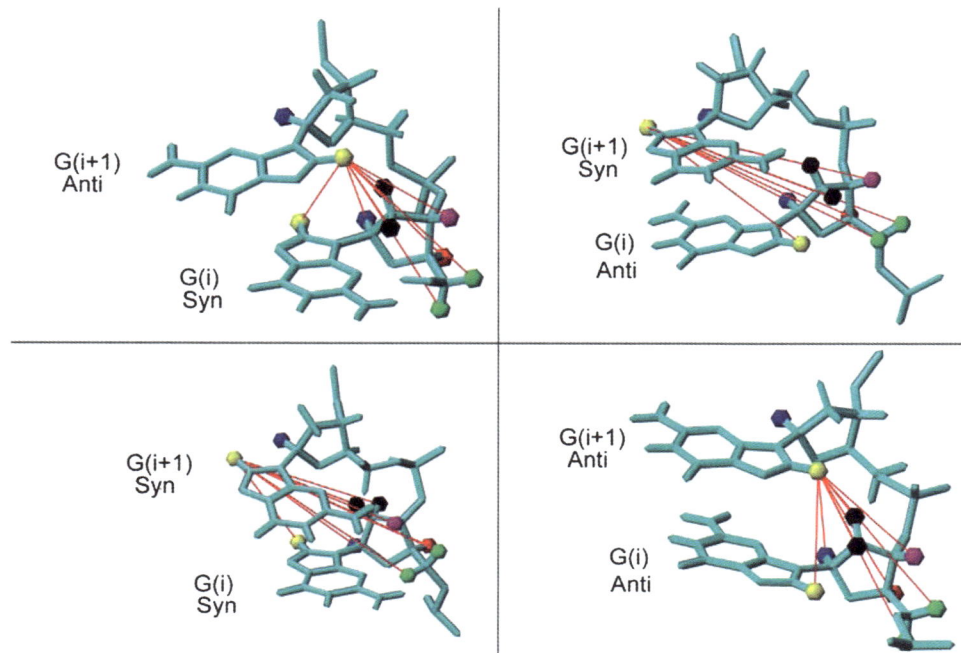

*Figure 6. Sequential steps of different GBA sequence in G-Quadruplexes. Distances between H8 (yellow) atom of residue (i+1) and H1' (blue), H2'/H2'' (black), H3' (magenta), H4' (red) and H5'/H5'' (green) atoms of the preceding (i) residue are indicated. H1' atom of residue (i) is indicated so the syn or anti disposition is easily identified. Examples are taken from PDB entries 201D (G1-G2, syn-anti and G2-G3, anti-syn) and 2GKU (G4-G5, anti-anti and G15-G16, syn-syn).*

*Table 2. Sequential distances between H8 of residue (i) and H8 plus sugar ring protons of residue (i+1). Distances are extracted from PDB entries 201D (G1-G2, syn-anti and G2-G3, anti-syn) and 2GKU (G4-G5, anti-anti and G15-G16. syn-syn)*

|  | **Syn-Anti** | **Anti-Syn** | **Anti-Anti** | **Syn-Syn** |
|---|---|---|---|---|
|  | **H8(i)** | **H8(i)** | **H8(i)** | **H8(i)** |
| **H8(i+1)**   | 5.0 | 9.0 | 5.3 | 5.8 |
| **H1'(i+1)**  | 4.3 | 8.4 | 8.7 | 5.2 |
| **H2'(i+1)**  | 5.8 | 8.2 | 7.5 | 6.8 |
| **H2''(i+1)** | 6.4 | 9.5 | 9.1 | 7.2 |
| **H3'(i+1)**  | 6.5 | 9.4 | 8.8 | 7.3 |
| **H4'(i+1)**  | 4.4 | 7.8 | 9.1 | 5.2 |
| **H5'(i+1)**  | 5.8 | 6.9 | 7.3 | 6.2 |
| **H5''(i+1)** | 5.2 | 7.9 | 8.4 | 4.8 |

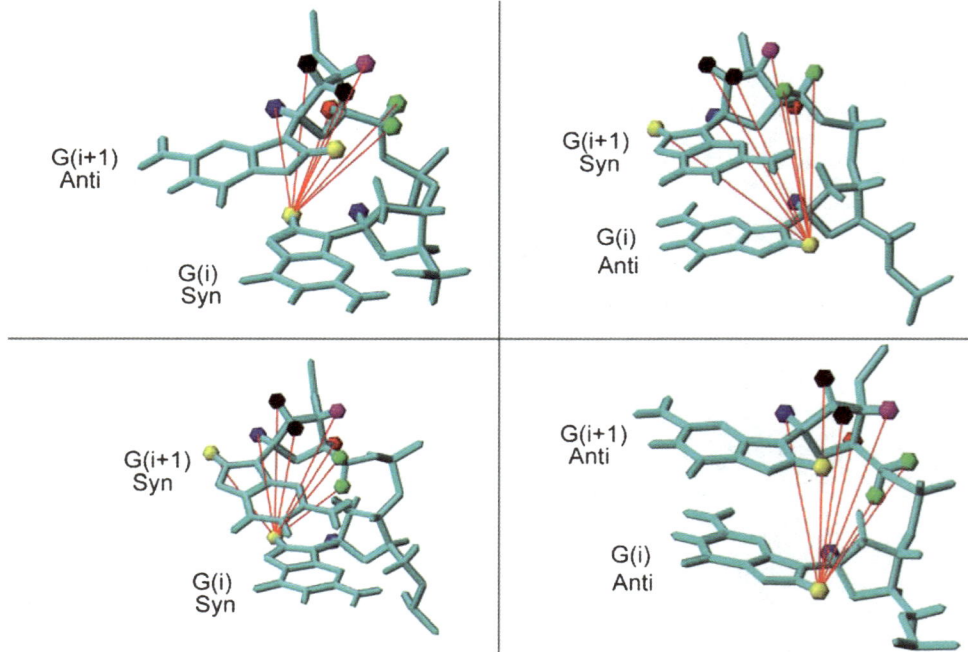

*Figure 7. Sequential steps of different GBA sequence in G-Quadruplexes. Distances between H8 (yellow) atom of residue (i) and H1' (blue), H2'/H2'' (black), H3' (magenta), H4' (red) and H5'/H5'' (green) atoms of the following (i+1) residue are indicated. H1' atom of residue (i) is indicated so the syn or anti disposition is easily identified. Examples are taken from PDB entries 201D (G1-G2, syn-anti and G2-G3, anti-syn) and 2GKU (G4-G5, anti-anti and G15-G16, syn-syn).*

## 9 IDENTIFYING THE TOPOLOGY UTILIZING THE TDS SIGNATURE

Mergny et al. have investigated the light absorption properties of quadruplexes(11,12) and identified a spectral pattern for UV-VIS thermal difference spectrum of quadruplexes.(12) The difference between the UV-VIS spectrum at a temperature in which the topology is folded and a temperature at which the topology is thermally unfolded shows maxima at approximately 240 nm, 255 nm, 275 nm and a trough at 295 nm. We reasoned that since quadruplexes may have three different intrastrand stacking conformations defined by the different types of quadruplexes (Types I through III), the light absorption properties of these have to be different. Indeed, the intensity at 240 nm weighted by the magnitude of the intensity of the trough gives factors that define the stacking sequence within the quadruplex stem.(13) Essentially, the $Q_{factor}$= 240 nm/295 nm may assume a magnitude between 0 and 0.9 for Type III quadruplexes, between 1.1 and 2.2 for Type II quadruplexes, and above 3.6 for Type I quadruplexes. This is thus an inexpensive method for topological characterization that is at least as useful as the characteristic CD signatures for parallel (Type I) and anti-parallel stranded (Type II and Type III) quadruplexes used until this paper was published.

## 10 IDENTIFYING THE TOPOLOGY UTILIZING CIRCULAR DICHROISM

However, since the CD signal is also primarily derived from light absorption we also hypothesized that there should be different signature for each of Type I, Type II, and Type

III quadruplexes.(13) We established that if the spectrum consists of a trough at 240 nm and a band at 264 nm it indicates the presence of stacked guanosines with the same GBA characteristic of Type I quadruplexes. We established that if a spectrum has at least a band at 290 nm it contains stacking guanosines of different GBA in the quadruplex stem. Thus a spectrum with the bands characteristic of Type I quadruplexes that also has the band at 290 nm will contain both stacking of guanosines with the same GBA, and stacking of guanosines of different GBA. Thus the spectrum for Type II quadruplexes consists of a trough at 240 nm, and bands at 264 and 290 nm. Type III quadruplexes have the band characteristic for stacking guanosines of different GBA, but also another band at 240 nm and a trough at 264 nm. These two are the inverse of the signals for Type I quadruplexes. Thus, it can be assumed a band at 240 nm coupled with a trough at 264 nm is an indication of absence of intrastrand stacking of guanosines of the same GBA. These assertions are significant since Type III quadruplexes may be further stacked by other hydrogen-bond alignments that may mask their correct characterization. This can occur when a quadruplex stem of Type III is further stacked by guanosines with the same GBA. Thus all of these assertions are of diagnostic value.

## 11 CONCLUDING REMARKS

An approach for understanding the folding of unimolecular quadruplex topologies is to combine geometric structural determinants with kinetic and thermodynamic data for folding of primary sequence. The Geometric Formalism of Quadruplex Folding attempts to define the interdependency of geometric structural determinants to allow for evolution of our understanding of their utility in defining the architecture. The Geometric Formalism of Quadruplex Folding is thus a system of self-consistent inferences that hold true through experiment. If, and when assumptions result in inferences that do not hold true the assumptions are reformulated in keeping with the self-consistent nature of the formalism. The formalism is thus a tool for describing quadruplexes, enables the directed self-assembly of quadruplexes known or predicted, it enables detailed characterization of anti-parallel quadruplexes utilizing CD, and allows for the characterization of quadruplex topologies utilizing UV-VIS. Both of these are due to the light absorption characteristics of the quadruplex stem. Thus, we should also expect subtle differences in light emission for the different types of quadruplexes. These still remain to be explored.

Currently, there are examples of lateral loops bridging narrow and wide grooves; each of these progressing either clockwise or anti-clockwise. There are also examples of propeller loops progressing either clockwise or anti-clockwise, and a few diagonal loops. However, only 9 of the 26 topologies have thus far been experimentally verified. In order to better understand the controlling features of quadruplex folding many more structures have to be determined.

A fundamental issue is of the definition of the universe of possible topologies of three loops in unimolecular quadruplexes. There could be more topologies amenable to experimental determination- thus extending the number of "feasible" topologies. There also may be topologies considered "feasible" for which experimental evidence will remain to be established.

## REFERENCES

1. Webba da Silva, M. (2007) Geometric formalism for DNA quadruplex folding. *Chemistry European Journal*, **13**, 9738-9745.

2. Cmugelj, M., Sket, P. and Plavec, J. (2003) Small change in a G-rich sequence, a dramatic change in topology: New dimeric G-quadruplex folding motif with unique loop orientations. *Journal of the American Chemical Society*, **125**, 7866-7871.
3. Kuryavyi, V. and Patel, D.J. (2010) Solution Structure of a Unique G-Quadruplex Scaffold Adopted by a Guanosine-Rich Human Intronic Sequence. *Structure*, **18**, 73-82.
4. Phan, A.T., Kuryavyi, V., Burge, S., Neidle, S. and Patel, D.J. (2007) Structure of an Unprecedented G-Quadruplex Scaffold in the Human c-kit Promoter. *Journal of the American Chemical Society*, **129**, 4386.
5. Kelsey, I., Nakayama, S. and Sintim, H.O. (2012) Diamidinium and iminium aromatics as new aggregators of the bacterial signaling molecule, c-di-GMP. *Bioorganic & Medicinal Chemistry Letters*, **22**, 881-885.
6. Webba da Silva, M., Trajkovski, M., Sannohe, Y., Ma'ani Hessari, N., Sugiyama, H. and Plavec, J. (2009) Design of a G-quadruplex topology through glycosidic bond angles. *Angew Chem Int Ed Engl*, **48**, 9167-9170.
7. Zhang, Z.J., Dai, J.X., Veliath, E., Jones, R.A. and Yang, D.Z. (2010) Structure of a two-G-tetrad intramolecular G-quadruplex formed by a variant human telomeric sequence in K+ solution: insights into the interconversion of human telomeric G-quadruplex structures. *Nucleic Acids Research*, **38**, 1009-1021.
8. Hu, L., Lim, K.W., Bouaziz, S. and Phan, A.T. (2009) Giardia Telomeric Sequence d(TAGGG)(4) Forms Two Intramolecular G-Quadruplexes in K+ Solution: Effect of Loop Length and Sequence on the Folding Topology. *Journal Of The American Chemical Society*, **131**, 16824-16831.
9. Setnicka, V., Novy, J., Boehm, S., Sreenivasachary, N., Urbanova, M. and Volka, K. (2008) Molecular structure of guanine-quartet supramolecular assemblies in a gel-state based on a DFT calculation of infrared and vibrational circular dichroism spectra. *Langmuir*, **24**, 7520-7527.
10. Marusic, M., Sket, P., Bauer, L., Viglasky, V. and Plavec, J. (2012) Solution-state structure of an intramolecular G-quadruplex with propeller, diagonal and edgewise loops. *Nucleic Acids Research*.
11. Mergny, J.L., Phan, A.T. and Lacroix, L. (1998) Following G-quartet formation by UV-spectroscopy. *FEBS Lett*, **435**, 74-78.
12. Mergny, J.L., Li, J., Lacroix, L., Amrane, S. and Chaires, J.B. (2005) Thermal difference spectra: a specific signature for nucleic acid structures. *Nucleic Acids Research*, **33**.
13. Karsisiotis, A.I., Hessari, N.M.a., Novellino, E., Spada, G.P., Randazzo, A. and Webba da Silva, M. (2011) Topological Characterization of Nucleic Acid G-Quadruplexes by UV Absorption and Circular Dichroism. *Angewandte Chemie-International Edition*, **50**, 10645-10648.

# GUANINE, XANTHINE AND URIC ACID ASSEMBLIES: COMPARATIVE THEORETICAL AND EXPERIMENTAL STUDIES

Gábor Paragi,[1] János Szolomájer,[2] Zoltán Kupihár,[2] Gyula Batta,[3] Zoltán Kele,[2] Petra Pádár,[2] Botond Penke,[1,2] Hester Zijlstra,[4] Célia Fonseca Guerra[4], F. Matthias Bickelhaupt,[4] and Lajos Kovács[2]

[1]Supramolecular and Nanostructured Materials Research Group of the Hungarian Academy of Sciences at the University of Szeged, Dóm tér 8, Szeged, Hungary. E-mail: paragi@sol.cc.u-szeged.hu
[2]Department of Medicinal Chemistry, University of Szeged, Dóm tér 8, Szeged, Hungary. E-mail: kovacs.lajos@med.u-szeged.hu
[3]Department of Organic Chemistry, University of Debrecen, Debrecen, Hungary. E-mail: batta@unideb.hu
[4]Department of Theoretical Chemistry and Amsterdam Center for Multiscale Modeling (ACMM), Scheikundig Laboratorium der Vrije Universiteit Amsterdam, De Boelelaan 1083, NL-1081 HV Amsterdam, The Netherland. E-mail: c.fonsecaguerra@vu.nl, f.m.bickelhaupt@vu.nl

## 1 INTRODUCTION

Higher-ordered structures, based on self-recognition and self-assembly of simpler molecules, are of potential interest in a variety of fields ranging from structural biology to medicinal chemistry, supramolecular chemistry, nanotechnology and molecular-scale devices. Nucleic acids are prominent representatives of such simple building blocks. The guanine base present in nucleic acids is well known for its ability to participate in the formation of tetrads/quartets and quadruplex structures, as well as more complex motifs, owing to its multiple hydrogen bonding pattern, stacking and ion-binding capacity.[1-5]

Recently, it was found that guanosine-based quadruplexes, occurring e.g. in telomere sequences, are heavily involved in switching on and off genes. Over the years several other quadruplex structures have been reported, based on experimental results or theoretical considerations (see e.g. refs.[6, 7] and references therein). These new structures cover a wide range of changes from small modification in the guanine base to the complete replacement of the whole monomeric building block. It was also shown that in most of the cases cations play a stabilizing role in the formation of quadruplex strands by intercalating into the stacking tetrads. Negatively charged molecules (*e.g.* picrates) can bind to the "lateral part" of stacked structure ensuring additional stabilization. The guanine quartet is one of the most suitable structures for stacking since guanine quartets are strongly bound (there are two H-bonds between neighbouring molecules) and already posses a planar equilibrium structure.[8] Other purine-based tetramers (e.g. adenine, A or inosine, I) have also been investigated,[8, 9] however these structures have been found less stable and they tend to adopt non-planar geometries. This non-planarity is understandable since the nucleobases in these tetramers are connected to each other by only one H-bond which yields a flexible structure.

In the present chapter, we report our computational and experimental results on tetrad/quadruplex forming ability of various xanthine and uric acid derivatives, in comparison with guanine. A highlight of our theoretical analyses is the finding that the enhanced stability of guanine tetrads originates from a cooperativity effect generated by the covalent character of the hydrogen bonds, contrary to other suggestions which are based on the concept of π-resonance assisted hydrogen bonding. The possibility of a special tetramer structure containing low barrier hydrogen bonds (LBHB) was also investigated by computational simulations.

## 2 METHODS AND RESULTS

### 2.1 Computational and experimental methods

All calculations were performed using the Amsterdam Density Functional (ADF) program developed by Baerends, Ziegler, and others,[10] and the QUantum-regions Interconnected by Local Descriptions (QUILD) program developed by Swart and Bickelhaupt.[11] The applied level of theory was density functional theory (DFT) using the BLYP functional[12,13] with TZ2P basis set and dispersion corrections were taken into account according to Grimme's suggestion[14,15] (BLYP-D). Equilibrium structures were optimized using analytical gradient techniques and stationary points were verified to be minima through vibrational analysis. Solvent effects in water have been estimated using the conductor-like screening model (COSMO)[16] and for settings, see ref.[17]. It is worth to note that according to the work by Riley et al.[18] the dispersion correction does not need to be modified for solvated systems. The bond energy $\Delta E_{bond}$ of a quartet is defined as [Eq. (1)]:

$$\Delta E_{bond} = E_{quartet} - 4E_{base} \tag{1}$$

in which $E_{quartet}$ is the energy of the quartet, and $E_{base}$ is the energy of the guanine or xanthine, optimized in $C_1$ symmetry, that is, without any geometrical constraint. The overall bond energy $\Delta E$ is made up of two major components [Eq. (2)]:

$$\Delta E_{bond} = \Delta E_{prep} + \Delta E_{int} \tag{2}$$

In this formula, the preparation energy $\Delta E_{prep}$ is the amount of energy required to deform the separate bases from their equilibrium structure to the geometry that they acquire in the quartet. The interaction energy $\Delta E_{int}$ corresponds to the actual energy change when the prepared bases are combined to form the quartet.
The interaction energy is examined in the hydrogen-bonded model systems in the framework of the Kohn–Sham MO model using a quantitative energy decomposition analysis (EDA) into electrostatic interaction, Pauli-repulsive orbital interactions, attractive orbital interactions and dispersion term [Eq. (3)]:[14,15,19-22]

$$\Delta E_{int} = \Delta V_{elstat} + \Delta E_{Pauli} + \Delta E_{oi} + \Delta E_{disp} \tag{3}$$

The detailed explanation of the different terms can be found in ref.[19] The orbital interaction energy can be further decomposed into the contributions from each irreducible representation of the interacting system [Eq. (4)] using the extended transition state (ETS) scheme developed by Ziegler and Rauk.[21] Our approach differs in this respect from the Morokuma scheme,[22] which instead attempts a decomposition of the orbital interactions into polarization and charge transfer. In systems with a clear $\sigma/\pi$ or A'/A'' separation (such as our planar DNA quartets), the symmetry partitioning in our approach proves to be most informative.

$$\Delta E_{oi} = \Delta E_{\sigma} + \Delta E_{\pi} \qquad (4)$$

The cooperativity in the hydrogen bonds in quartets is quantified by comparing $\Delta E_{int}$ (that is, formation of the quartet from four bases in the geometry of the former) with the sum ($\Delta E_{sum}$) of the individual pairwise interactions for all possible pairs of bases in the quartet, defined as [Eq. (5)]:

$$\Delta E_{sum} = 4\ \Delta E_{pair} + 2\ \Delta E_{diag} \qquad (5)$$

Here $\Delta E_{pair}$ is the interaction between two neighboring bases (that is, the interaction between two doubly hydrogen-bonded bases in the geometry of the quartet), and $\Delta E_{diag}$ is the interaction between two mutually diagonally oriented bases (that is, the interaction between two non-hydrogen bonded bases in the geometry of the quartet). The synergy that occurs in the quartet is then defined as the difference [Eq. (6)]:

$$\Delta E_{syn} = \Delta E_{int} - \Delta E_{sum} \qquad (6)$$

Thus, a negative value of $\Delta E_{syn}$ corresponds to a stabilizing cooperative effect, that is, a reinforcement of the quartet stability due to the occurrence of all hydrogen bonds simultaneously.

To follow the change in the charge distribution, the electron density distribution is analyzed using the Voronoi deformation density (VDD) method introduced in ref.[23] The VDD charge $Q_A$ is computed as the (numerical) integral of the deformation density $\Delta\rho(r) = \rho(r) - \Sigma_B \rho_B(r)$ associated with the formation of the molecule from its atoms in the volume of the Voronoi cell of atom A. The Voronoi cell of an atom A is defined as the compartment of space bounded by the bond midplanes on and perpendicular to all bond axes between nucleus A and its neighboring nuclei (cf. the Wigner–Seitz cells in crystals). The mass spectrometric and NMR methods are described elsewhere.[24]

## 2.2 Neutral and positively charged xanthine and uric acid derivatives as tetrad-forming agents[25]

The replacement of guanine with xanthine instead of hypoxanthine (which was used previously to model inosine nucleoside[26]), can also produce two H-bonds between neighbouring xanthines in special circumstances by low-barrier hydrogen bond to yield strongly bonded, planar structures. A possible implementation is to have a proton positioned between heteroatoms that face each other at nearly H-bond distance. Consequently, the system will have an additional positive charge and this provides a very strong interaction. Moreover, naturally occurring uric acid can be also involved in tetramer formation with or without xanthine. The low-barrier hydrogen bond is well characterized experimentally as well as theoretically[27] and according to our expectation; these systems are capable to bind anions naturally. In our computational studies we have investigated tetramer systems based on 9-methylguanine (G), 9-methylxanthine (Xa) and 9-methyluric acid (Ua) and due to the pK$_a$ values these protonated systems can exist only under strongly acidic conditions.[25]

Homo- and heterodimer structures from the above bases were constructed and evaluated (Fig. 1). The calculated bond energies are as follows in kcal/mol: −11.4 (G-G), −10.6 (Ua-Ua), 19.2 (XaH$^+$-XaH$^+$), −15.8 (Ua-XaH$^+$), −22.1 (XaH$^+$-Ua), and −23.6 (XaH$^+$-Xa). These dimer bonding energies are very promising, compared to the G-G dimer, especially in the presence of LBHB, with the exception of XaH$^+$-XaH$^+$ situation, where the bond energy (and the interaction energy as well) is positive. Regarding the large binding energy difference between the XaH$^+$-Xa and XaH$^+$-XaH$^+$ dimers, a logical explanation of the results is that the H-bonds cannot compensate the electrostatic repulsion in the XaH$^+$-XaH$^+$ case. All the structures discussed in this work are real minima, in particular also the XaH$^+$-XaH$^+$ dimer complex. The new dimers contain alternating H-bonds (proton donor and acceptor occur in each monomer), contrary to the G-G dimer, where all the proton donors belong to the G moiety. This pattern of alternating hydrogen bonds enhances the planarity of the new arrangements which is important in the formation of quadruplexes because planar tetramers establish stacking structures more efficiently. According to the binding energy values, the most stable quadruplex can be formed by XaH$^+$ and Ua molecules, but the homo Ua tetramer structure is also promising. In the former case, the quadruplex will be doubly positively charged, while the latter one will be neutral. Since all the dimer structures are true minima, the quadruply charged tetramer can be also constructed by XaH$^+$ molecules in principle but positive interaction energy is expected without additional circumstances. It is an open question whether the (XaH$^+$)$_4$ structure can eventually be stabilized by properly placed negative ions. Finally, although the XaH$^+$-Xa dimer provides the strongest binding energy, it is not expected to form the most stable quartet since two XaH$^+$-Xa units can be connected only through two H-bonds, i.e., one on each side.

The calculated bond energies (Figure 1, in kcal/mol) for tetramers with C$_s$ symmetry, comprising the above units with +4, +2 and zero net charges, are listed hereunder: −41.0 (Ua)$_4$, −24.7 (Xa-Ua)$_2$, −3.1 (Xa)$_4$, −41.9 (XaH$^+$-Ua)$_2$, 129.0 (XaH$^+$)$_4$, −42.9 (XaH$^+$-Xa)$_2$. Considering the total binding energies the quadruply charged (XaH$^+$)$_4$ system has a positive $\Delta E_{bond}$ value in accordance with the dimer consideration (see above) but the bond energy of the (Xa)$_4$ structure (−3.1 kcal/mol) is unexpectedly weak compared to that of A$_4$ (−33.1 kcal/mol at BLYP-D/TZ2P).[8] This is a consequence of that the relatively long (1.9 Å) sole H-bond cannot compensate the steric repulsion between the N7 and the O2 atoms in the C$_s$ symmetric case. Moreover, frequency analyses show that the (Xa)$_4$ optimum is a transition state in C$_s$ symmetry similar to the A$_4$ structure[8] and lone pair − lone pair repulsion between the N and O atoms is also possible. All other complexes are in real equilibrium, even in the case of (XaH$^+$-Xa)$_2$, in which steric repulsions can be found between the N7 and the O2 atoms, analogous to the situation in (Xa)$_4$.

For comparison, we optimized the G$_4$ tetrad at the same level of theory, and the calculated total binding energy of the guanine monomers is −64.6 kcal/mol. This value is in good agreement with other calculations (see e.g. ref.[8]) but a bit surprising regarding the dimer calculations. According to the preliminary dimer studies similar or stronger interactions were expected in certain cases [(Ua)$_4$, (XaH$^+$-Ua)$_2$, (XaH$^+$-Xa)$_2$] but the G$_4$ tetrad shows an extra strong binding compared to our new cases. This is the consequence of a cooperative bond effect in G$_4$ which is explained in detail in Section. 2.4. Here, we find that even the strong LBHB cannot compensate its absence in the present tetrads. In Fig. 1, we present the results of an alternative fragmentation scheme, namely (AB)+(CD)

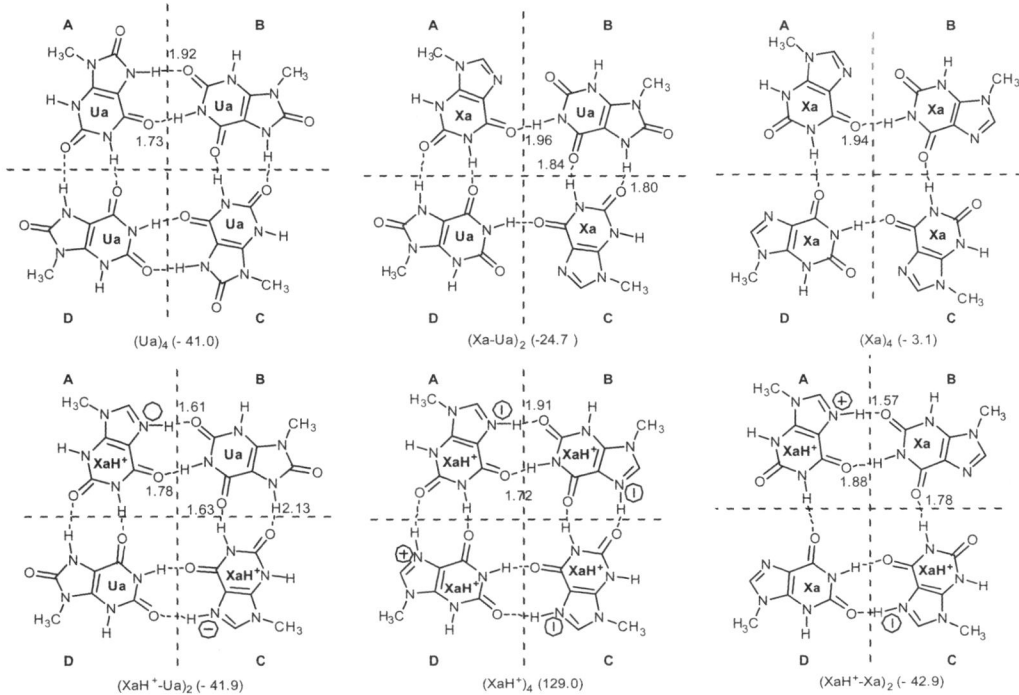

**Figure 1** *Tetramer structures investigated. The numbers in the parentheses show the calculated bond energy (in kcal/mol) referring to the monomer fragmentation in C(s) symmetry. Optimal bond distances are also presented (in Å). Dimer fragmentation is designated with dotted lines and letters A, B, C and D.*

and (AD)+(BC) according to the introduced areas A, B, C, and D in Fig. 1, in which the interaction between differently chosen dimer fragments have been calculated.

Concerning the optimized structures we found that the tetramer structures show universal distortion in the presence of the LBHB. Namely, taking the optimized dimers as reference systems, the N7-O2 distance is more elongated regarding the B-C (or A-D) fragments comparing to that of the LBHB containing A-B (or C-D) fragments. The consequence is an asymmetric dimer-dimer interaction. In particular, the fragment analyses provide 1.4 kcal/mol (AB+CD) and –19.9 kcal/mol (AD+BC) interaction energy in the case of $(XaH^+\text{-}Xa)_2$. The positive interaction value shows that the formation of this quartet is unlikely, contrary to the interaction in the dimer. This fact demonstrates that the interaction between the neutral and the protonated xanthine is still strong between the AD and BC dimer fragments but the H-bonds cannot compensate for the strong electrostatic repulsion between AB and CD units which have both a net positive charge. Likewise, in the case of $(XaH^+\text{-}Ua)_2$, we have –3.5 kcal/mol and –18.6 kcal/mol interaction energies for the AB+CD and AD+BC decompositions, respectively. Thus, the new H-bond is still weak between the N7-H group of the Ua and the O2 atom of the $XaH^+$ fragment, which is also confirmed by the relatively long H7···O2 bond length of 2.13 Å.

The lateral and central H-bonds in the tetrad are elongated and shortened, respectively, if compared to the corresponding ones in the dimer. There is only one exception in the $(XaH^+\text{-}Ua)_2$ complex, where the O6···H1 connection between the Ua and the $XaH^+$ was 1.61 Å in the dimer and 1.63 Å in the tetramer. Two effects are worth to mention in connection with this finding. First, the LBHB tends to form as short bond distance as possible while the electrostatic repulsion between the two positively charged dimer fragments tries to shift the dimers as far away as possible. These effects can cause together the different behaviour of this system. Moreover, this fragmentation proves that, in the present set of model systems, the LBHB cannot yield the same amount of energy gain as the cooperative effect (see e.g. the (AD)-(BC) interaction energies). From an electrostatic point of view it is understandable since the LBHB systems are divided into two positively charged fragments, where the interaction energy contains the electrostatic repulsion effect, too.

Finally, we have performed Voronoi Deformation Density (VDD) charge analyses[23a] for the $(Ua)_4$, $(XaH^+-Ua)_2$, $(XaH^+-Xa)_2$ and $(G)_4$ systems. According to the atomic VDD charges (data not shown), an additional anion cannot be bound in the central position, since the four oxygen atoms carry a large negative partial charge in all the cases. However, this situation opens the possibility that if the +2 total charge of the system is compensated by anions, an additional cation can still be bound. Moreover, the largest negative charges are present at the oxygen atom in the LBHB bonds. Thus, the binding position of the anion is expected to be slightly shifted away from the LBHB region.

We have also investigated the ion binding capacity of the new assemblies. Stacked quartets can form quadruplexes that are often further stabilized by cations (typically by potassium, sodium or ammonium), since they coordinate to the negatively charged carbonyl groups at the center of the guanine tetrad. The preferred guanine quadruplex structure adopted by a guanine-rich sequence in a DNA or RNA chain depends on the nature of the coordinating cations. VDD charge analyses point out that the central region of the new tetrads is similarly negatively charged as the G-tetramer. Hence, a $Na^+$ cation was placed into the center of the tetrads. Furthermore, when LBHB was assumed in the systems, it carries a positive charge (which according to the charge analyzes spreads around the neighboring atoms) and the anion binding is expected near this area. We have to note that two anions can be present in the doubly charged systems at the same time, and we could distinguish two arrangements: the anions are on the same or on the opposite sides of a tetrad.

Taking into account these facts, we have investigated the $(XaH^+-Ua)_2$ system with a $Na^+$ and two $F^-$ ions and the neutral $(Ua)_4$ system with a sodium cation. The $Na^+$ was always constrained to be on the $C_4$ symmetry axis, and the $F^-$ ions were placed close to the LBHB region at the $XaH^+$ molecule. The important role of the central cation at the $(XaH^+-Ua)_2$ tetrad is shown by the observation that when we tried to optimized the structures without the sodium ion, the two anions always destroyed the tetramer structures. For comparison the G-tetrad was also calculated with $Na^+$ in the center and we introduced a different fragmentation of the ion binding systems in contrast to the ion-free ones. The fragmentation schemes are as follows: (i) we calculated the interactions of the $Na^+$ ion with the remaining part of a complex to reveal the strength of $Na^+$ ion binding; (ii) we defined (dimer+anion+cation) and (dimer+anion) fragments in the complexes (where anions were present as well) in two ways, analogously to the corresponding situations without extra ions. This allows us to investigate how coordination of the cation affects the interactions between fragments in AB+CD and AD+BC decomposition.

We can conclude that the $G_4$ tetramer binds most strongly the sodium ion and the compensated $(XaH^+-F^--Ua)_2$ complexes are capable to bind $Na^+$ with nearly similar strength. Taking into account that the four oxygens around the sodium have very similar VDD charges in the ion free situation, this shows that the metal-tetramer interaction is mainly based on the ionic interaction between $Na^+$ and the four oxygen atoms. Regarding fragmentation (ii), we can see that the interaction energy differences were eliminated in the presence of a cation and anions, (the $(XaH^+-Ua)_2+2F^-+Na^+$ system) which ones were found between the AB+CD and AD+BC decompositions in the situation without ion. The different interaction energies between the fragments are nearly equal in the ion-containing systems and the values show a similar tendency regarding the different tetramer systems.

## 2.3 The behaviour of 3-methylxanthine in quadruplex structures[24]

The dominant 7H tautomeric form of a 3-substituted xanthine can bind to each neighboring xanthine moiety with two H-bonds in a tetrad, similarly to guanine. This is different from the usual 9-substituted xanthines in which additional protonation is required to form efficiently tetrads (cf. Section 2.2). Therefore, we have anticipated that these assemblies should yield strong interactions and a planar geometry in gas phase. As the simplest model, the naturally occurring 3-methylxanthine (3MX) was chosen for our studies (Fig. 2) with or without additional cation.

First, computational investigations have been performed for 3MX, as the simplest representative of the family of 3-substituted xanthines, at BLYP-D/TZ2P level of theory. It is worth to note that the inclusion of the dispersion correction shifts the absolute value of the interaction energies of the H-bonds, thus direct comparison with previous chapter values is not possible. The optimized tetrameric and octameric structures are presented in Fig. 2 in the absence (a) and the presence (b-d) of cations. The calculated hydrogen bond energy of the four 3MX monomer in the tetrad amounted to –66.1 kcal/mol.

The optimized tetrad structures have been found to be bent even in the ion-free case but cations show similar behavior to guanine tetrads.[28, 29] Namely, sodium ion tends to stay in the plane of the tetrad while potassium and ammonium ions occupy an out-of-plane position. According to the different ion radii of $Na^+$ (1.02 Å) and $K^+$ (1.38 Å), the results are reasonable and the similar out-of-plane optimums in the $K^+$ and $NH_4^+$ cases explain the ion bonding preference order in gas phase. Thus, the stability order is $Na^+ > NH_4^+ \approx K^+$ and the binding energy values between the $(3MX)_4$ tetrad and the cations are –96.6, –71.8 and –70.3 kcal/mol for $Na^+$, $NH_4^+$ and $K^+$ ions, respectively.

Regarding the octads, the most planar structure was found with the sodium ion, and the structure without any intercalating ion has the most bent optimum. In all cases the $(3MX)_4$ complex turned to be more planar in the octad structure. Cations have been found close to the midpoint between the layers and the rotation of the two layers in octads is ca. 17° in each aggregate. The distances between the layers have also been found to be very similar in each complex, thus neither the metallic ions nor the $NH_4^+$ affect the optimum stacking distance. The „external" H-bonds are always very close to the ideal linear bond orientation in the slightly bent structures (172-178°). These „external" bonds should be less linear in a planar geometry which can be also interpreted through the C=O and N-H bond angles.

The orders of the strength of ion-binding interaction in the octameric bilayer structure and the tetrad were the same ($Na^+ > NH_4^+ \approx K^+$). Note that the pure stacking bond energy between two $(3MX)_4$ tetrads is –47.0 kcal/mol. This is stronger than the corresponding stacking energy of the octameric 9-methylguanine cluster $G_8$ (–35.4 kcal/mol). Together with the theoretical tetrads, this prompted us to investigate the existence of $(3MX)_4$ and $(3MX)_8$ structures with or without intercalating cations experimentally as well.

First, the tetrad formation of 3MX was examined by mass spectrometry using a Q-TOF instrument equipped with a nano-ESI ion source. Although only $NH_4^+$ was intentionally added as an adduct forming ion in aq. methanol, the most intense peak series in the mass spectrum (Fig. 3) contains $Na^+$, possibly due to an ion-exchange on the surface

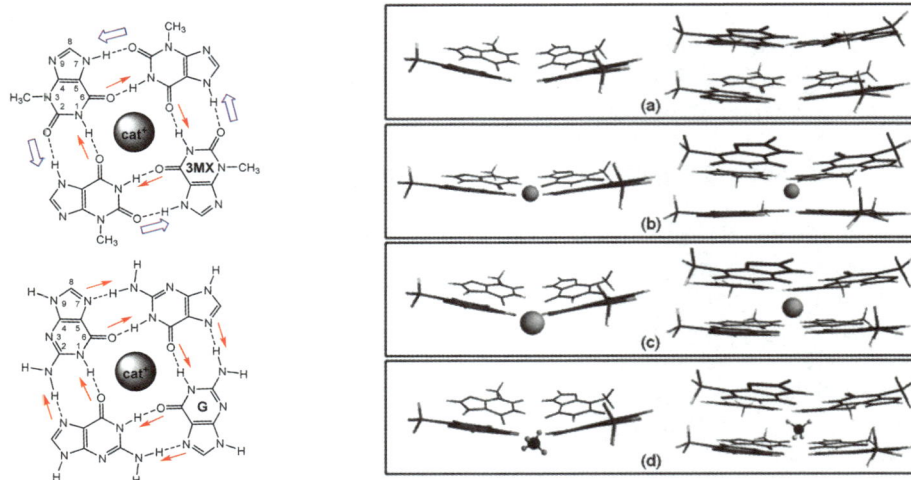

**Figure 2** *Top view of 7H-3-methylxanthine (3MX) (upper left panel), 9H-guanine (G) tetrads (lower left panel) and optimized lateral views of tetrad and octad structures of 3MX (right panels) without (a) and with cations (cat$^+$: b: Na$^+$, c: K$^+$, d: NH$_4^+$). The arrows (left panel) indicate the hydrogen bond orientation.*

of the borosilicate capillary. This is in good accordance with the calculations, where the sodium ion is bound most strongly to the tetrad. Interestingly, the peak intensities increased along the series of adduct ions of [3MX+Na]$^+$, [(3MX)$_2$+Na]$^+$, [(3MX)$_3$+Na]$^+$, and [(3MX)$_4$+Na]$^+$, in contrast to the "normal" case for compounds not forming tetrads. Furthermore, beyond the [(3MX)$_4$+Na]$^+$ peak, the next most intense peak series in the upper mass region belongs to the [(3MX)$_8$+cat]$^+$ (cat=K$^+$, Na$^+$, NH$_4^+$) peaks suggesting an increased stability of the tetrad constructed by four 3MX molecules. These results are in line with the findings of Mezzache et al. on metal ion adducts of the parent xanthine.[28]

Second, multinuclear NMR studies have performed. Thus, full $^1$H, $^{13}$C and $^{15}$N NMR assignment of 3MX has been accomplished using standard HMBC techniques in dilute DMSO-d$_6$ solution at 300 K. The obtained data (not shown)[24] support the 7H-tautomeric form as shown in Fig. 3. Since the MS study suggests (3MX)$_n$ · cat$^+$ aggregates (n = 4, 8; cat$^+$ = NH$_4^+$, Na$^+$, K$^+$), several additional NMR experiments were carried out to disclose these features. Diffusion ordered spectroscopy (DOSY)[30-32] yielded an apparent MW of 920 Da. For a (3MX)$_4$ aggregate MW = 664 Da is expected and the difference between the experimental and theoretical values may be attributed to solvent molecules associated to the clusters of 3MX. When 0.4 equiv. of potassium picrate was added to the solution, the apparent MW increased to 1130 Da, with 1.04 equiv. up to 1300 Da. To the contrary, increasing the temperature to 315 K decreased the apparent MW to 710. In other solvents (water, DMF, methanol, chloroform) the solubility was generally poor. We did not observe association e.g. in dilute water-DMSO solution. This is in accordance with the observation that co-solvents or co-solutes with relatively low dielectric constant are known to stabilize quadruplexes.[33] To observe the intermolecular H-bonding in the self-assemblies we measured the deuteration isotope shifts[34] on the acceptor carbonyls but this was not indicative because of the concurrent two- and three-bond intramolecular effects. Similarly, NH proton chemical shift temperature dependences were not well above the accepted limits for H-bonding. In selective transient 1D NOESY experiments[35] we observed

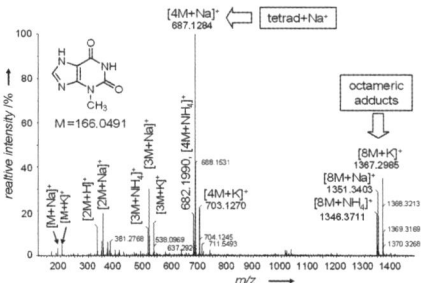

**Figure 3** *Nano-ESI-Q-TOF MS spectrum of 3-methylxanthine (3MX).*

significant magnetization transfer between N7H and water due to exchange while in the case of N1H this transfer was negligible. Consequently, this proves that, in accordance with the gas phase theoretical results, the „internal" H-bonds (N1H···O6) in the (3MX)$_4$ structures must be stronger than the „external" H-bonds (N7H···O2). Ribbon-like structures in solution with indefinite MW would involve equally both N1H and NH7 and therefore can be excluded as the two protons behave very differently (see above). Concerning the homonuclear NOEs from the same 1D NOESY experiments, only a small NOE is seen between N1H and the N3CH$_3$ group, and a stronger intramolecular effect between N7H and C8H can be observed. Intermolecular effects of N7H may possibly be reduced by the water exchange and were not observed. Using steady state $^{13}$C{$^{1}$H} heteronuclear NOE,[36, 37] we observed 14% and 44% heteronuclear NOE at C6 and C2, respectively, when N1H was saturated, while these carbons were „silent" when N7H was irradiated though at C-5 the two-bond hetero NOE yielded 17% enhancement. Furthermore, no couplings via intermolecular H-bonds were detected in either $^{15}$N or $^{13}$C HMBC experiments. In summary, the DOSY experiments seem to support the presence of (3MX)$_4$ · cat$^+$ and possibly of (3MX)$_8$ · cat$^+$ (cat$^+$ = none or K$^+$) clusters in DMSO-d$_6$ solution but it is difficult to prove unequivocally the H-bond pattern by direct NMR methods, most likely because of the dynamic nature of the weak self-assemblies.

## 2.4 The nature of cooperativity in 9*H*-guanine and 7*H*-xanthine tetrads[38]

The nature of hydrogen bonds in DNA base pairs has been the subject of numerous theoretical studies.[39, 40] Gilli et al.[41] proposed that these hydrogen bonds were reinforced by π assistance, the so-called resonance-assisted hydrogen bonding (RAHB). In previous work,[39] it was established theoretically that, for these hydrogen bonds, the electrostatic interactions and orbital interactions are of equal importance and that indeed the π electrons provide an additional stabilizing component. This finding was recently reconfirmed by Ziegler et al.[40] However, it was also shown computationally that the synergetic interplay between the delocalization in the π-electron system and the donor-acceptor interactions in the σ-electron system was small, that is, the simultaneous occurrence of the π and σ interactions is only slightly stronger than the sum of each of these interactions occurring individually.

Recently, cooperative effects were computationally established to occur also in the hydrogen-bonded quartet of guanine (G) bases (Fig. 2, left panel)[42] and in hydrogen-bonded supramolecular polymers. The computed hydrogen-bond energy of the guanine quartet (G$_4$) was found to be substantially more stabilizing than four times the hydrogen bond energy of one guanine pair (G$_2$) although the former has four times the number of hydrogen bonds as the latter. This phenomenon of cooperativity was ascribed to RAHB, i.e., resonance-assistance of the hydrogen bonds by the π electrons in the guanine quartet.[42] No such cooperativity was found in the structurally related xanthine quartet $X_4$ although this quartet also has a π-electron system at its disposal.[24, 25] The main difference between the guanine and xanthine quartets is that $G_4$ has all hydrogen bonds pointing in the same direction (Fig. 2, left panel: red arrows), whereas $X_4$ has its hydrogen bonds between two bases pointing in opposite directions (Fig. 2, left panel: red versus blue arrows). In this chapter, we show that the cooperative reinforcement between hydrogen bonds in guanine quartets is related to its special H-bond pattern and not caused by resonance assisted hydrogen bonding (RAHB).

The structures and energies of the guanine and xanthine tetrads have been computed at the BLYP-D/TZ2P level of theory. We have verified that $7H$-xanthine is the lowest-energy tautomer of xanthine, both in the gas phase and in water. Then, we have examined the cooperativity in $G_4$ and $X_4$ in the gas phase, in water, and within stacking structures. In the gas phase, $G_4$ is significantly more strongly bound than $X_4$, in aqueous solution, this difference disappears but in stacks, with or without cations, it appears again.

*Gas phase*: Appling [Eq. (6)] we found that there is indeed a large synergy of –20.9 kcal/mol in $G_4$ whereas in $X_4$ this term is essentially zero, i.e., only –1.5 kcal/mol (see Table 1). The large cooperative effect in the hydrogen bonds in the $G_4$ quartet has been ascribed by Otero et al.[42] to RAHB. This, according to the definition of Gilli et al.,[41] implies that the π electrons are responsible for this reinforcement and thus the high stability of $G_4$. To scrutinize this statement,[42] we calculated the synergy in the $G_4$ and $X_4$ quartet with and without the assistance by the π electrons. Switching off the π interactions is achieved by removing all π virtuals (i.e., unoccupied π orbitals) of the DNA bases. The removal of π virtuals precludes the mixing of occupied and unoccupied π orbitals as the DNA bases are forming the quartet which otherwise causes any resonance-assistance to the hydrogen bonds.

Switching off π assistance in the case of $X_4$ has little effect: the synergy was small (-1.5 kcal/mol) and it remains small (–2.5 kcal/mol). On the other hand, switching off π assistance in the case of $G_4$, causes the synergy to change only moderately, from –20.8 to still –14.3 kcal/mol (cf. Table 1). However it is still larger with an order of magnitude than the former one. Importantly, we can conclude that the synergy in the $G_4$ hydrogen bonds is caused primarily by another mechanism than π assistance.

*Aqueous solution*: Solvation reduces the interaction $\Delta E_{int}$ between DNA bases in both $G_4$ and $X_4$ but the weakening is 15.9 kcal/mol more pronounced in the former (from –89.1 to –37.1 kcal/mol, i.e., by 52 kcal/mol) than in the latter (from –72.6 to –36.5 kcal/mol, i.e., by 36.1 kcal/mol). Interestingly, this equalizing effect that solvation has on the quartet stabilities is largely caused by the disappearance of the synergy $\Delta E_{syn}$ in $G_4$. In the case of $G_4$, going from the gas phase to water causes the synergy to drop from –20.9 to 1.3 kcal/mol (see Table 2). At variance, in the case of $X_4$, the synergy was negligible in the gas phase (1.3 kcal/mol) and it remains small in water (2.2 kcal/mol).

*Stacks*: We have modelled this situation by stacks of three quartets, i.e., $(G_4)_3$ and $G_4X_4G_4$, optimized in $C_{4h}$ symmetry, in the absence of any further sodium ions. Interlayer distances (between respective planes of O6 atoms, Fig. 2) are 3.25 - 3.30 Å and the central quartet ends up being rotated with respect to the top and bottom ones by some 30° - 45° with respect to an eclipsed conformation of the O6 atoms in the respective layers. In the case of $(G_4)_3$, we have examined two different stacking patterns: either, all three layers have their hydrogen bonds pointing in the same direction (designated conrotatory: con-$(G_4)_3$) or the middle layer has its hydrogen bonds pointing in the opposite direction as compared to the lower and upper layer (designated disrotatory: dis-$(G_4)_3$). Thus, all together, we have three model stacks, namely, con-$(G_4)_3$, dis-$(G_4)_3$, and $G_4X_4G_4$ (Table 1).

As one can see, stacking effect has much less influence than solvation in water on the tetramer complexes. Stacking modifies the interaction $\Delta E_{int}$ between DNA bases only very slightly in both $G_4$ (from −89.1 to −87.0 kcal/mol, i.e., by 2.1 kcal/mol in both con-$(G_4)_3$ and dis-$(G_4)_3$) and $X_4$ (from −72.6 to −73.4 kcal/mol, i.e., a slight increase of less than 1 kcal/mol). Note that the effect of stacking does not differ for conrotatory and disrotatory orientation of the embedded guanine quartet. A slight influence of stacking interactions on the hydrogen bonds in DNA-base quartets is consistent with similar findings for stacked DNA and artificial base pairs.[43]

Importantly, the synergy in the $G_4$ quartet in a stacking environment ($\Delta E_{syn} \approx -15$ kcal/mol) remains 14 kcal/mol more stabilizing than that in the corresponding $X_4$ quartet ($\Delta E_{syn} \approx -1$ kcal/mol). Consequently, the interaction energy $\Delta E_{int}$ for formation of $G_4$ in a stack is also ca. 14 kcal/mol more binding than that for $X_4$. Thus, cooperativity between hydrogen bonds reinforces guanine quartets as compared to xanthine quartets also in stacked arrangements.

*Stacks with sodium*: The dicationic model stacks con-$G_4Na^+[G_4]Na^+G_4$, dis-$G_4Na^+[G_4]Na^+G_4$ and $G_4Na^+[X_4]Na^+G_4$ (optimized again in $C_{4h}$ symmetry) are bound with respect to the three separate (and optimized) quartets and two sodium ions by −318.8, -316.4, and -309.5 kcal/mol, respectively. This corresponds to average inter-layer stacking interactions of -159.4 ($G_4$–$G_4$), −158.2 ($G_4$–$G_4$), and −154.8 kcal/mol ($G_4$–$X_4$). Note that in the presence of inter-layer sodium ions, it is somewhat less favorable to stack $X_4$ than $G_4$ in between two $G_4$ quartets, just the other way around as compared to the stacks without sodium ions, in which case $X_4$ yields somewhat stronger stacking interactions (vide supra). Furthermore, interlayer distances (between respective planes of O6 atoms) are 2.67 - 2.78 Å and the central quartet ends up being rotated with respect to the top and bottom ones by some 35° - 45°. The sodium cations are involved in electrostatic as well as donor–acceptor orbital interactions with the lone-pairs on the O6 oxygen atoms. This results are in tight O–Na coordination bonds of 2.53 - 2.68 Å and slightly bowl-shaped outer quartets, curling away from the central layer.

Interestingly, as mentioned above, the "telomere-like" environment in the dicationic stacks involving sodium cations significantly reduces the interaction $\Delta E_{int}$ within $G_4$ and $X_4$ quartets but, at the same time, the cooperativity $\Delta E_{syn}$ between hydrogen bonds essentially retains the high value of the corresponding isolated quartet in the gas phase. The interaction $\Delta E_{int}$ and cooperativity $\Delta E_{syn}$ are computed again applying the approach as introduced above for the neutral stacks where now, in Eqs. 8 and 9, "$G_4[\ ]G_4$" is to be substituted by "$G_4Na^+[\ ]Na^+G_4$" (see also Eqs. 4 - 5). Thus, the interaction $\Delta E_{int}$ within the central quartets amounts to only −59.0 ($G_4$), −62.0 ($G_4$) and −53.2 kcal/mol ($X_4$) in con-$G_4Na^+[G_4]Na^+G_4$, dis-$G_4Na^+[G_4]Na^+G_4$ and $G_4Na^+[X_4]Na^+G_4$, respectively (see Table 2). This interaction is roughly 20 - 30 kcal/mol less stabilizing than the corresponding ones within the quartets in the neutral stacks or in the gas phase, but it is still 17 - 25 kcal/mol more stabilizing than the corresponding interactions within the quartets in aqueous solution (see Table 2). On the other hand, the cooperativity $\Delta E_{syn}$ between the hydrogen bonds within the central quartets is −20.5 ($G_4$), −22.3 ($G_4$) and −5.5 kcal/mol ($X_4$) in con-$G_4Na^+[G_4]Na^+G_4$, dis-$G_4Na^+[G_4]Na^+G_4$ and $G_4Na^+[X_4]Na^+G_4$, respectively (see Table 2). This is within a few kcal/mol the same value as in the gas-phase situation. Thus, guanine quartets still benefit from additional stabilization through a cooperativity effect in a telomere-like environment, whereas xanthine quartets do not.

**Table 1** Analysis of interaction energy (in kcal/mol) of 9H-guanine and 7H-xanthine tetrads.[a]

| Environment | Quartet | $\Delta E_{int}$[b] | $\Delta E_{diag}$[b] | $\Delta E_{pair}$[b] | $\Delta E_{sum}$[b] | Synergy[b] |
|---|---|---|---|---|---|---|
| Gas phase | $G_4$ | −89.1 | −1.9 | −16.1 | −68.2 | −20.9 |
| | $G_4$ no π | −75.4 | −1.9 | −14.4 | −61.2 | −14.3 |
| | $X_4$ | −72.6 | −0.2 | −17.7 | −71.1 | −1.5 |
| | $X_4$ no π | −64.6 | −0.2 | −15.5 | −62.2 | −2.4 |
| Water | $G_4$ | −37.1 | −0.3 | −9.5 | −38.4 | 1.3 |
| | $X_4$ | −36.5 | −0.4 | −9.5 | −38.7 | 2.2 |
| $G_4$–[ ]–$G_4$ | $G_4$ con | −87.0 | −1.9 | −17.0 | −71.6 | −15.3 |
| $G_4$–[ ]–$G_4$ | $G_4$ dis | −87.0 | −1.6 | −17.0 | −71.0 | −16.0 |
| $G_4$–[ ]–$G_4$ | $X_4$ | −73.4 | −0.4 | −17.9 | −72.3 | −1.1 |
| $G_4$–$Na^+$–[ ]–$Na^+$–$G_4$ | $G_4$ con | −59.0 | 1.5 | −10.4 | −38.4 | −20.5 |
| $G_4$–$Na^+$–[ ]–$Na^+$–$G_4$ | $G_4$ dis | −62.0 | 1.1 | −10.5 | −39.7 | −22.3 |
| $G_4$–$Na^+$–[ ]–$Na^+$–$G_4$ | $X_4$ | −53.2 | 1.8 | −12.8 | −47.7 | −5.5 |

[a] Computed at BLYP-D/TZ2P in $C_{4h}$ symmetry. See Eq. 1. [b] See Eqs. 2 - 6.

*Origin of cooperativity in guanine quartets.* To trace the appearance of cooperativity in guanine quartets, we have constructed $G_4$ by taking one of the guanine bases in the quartet and stepwise adding the other three guanine bases (always in the geometry of $G_4$), i.e., $G + G = G_2$, $G_2 + G = G_3$ and $G_3 + G = G_4$. This stepwise approach enables us to examine accurately why and at which point cooperativity begins to appear. The same approach was applied to the xanthine quartet. The synergy ($\Delta E_{syn}$), defined by Eq. (6), in the interaction energy amounts to −20.9 kcal/mol for $G_4$ (Table 2). The energy decomposition analyses in this section show that there are two main contributions to this synergy which are of the same order of magnitude: (i) the synergy in the electrostatic attraction of −8.7 kcal/mol; and (ii) the somewhat larger synergy in the orbital interactions of −10.8 kcal/mol. The latter originates mainly from a synergy of −7.9 kcal/mol in the orbital interactions in the σ-electron system, with a smaller contribution of −2.9 kcal/mol stemming from synergy in the resonance assistance of the π-electron system (Table 2). The corresponding terms in the energy decomposition analyses for $X_4$ are all close to zero, in line with cooperativity being essentially absent in the xanthine quartet (Table 2). Therefore, one can conclude that the cooperativity leading to the enhanced stability of the guanine quartet does not stem from resonance assistance.

**Table 2** Energy decomposition for the formation of 9H-guanine and 7H-xanthine tetrads (kcal/mol). a. Computed at BLYP-D/TZ2P for frozen (fragments of) $C_{4h}$-symmetric $Z_4$ (Z = 9H-guanine or 7H-xanthine). b. Overall synergy in each energy term is defined as $\Delta E(Z_2) + \Delta E(Z_3) + \Delta E(Z_4) - 4 \cdot \Delta E(Z_2) - 2 \cdot \Delta E(Z_2^{diag})$.

| | $\Delta E_\sigma$ | $\Delta E_\pi$ | $\Delta E_{oi}$ | $\Delta E_{Pauli}$ | $\Delta V_{elstat}$ | $\Delta E_{disp}$ | $\Delta E_{int}(Z_4)$ |
|---|---|---|---|---|---|---|---|
| Guanine synergy[b] | −7.9 | −2.9 | −10.8 | −1.4 | −8.7 | 0 | −20.9 |
| Guanine synergy[b] | −7.9 | −2.9 | −10.8 | −1.4 | −8.7 | 0 | −20.9 |

What is the mechanism behind the synergy in the electrostatic attraction and charge-transfer orbital interactions in the $G_4$ hydrogen bonds? The bonding analyses lead to a clear answer: the donor–acceptor orbital interactions associated with the $\Delta E_\sigma$ term induce a charge separation which in turn enhances both the orbital interactions and the electrostatic attraction with an additional guanine base. In the $\pi$-electron system, there is no charge transfer from one base to the another. The reason is that the $\pi$-electron system is effectively interrupted across two bases by the bridging proton in the hydrogen bonds. Therefore, the overlap between $\pi$-frontier orbitals is one to two orders of magnitude smaller than that between $\sigma$-frontier orbitals.

## 3 SUMMARY

First, dimer and tetramer structures based on 9-methylxanthine and 9-methyluric acid have been investigated by high level quantum-chemical calculations. We have found that while dimers show very promising stabilities, the new tetramer structures are in all cases more weakly bound than the well-known $G_4$ tetrad. This is the consequence of the absence of the cooperative H-bonding in the new systems, which provides additional stabilization in the $G_4$ complex. Still, the binding energies between the monomers in the new quartets are acceptable and the planarity of the latter ones is also good compared to the $G_4$ case. The absence of a cooperative effect is partly compensated by strong low-barrier hydrogen bonds (LBHB) in particular systems with net positive charges. The LBHB provides an extra strong interaction between dimer fragments in the tetramers and because of the positive net charge the structures are capable of binding anions. This latter character of the new systems is very important because so far quadruplexes or higher ordered structures with anion intercalation, to our knowledge, has been investigated only in one case.[44] The new positively charged tetramers most probably can exist only in presence of ions, and the monomers without ions easily decompose into dimer fragments. This decomposition is prevented by coordination of ions. We would like to emphasize that either the LBHB or the cooperative effects cannot be described and thus not uncovered by any of the commonly used nolecular mechanics (MM) methods. Only quantum chemical calculation or a properly modified molecular mechanical method is suitable to investigate these systems.

Furthermore, gas-phase computational studies suggest that 3-methylxanthine (3MX) is a good candidate for tetrad and quadruplex structures, Indeed, the existence of $(3MX)_4 \cdot cat^+$ and $(3MX)_8 \cdot cat^+$ ($cat^+ = NH_4^+$, $Na^+$, $K^+$) aggregates in the gas phase has been experimentally observed (MS). Detailed NMR studies have also verified that the „internal" H-bonds (N1H···O6) in the $(3MX)_4$ structures must be stronger than the „external" H-bonds (N7H···O2). Clearly, to achieve stronger interactions in tetrads/quadruplexes composed of 3-substituted xanthines, substituents in position 3 other than methyl should be present to allow for further stabilization and this work is in progress.

In general, we found that guanine quartets ($G_4$) are more strongly bound than xanthine quartets ($X_4$), despite the fact hat they have the same number of hydrogen bonds. It is due to a cooperativity effect in the former case, as follows from our computational study. This is true not only in the gas phase, but also in telomere-like structures where the quartet coordinates to two sodium ions and this complex is stacked in between two other quartets. In aqueous solution and in the absence of stacking quartet partners, however, the cooperativity is quenched and $G_4$ and $X_4$ are equally strongly bound.

The cooperativity in $G_4$ originates from the charge separation that goes with donor–acceptor orbital interactions in the $\sigma$-electron system from N and O lone-pair orbitals on one G base to $\sigma^*$N–H acceptor orbitals on the other G base. This picture emerges from detailed bonding analyses based on quantitative Kohn-Sham MO theory in combination with energy decomposition analyses (EDA). Resonance assistance by the π-electron system (i.e., RAHB) is shown to slightly contribute to the hydrogen bond strength but not to the cooperativity. Thus, the cooperativity in guanine quartets, which also contributes to the stability of telomere structures, is a direct consequence of the covalent (charge transfer) component in the hydrogen bonds.

## Acknowledgements

This study was supported by grants OTKA NK 73672, NK 68578, CK 61577, CK 77515, COST Action MP0802, TÁMOP 4.2.1.B 09/1/KONV, University of Szeged, HPC-Europa2 (project number 228398, Hungarian Eötvös Fellowship, Netherlands Organization for Scientific Research (NWO-CW and NWO-NCF), the National Research School Combination - Catalysis (NRSC-C) and the Netherlands Organization for Scientific Research (NWO-CW and NWO-NCF) for financial support.

## References

1. S. Burge, G. N. Parkinson, P. Hazel, A. K. Todd and S. Neidle, *Nucleic Acids Res.*, 2006, **34**, 5402.
2. J. T. Davis and G. P. Spada, *Chem. Soc. Rev.*, 2007, **36**, 296.
3. S. Lena, S. Masiero, S. Pieraccini and G. P. Spada, *Chem. Eur. J.*, 2009, **15**, 7792.
4. S. Neidle and S. Balasubramanian, eds., *Quadruplex nucleic acids*, The Royal Society of Chemistry, Cambridge, 2006.
5. A. Taylor, J. Taylor, G. W. Watson and R. J. Boyd, *J. Phys. Chem. B*, 2010, **114**, 9833.
6. M. Franceschin, *Eur. J. Org. Chem.*, 2009, 2225.
7. S. Martic, X. Y. Liu, S. N. Wang and G. Wu, *Chem. Eur. J.*, 2008, **14**, 1196.
8. C. Fonseca Guerra, T. van der Wijst, J. Poater, M. Swart and F. M. Bickelhaupt, *Theor. Chem. Acc.*, 2010, **125**, 245.
9. B. C. Pan, K. Shi and M. Sundaralingam, *J. Mol. Biol.*, 2006, **363**, 451.
10. Computer code ADF2008.01: E. J. Baerends, *et al.*, http://www.scm.com.
11. (a) M. Swart and F. M. Bickelhaupt, *Int. J. Quantum. Chem.*, 2006, **106**, 2536. (b) M. Swart and F. M. Bickelhaupt, *J. Comput. Chem.*, 2008, **29**, 724.
12. A. D. Becke, *Phys. Rev. A*, 1988, **38**, 3098.
13. C. Lee and W. Yang, R. G. Parr, *Phys. Rev. B*, 1988, **37**, 785.
14. S. Grimme, *J. Comput. Chem.*, 2004, **25**, 1463.
15. S. Grimme, *J. Comput. Chem.*, 2006, **27**, 1787.
16. (a) A. Klamt and G. Schüürmann, *J. Chem Soc. Perkin Trans.*, 1993, 799. (b) A. Klamt, *J. Phys. Chem.*, 1995, **99**, 2224.
17. M. Swart, E. Rösler and F. M. Bickelhaupt, *Eur. J. Inorg. Chem.*, 2007, 3646.
18. K. E. Riley, J. Vondrasek and P. Hobza, *Phys. Chem. Chem. Phys.*, 2007, **9**, 5555.
19. F. M. Bickelhaupt and E. J. Baerends, In *Reviews in Computational Chemistry*; K. B. Lipkowitz and D. B. Boyd, Eds.; Wiley-VCH: New York, 2000, Vol. 15; p 1-86.
20. C. Fonseca Guerra and F. M. Bickelhaupt, *J. Chem. Phys.*, 2003, **119**, 4262.
21. (a) T. Ziegler and A. Rauk, *Inorg. Chem.*, 1979, **18**, 1755. (b) T. Ziegler and A. Rauk, *Inorg. Chem.*, 1979, **18**, 1979, 1558. (c) T. Ziegler and A. Rauk, *Theor. Chim. Acta*, 1977, **46**, 1
22. (a) K. Morokuma, *J. Chem. Phys.*, 1971, **55**, 1236. (b) K. Kitaura and K. Morokuma, *Int. J. Quantum. Chem.*, 1976, **10**, 325.

23. (a) C. Fonseca Guerra, J.-W. Handgraaf, E. J. Baerends and F. M. Bickelhaupt, *J. Comput. Chem.*, 2004, **25**, 189. (b) F. M. Bickelhaupt, N. J. R. van Eikema Hommes, C. Fonseca Guerra and E. J. Baerends, *Organometallics*, 1996, **15**, 2923.
24. J. Szolomájer, G. Paragi, G. Batta, C. Fonseca Guerra, F. M. Bickelhaupt, Z. Kele, P. Pádár, Z. Kupihár and L. Kovács, *New J. Chem.*, 2011, **35**, 476.
25. G. Paragi, L. Kovács, Z. Kupihár, J. Szolomájer, B. Penke, C. Fonseca Guerra and F. M. Bickelhaupt, *New J. Chem.*, 2011, **35**, 119.
26. F. C. Meng and X. Zhao, *J. Mol. Struct.-Theochem*, 2008, **869**, 94.
27. F. W. Smith and J. Feigon, *Biochem.*, 1993, **32**, 8682.
28. S. Mezzache, S. Alves, J. P. Paumard, C. Pepe and J. C. Tabet, *Rapid Commun. Mass Spectrom.*, 2007, **21**, 1075.
29. J. D. Gu and J. Leszczynski, *Chem. Phys. Lett.*, 2002, **351**, 403.
30. C. S. Johnson, *Prog. Nucl. Magn. Reson. Spectrosc.*, 1999, **34**, 203.
31. D. H. Wu, A. D. Chen and C. S. Johnson, *J. Magn. Reson., Ser. A*, 1995, **115**, 260.
32. G. A. Morris and H. Barjat, in *Methods for structure elucidation by high-resolution NMR*, G. Batta, K. E. Kövér and C. Szántay Jr., eds., Elsevier, Amsterdam, 1997, pp. 209.
33. I. V. Smirnov and R. H. Shafer, *Biopolymers*, 2007, **85**, 91.
34. T. Dziembowska, P. E. Hansen and Z. Rozwadowski, *Prog. Nucl. Magn. Reson. Spectrosc.*, 2004, **45**, 1.
35. K. Stott, J. Stonehouse, J. Keeler, T. L. Hwang and A. J. Shaka, *J. Am. Chem. Soc.*, 1995, **117**, 4199.
36. K. E. Kövér and G. Batta, *Prog. Nucl. Magn. Reson. Spectrosc.*, 1987, **19**, 223.
37. H. B. Seba, P. Thureau, B. Ancian and A. Thevand, *Magn. Reson. Chem.*, 2006, **44**, 1109.
38. C. Fonseca Guerra, H. Zijlstra, G. Paragi and F. M. Bickelhaupt, J. G. Snijders and E. J. Baerends, *Chem. Eur. J.*, 2011, **17**, 12612.
39. (a) C. Fonseca Guerra, F. M. Bickelhaupt, J. G. Snijders and E. J. Baerends, *Chem. Eur. J.*, 1999, **5**, 3581. (b) C. Fonseca Guerra and F. M. Bickelhaupt, *Angew. Chem. Int. Ed.*, 1999, **38**, 2942. (c) C. Fonseca Guerra, F. M. Bickelhaupt, J. G. Snijders and E. J. Baerends, *J. Am. Chem. Soc.*, 2000, **122**, 4117. (d) C. Fonseca Guerra, F. M. Bickelhaupt and E. J. Baerends, *Cryst. Growth Des.*, 2002, **2**, 239. (e) C. Fonseca Guerra, Z. Szekeres and F. M. Bickelhaupt, *Chem. Eur. J.*, 2011, **17**, online.
40. R. Kurczab, M. P. Mitoraj, A. Michalak and T. Ziegler, *J. Phys. Chem. A*, 2010, **114**, 8581.
41. G. Gilli, F. Belluci, V. Ferretti and V. Bertolasi, *J. Am. Chem. Soc.*, 1989, **111**, 1023.
42. R. Otero, M. Schöck, L. M. Molina, E. Loegsgaard, I. Stensgaard, B. Hammer and F. Besenbacher, *Angew. Chem. Int. Ed.*, 2005, **44**, 2270
43. (a) A. Gil, V. Barbchadell, J. Bertran and A. Oliva, *J. Phys. Chem. B*, 2010, **113**, 4907. (b) D. Guo, R. P. Sijbesma and H. Zuilhof, *Org. Lett.*, 2004, **6**, 3667. (c) K. Vanommeslaeghe, P. Mignon, S. Loverix, D. Tourwé and P. Geerlings, *J. Chem. Theory Comput.*, 2006, **2**, 1444.
44. T. van der Wijst, B. Lippert, M. Swart, C. F. Guerra and F. M. Bickelhaupt, J. Biol. Inorg. Chem., 2010, 15, 387.

# COMPUTATIONAL METHODS FOR STUDYING G-QUADRUPLEX NUCLEIC ACIDS

B. Islam[1], V. D'Atri[1], M. Sgobba[1], J. Husby[2] and S. Haider[1]*

[1] Centre for Cancer Research and Cell Biology, Queen's University of Belfast, Belfast BT9 7BL, UK
[2] UCL School of Pharmacy, Brunswick Square, London WC1N 1AX
*Corresponding Author, Email: s.haider@qub.ac.uk

## 1 INTRODUCTION

Guanine repeat nucleic acid sequences in the presence of cations can self-associate to form a four-stranded arrangement called G-quadruplex[1]. It has garnered interest in the past decade due to abundance of potential quadruplex structures in several biologically relevant sequences such as telomere, immunoglobulin switch regions, promoter regions and other disease associated regions[2]. G-quadruplex forming regions have also been explored as promising anti-cancer targets[3]. Along with the experimental techniques, computational methods including molecular dynamics have been used to appreciate the biological relevance of quadruplexes [4, 5]. A quadruplex is stabilized via cyclic arrangement of eight Hoogsteen hydrogen bonds between four planar guanine bases to form a G-quartet (or G-tetrad)[5]. Several quartets can stack over one another to form the stem of the quadruplex. The cations are located in the central helical cavity and forms coordination with carbonyl oxygen from guanines. Quadruplexes are highly diverse with regard to the origin, the cations involved in the stabilization, orientation of the strands and the glycosidic conformation of the bases. They can be monomeric (single strand), dimeric (two strands) and tetrameric (four strands) in origin. The monomeric quadruplex are formed in strands with general sequence $G_aXG_bYG_cZG_d$, where a, b, c and d are in the range 3-5. X, Y and Z represent the loop sequences and comprise of 1-8 non-guanine nucleotides. Dimeric quadruplexes are formed in two seperate strands of sequence $G_aXG_b$. A quadruplex is called tetramolecular when four separate strands interact to form the quadruplex structure.

## 2 ROLE OF CATIONS

The formation of quadruplex requires coordination of cations for its stability[6]. The experimental observations of quadruplex crystal structures confirm that cations are an integral part of all quadruplexes[6]. The quartets undergo hydrophobic stacking to form a central channel contributed by carbonyl oxygen of the four guanines. The cations interact with the carbonyl oxygen by electrostatic forces and neutralize the repulsion between them. The role of $K^+$, $Na^+$, $Rb^+$, $Cs^+$, $NH_4^+$, $Tl^+$, $Sr^{2+}$, $Ba^{2+}$ and

Pb$^{2+}$ ions in stabilizing the quadruplex has been elucidated[6]. Ca$^{2+}$ and Mg$^{2+}$ ions do not promote the formation of G-quadruplex structures. K$^+$ is the most stabilizing cation and coordinates with O6 atoms of guanine bases in bipyramidal antiprismatic geometry. The ionic radius of K$^+$ ions (1.33Å) is too large to coordinate in the plane of G-quartet. Therefore, they occupy interstitial space in between two quartets. The adjacent K$^+$ ions in the ion channel are separated by an average distance of 3.38Å[7]. The order of stability provided by cation is K$^+$ > Rb$^+$ > Na$^+$ > Li$^+$ = Cs$^+$ and Sr$^{2+}$ > Ba$^{2+}$ > Ca$^{2+}$ > Mg$^{2+}$ [8]. In addition to their role in stabilizing quadruplexes, cations also play an important role in determining their structure. Quadruplexes with multiple cation coordination geometries have also been reported [9]. Quadruplexes with K$^+$ ions have larger G-quartets and hence provide greater loop flexibility than with those with Na$^+$ ions [10]. K$^+$ ions favour formation of anti-parallel quadruplex and the interconversion between parallel and anti-parallel structures can be controlled by Na$^+$/K$^+$ balance[11]. The cations stabilize as well as direct the structural conformation of thrombin binding aptamer (TBA)[12]. Spontaneous exchange of cation between the quadruplex structure and bulk solvent during the course of simulation has also been evidenced. The exchange is however very rapid so as to minimize any destabilization to quadruplex structure[12].

## 3 MOLECULAR DYNAMICS SIMULATIONS: CONSIDERATIONS AND LIMITATIONS

The interactions of G-quadruplexes have been widely studied by Molecular Dynamics (MD) carried out in explicit solvent[4, 5, 12, 13]. MD simulation consists of solution of the classical equations of motion, which for the force on a simple atomistic system may be represented as $F=ma$. The force on an individual atom is contributed by its bonded and non-bonded terms. The interatomic parameters like bond length, bond angle, dihedral angles, coulombic and van der Waals interactions all determine the effective force on an atom. The acceleration term in the equation is used to determine the velocity and position of each atom at a course of time. A trajectory is then calculated by tracing the course of position vectors as a function of time. The determination of suitable potential function is extremely important for a MD simulation[14]. It is assumed that atoms attract each other when they are at long distance while they repel at short interatomic distance. This is represented by Lennard-Jones potential. MD simulations are usually carried out in canonical (NVT) or Isobaric-isothermal (NPT) ensemble.

### 3.1 Force Fields

Force fields in molecular mechanics are used to include all the quantum mechanical parameters that describe the molecular behaviour. The total energy of a molecule is calculated by taking into account all atom-atom contributions such as Coulombic, polarization, dispersion and repulsive energies[15]. There is no "universal force-field" and preference of one force field over the other may vary based on the target application. Accurate parameterization requires set of input data and significant physical approximations to describe the intermolecular interactions. Atomistic molecular dynamics (MD) computer simulations for B-DNA have been successfully carried for several years and show good agreement with the experimental results[16]. Haider *et al.* performed a 20ns simulation of quadruplex DNA using GROMOS force field. The four-stranded structure was lost within the first 10ns of the simulation[17]. The CHARMM27 force field when used on folded RNA structures reveals unstable trajectories[18]. However, Cornell parameters that are normally associated with AMBER have been successfully used with CHARMM suite of programs for hammerhead ribozyme and guanine riboswitch[18]. MD simulations of G-quadruplex loops using Cornell force field, parm94 and parm98/99 on a nanosecond-scale are reported to show significant variations although they work properly for G-stems[19]. An α/γ torsional term was re-parameterized and presented in parmbsc0 force field to resolve various inaccuracies of previous force fields[20]. The benchmarking exercise of MD simulations of G-quadruplex for 1.5µs in parmbsc0 force fields shows reasonable agreement with the experimental values. The force fields were tested on the $d(G_4T_4G_4)_2$ dimeric quadruplex with diagonal loops and also the parallel stranded human telomeric monomolecular quadruplex $d[AG_3(TTAG_3)_3]$ with three propeller loops[21]. The parmbsc0 force field have also been used for 10ns simulation for free-energy analysis of G-quadruplex DNA. The estimates based on MD simulations are in close accord with the experimental results proving the validity of force field for quadruplex analysis[13]. A combination of parm99 and parmbsc0 has also been used for a longer microsecond scale simulation of quadruplex[21].

The binding of ligands to G-quadruplex has also been explored by MD using different force fields of AMBER. AMBER force field ff03 was used for quadruplex, water and ions while parameters for the ligands were assigned using generalised AMBER force field (GAFF) method for MD simulations of G-quadruplex with ligands carried for 10-30ns. The backbone of the quadruplex was reasonably stable but the flanking nucleotides show fluctuations during the course of simulation due to the limitations of the force fields[22]. MD simulations of Azatrux and parallel quadruplex DNA using parmbsc0 force field has been carried out for 15ns. The additional parameters for the Azatrux were assigned using AMBER-GAFF method. The experimental results corroborated with the molecular dynamics data implying that parmbsc0 force field is suitable for MD simulations of quadruplex-ligand complex along with the GAFF method for ligand parameterisation[23]. It has to be mentioned that quantitative derivations like free-energy calculations should be avoided using ligand force-fields. MD simulations of ligand-quadruplex interactions can be best used to augment the experimental predictions or interpretations.

**3.2 Long-range Electrostatic Interactions**

It is computationally ineffective to sum all the non-bonded interactions in a MD simulation. Therefore, "cutoff" procedures are employed to neglect long-range electrostatic interactions beyond some distance without altering the quality of simulations. The long range electrostatic forces play an important role in stability of quadruplex DNA as they represent the quadruplex-cation interactions. Thus, MD simulations of quadruplex DNA structures require accurate treatment of electrostatic interactions for achieving stable trajectories. Initial report of quadruplex analysis used spherical cutoff methods for handling electrostatic interactions[24]. The trajectories of these simulations were unstable and cations from the quadruplex were also dislodged during the course of simulation eventually leading to collapse of structure[24]. With the introduction of Particle Mesh Ewald (PME) truncation method a stable DNA simulation is achievable[25]. The computational time is longer for the PME method than for the atom-based truncation method due to the implicit requirement of periodic boundary conditions for Ewald summation. Inspite of time consummation factor, PME is universally used for accurate and stable simulations of quadruplex DNA[26]. Quadruplex MD simulations with PME summation method and 10Å cut-off have yielded fairly stable trajectories[5].

## 3.3 Inter-base Interactions and Backbone Descriptions

The force field parameters of nucleic acid molecules are also influenced by interactions between the bases. The amino groups of nucleobases tend to be non-planar due to a partial $sp^3$ pyramidal hybridization[27]. This affects the stabilization of bifurcated H-bonds, close amino group contacts, non-planar G/A base pairs and some other specific interactions[27]. The force fields assume purely $sp^2$ amino nitrogen[27]. This should be sufficient for most interactions, as primary H-bonds stabilize the $sp^2$ electronic structure. However, the force fields support neither out-of-plane H-bonds nor amino–acceptor interactions. Not surprisingly, neither bifurcated H-bonds nor amino-group contacts are properly reproduced by contemporary MD simulations.

Assignment of charge on a base for quantum mechanics is purely hypothetical as there is no quantum mechanical operator for atomic charges[27]. Therefore, treatment of molecular interactions based on charge distributions like hydrogen bonding should be done with care. The arbitrary charges on the bases can be assigned by fitting to molecular electrostatic potential (MEP, ESP charges). The popular AMBER fields use this method for the assignment of charges for calculation of electronic interactions[28]. The backbone of nucleic acid is a fairly difficult molecule to deal for MD simulations due to two major reasons. Firstly, the backbone is a highly flexible structure and requires geometry-dependent charges rather than constant point charges. Secondly, the highly negative charge on the nucleic acid changes with conformation and solvation dynamics. This is not handled by the non-polarizable atom-atom pair additive force fields. Failure of parm94 and parm98/99 AMBER force fields for simulation of DNA is reported to be primarily due to accumulation of α/γ substrates resulting in distortions of the double helix[29-33]. An improvised force-field parmbsc0 has correction term for α/γ DNA backbone[34]. Recent simulations on nanoscale and microscale report that G-stems remain stable while the loops may show slight but acceptable fluctuations over the entire course of simulation[20, 35]. Parmbsc0 is therefore preferred force field for the simulation of quadruplex DNA.

## 4 SIMULATION-BASED METHODS

### 4.1 Classical Molecular Dynamics

With the recent advances in force field parameterization and PME technique for treatment of long range electrostatic interactions, Classical MD has been the most common method to investigate the conformational flexibility and dynamic behaviour of quadruplex structures. Initial MD of G-quadruplex employed AMBER4.1 force fields for simulations[36]. Parallel and Antiparallel quadruplex formed by the sequences $d(G_4)$ and $d(G_4T_4G_4)$ respectively, with $Na^+$ reveal exceptionally stable trajectories over 25ns of simulation. The study revealed that G-stems can exchange cations from the solvent without significant perturbation. Complete removal of cations from the central ion channel of quadruplex has also been observed in this simulation. It may be due to inappropriate force field calculations of ion-guanine interactions. The removal of cations from the quadruples results in immediate collapse underscoring their importance in maintaining the stability of quadruplex[36]. During the simulation of parallel quadruplex, deformation of inner quartets has also been reported. This implicates that radius of ion is too large to optimally describe the ion interactions inside the stem. The trajectories associated with loops show more fluctuations compared to the G-stem[36]. Recently Classical MD has also been used to study the cation binding to Thrombin Binding Aptamer (TBA) quadruplex structure using parmbsc0 force field. The study reveals that cations are sucked into the quadruplex structure for its stabilization. The penetration of cation and binding events are modulated by the loop structure and depends on the type of cation. The spontaneous exchange of cations between the solvent and quadruplex molecule at the atomistic resolution is also evidenced in this study[12].

The presentation of crystal structure of quadruplex from human telomeric DNA attracted many molecular dynamics studies to provide an insight in this assembly[7,37]. Classical MD simulations have also been used to investigate the ligand binding to telomeric quadruplex DNA[22,38-41]. The stability of quadruplex increases with the number of quartets in the structure. An empty pseudo-intercalation ligand binding site created between two quadruplex units (distance >7Å) is not tolerated and the units separate within the first 2 ns. However, the presence of acridine in this binding site is sufficient to stabilise the quadruplex and maintain its topology[5]. MD for investigation of porphyrin binding with two antiparallel human telomeric quadruplex (d[(GGGTTA)$_3$GGG]) in the presence of $K^+$ show stable trajectories. The ligand binds to quadruplex in 1:1 stoichiometry and shows binding modes of end stacking, intercalation, external groove binding, and external loop stacking[42]. MD has also been used to study interaction of quadruplex with selective stabilizing agents based on bisquinolinium and bispyridinium derivatives of 1,8-naphthyridine. The study reveals that derivatives with low binding free energy bind by end-stacking binding mode[43]. MD simulations show that telomeric G-quadruplex interaction with a single molecule of 4,11-bis[(2-ethyl)amino]anthra[2,3-b]thiophene-5,10-dione disrupt the G-quartets even in the presence of $Na^+$ and $K^+$ ions. Also, despite the loss of G-quartet the sugar-phosphate backbone of quadruplex remained stable[44].

The conformational variability of quadruplexes is enhanced by intercalated four-stranded DNA structures called i-motifs[45]. The DNA-tetrameric structure of i-motif is formed of two parallel duplexes that are stabilised by hemi-protonated cytosine/cytosine$^+$ (C/CH$^+$) base pairs that intercalate into each other in a head-to-tail orientation[36]. Each base pair carries a +1 charge, which is distributed over the two cytosines. The MD simulation of i-motif on a nanosecond scale was stable. The attractive interactions between C4–N4 and C2–O2 dipoles of stacked C/C+ pairs and between the N3 and H3+ atoms of the cytosines forming a hemiprotonated C/C+ pair can thus compensate for the C-imino proton repulsion[46]. Also, solvent screening that is modulated by the over-all topology of i-motif structure counterbalances the base pair repulsion. Unlike other quadruplex structures, i-motifs have intrinsic repulsive energy stacking terms and the stability in simulations is due to common electrostatics. The AMBER force field does not include any polarisation, exocyclic-group aromatic ring interactions or resonance contribution terms[46]. Thus, simulations of i-motifs give an excellent account of how force fields deal with such unusual vertically stabilised structures.

## 4.2 Enhanced Sampling Methods

The data generated by classical MD for quadruplex stems are usually in agreement with the experimental results. However, the modeling of single-stranded loops is challenging as the electronic structure of loops is different from the backbone[19, 21]. Furthermore, the loops are more exposed to the solvent and cations which makes the presentation of their interactions difficult. Localised enhanced sampling (LES) splits the loop regions of the simulated molecules in N independently moving copies[47, 48]. This reduces the energy barriers between different substrates of the loop regions, proportionate to the number of copies generated (1/N). The LES technique mediates faster relaxation of the structure and allows comparison between the different loop geometries. Besides conventional MD simulations, Sponer et al. also implemented LES to study the loop dynamics[19]. The diagonal loop arrangement of d($G_4T_4G_4$)$_2$ is highly unstable and different from the well-characterized crystal structure[7]. The subsequent free energy analysis also introduces a discrepancy as it predicts the incorrect structure to be more stable than crystal structure[19]. It has been derived that the instability of diagonal loop arrangement and imbalances in solute-cation and solvent-cation interactions observed in this study are due to the pair additivity of the Cornell force-field parm94 and parm98/99 force-field[36]. LES modelling of loops in *Oxytricha nova* d($G_4T_4G_4$)$_2$ quadruplex structure was done by Neidle and coworkers[5]. The lateral loops remained stable over nanosecond scale during the course of simulation. The G-quadruplex stem was unaffected by the loop dynamics. They also demonstrated that dimeric structures with T2 and T3 loops depend on loop lengths and not only on quartet stability. LES is capable of achieving major conformational changes within flexible loop region therefore multiple LES runs should be done for checking the reproducibility of the results.

## 4.3 Principal Component Analysis

The Principal Component Analysis (PCA) is a mathematical algorithm with fixed parameters to find a simplified coordinate representation for data so as to separate large amplitude motions from irrelevant fluctuations[49]. To find the representation, eigen value is segregated such that resulting eigen vectors provide an orthonormal basis for the data and the corresponding eigen values provide information of each individual vector. PCA treats the trajectory to extract the dominant representations over the course of simulation. The simulated structure is translated to the geometrical centre of the molecule by least-square fit superimposition onto a reference structure. This reduces the overall rotation and translation of the structure over the course of trajectory. The configurational space is then constructed over a simple linear transformation in Cartesian coordinate space to generate a 3N X 3N covariance matrix. The matrices are summed and averaged over the whole trajectory. The resulting matrix is then diagonalized generating a set of eigenvectors that gives a vectoral description of each component of the motion by indicating the direction of the motion. Each eigenvector describing the motion has a corresponding eigenvalue that represents the energetic contribution of the particular component to the motion. The eigenvalue is the average square displacement of the structure in the direction of the eigenvector. Projection of a trajectory on a particular eigenvector highlights the time dependent motions that the component performs in the particular vibrational mode. The time average of the projection shows the contribution of the components of the atomic vibrations to this mode of concerted motion[50]. The eigen values are placed in descending order where the first eigenvector and eigenvalue describes the largest internal motion of the structure. On average, only about 5% of eigenvectors are necessary to describe 90% of the total dynamics. Although PCA is a convenient method to visualise trajectories, its limitations should be taken into consideration when interpreting results[51]. PCA is most suited to analyse trajectories of systems that undergo transitional changes instead of trajectories that highlight thermal fluctuations of flexible molecules.

PCA has been applied to study the dynamic behaviour of human telomeric quadruplex dimeric and multimeric structures by Haider et al[5]. The most prominent motions observed are the movements of the loops with a thymine adenine stack maintaining its adopted conformation and moving as a single unit in a concerted manner instead of wobbling of bases. However, the motions of different loops are independent of each other. The presence of ligand in the multimer changes the internal motion of the model. The most dominant motion in the quadruplex-ligand model is not the motion of the loops. This may suggest that the ligand in the pseudo-intercalation site is able to stabilise the model by reducing the motion of the loops to a lower component. PCA has also been employed to study interaction of dimeric quadruplex with perylene derivative (Tel03)[52]. In corroboration with the previous studies, it was observed that Tel03 binding to quadruplex reduces the flexibility of the loop regions[52].

## 4.4 Gas-phase Simulations

It has been observed that parallel and antiparallel quadruplex maintain their structural integrity in the gas-phase in the presence of cation[53]. MD simulations of quadruplexes in gas phase covering 1μs yield stable trajectories[54]. The first principal components obtained from the trajectory of analogous simulations of gas phase and water is similar, implying that dominant motions in DNA are not altered by vaporization. During the course of gas-phase simulation of quadruplex DNA different structures are obtained depending on the central cation and the temperature. Similar to the explicit solvent conditions, expulsion of cation from the central channel of quadruplex leads to instantaneous fluctuations in the trajectory[54]. The G-quadruplex observed in the simulations with $Li^+$ as the coordinating ion appears distorted and collapsed while $Na^+$ and $K^+$ ions yield stable quadruplexes. A distinct feature of gas phase simulation is that even when the structure is partially lost large numbers of guanine-guanine interactions are preserved. The interactions of a quadruplex in the gas phase are more complex than in the solvent phase. Along with MD, ESI-MS has been successfully used to gain further insight into quadruplex structure[55]. Mazzitelli et al. performed 50ns MD simulations for twelve quadruplex models to complement ESI-MS[55]. They used AMBER restrained electrostatic potential charge (RESP) procedure to reassign the charges of the neutral phosphate group and parm99 force-field for gas-phase simulations.

## 4.5 Continuum (Implicit) Solvent Methods

Assessment of structure and free energies during course of a simulation can be done by continuum solvent method[56]. The approach is to simplify the energy calculations by implicitly integrating out all the solvent coordinates. The electrostatic contribution to the solvation energies is calculated by Poisson Boltzmann (PB) or Generalized

Born (GB) approaches. The hydrophobic contribution is represented by a surface area (SA) term obtained by scaling the solvent accessible surface area by an appropriate surface tension. The PB/GB solution along with SA term is used to calculate the absolute free energies of the molecules in solution[57]. The MM-PBSA method extracts estimates of the free-energies from the MD trajectories based on averages of the gas phase molecular mechanical energy of the solute with an estimate of solvation free-energies from a PB continuum solvation model. The solvent and periodicity are removed in the post processing followed by the averaging of the energies over structural snapshots taken from the simulations. Both single and multiple trajectories can be used for MM-PBSA calculations however, single trajectories are preferred as it cancels sampling errors in the intra-molecular terms[58]. The PB model is theoretically more complex than the GB model and MM/PBSA is often estimated better than MM/GBSA for predicting binding free energies. MM-PBSA has been used to calculate free-energies in G-quadruplex simulations including those involving quadruplex-ligand interactions. Hazel *et al.* used MM-PBSA contributions from loops and stem for analysis of quadruplex sub-structures[4]. They reported that energy values estimated from MM-PBSA do not reflect the stability of loops. The local fluctuations in the loop significantly affect the free-energy calculation at each step. This overshadowed the free-energies difference between distinct loops and no inference could be drawn based on it. However, energy difference between parallel and antiparallel quadruplex reflected that antiparallel structure is more stable than parallel structure. Similar approach was used by Haider *et al.* for use of MM-PBSA to compare the different quadruplex conformations[5]. The free-energy calculations in this study were well in agreement with the structural stability of quadruplex conformations. Ishikawa *et al.* have used MM-PBSA results to analyse the binding forces of quadruplex and porphyrins interactions[59, 60].

## 4.6 Free Energy Perturbation, Thermodynamic Integration, Potential of Mean Force and Umbrella Sampling

Free energy analysis between two states observed during the course of explicit MD simulation can be done by Free-energy pertubation (FEP) and thermodynamic integration (TI). FEP provides a quantitative method for studying the behaviour of DNA in response to a particular perturbation like base flipping[61]. TI allows accurate calculations of binding parameters of bis-intercalating molecules interaction with DNA[62]. As TI is very sensitive to the structural changes, it should only be used in the cases where ligand binding to DNA does not significantly affect the double helical nature. TI methods have also been used by Clark *et al.* to study the DNA-protein interactions. The work suggests the specificity of DNA base pair in recognition and binding of DNA[63].

The potential of mean force (PMF) is described as the change in free energy between the initial and final states as a function of Cartesian coordinates. PMF of simulation provides the thermal probability of finding a system at different points along the coordinates. The thermal probability is directly related to the free energy of the conformation. The widely used algorithm for calculating PMF is umbrella sampling[64]. PMF profile has been used along with explicit atomistic MD simulation to investigate the folding and unfolding of TBA in the presence of $Sr^{+2}$. The analysis suggests that unfolding of TBA in presence of cation is a distinct multiple stepwise process and that the interplay of guanine, water molecules and cation govern the behaviour[65].

### 4.7 Quantum Mechanics/Molecular Mechanics (QM/MM)

Quantum Mechanical methods are based on the solution of Schrodinger equation. The equation describes the motions of electrons and nuclei in a molecular system. The *ab initio* method molecular orbitals are approximated by a linear combination of atomic orbitals[66]. *Ab initio* quantum mechanical (QM) calculation is an important tool to investigate functional mechanisms of biological macromolecules based on their three-dimensional and electronic structures[67]. In theory, the system size, which QM calculations can treat, is usually up to a few hundred atoms despite of huge sizes of biomacromolecules including solvent water molecules. The QM calculations are computationally very demanding. The largest system on which such calculations are available is a guanine dimer[68]. The QM calculations deal with only intra-quartet interaction and do not take into account the effect of long-range electrostatic forces or solvation effects. The accuracy of the calculated molecular properties depends on the number of atomic orbitals used and quality of the basis set. Density functional theory (DFT) can be formulated as a variant of *ab initio* methods where correlation functionals are used to represent electron correlation energy[69]. DFT can account for hydrogen bonding effects and can therefore calculate interactions within quartet and guanine-cation interactions. DFT does not assimilate base stacking parameters and therefore cannot describe inter-quartet interactions. Hybrid QM/MM methods rely on dividing the bimolecular systems into two regions: a smaller region which is treated quantum mechanically and the remaining region which is modelled by classical molecular mechanics force fields. QM/MM has been used to explore cation binding to TBA using $Ba^{+2}$ as the coordinating cation[12].

## 5 EXAMPLES OF MICROSECOND SCALE MOLECULAR DYNAMICS SIMULATIONS

### 5.1 Propeller-type Human Telomeric Repeat Quadruplex

Haider *et al.* have performed 1.5μs long simulation of 22mer sequence [$AG_3(T_2AG_3)_3$] (PDB id:1KF1) from human telomeric repeat that folds in a propeller-topology (unpublished results). The molecular dynamics simulations were carried out using Amber11 software employing the parmbsc0 force field. Additional $K^+$ counterions were added to system to neutralize the charge on the quadruplex backbone. The $K^+$ ions are arranged in a square antiprismatic coordination. They are sandwiched between the quartets and were retained in positions as in the crystal structure. The protocol for energy minimization and molecular dynamics in explicit solvent was adopted from Haider *et al.*[5]. The rmsd of all-atom structure is significantly greater than the backbone and is a result of flexible nature of TTA loops. The G-quartet stem was found to be the most stable sub-segment of the structure.

During the dynamics run, the central core geometry constituted by the three G-quartet planes pairing through there Watson-Crick and Hoogsteen edges, was retained along with its characteristic square planar arrangement. The grooves also retained their width and depth such that the same pattern of hydrogen donors and acceptors is observed throughout the course of the simulation. This pattern when compared with the crystal structure reflects the four-fold symmetry of the quadruplex. The geometry of the stem stacking as observed in the crystal structure, measured by the rise and twist, was also maintained throughout the course of the simulation. The stacking geometry of 12 G-quartets per turn is retained, as observed in other intramolecular parallel stranded G-quartets. There was no movement of the potassium ions out of the central quadruplex channel as observed in some simulations of quadruplexes[12, 36].

We observed small fluctuations within the first 300ns, which eventually stabilized over the course of the simulation. Examination of the trajectory in short bins revealed that the structure adopted stable conformations that could last several nanoseconds. While this stability might have sufficed for short multi-nanosecond scale simulation, however, placing it in the context of a long microsecond or longer simulation, these structures are transitionary. Therefore the first 300ns were treated as equilibration and all analysis was carried on the trajectory beyond this point. The cutoff point at 300ns was determined via principal components analysis, until which no convergence was observed. The microscale simulation of quadruplex highlighted the importance of timescale on which these propeller-type structures equilibrate, which would otherwise be thought as stable conformations. Following the flexibility of the loop structure over the course of the simulation and comparing them with the X-ray structure reveals significant conformational changes. Moreover, the mobility of the simulated loops is independent of each other. Loop1 and loop2 shows high rmsd reflecting more movement while loop3 shows is more stable as reflected by less rmsd. However, the loop conformational rearrangement did not have any great impact on the structure of the central G-quartet stem. A cluster analysis over the microsecond trajectory resulted in eight clusters. It is noteworthy that three of these can be classified as clusters that contain metastable intermediates that resemble conformations to that observed in various NMR and X-ray solved structures.

## 5.2 Thrombin Binding Aptamer

Aptamers are short DNA or RNA fragments that can bind, with high affinity and selectivity to defined molecular targets, including biological proteins[70]. Among the different conformational arrangements that can be assumed by aptamers, a quadruplex represents suitable nucleic acid architecture of noteworthy importance, due to its dramatic thermal stability. Contextually, potentially therapeutic aptamers like those endowed with anti-HIV and anti-thrombin activity have driven great interests[10, 71-73]. In the study of both these classes of aptamers, computational methods have proved to be excellent research tools, able to provide a deep insight into the aptamer-protein interactions[35, 74, 75]. Reshetnikov *et al.* investigated the TBA-thrombin interaction by MD simulation using AMBER-99φ and parmbsc0 force-field on GROMACS software package[35]. The simulations were conducted in a range time from 600 to 900ns in individual runs, reaching a total simulation time that exceeds 12µs. Separate 900ns MD simulation runs were performed, in both parm99 and parmbsc0 force fields, to compare the viability of NMR- and X-ray-based structures of free TBA. Data from this study suggest that the X-ray-based conformation is unstable, while the NMR-based structure is the only viable TBA conformation. Furthermore, this study highlighted all the structural features that influence the stability of the structure, and the parmbsc0 force field was elected as the most suitable force field to conduct the subsequent simulations. To analyse the precise TBA residues that interact with the thrombin structure, 600ns MD simulation runs were performed in parmbsc0 force field to compare the NMR- and X-ray-based models of the thrombin-TBA 1:1 complex. It could be noted that the two starting complex structures showed different binding modes. In the NMR-based complex, aptamer interacts by the TT-loops, involving the Exosite I of the thrombin, while in the X-ray-based complex the interaction involves the TGT-loop and the Exosite II. Also in this case, since the X-ray-based complex collapses during the simulation, data show that the NMR-based complex was favoured, resulting in the only viable complex. The same studies were performed for the thrombin-TBA 1:2 complex. In this case both the complexes survived during the simulations, even if the G-quartet planarity of the aptamer is disrupted in the X-ray-based complex. In summary, all the simulations performed by Reshetnikov et al. definitely support the NMR-based model of TBA, where the TGT-loop has a stabilizing influence on the TBA molecule and TT-loops mediate the thrombin binding and its inhibition.

In agreement with these results, the changes made to the TT-loops can lead to different consequences. An example is provided by the work of Pagano et al., in which a modified TBA (mTBA), containing a 5'-5' site of polarity inversion up to TT-loop, shows a better stability and a higher thrombin affinity but a less inhibitory activity than the unmodified TBA[75]. To better understand the biological results, the interaction patters for the mTBA-thrombin 1:2 complex have been investigated by 5ns MD simulation, performed with the GROMACS package software in parm98 force field. The results highlight that Exosite I interacting with mTBA makes less contact when compared with the unmodified TBA but, simultaneously, the Exosite II of a second thrombin makes more contacts with mTBA. In this case, the MD simulation data corroborate the hypothesis that the binding with the Exosite I is the only one with an inhibitory effect[76, 77]. In fact, even if mTBA binds to thrombin with an increased affinity, this enhancement is due to better interactions with the non-inhibiting Exosite II of the protein, resulting in a decreasing of inhibitory activity. Furthermore, it is fair to emphasize that the deep understanding of biological data has been possible only through the use of MD simulation studies.

As previously mentioned, a solved structure of the aptamer-protein complex may not be necessary to conduct computational studies when the structures are independently solved. A suitable example is the work of Sgobba et al., in which *93del*-HIV1-integrase complex has been obtained through docking studies and then subjected to MD simulations to obtain a structural perspective on the mechanism of inhibition[74]. In this case, the binding interactions between aptamer and protein have been investigated by 35ns MD simulation, performed with the AMBER11 package software, and parmbsc0 and ff99SBildn force fields. The energy minimization and MD simulation in explicit solvent have been performed following the protocol reported by Perryman et al.[78]. Moreover, the Principal Component Analysis (PCA) and the Elastic Network Models (ENM) have been performed to identify the motions of HIV1-Integrase (HIV1-IN) in complex with *93del* aptamer and to describe structural flexibility in order to correlate the relationship between dynamics and function of the protein. The reported data have highlighted the key residues of HIV1-IN essential for the binding with the aptamer, capable to disrupt the HIV1-IN-DNA interactions. Furthermore, data collected by PCA and ENM have confirmed that interaction involving the catalytic loops are responsible of inhibition activity of *93del,* since they conduct to a conformational rearrangement that hold the protein in an inactive conformation.

## 6 VIRTUAL SCREENING OF G-QUADRUPLEXES

The expression 'virtual screening' describes the use of computational algorithms and models for the identification of novel bioactive molecules. Over the years, virtual screening has become an important component of the *in silico* search for the hit and lead compounds and their optimization[79]. It can be divided into two main categories, namely structure-based virtual screening which utilizes the 3D structure of biological target and ligand-based virtual screening where the structure activity data from a set of known active compounds are employed[80]. Despite significant progress in generation of potential ligand 'poses' by automated molecular docking, there are a number of pitfalls in the existing virtual screening methods that needs to be addressed[79, 81]. The scoring functions used for prioritization of various suggested binding poses are still very much inaccurate[79, 82]. To this date, there are over 150 quadruplex structures deposited in the Protein Data Bank (http://www.rscb.org/pdb). However, the structure of quadruplex may vary with the experimental methods and conditions. Therefore, choosing a suitable model for modelling and structure-based virtual screening is critical.

With the advent of computational techniques, experimental biophysical assays are complemented by molecular docking and ligand-based pharmacophore studies to explore plausible binding modes of quadruplex-ligand interaction, optimize the lead compounds and rationalize their selectivity[5, 83]. Interaction of cationic tetraolylporphyrin derivatives with the NMR intramolecular antiparallel telomeric G-quadruplex d(AG3[T2AG3]3)(PDB code: 143D) was studied by molecular docking using DOCK6[60]. A model of *c-myc* quadruplex was generated for molecular docking study with platinum (II) Schiff base complexes. The first virtual screen of library of FDA-approved drugs (3000 compounds) for *c-myc* quadruplex stabilizing ligands was done employing the ICM (molsoft) program[84]. Selectivity of a group of napthalene diimide ligands for telomeric G4-RNA over the G4-DNA was explained by docking using the Accelrys AFFINITY program[85]. Two quadruplex stabilizing alkaloid compounds were selected by screening a Chinese herbal database of 10000 compounds with previously generated pharmacophores using Catalyst, Accelrys[86]. Chen *et al.* on the basis of acridine derivatives, developed ligand-based pharmacophore model and identified triaryl-substituted imidazole derivative TSIZ01 as potent G-quadruplex ligand[87].

The biggest challenge in virtual screening of G-quadruplex has been the selection of most favourable pose of quadruplex for *in silico* studies. This is due to highly charged backbone, presence of stabilizing alkali metal cations and in particular the flexibility of the loops. Quality of the quadruplex docking, and subsequently the binding affinity of the ligands might be strongly affected by the flexibility of the loop regions[88]. To address the receptor and ligand flexibility issue and subsequently conformational change upon binding, a novel form of 'dynamic docking' has been developed by Neidle and co-workers (unpublished results). Over 250 short explicit solvent molecular dynamics (MD) simulations were employed to explore binding, transition states and preferred conformations of the quadruplex DNA end-stacking ligand *pyridostatin* and *RHPS4* with the parallel native X-ray structure of telomeric quadruplex. Diverse conformations of the ligand were placed in a grid-like manner and in multiple orientations parallel to the 3' and 5' site of the quadruplex structure. Multiple starting positions for subsequent MD simulations were generated, allowing large conformational space for both ligand and its target to be explored (unpublished data).

Virtual screening of large chemical libraries to aid the selection of new quadruplex-binding ligands has recently been explored[89]. A synthetic substituted indole was identified by screening ~100,000 drug-like compounds[90]. Fonsecin B, a natural napthopyrone compound was identified by screening 20,000 compound in natural product database[91]. *In silico* studies along with NMR experiments in tandem provide better details of interactions of ligands with G4-DNA[92]. A relatively small but structurally diverse commercially available database (6000 compounds) was screened against the parallel quadruplex [d(TG$_4$T) $_4$] using Autodock v4. Subsequent NMR screening of the top 30 hits identified six G4-groove binding molecules that were further studied in detail by NMR, ITC measurements and molecular docking with modified quadruplexes. These six molecules are the most potent G-quadruplex groove-binders identified so far[93].

Large-scale integrated *in silico* and *in vitro* screening platforms to discover novel small molecules binding to specific nuclei acids were developed in the Chaires group, using the analogy of a 'funnel'[94]. The ZINC 'drug-like' virtual database of 1.3 million compounds was screened against the antiparallel quadruplex target (pdb id: 2HY9) employing the Surflex-Dock molecular docking software. An array of potential nucleic acids competing sites, as well as all possible binding sites on the target nucleic acid itself were considered for the virtual screening, which was performed on a grid of more than 10000 computer processors. This approach and the appropriate choice of the software, was previously validated [95, 96]. The top 160 hits that emerged after scoring, for selective binding to the quadruplex target, were tested by high-throughput melting assay followed by a secondary screening of the top compounds. Characterization of their binding behaviour with the quadruplex target, by means of rigorous binding studies using calorimetry, spectroscopy, competition dialysis, molecular dynamics simulations and functional assays, identified a substantially stabilizing quadruplex binding ligand, which in turn suggests that the proposed *in silico* and in vitro platform may be used to discover new G-quadruplex binding scaffolds[94].

At present, virtual screening alone is rather unable to adequately predict the selectivity of a particular ligand for different G-quadruplexes, because of the inherent inaccuracies in comparing binding energies between multiple G-quadruplex structures[88]. However the G-quadruplex structures display a great diversity in their loops and grooves geometries, which in particular could be used to enhance the selectivity of ligands, allowing for their excellent structure-specific recognition, affinity and specificity. For instance, in the case of the promoter *KIT1* quadruplex structure[97, 98], the presence of a distinct cleft may be suitable for ligand binding and virtual high-throughput screening[99].

## 7 CONCLUSIONS

There are over 150 NMR or X-ray derived structures present in the PDB that detail atomistic information. However, it should be emphasised that these are time-average static representations and need to be simulated in before any realistic information can be extracted. Such simulations carried out at ambient temperatures can provide information on interactions between solvent, solute and ions, cross energy barriers to access conformations, folding intermediates and metastable structures for which there is no experimental data available. There is no other technique that can provide time coursed dynamic details of interactions and therefore molecular dynamics simulations have been a chosen method to study quadruplexes, both native and in complex with ligands and proteins. The best usage of simulations is envisaged to complement the explanation and interpretation of experimental data while also supplements it with information that cannot be accessed by experimental measurements.

**References**

1. T. Simonsson, *Biological chemistry*, 2001, **382**, 621-628.
2. K. J. Neaves, J. L. Huppert, R. M. Henderson and J. M. Edwardson, *Nucleic acids research*, 2009, **37**, 6269-6275.
3. R. Rodriguez, K. M. Miller, J. V. Forment, C. R. Bradshaw, M. Nikan, S. Britton, T. Oelschlaegel, B. Xhemalce, S. Balasubramanian and S. P. Jackson, *Nat Chem Biol*, 2012, **8**, 301-310.
4. P. Hazel, G. N. Parkinson and S. Neidle, *Nucleic acids research*, 2006, **34**, 2117-2127.
5. S. Haider, G. N. Parkinson and S. Neidle, *Biophysical journal*, 2008, **95**, 296-311.
6. N. V. Hud and J. Plavec, *The role of cations in determining quadruplex structure and stability*, Royal Society of Chemistry, U.K., 2006.
7. S. Haider, G. N. Parkinson and S. Neidle, *Journal of molecular biology*, 2002, **320**, 189-200.
8. E. A. Venczel and D. Sen, *Biochemistry*, 1993, **32**, 6220-6228.
9. N. V. Hud, F. W. Smith, F. A. Anet and J. Feigon, *Biochemistry*, 1996, **35**, 15383-15390.
10. P. Schultze, R. F. Macaya and J. Feigon, *Journal of molecular biology*, 1994, **235**, 1532-1547.
11. G. D. Strahan, M. A. Keniry and R. H. Shafer, *Biophysical journal*, 1998, **75**, 968-981.
12. R. V. Reshetnikov, J. Sponer, O. I. Rassokhina, A. M. Kopylov, P. O. Tsvetkov, A. A. Makarov and A. V. Golovin, *Nucleic acids research*, 2011, **39**, 9789-9802.
13. X. Cang, J. Sponer and T. E. Cheatham, 3rd, *Journal of the American Chemical Society*, 2011, **133**, 14270-14279.
14. S. Haider and S. Neidle, *Molecular Modeling and Simulation of G-Quadruplexes and Quadruplex-Ligand Complexes*, Springer Protocols, 2008.
15. P. Maurer, A. Laio, H. W. Hugosson, M. C. Colombo and U. Rothlisberger, *J. Chem. Theory Comput.*, 2007, **3**, 628–639.
16. T. A. Soares, P. H. Hunenberger, M. A. Kastenholz, V. Krautler, T. Lenz, R. D. Lins, C. Oostenbrink and W. F. van Gunsteren, *J Comput Chem*, 2005, **26**, 725-737.
17. S. M. Haider and S. Neidle, *Molecular Dynamics and Force Field Based Methods for studying Quadruplex Nucleic Acids*, The Royal Society of Chemistry, U.K., 2012.
18. A. W. Van Wynsberghe and Q. Cui, *Biophysical journal*, 2005, **89**, 2939-2949.
19. E. Fadrna, N. Spackova, R. Stefl, J. Koca, T. E. Cheatham, 3rd and J. Sponer, *Biophysical journal*, 2004, **87**, 227-242.
20. A. Perez, F. J. Luque and M. Orozco, *Journal of the American Chemical Society*, 2007, **129**, 14739-14745.
21. E. Fadrná, N. a. Špačková, J. Sarzyňska, J. Koča, M. Orozco, T. E. Cheatham, T. Kulinski and J. i. Šponer, *Journal of Chemical Theory and Computation*, 2009, **5**, 2514-2530.

22. J. Q. Hou, S. B. Chen, J. H. Tan, T. M. Ou, H. B. Luo, D. Li, J. Xu, L. Q. Gu and Z. S. Huang, *The journal of physical chemistry. B*, 2010, **114**, 15301-15310.
23. L. Petraccone, I. Fotticchia, A. Cummaro, B. Pagano, L. Ginnari-Satriani, S. Haider, A. Randazzo, E. Novellino, S. Neidle and C. Giancola, *Biochimie*, 2011, **93**, 1318-1327.
24. W. S. Ross and C. C. Hardin, *Journal of the American Chemical Society*, 1994, **116**, 6070-6080.
25. T. E. Cheatham, III, J. L. Miller, T. Fox, T. A. Darden and P. A. Kollman, *Journal of the American Chemical Society*, 1995, **117**, 4193-4194.
26. J. Norberg and L. Nilsson, *Biophysical journal*, 2000, **79**, 1537-1553.
27. J. Sponer, J. Leszczynski and P. Hobza, *Biopolymers*, 2001, **61**, 3-31.
28. P. Cieplak, W. D. Cornell, C. Bayly and P. A. Kollman, *J Comput Chem*, 1995, **16**, 1357-1377.
29. P. Varnai and K. Zakrzewska, *Nucleic acids research*, 2004, **32**, 4269-4280.
30. W. Cornell, P. Cieplak, C. Bayly, I. Gould, K. Merz, D. Ferguson, D. Spellmeyer, T. Fox, J. Caldwell and P. Kollman, *J. Am. Chem. Soc.*, 1995, **117**, 5179-5197.
31. D. L. Beveridge, G. Barreiro, K. S. Byun, D. A. Case, T. E. Cheatham, 3rd, S. B. Dixit, E. Giudice, F. Lankas, R. Lavery, J. H. Maddocks, R. Osman, E. Seibert, H. Sklenar, G. Stoll, K. M. Thayer, P. Varnai and M. A. Young, *Biophysical journal*, 2004, **87**, 3799-3813.
32. F. Barone, F. Lankas, N. Spackova, J. Sponer, P. Karran, M. Bignami and F. Mazzei, *Biophysical chemistry*, 2005, **118**, 31-41.
33. T. E. Cheatham, P. Cieplak and P. A. Kollman, *Journal of Biomolecular Structure and Dynamics*, 1999, **16**, 845-862.
34. A. Perez, I. Marchan, D. Svozil, J. Sponer, T. E. Cheatham, 3rd, C. A. Laughton and M. Orozco, *Biophysical journal*, 2007, **92**, 3817-3829.
35. R. V. Reshetnikov, A. V. Golovin, V. Spiridonova, A. M. Kopylov and J. Sponer, *J Chem Theory Comput*, 2010, **6**, 3003-3014.
36. N. a. Špačková, I. Berger and J. Šponer, *Journal of the American Chemical Society*, 1999, **121**, 5519-5534.
37. G. N. Parkinson, M. P. Lee and S. Neidle, *Nature*, 2002, **417**, 876-880.
38. F. Cuenca, O. Greciano, M. Gunaratnam, S. Haider, D. Munnur, R. Nanjunda, W. D. Wilson and S. Neidle, *Bioorganic & medicinal chemistry letters*, 2008, **18**, 1668-1673.
39. S. Agrawal, R. P. Ojha and S. Maiti, *The Journal of Physical Chemistry B*, 2008, **112**, 6828-6836.
40. M. Cavallari, A. Garbesi and R. Di Felice, *The Journal of Physical Chemistry B*, 2009, **113**, 13152-13160.
41. D.-Y. Yang and S.-Y. Sheu, *The Journal of Physical Chemistry A*, 2007, **111**, 9224-9232.
42. M.-H. Li, Q. Luo and Z.-S. Li, *The Journal of Physical Chemistry B*, 2010, **114**, 6216-6224.
43. V. Dhamodharan, S. Harikrishna, C. Jagadeeswaran, K. Halder and P. I. Pradeepkumar, *The Journal of Organic Chemistry*, 2011, **77**, 229-242.
44. D. Kaluzhny, N. Ilyinsky, A. Shchekotikhin, Y. Sinkevich, P. O. Tsvetkov, V. Tsvetkov, A. Veselovsky, M. Livshits, O. Borisova, A. Shtil and A. Shchyolkina, *PLoS ONE*, 2011, **6**, e27151.

45. T. A. Brooks, S. Kendrick and L. Hurley, *The FEBS journal*, 2010, **277**, 3459-3469.
46. N. Špacková, I. Berger, M. Egli and J. Šponer, *Journal of the American Chemical Society*, 1998, **120**, 6147-6151.
47. C. Simmerling, J. L. Miller and P. A. Kollman, *Journal of the American Chemical Society*, 1998, **120**, 7149-7155.
48. R. Elber and M. Karplus, *Journal of the American Chemical Society*, 1990, **112**, 9161-9175.
49. A. Amadei, A. B. Linssen, B. L. de Groot, D. M. van Aalten and H. J. Berendsen, *Journal of biomolecular structure & dynamics*, 1996, **13**, 615-625.
50. D. M. van Aalten, A. Amadei, A. B. Linssen, V. G. Eijsink, G. Vriend and H. J. Berendsen, *Proteins*, 1995, **22**, 45-54.
51. K. Reblova, Z. Strelcova, P. Kulhanek, I. Besseova, D. H. Mathews, K. V. Nostrand, I. Yildirim, D. H. Turner and J. Sponer, *J Chem Theory Comput*, 2010, **2010**, 910-929.
52. M.-H. Li, Q. Luo, X.-G. Xue and Z.-S. Li, *Journal of Molecular Modeling*, 2011, **17**, 515-526.
53. M. Rueda, F. J. Luque and M. Orozco, *Journal of the American Chemical Society*, 2006, **128**, 3608-3619.
54. V. Gabelica, F. Rosu, M. Witt, G. Baykut and E. De Pauw, *Rapid communications in mass spectrometry : RCM*, 2005, **19**, 201-208.
55. C. L. Mazzitelli, J. Wang, S. I. Smith and J. S. Brodbelt, *Journal of the American Society for Mass Spectrometry*, 2007, **18**, 1760-1773.
56. T. Rodinger and R. Pomes, *Current opinion in structural biology*, 2005, **15**, 164-170.
57. M. Zacharias, eds. J. Šponer and F. Lankaš, Springer Netherlands, 2006, pp. 95-119.
58. N. Spackova, T. E. Cheatham, 3rd, F. Ryjacek, F. Lankas, L. Van Meervelt, P. Hobza and J. Sponer, *Journal of the American Chemical Society*, 2003, **125**, 1759-1769.
59. Y. Ishikawa, Y. Tomisugi and T. Uno, *Nucleic acids symposium series*, 2006, 331-332.
60. Y. Ishikawa, E. Higashi and H. Morioka, *Nucleic acids symposium series*, 2007, 247-248.
61. N. K. Banavali and A. D. Mackerell, Jr., *PLoS One*, 2009, **4**, e5525.
62. E. Marco, A. Negri, F. J. Luque and F. Gago, *Nucleic acids research*, 2005, **33**, 6214-6224.
63. Frank R. Beierlein, G. G. Kneale and T. Clark, *Biophysical journal*, 2011, **101**, 1130-1138.
64. B. Roux, *Computer Physics Communications*, 1995, **91**, 275-282.
65. C. Yang, S. Jang and Y. Pak, *J Chem Phys*, 2011, **135**, 225104.
66. R. Iftimie, P. Minary and M. E. Tuckerman, *Proceedings of the National Academy of Sciences of the United States of America*, 2005, **102**, 6654-6659.
67. B. Kirchner, F. Wennmohs, S. Ye and F. Neese, *Current Opinion in Chemical Biology*, 2007, **11**, 134-141.
68. J. Sponer, P. Jurecka, I. Marchan, F. J. Luque, M. Orozco and P. Hobza, *Chemistry*, 2006, **12**, 2854-2865.
69. R. G. Parr and W. Yang, *Annual Review of Physical Chemistry*, 1995, **46**, 701-728.

70. S. M. Nimjee, C. P. Rusconi and B. A. Sullenger, *Annual review of medicine*, 2005, **56**, 555-583.
71. K. Padmanabhan and A. Tulinsky, *Acta crystallographica. Section D, Biological crystallography*, 1996, **52**, 272-282.
72. J. D'Onofrio, L. Petraccone, E. Erra, L. Martino, G. Di Fabio, L. De Napoli, C. Giancola and D. Montesarchio, *Bioconjugate Chemistry*, 2007, **18**, 1194-1204.
73. A. T. Phan, V. Kuryavyi, J. B. Ma, A. Faure, M. L. Andreola and D. J. Patel, *Proceedings of the National Academy of Sciences of the United States of America*, 2005, **102**, 634-639.
74. M. Sgobba, O. Olubiyi, S. Ke and S. Haider, *Journal of biomolecular structure & dynamics*, 2012, **29**, 863-877.
75. B. Pagano, L. Martino, A. Randazzo and C. Giancola, *Biophysical journal*, 2008, **94**, 562-569.
76. M. Tsiang, A. K. Jain, K. E. Dunn, M. E. Rojas, L. L. Leung and C. S. Gibbs, *The Journal of biological chemistry*, 1995, **270**, 16854-16863.
77. K. Y. Wang, S. H. Krawczyk, N. Bischofberger, S. Swaminathan and P. H. Bolton, *Biochemistry*, 1993, **32**, 11285-11292.
78. A. L. Perryman, S. Forli, G. M. Morris, C. Burt, Y. Cheng, M. J. Palmer, K. Whitby, J. A. McCammon, C. Phillips and A. J. Olson, *Journal of molecular biology*, 2010, **397**, 600-615.
79. G. Schneider, *Nature reviews. Drug discovery*, 2010, **9**, 273-276.
80. A. Jahn, G. Hinselmann, N. Fechner and A. Zell, *Journal of cheminformatics*, 2009, **1**, 14.
81. T. Scior, A. Bender, G. Tresadern, J. L. Medina-Franco, K. Martinez-Mayorga, T. Langer, K. Cuanalo-Contreras and D. K. Agrafiotis, *Journal of chemical information and modeling*, 2012.
82. M. Totrov and R. Abagyan, *Proteins*, 1997, **Suppl 1**, 215-220.
83. S. M. Haider, I. Autiero and S. Neidle, *Biochimie*, 2011, **93**, 1275-1279.
84. D. S. Chan, H. Yang, M. H. Kwan, Z. Cheng, P. Lee, L. P. Bai, Z. H. Jiang, C. Y. Wong, W. F. Fong, C. H. Leung and D. L. Ma, *Biochimie*, 2011, **93**, 1055-1064.
85. G. W. Collie, S. M. Haider, S. Neidle and G. N. Parkinson, *Nucleic acids research*, 2010, **38**, 5569-5580.
86. Q. Li, J. Xiang, X. Li, L. Chen, X. Xu, Y. Tang, Q. Zhou, L. Li, H. Zhang, H. Sun, A. Guan, Q. Yang, S. Yang and G. Xu, *Biochimie*, 2009, **91**, 811-819.
87. S. B. Chen, J. H. Tan, T. M. Ou, S. L. Huang, L. K. An, H. B. Luo, D. Li, L. Q. Gu and Z. S. Huang, *Bioorganic & medicinal chemistry letters*, 2011, **21**, 1004-1009.
88. D. L. Ma, V. P. Ma, D. S. Chan, K. H. Leung, H. J. Zhong and C. H. Leung, *Methods*, 2012.
89. S. Neidle, in *Therapeutic Applications of Quadruplex Nucleic Acids*, Academic Press, Boston, 2012, pp. 151-174.
90. D. L. Ma, T. S. Lai, F. Y. Chan, W. H. Chung, R. Abagyan, Y. C. Leung and K. Y. Wong, *ChemMedChem*, 2008, **3**, 881-884.
91. H. M. Lee, D. S. Chan, F. Yang, H. Y. Lam, S. C. Yan, C. M. Che, D. L. Ma and C. H. Leung, *Chemical communications*, 2010, **46**, 4680-4682.
92. S. Cosconati, L. Marinelli, R. Trotta, A. Virno, L. Mayol, E. Novellino, A. J. Olson and A. Randazzo, *Journal of the American Chemical Society*, 2009, **131**, 16336-16337.

93. R. Trotta, S. De Tito, I. Lauri, V. La Pietra, L. Marinelli, S. Cosconati, L. Martino, M. R. Conte, L. Mayol, E. Novellino and A. Randazzo, *Biochimie*, 2011, **93**, 1280-1287.
94. P. A. Holt, R. Buscaglia, J. O. Trent and J. B. Chaires, *Drug development research*, 2011, **72**, 178-186.
95. P. A. Holt, J. B. Chaires and J. O. Trent, *Journal of chemical information and modeling*, 2008, **48**, 1602-1615.
96. P. A. Holt, P. Ragazzon, L. Strekowski, J. B. Chaires and J. O. Trent, *Nucleic acids research*, 2009, **37**, 1280-1287.
97. A. T. Phan, V. Kuryavyi, S. Burge, S. Neidle and D. J. Patel, *Journal of the American Chemical Society*, 2007, **129**, 4386-4392.
98. D. Wei, G. N. Parkinson, A. P. Reszka and S. Neidle, *Nucleic acids research*, 2012, **40**, 4691-4700.
99. S. Balasubramanian, L. H. Hurley and S. Neidle, *Nature reviews. Drug discovery*, 2011, **10**, 261-275.

# Chapter 4

## Recognition of Quadruplexes

Chapter Editor: **Mateus Webba da Silva**

'Recognition of quadruplexes' is a wide-ranging theme in the field of nucleic acid quadruplexes. Quadruplexes of DNA or RNA have four grooves with three distinct dimensions. They may be populated by single stranded architectures that have the potential for hydrogen bonding and base-stacking interactions. Their backbones adopts a variety of conformations characterizing greater diversity of structural motifs as compared, for example, to double-stranded DNA. Thus quadruplex nucleic acid structures are a richly provided with structural motifs that favor interactions with other molecules.

In this section authors from a variety of disciplines describe a range of interactions involving quadruplex nucleic acids from a variety of perspectives. The section starts with an analysis of the selective pressure G4 motifs are associated with. It is described and contextualized in terms of specific chromosomal domains, specific regions within genes, and specific classes of genes. The current belief that quadruplex structures of DNA and RNA play roles in the regulation of gene expression is then scrutinized. There is specific emphasis on current findings and ideas on the proposed roles of quadruplex structures of DNA in the regulation of gene transcription. The section than moves into the issue of quadruplex DNA as a viable therapeutic target. Are quadruplex topologies suitable targets for the rational design and development of therapeutics? Here a discussion identifies both difficulties and opportunities in relation to this question. Next, a discussion ensues on the utility of high-throughput screening in the discovery of highly specific quadruplex ligands; including those that may not obey the rules dictated by drug design. The features that make metal complexes suitable quadruplex DNA binders are explored next. In particular the interaction of metal complex intercalators with quadruplex DNA is discussed including those complexes that have been studied as optical probes for DNA.

# BIOLOGICAL FUNCTIONS OF G-QUADRUPLEXES

Nancy Maizels

Departments of Immunology and Biochemistry, University of Washington, Seattle, WA, 98195 USA

## 1 INTRODUCTION

In the human genome, the G4 motif is enriched at specific chromosomal domains, in specific regions within genes, and in specific classes of genes. This strongly suggests that this motif is under selective pressure, particularly because G4 motifs are associated with genomic instability. Examples of specific functions of G4 motifs are accumulating, and provide persuasive evidence both of formation of these alternative structures in vivo, and of their functional importance.

## 2 RESULTS

### 2.1. G4 Structures Form in Vivo

The G4 motif, $G_{\geq 3}N_xG_{\geq 3}N_xG_{\geq 3}N_xG_{\geq 3}$, confers intrinsic potential to form structures, referred to as G4 (or G-quadruplex) DNA. The essential unit of a G4 structure is the G-quartet, in which four guanines interact by pairwise hydrogen bonds between the N7 of one guanine and the extracyclic amine of another[1,2]. Hydrogen bonds between guanines and stacking of the G-quartets upon one another confer very high thermodynamic stability upon G4 structures in vitro.

Sequences that contain G4 motifs can be easily discussed in terms of the numbers and lengths of "G-runs" (adjacent guanines) and "loops" (non-guanine nucleotides between the G-runs). However, structural analysis has shown that even relatively simple sequences bearing G4 motifs can form three-dimensional structures of great variety[3]. This contributes to the utility of G-quadruplexes in technical applications. It also makes it difficult (or even impossible) to predict the actual structure that a G4 motif will form in genomic DNA, where the motifs themselves are not simple but instead consist of G-runs that are heterogeneous in length and variable in number; and where a variety of proteins can stabilize or resolve G structures. The potential polymorphism of G4 motifs is illustrated by example in Figure 1, which presents just two of the many structures that can in principle be formed by a G4 motif bearing the sequence $G_3N_3G_3N_4G_4N_2G_5$.

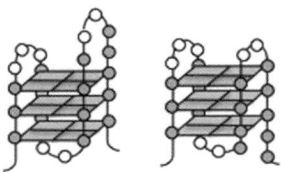

**Figure 1**   *Structural polymorphism of G4 motifs. The figure shows two of the many structures that can in principle be formed by a G4 motif bearing the sequence $G_3N_3G_3N_4G_4N_2G_5$. Circles denote bases; guanines, filled; other bases, open. Planes of G-quartets are shaded.*

Motifs with structural potential pose a risk for the genome, because they can promote genomic instability. The first evidence that G4 structures form in vivo was indirect, but it has proven to be very revealing about a fundamental property of G4 motifs and the human genome. Early analysis showed that some of the most polymorphic variable number tandem repeats (VNTRs) in the human genome bear the G4 motif. These include D4S43, GGGGAGGGGGAAGA; the insulin-linked hypervariable repeat, ACAGGGGTGTGGGG; and the CEB1 minisatellite[4]. Unstable G4 motifs are also associated with neurological diseases, including Fragile X syndrome, due to CGG expansion in the *FRAX* gene[5]; mental retardation[6]; and progressive myoclonus epilepsy[7].

DNA is normally a Watson-Crick duplex, which prevents formation of non-B structures such as G-quadruplexes. In vivo, G4 structures can form within single-stranded regions that form transiently in the course of denaturation that accompanies DNA replication or transcription, as illustrated in Figure 2.

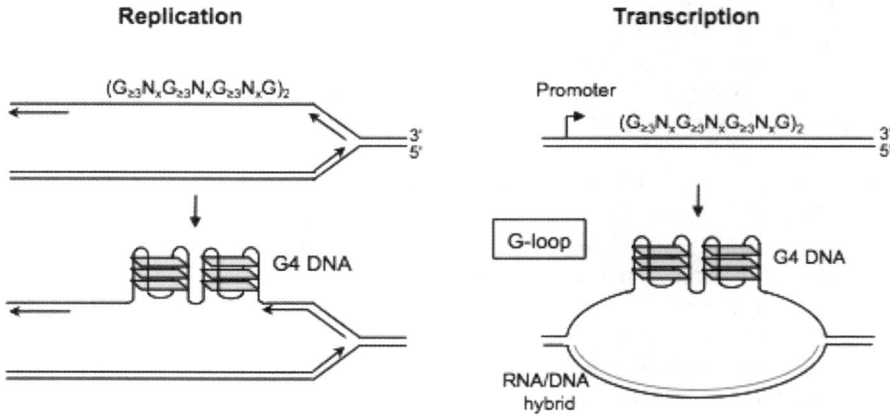

**Figure 2**   *Formation of G4 DNA during transcription or replication. Left, replication results in formation of G4 DNA on the lagging strand. Right, transcription results in formation of a G-loop, containing a stable RNA/DNA hybrid and G4 DNA on the nontemplate strand.*

We showed that G4 structures do form in transcribed regions by single molecule imaging (Figure 3). Templates from G4$^{hi}$ chromosomal domains, including the immunoglobulin switch regions and the telomeres, were transcribed either in vitro or intracellularly, and then examined by transmission electron microscopy. Transcription promoted formation of large, characteristic large loops, hundreds of bp in length[8-10]. These loops, which we called "G-loops", map to the G-rich region. They contain a cotranscriptional RNA/DNA hybrid on the C-rich template strand, and G4 DNA interspersed with single-stranded regions on the G-rich strand[8,9]. The stability of the cotranscriptional hybrid reflects two chemical properties: the unusual stability of the rG:dC base pair (the most stable of the 16 natural ribo/deoxy base pair combinations); and stacking of adjacent guanines.

We identified G-loops in G4$^{hi}$ regions of transcribed immunoglobulin switch regions, telomeres, and oncogenes. G-loops are also predicted to form at other regions bearing the G4 consensus, including many single copy genes. Notably, G-loops form only if transcription is in the physiological orientation, with a G-rich nontemplate strand. The G4 structures in G-loops are recognized by proteins that bind G4 DNA with high affinity in vitro, including the DNA repair proteins BLM helicase[10] and MutSα[11].

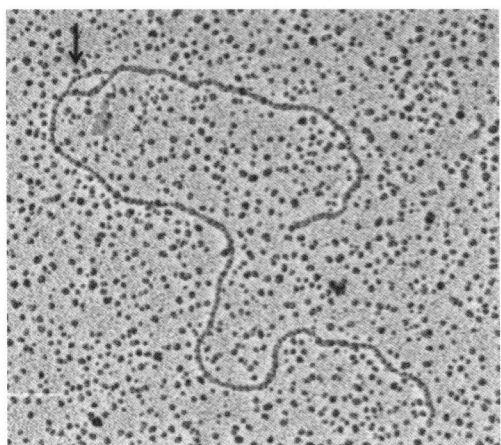

**Figure 3**  *G-loop formed upon transcription of the c-MYC gene. The human c-MYC gene, cloned on a plasmid, was transcribed in vitro, free RNA transcripts degraded, and the DNA template examined by transmission electron microscopy. The arrow points to the G-loop. Note the thick (lower) filament, corresponding to the RNA/DNA hybrid, and the then (upper) filament, which contains single-stranded DNA interspersed with G4 structures.*

### 2.2. G4 Motifs Are Under Selection

The G4 motif characterizes specific functional chromosomal domains in human cells. Telomeres, at the two ends of each chromosome, consist of 10-18 kb repeats of (TTAGGG)$_n$; the repetitive ribosomal DNA (200 copies/genome), which is transcribed to generate ribosomal RNAs, carries G4 motifs throughout its 35 kb length; and the immunoglobulin switch regions, sites of recombination essential for the immune response, consist of highly degenerate and repetitive G4 motifs.

We searched the human genome to identify other regions that carry G4 motifs and would thus have the potential to form G-quadruplex structures. We used sliding windows 100 nt in length to analyze genomic sequence, scored each region that contained a G4 motif as a "hit", and then quantified the percentage of hits in the total numbers of windows searched[12]. This analysis showed that G4 motifs exhibit a very uneven distribution among genes and gene regulatory regions[12-15]. These differences in abundance of G4 motifs do not reflect coding capacity, but apply over entire genes, including both exons and introns.

Enrichment or depletion in G4 motifs ($G4^{hi}$ or $G4^{lo}$) correlates with gene function[12]. This was apparent upon evaluating the distribution of Gene Ontology (GO) terms for each human gene across the spectrum of G4 DNA potentials. $G4^{lo}$ genes function in nucleic acid binding, nucleosome assembly, ubiquitin-dependent proteolysis, cell adhesion, and regulation of cell division. This class of genes includes essentially all genes that encode proteins necessary for DNA repair and replication. In contrast, $G4^{hi}$ genes function as transcription factors, growth factors and cytokines, or in development, cell signalling and muscle contraction. This class encodes essentially all genes that function as oncogenes.

The correlation of gene function with $G4^{hi}$ or $G4^{lo}$ status suggests that G4 motifs are under evolutionary selection. The $G4^{lo}$ status of a gene may be explained as providing protection from genomic instability conferred by formation of G4 structures, but there is no obvious explanation for $G4^{hi}$ status.

## 2.3 G4 Motifs in Regulatory Regions of Human Genes

A generic or idealized gene contains a transcription start site (TSS), an upstream regulatory region, and downstream exons and introns. G4 motifs are enriched both upstream and downstream of the TSS[13,16,17]. However, it is important to distinguish between the presence of a G4 motif in DNA sequence and the actual formation of a G4 structure in the DNA.

The regions upstream and downstream of the TSS have very distinct potential to form non-duplex structures in the course of transcription. The downstream region undergoes transient denaturation to enable passage of RNA polymerase along the template DNA strand. The nontemplate strand is thereby free to form G4 DNA. This suggested that there might be strand bias of enrichment of G4 motifs near the TSS. We found that upstream of the TSS, G4 motifs are comparably enriched on the template and nontemplate DNA strands (Figure 4)[13]. Downstream of the TSS, G4 motifs are enriched on the nontemplate DNA strand.

Most of the $G4^{hi}$ character of the region upstream of the TSS can be eliminated upon masking sequence motifs for well-defined regulatory processes, such as CpG dinucleotides that are sites of methylation, and G-rich recognition motifs for common transcription factors, such as SP1 or KLF[13]. This suggests that these well-defined processes are likely to account for the $G4^{hi}$ character of the region upstream of the TSS.

The strand asymmetry of G4 motifs downstream of the TSS suggests that, in a fraction of genes, transcription of regions bearing G4 motifs may promote formation of nucleic acid structures that could serve as targets for regulation. Three kinds of unusual structures are associated with G-loop formation (see Figure 2). (1) G4 DNA structures in the nontemplate strand of a G-loop, which may be recognized by regulatory factors. (2) G4 structures in the pre-mRNA or mRNA, which may bind specific factors to determine mRNA processing or functions. (3) Cotranscriptional RNA/DNA hybrids, which place special demands on RNA processing[18] and may thus be subject to unusual regulation. As described below, examples are emerging that illustrate the functional importance of G4 structures in both DNA and RNA.

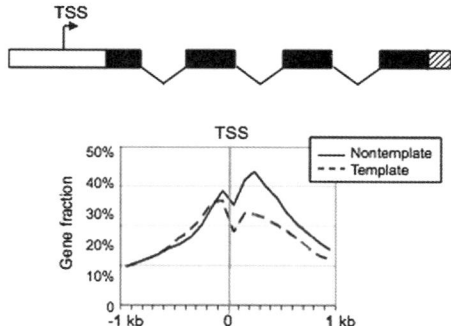

**Figure 4**  Enrichment of genes with G4 motifs near the TSS. Above, diagram of a generic gene, showing transcription start site (TSS), exons (dark), introns (lines), and 3'-untranslated region (striped). Below, graph of fraction of genes containing one or more G4 motifs within the 2 kb region spanning the TSS, on the nontemplate (solid line) or template (dotted line) strand.

### 2.4 Conserved Elements Carrying G4 Motifs in the First Intron

We found that nearly half of human genes carry one or more G4 motifs at the very 5' end of the first intron, on the nontemplate DNA strand[13]. Enrichment displays clear strand bias, and is evident predominately on the nontemplate strand. These G-rich intron 1 elements are conserved in potential to form G4 structures, though not in sequence, from humans to frogs. Many of these elements are quite long, and contain eight or more G-runs.

This observation is especially tantalizing because the 5'-end of the first intron represents a privileged position within genomic sequence. It is very near the TSS but does not appear in the mature mRNA or protein coding sequence and is therefore not under selection determined by codon use or translation. Proximity to the TSS could in principle enable an element to load factors critical for regulation of gene expression and enable their access to the core transcription complex.

### 2.5 G4 Motifs Are Not Enriched in Coding Regions or Untranslated Regions

G4 motifs are not enriched within coding regions: 85% of human exons contain no G4 motifs[14]. Relatively few genes carry G4 motifs within 1 kb downstream of ATG initiator codons and within 1 kb upstream of termination codons, and where they do occur they predominate on the template rather than nontemplate DNA strand. There are also few genes with G4 motifs in the 5'- and 3'-untranslated regions. The paucity of G4 motifs within genes provides further evidence for evolutionary selection against these motifs within specific genomic regions.

### 2.6 G4 Motifs Contribute to Genomic Instability

G4 motifs promote genomic instability, and this may lead to their loss on an evolutionary time scale. G4 structures that form during replication can promote DNA deletion if not resolved. G4 helicases, which unwind G-quadruplex structures, are important in maintaining stability of $G4^{hi}$ regions. Among these are human helicases BLM[19], WRN[20] and FANCJ[21,22].

Deficiencies in G4 helicases cause human diseases associated with genomic instability and cancer predisposition: Bloom syndrome (BLM), Werner syndrome (WRN), and Fanconi anemia (FANCJ)[23-27].

The importance of G4 helicases is evident in human disease, where deficiency in G4 helicases may have particular impact on a G4$^{hi}$ chromosomal domain. For example, WRN helicase is especially important at the telomeres (repeat TTAGGG), and cells from individuals with Werner syndrome exhibit telomere shortening, which contributes to the premature aging characteristic of this disease[28].

The role of G4 helicases in genomic stability has been dramatically demonstrated by experiments with model organisms. Nematode worms lacking the FANCJ homolog, *dog-1*, exhibit genomic instability mapping to regions bearing G4 motifs[29]. In the bakers yeast, *S. cerevisiae*, absence of the G4 helicase Pif-1 results in instability at a G4$^{hi}$ reporter sequence, which is exacerbated by small molecule ligands selective for G4 structures[30,31].

## 2.7 G4 Structures Can be Targets of Regulation

While there is extensive evidence for how G4 motifs may contribute to genomic instability, and thus be under negative selection, we are only beginning to understand how positive selection might maintain G4 motifs in the genome. Thus far, a few specific functional roles for G4 motifs have been identified in diverse and unanticipated contexts. Several examples in which G4 structures have been unambiguously correlated with gene regulation illustrate this functionality. Most notably is a G-quadruplex structure that controls antigen switching in *N. gonorrhea*[32]. In this pathogenic microorganism, the pilE gene encodes a component of the cell surface pilin, a target for immune recognition. To evade the immune response, the expressed pilin gene undergoes regulated variation, by a process that alters the identity of a cassette at a fixed chromosomal expression site. The expression site carries a G4 motif just downstream of the promoter, and formation of a structure at that site enables recombination (Figure 5). G4 DNA structures have also been implicated in epigenetic regulation of human genes by the SWI/SNF family member, ATRX[33].

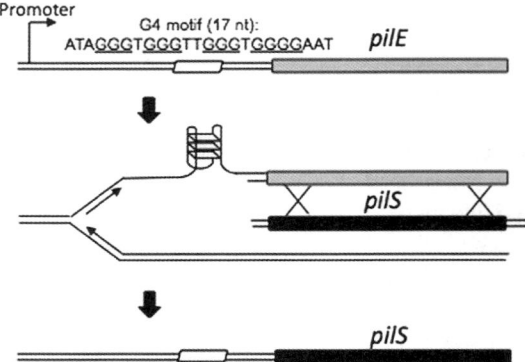

**Figure 5** *A G4 structure controls antigen switching in N. gonorrheae. Above, the regulatory region for expression of a component of the cell surface pilin, shown expressing the pilE cassette. Formation of a G4 structure is necessary for recombination that swaps in the pilS casette, which alters the pilin protein sequence and antigen recognition to enable evasion of the host immune response.*

G4 structures form readily in RNA in vitro, and G4 RNA structures have also been identified with biological function in vivo. The mRNA transcript encoded by the P53 gene, the "guardian" of genomic stability, undergoes alternative 3'-processing in human cells subjected to environmental stress, and processing is determined by a G4 structure formed within a 100 nt region of the P53 gene transcript[34]. G4 RNA structures can also contribute to the pathology of human disease. Transcripts of FRAX genes carrying CGG repeat expansions titrate essential RNA processing factors that recognize G4 structures, and thereby alter RNA processing throughout the cell and contribute to the disease pathology of Fragile X syndrome[5].

## 3 CONCLUSION

G4 motifs are clearly under selection in the human genome. Depletion of G4 motifs is likely to contribute to genomic stability. Moreover, recent examples of how G4 motifs contribute to genome functions suggests why these motifs may be selected.

**References**

1. M. Gellert, M.N. Lipsett and D.R. Davies, *Proc. Natl. Acad. Sci. USA,* 1962, **48**, 2014-2018.
2. D. Sen and W. Gilbert, *Nature,* 1988, **334**, 364-366.
3. A.T. Phan, V. Kuryavyi and D.J. Patel, *Curr. Opin. Struct. Biol.,* 2006, **16**, 288-298.
4. M.N. Weitzmann, K.J. Woodford and K. Usdin, *J. Biol. Chem.,* 1997, **272**, 9517-9523.
5. M.R. Santoro, S.M. Bray and S.T. Warren, *Annu. Rev. Pathol.,* 2011.
6. M.C. Bonaglia, R. Giorda, E. Mani, G. Aceti, B.M. Anderlid, A. Baroncini, T. Pramparo and O. Zuffardi, *J. Med. Genet.,* 2006, **43**, 822-828.
7. T. Saha and K. Usdin, *FEBS Lett.,* 2001, **491**, 184-187.
8. M.L. Duquette, P. Handa, J.A. Vincent, A.F. Taylor and N. Maizels, *Genes Dev.,* 2004, **18**, 1618-1629.
9. M.L. Duquette, P. Pham, M.F. Goodman and N. Maizels, *Oncogene,* 2005, **24**, 5791-5798.
10. M.D. Huber, M.L. Duquette, J.C. Shiels and N. Maizels, *J. Mol. Biol.,* 2006, **358**, 1071-1080.
11. E.D. Larson, M.L. Duquette, W.J. Cummings, R.J. Streiff and N. Maizels, *Curr. Biol.,* 2005, **15**, 470-474.
12. J. Eddy and N. Maizels, *Nucleic Acids Res.,* 2006, **34**, 3887-3896.
13. J. Eddy and N. Maizels, *Nucleic Acids Res.,* 2008, **36**, 1321-1333.
14. J. Eddy and N. Maizels, *Mol. Carcinog.,* 2009, **48**, 319-325.
15. J. Eddy, A.C. Vallur, S. Varma, H. Liu, W.C. Reinhold, Y. Pommier and N. Maizels, *Nucleic Acids Res.,* 2011, **39**, 4975-4983.
16. J.L. Huppert and S. Balasubramanian, *Nucleic Acids Res.,* 2007, **35**, 406-413.
17. Y. Zhao, Z. Du and N. Li, *FEBS Lett.,* 2007, **581**, 1951-1956.
18. A. Aguilera, *Nat. Struct. Mol. Biol.,* 2005, **12**, 737-738.
19. H. Sun, J.K. Karow, I.D. Hickson and N. Maizels, *J. Biol. Chem.,* 1998, **273**, 27587-27592.
20. M. Fry and L.A. Loeb, *J. Biol. Chem.,* 1999, **274**, 12797-12802.

21. T.B. London, L.J. Barber, G. Mosedale, G.P. Kelly, S. Balasubramanian, I.D. Hickson, S.J. Boulton and K. Hiom, *J. Biol. Chem.,* 2008, **283**, 36132-36139.
22. Y. Wu, K. Shin-ya and R.M. Brosh, Jr., *Mol. Cell. Biol.,* 2008, **28**, 4116-4128.
23. C. Sissi, B. Gatto and M. Palumbo, *Biochimie,* 2010, **93**, 1219-1230.
24. R.J. Monnat, Jr., *Semin. Cancer Biol.,* 2010, **20**, 329-339.
25. M.L. Rossi, A.K. Ghosh and V.A. Bohr, *DNA Repair (Amst),* 2010, **9**, 331-344.
26. A.J. Deans and S.C. West, *Nat. Rev. Cancer,* 2011, **11**, 467-480.
27. J.S. Deakyne and A.V. Mazin, *Biochemistry (Mosc),* 2011, **76**, 36-48.
28. L. Crabbe, A. Jauch, C.M. Naeger, H. Holtgreve-Grez and J. Karlseder, *Proc. Natl. Acad. Sci. USA,* 2007, **104**, 2205-2210.
29. E. Kruisselbrink, V. Guryev, K. Brouwer, D.B. Pontier, E. Cuppen and M. Tijsterman, *Curr. Biol.,* 2008, **18**, 900-905.
30. C. Ribeyre, J. Lopes, J.B. Boule, A. Piazza, A. Guedin, V.A. Zakian, J.L. Mergny and A. Nicolas, *PLoS Genet.,* 2009, **5**, e1000475.
31. A. Piazza, J.B. Boule, J. Lopes, K. Mingo, E. Largy, M.P. Teulade-Fichou and A. Nicolas, *Nucleic Acids Res.,* 2010, **38**, 4337-4348.
32. L.A. Cahoon and H.S. Seifert, *Science,* 2009, **325**, 764-767.
33. M.J. Law *et al., Cell,* 2010, **143**, 367-378.
34. A. Decorsiere, A. Cayrel, S. Vagner and S. Millevoi, *Genes Dev.,* 2011, **25**, 220-225.

# REGULATION OF GENE TRANSCRIPTION BY DNA G-QUADRUPLEXES

Michael Fry

Department of Biochemistry, Rappaport Faculty of Medicine, Technion - Israel Institute of Technology, POB 9649 Bat Galim, Haifa, 31096 ISRAEL

## 1 INTRODUCTION

Local sequences of DNA and RNA can form various secondary structures such as hairpins, slipped hairpins, left-handed Z-DNA, triplex DNA, cruciforms and four-stranded G-quadruplexes and i-quadruplexes. Of prime interest are potential biological roles that such secondary structures of nucleic acids may have. Four-stranded G-quadruplexes arguably represent the most extensively documented and convincing case of biologically important non-canonical structure of nucleic acids. DNA and RNA tetraplexes were primarily implicated in telomere metabolism and in the regulation of gene expression. As was extensively reviewed elsewhere, folding of the guanine-rich (G-rich) single-strand overhang of telomeric DNA[1, 2] and of telomerase RNA[3-6] into tetrahelical structures contribute to both the stability and to the regulated elongation of telomeres. A parallel body of evidence indicated that quadruplex structures of both DNA and RNA contribute to the regulation of gene expression. To establish any biological role DNA or RNA quadruplex such structures must be detected and manipulated in cells. Indeed, use of quadruplex-specific antibodies and of fluorescent probes provided evidence for the existence *in vivo* of tetrahelical structures in telomeric DNA.[7-9] However, the limited resolution of current technologies does not enable yet direct detection and manipulation *in vivo* of quadruplexes that putatively exist in non-telomeric chromatin or in cellular RNA. The current belief that quadruplex structures of DNA and RNA play roles in the regulation of gene expression is therefore based by necessity on inferred rather than direct evidence.

Data suggest that G-quadruplex domains in genomic DNA and in RNA transcripts affect gene expression by acting at several critical junctions. At one level, G-rich sequences in regulatory regions of protein-encoding genes may fold into quadruplex conformations that repress or enhance gene transcription. At other levels, quadruplex structures formed in RNA transcripts are thought to play regulatory roles in transcription termination and polyadenylation, in pre-mRNA splicing and in the targeting and translation of mRNA. Evidence for the involvement of RNA quadruplexes at these junctions was reviewed in a number of excellent recent articles.[2, 10, 11] The scope of this chapter is thus limited to a survey of current findings and ideas on the proposed roles of quadruplex structures of DNA in the regulation of gene transcription. Several lines of indirect evidence implicated DNA quadruplexes in transcription regulation. First, G-rich tracts that potentially fold into G-quadruplex structures were found to be overrepresented in regulatory regions of protein-coding genes. Second, gene transcription could be modulated by mutational manipulation of these quadruplex-forming sequences and last, transcription was shown to be affected by quadruplex-interacting small molecules and proteins.

## 2 REGULATION OF TRANSCRIPTION BY UPSTREAM G-QUADRUPLEXES

### 2.1 Upstream gene regulatory regions are enriched in potential G-quadruplex forming motifs

Early global bioinformatics analyses detected about 375,000 G-rich tracts in the human genome that can potentially fold into G-quadruplex structures.[12,13] A subsequent survey of 16,654 genes revealed an unequal distribution of quadruplex-generating motifs among different functional classes of genes. Such elements were particularly abundant in exons and introns of genes that encode kinases and transcription factors and in genes that affect neurogenesis and developmental processes. Remarkably, nearly 70% of proto-oncogenes contained potential quadruplex-generating motifs whereas tumour suppressor genes had very low content of such elements.[14] Additionally, genes that are involved in the immune response, in protein synthesis and in G-protein signalling and in nucleic acids binding also had low content of quadruplex-generating sequences.[14] Computational analyses and biochemical and biophysical inspection of the 5'-untranslated regions (5'-UTR) and of core promoters of protein-coding genes in diverse species detected G-rich motifs that can potentially fold into quadruplex structures in a large percentage of regions proximal to transcription start sites (TSS). Bioinformatics survey identified in 65% of 2892 inspected chicken genes sequences upstream to the TSS, mostly around the core promoter region, that could potentially form quadruplexes.[15] This analysis was extended to demonstrate enrichment in potential quadruplex-generating motifs in regulatory regions of genes of warm-blooded species.[16] Bioinformatics analysis of human genes showed that 43% of the genes had G-rich tracts within 1 Kb upstream to the TSS that could potentially generate at least one G-quadruplex structure.[17] Comparative analysis of human, chimpanzee, rat and mouse genomes revealed that at least 40% of the regions within 1 Kb of TSS contained potential quadruplex forming motifs and that such sequences were conserved in over 700 orthologous promoter regions.[18] Indeed, analysis revealed a consistent enrichment in G-quadruplex-forming motifs in seven different types of gene regulatory regions. It was speculated that due their purported contribution to gene regulation, once quadruplex-generating elements became localized in gene regulatory regions they enjoyed selective advantage during evolution.[19] Remarkably, nuclease hypersensitive sites were 230-fold richer in potential quadruplex-forming motifs than the rest of the genome.[17] This association of G-quadruplex-generating sequences with transcription-induced open structure of chromatinised DNA was in line with the idea that negatively supercoiled G-rich single strands in transcriptionally active DNA can fold into tetraplexes.[20,21] Yet, folding of genomic G-rich sequences into G-quadruplex structures may be largely hampered by their wrapping around nucleosomes in chromatin. Notably, however, computerized analyses of upstream regulatory regions of *C. elegans* and human genes revealed that quadruplex-forming sequences were located outside nucleosome-bound tracts.[22] An independent genome-wide examination of reported nucleosome positions in *Saccharomyces cerevisiae* and in humans also showed that nucleosomes were excluded from regions that were enriched in potential quadruplex-forming sequences.[23] It was argued that by not being wrapped around nucleosomes, G-rich tracts are better able to fold *in vivo* into stable tetraplexes.[22] Also, absence of nucleosomes from G-rich regions could indicate that quadruplexes regulate gene expression by serving as signals for nucleosome exclusion.[23]

The identification of potential G-quadruplex-generating motifs in regulatory regions of genes by genome-wide analyses was complemented by examination of such elements in promoters of specific genes. The most comprehensively investigated case is that of a G-quadruplex-forming element in a *c-MYC* promoter. Following its initial detection,[24] *c-MYC* promoter quadruplex was investigated in detail (see [25-27] for reviews). The c-MYC gene encodes a transcription factor that takes part in the regulation of the expression of about 15% of human genes and whose overexpression is associated with a variety of malignancies.[28-31] About 80-90% of *c-MYC* transcription is controlled by an upstream (-142 to -115) nuclease hypersensitive element III$_i$ (NHE III$_i$). Studies showed that a far upstream element (FUSE) and NHE III$_i$ control *cMYC* expression through a combination of the action of transcription factors and transcription-induced negative superhelicity of the DNA.[32] An NHE III$_i$ core sequence of 27 base-pairs contains in the non-coding strand five consecutive tracts of contiguous guanines that are paired by complementary cytosines in the coding strand. Folding *in vitro* of this sequence into parallel-stranded intramolecular G-quadruplex structure was demonstrated biochemically and biophysically. Data also suggested that the C-rich coding strand could fold back to generate an i-motif structure of two parallel duplexes with intercalated hemiprotonated cytosine$^+$-cytosine base pairs (see [27, 33] for comprehensive reviews). As is discussed below, indirect evidence implied that the G-rich NHE III$_i$ core sequence is likely to fold during transcription *in vivo* into a unimolecular quadruplex structure that represses *c-MYC* transcription.

The study of the *c-MYC* tetraplex guided parallel and subsequent identification of quadruplex-generating motifs in promoters of many other genes. G-rich sequences were identified in regulatory regions of genes such as *c-kit*,[34-36] *VEGF*,[37, 38] *PDGF-A*[39], *c-myb*,[40] *KRAS*,[41] *HRAS*,[42, 43] *hif-1a*,[44] *RET*,[45] *bcl-2*,[46, 47] *Rb*,[48] *bcl-2*,[49, 50] *TK1*,[51] in several muscle-specific genes[52] and in the insulin regulating polymorphic region ILPR.[53-55] These motifs folded *in vitro* into diverse tetraplex structures whose putative formation *in vivo* was shown to affect transcription (*vide infra*).

## 2.2 Transcription modulation by putative G-quadruplex structures in gene promoters

The profusion of potential G-quadruplex forming sequences in gene promoters and their co-localization with nuclease hypersensitive sites in DNA[17] suggested that they may contribute to the regulation of gene transcription. This idea gained support from different lines of indirect evidence. Essentially, transcription in diverse systems was found to be affected by mutational disruption of G-quadruplex formation and by the action of quadruplex-interacting small molecules and proteins. Data thus gathered indicated that putatively formed tetrahelices in different promoter systems may act alone or in collaboration with interacting proteins to either enhance or suppress gene transcription.

*2.2.1 Mutations in G-quadruplex forming promoter sequences affect gene transcription.*
Potential contribution of G-quadruplexes to gene transcription was investigated by the introduction of quadruplex destabilizing mutations into G-rich tracts in gene promoters and studying their effects on transcription. Results indicated that such mutations significantly enhanced or diminished the transcription of different genes. Yet, caution is called for when interpreting these results. One reading is that mutation-induced unfolding of quadruplex structures modulates gene transcription because the transcription machinery interacts directly with the tetraplexes. An alternative possibility is that stable or unfolded quadruplexes affect transcription indirectly by interfering or enhancing protein binding to gene promoters.

The possibility that promoter quadruplex structures may augment transcription was first raised in early studies on a polymorphic region (ILPR) in the promoter of the human insulin gene. This region comprises a variable number of tandemly repeating G-rich sequences that can fold *in vitro* into various inter- and intramolecular G-quadruplex structures.[44-46] Results indicated that the *in vivo* transcription of the insulin gene was positively correlated to tetraplex formation within the ILPR. Specifically, single nucleotide differences in the ILPR that enhanced or decreased G-quadruplex formation respectively increased or lowered insulin transcription.[54] An analogous example for potential action of a promoter quadruplex structure as a facilitator of gene transcription was provided in studies on the murine *KRAS* gene. A G-rich nuclease hypersensitive element (GA-element) upstream of the *KRAS* TSS that was essential for transcription could fold *in vitro* into a quadruplex structure. Destabilization of this tetraplex by G→T or G→A point mutations entailed depressed of *KRAS* expression, consistent with the notion that the G-quadruplex structure operated as a positive regulator of transcription.[42] Expression of a reporter luciferase gene that had an upstream synthetic quadruplex motif or a mutated unfolded counterpart was measured in transfected cells. Transcription in this system was found to be higher when the luciferase gene was preceded by an intact G-quadruplex compared to its destabilized form.[56]

Mutational manipulation of the stability of G-quadruplex structures also indicated that in some systems tetraplex structures in gene promoters acted as negative rather than positive regulators of gene transcription. The capacity of a G-quadruplex structure to arrest transcription was directly demonstrated in T7 RNA polymerase-catalysed *in vitro* system where a tetrahelical form of a human *c-myb* $(GGA)_4$ repeat sequence blocked transcription.[57] A series of other studies indicated that mutational destabilization of some G-quadruplexes enhanced gene transcription in cells. A single G→A transitions within a four guanines cluster in the *c-MYC* NHE $III_i$ core sequence destabilized an intramolecular G-quadruplex structure that was putatively formed *in vivo*. Using a luciferase reporter assay, this quadruplex-disrupting single-base substitution resulted in a 3-fold increase in the basal *in vivo* transcription of *c-MYC*, suggesting that the quadruplex motif in the *c-MYC* promoter acted to suppress transcription.[58] A mixture of four biologically relevant parallel-loop isomeric quadruplexes was subsequently identified in the *c-MYC* NHE $III_i$ region. Mutating all four guanines that were involved in transitions of the quadruplex isomers produced a 5-fold increase in the basal gene expression.[59] Introduction of quadruplex-disrupting mutations into promoters of other genes similarly enhanced transcription. A G-rich sequence in the *thymidine kinase 1* (*TK1*) gene promoter formed *in vitro* an intramolecular quadruplex. A G → A substitution that destabilized *in vitro* this quadruplex entailed a 3-fold increase in the transcription in transfected cells of mutant relative to wild type *TK1*-luciferase reporter gene.[51] In another case, two G-rich *HRAS* promoter sequences, designated *hras-1* and *hras-2* folded *in vitro* into antiparallel and parallel quadruplexes, respectively. Destabilization of these tetrahelical structures by two point mutations resulted in up to 5-fold increase in the transcription of *HRAS* in cells that were transfected by *HRAS* promoter-luciferase reporter plasmid.[43] A G-rich tract located upstream to the P1 promoter of the human *bcl-1* oncogene folded *in vitro* into several quadruplex structures.[46] A G→A mutation that partially disrupted the *bcl-1* quadruplex structure resulted in elevated expression of a *bcl-1* promoter-driven luciferase gene in transfected HL-60 cells. Thus, in this case also, a promoter quadruplex motif was implicated in repression of gene transcription.[60] Modulated expression of the Ying Yang 1 (YY1) transcription factor provided an additional example of negative transcriptional control by a potential quadruplex structure in a gene promoter. A G-rich tract in the YY1 promoter folded *in vitro* into a monomolecular quadruplex structure.[61] When placed in the

5'-UTR of a *Gaussia* luciferase (Gluc) reporter gene the purported quadruplex repressed Gluc expression in transfected 293T cells. Conversely, introduction of tetraplex-disrupting mutations into the G-rich 5'-UTR sequence resulted in increased expression of YY1.[61]

*2.2.2 Transcription is modulated by small molecules that alter the stability of promoter G-quadruplexes.* Support of the notion that promoter G-quadruplex structures regulate gene transcription was additionally gained by demonstrations of modulation of transcription by small molecule ligands that altered the stability of putative quadruplexes in gene promoters (see [62] for a recent review). Some studies focused on the effect of certain quadruplex-interacting ligands on global gene expression. The cationic porphyrin 5,10,15,20-tetra(N-methyl-4-pyridil) porphine chloride (TMPyP4) was shown to bind tightly to[63] and to stabilize[58, 64, 65] or destabilize[66, 67] different G-quadruplex structures. By contrast, TMPyP2, an isomer of TMPyP4, bound poorly to quadruplex structures due to steric interference in its capacity to insert itself or to stack externally to guanine quartets.[64] An early cDNA microarray analysis of TMPyP4-treated cells indicated that this cationic porphyrin altered the expression of several gene clusters.[68] A more recent genome-wide comparison of gene expression profiles in TMPyP4-treated and in untreated HeLa cells revealed drug-induced differential expression of genes that harboured putative quadruplexes in their promoters.[18] Yet, these results should be viewed with caution because of the relatively low selectivity of binding of TMPyP4 to quadruplex versus duplex DNA. More recently, however, the bisquinolinium derivatives PhenDC3 and 360A that have higher binding specificity for quadruplex DNA were reported to exert wide-ranging changes in the transcription of genes that contained potential G-quadruplex motifs upstream to their TSS.[69]

Effects of quadruplex-interacting ligands on the expression of a specific gene were most exhaustively studied in the *c-MYC* model system. TMPyP4 that bound and stabilized a putative quadruplex structure in the *c-MYC* promoter suppressed its expression. Analysis of cDNA microarrays revealed that *c-MYC* expression was down-regulated in cells that were exposed to TMPyP4 but not to its inactive isomer TMPyP2.[68] Stabilization by TMPyP4 of a *c-MYC* promoter quadruplex structure decreased the expression of a linked luciferase gene in transfected Burkitt lymphoma cells.[58, 70] Also, measurements of RNA and protein levels in several transfected cell types showed that TMPyP4 but not TMPyP2 diminished the transcription of a reporter luciferase gene driven by a *c-MYC* promoter.[68, 70] A different class of quadruplex binding and stabilizing ligands, quindoline derivatives[71, 72] was shown to similarly down-regulate *c-MYC* expression in cultured Hep G2[71] and HL60 and K562[73] cells. Berberine derivatives, another group of G-quadruplex binding[74] and stabilizing[75] ligands also lowered *c-MYC* expression in HL60 cells.[76] Last, the quadruplex binding and stabilizing ellipticine analogue GQC-05 decreased *c-MYC* transcription in a CA46 Burkitt lymphoma cell line.[77] The described outcomes of exposure of cells to the various *c-MYC* quadruplex-interacting ligands as well as the above cited independent data[21, 24, 25, 27, 58] lent weight to the idea that the *c-MYC* promoter G-quadruplex operated as a negative regulator of transcription.

G-quadruplex-interacting ligands were also used to demonstrate suppression of transcription by tetraplex structures in promoters of genes other than *c-MYC*. A promoter nuclease hypersensitive element (NHE) is essential for the transcription of platelet-derived growth factor A (PDGF-A). A G-rich run within this NHE folded *in vitro* into a unimolecular quadruplex structure. Stabilization *in vivo* of the putative quadruplex by TMPyP4 was reported to diminish by 40% the basic transcription of the PDGF-A gene.[39] A proximal G-rich S1 nuclease sensitive element in the promoter of platelet-derived growth factor receptor beta (PDGFR-β) is critical for basal promoter activity. This region was capable of assuming *in vitro* several different G-quadruplex conformations. Exposure of Daoy cells to the quadruplex stabilizing ligand telomestatin significantly reduced the level of PDGFR-β mRNA molecules.[78] In an analogous case, a G-rich tract in the promoter of the human vascular endothelial growth factor (VEGF) gene was shown to fold *in vitro* into a monomolecular quadruplex structures.[37] The ligand Se2SAP that bound and stabilized quadruplexes more potently than TMPyP4, was also a more efficient inhibitor of VEGF mRNA synthesis, suggesting that the stabilized tetraplex structure in the VEGF promoter suppressed transcription.[79] Consistent with these results, it was reported recently that VEGF expression was reduced in A549 lung cancer cells that were exposed to two perylene monoimide derivatives that induced formation of a unimolecular quadruplex structure of a VEGF promoter sequence.[80] A class of 3,8,10-trisubstituted isoalloxazines that were identified as G-quadruplex binding and stabilizing ligands have been shown to depress the expression of the *c-KIT* proto-oncogene in cells.[81] Analogously, endogenous expression of *c-KIT* was suppressed in human gastric carcinoma cells that were exposed to benzo[α]phenoxazine that stabilized the *c-KIT* promoter quadruplex.[82] Complementary experiments conversely showed that *c-KIT* expression was elevated in cells that were exposed to triarylpyridine, a selective G-quadruplex interacting ligand that disrupted two quadruplex structures derived from *c-KIT* promoter sequence motifs.[83] Results obtained by mutagenesis that suggested a role for *bcl-1* promoter quadruplex structure in transcription suppression were supplemented by a demonstration that the quadruplex-stabilizing quindoline derivatives significantly reduced *bcl-1* transcription in cells.[60]

Use of quadruplex-stabilizing ligand also indicated that in some cases tetraplexes in gene promoters could function as transcription facilitators rather than suppressors. In an early report TMPyP4 reduced significantly the expression of *KRAS* promoter-driven CAT reporter gene in transfected 293 cells.[41] However, this result was later revised by using a ligand with higher specificity for quadruplex DNA and by adjusting more precisely its molar ratio to the DNA. Use of these better attuned conditions revealed that the quadruplex-specific stabilizing agent guanidine-modified phthalocyanines enhanced by 2-fold the expression of *KRAS* promoter-driven luciferase reported gene in transfected NIH 3T3 cells. This result was thus consistent with the notion that the *KRAS* quadruplex element functioned as an activator rather than suppressor of gene transcription.[42]

*2.2.3 Transcription is modulated by G-quadruplex interacting proteins.* Some proteins that interact with G-quadruplex structures were suggested to contribute to the regulation of transcription in cells by altering the equilibrium between duplex and quadruplex conformations of DNA in gene regulatory regions. Possible involvement of such proteins in the regulation of gene expression was inferred from global computerized identification of association between potential tetraplex motifs and binding sites for transcription factor. Additional evidence was obtained in direct studies of specific proteins that bind, stabilize or resolve tetraplex DNA structures.

2.2.3.1 <u>Potential G-quadruplexes are located in close proximity to binding sites for transcription factors</u>: Computerized genomic analyses detected close juxtaposition between putative G-quadruplex forming sequences in gene promoters and binding sites for diverse transcription factors. Survey of 0.5 kb tracts upstream to the TSS of human genes revealed a significant overlap between the binding site d(GGGCGG) for the zinc-fingers transcription factor SP1 and motifs that potentially folded into G-quadruplex structures.[84,85] Although SP1 was known to bind to its consensus sequence in double-stranded DNA, recent biochemical and biophysical results showed that it bound with comparable affinity to a *c-KIT* antiparallel quadruplex structure.[85] Proximity in gene promoters between potential quadruplex structures and binding sites of additional zinc-finger transcription factors was identified in a bioinformatics analysis of human, chimpanzee, mouse and rat cells. A significant proportion of the binding sites for the transcription factors AP-2, SP1, MAZ and VDR in multiple gene promoters were localized within 100 bases of potential quadruplex motifs.[86] This juxtaposition might be biologically significant as was suggested by an observed co-expression of genes that harbour in their promoters both potential quadruplex motifs and adjacent transcription factor binding sites.[86]

2.2.3.2 Quadruplex binding proteins: Several lines of evidence indicated that transcription of specific genes could be modulated by proteins that interacted with putative quadruplex structures in their promoters. These observations lent support to the notion that DNA tetraplexes contribute to the regulation of transcription and also provided insights into possible roles of G-quadruplex-interacting proteins in gene regulation.

Interaction of members of the myogenic regulatory factors (MRFs) family of muscle-specific transcription factors with G-quadruplex structures of 5'-UTR sequences of muscle-specific genes was linked to gene transcription. Sequences with high content of guanine clusters that were detected in promoters and enhancers of muscle-specific genes, readily folded into hairpin and mono- and bi-molecular quadruplex structures. Homodimers of the MRF transcription factor MyoD bound to some of these tetraplex structures more tightly than to their E-box duplex DNA target motif d(CANNTG)•d(GTNNAC).[52, 87] Whereas homodimeric MyoD is a poor activator of transcription, heterodimers of MyoD and E12 or E47 proteins are potent activators of muscle gene transcription. Indeed, MyoD-E47 dimers associated more tightly with E-box motifs than with the G-quadruplexes.[52] Molecular dissection indicated that MyoD employed partially overlapping but distinct elements to bind the E-box or quadruplex motifs of muscle-specific gene promoters.[88] Homodimers of two other members of the MRF family; Myogenin and MRF4 did not display a similar preferential binding to muscle-gene quadruplex structures. However, upon grafting of the MyoD tetraplex binding domain into Myogenin, its homodimers acquired MyoD-like tight and preferential binding to G-quadruplexes.[89] Muscle-gene promoter G-quadruplexes that were transfected into HEK293 cells enhanced significantly MyoD and E-box-driven expression of a luciferase reporter gene.[90] Based on these experimental data it was speculated that the tightly scheduled expression of muscle genes during myogenesis involves a switch between tetraplex-trapped transcriptionally inactive homodimeric MyoD and E-box-bound active heterodimeric MyoD. Confinement of MyoD homodimers to the quadruplex motifs was proposed to prevent their ineffective occupation of E-boxes and yet to maintain them in close proximity to these motifs. Following heterodimerization, the high affinity of MyoD for the quadruplexes is lost and the gained tight binding to E-boxes initiates muscle gene transcription.[90]

Some of the proteins that modulate *c-MYC* transcription by binding to its promoter NHE III$_i$ region[27] appear to do so by affecting the formation or stability of a putative quadruplex in this region. The hexameric human non-metastatic 23 isoform 2 protein (NM23-H2) activates *c-MYC* transcription *via* interaction with its NHE III$_i$ promoter domain.[91, 92] This protein binds to the *cMYC* NHE III$_i$ cytosine- and G-rich single strands but not to the duplex form of DNA tract. TMPyP4-mediated stabilization of the NHE III$_i$ G-quadruplex decreased NM23-H2 binding, suggesting that the presence of a stabilized tetrahelix prevented association of NM23-H2 with the DNA and inhibited activation of *c-MYC* transcription.[93] The nucleolar protein nucleolin bound preferentially to a parallel-stranded G-quadruplex structure in the *c-MYC* NHE III$_i$ region and induced its formation.[94] Specifically, interaction of the RNA binding domains of nucleolin and of its arginine-glycine-glycine (RGG) domain with the NHE III$_i$ G-quadruplex resulted in repression of *c-MYC* transcription.[94, 95] Another protein, cellular nucleic acid binding protein (CNBP), that also bound and promoted the formation of a parallel-stranded G-quadruplex structure in the *c-MYC* NHE III$_i$,[96] region enhanced rather than depressed the transcription of *c-MYC*.[97] Interestingly, binding of nucleolin to a quadruplex structure of a G-rich tract in the VEGF promoter together with the binding of hnRNP K to the complementary cytosine-rich single-strand similarly enhanced the transcription of the VEGF gene.[98] Considered together, these results are somewhat puzzling. Binding and promotion of the formation of a G-quadruplex structure in the *c-MYC* NHE III$_i$ region by nucleolin or CNBP resulted in contrasting respective repression or enhancement of *c-MYC* transcription. Also, association of nucleolin with putative quadruplex structures in the *c-MYC* or VEGF promoters, respectively, suppressed or enhanced transcription. These presently unexplained contradictions signal our incomplete understanding of the relations at the molecular level between specific quadruplex structures, their interacting proteins and gene transcription.

A role of a quadruplex binding protein in the promotion of gene transcription is illustrated by the interaction of insulin with a putative quadruplex structure of a G-rich tract in the insulin-linked polymorphic region (ILPR). ILPR that consists of an array of polymorphic G-rich tandem repeats is thought to regulate the expression of the human insulin gene. An ILPR repeat "a" that formed a highly stable monomolecular quadruplex structure *in vitro*[54, 55, 99] was found to enhance transcription.[54, 100] Importantly, insulin and insulin-like growth factor 2 (IGF-2) were shown to bind to the ILPR quadruplex "a" in preference over quadruplex structures of other G-rich sequences.[100, 101] The specificity of insulin and IGF-2 binding was underscored by demonstrations that it formed a tighter complex with the "a" repeat quadruplex than with quadruplex forms of other G-rich ILPR repeats.[102-104] Based on these observations it was conjectured that the association of insulin, IGF-2 or peptides derived thereof with the ILPR quadruplex "a" may regulate the expression of insulin.[100]

2.2.3.3 <u>Quadruplex destabilizing proteins</u>: Diverse proteins were found to possess quadruplex-destabilizing activity. By virtue of their potential capacity to affect the equilibrium between quadruplex and duplex conformations of DNA some of these proteins were implicated in the regulation of gene transcription. Two members of the heterogeneous nuclear ribonucleoprotein (hnRNP) super family; CBF-A[105, 106] and uqTBP25[107] bound and stabilized quadruplex forms of the d(TTAGGG)$_n$ telomeric repeat and of a G-rich sequence of the IgG switch region. However, these proteins employed specific hnRNP conserved motifs[108] to disrupt rather than stabilize bimolecular quadruplex structures of the fragile X d(CGG)$_n$ repeat sequence.[109] UP1, a product of partial proteolysis of hnRNP A1,[110, 111] was reported to similarly disrupt quadruplex structures of d(CGG)$_n$, d(GGCAG)$_5$, and d(TTAGGG)$_4$. Several Panc-1 cell proteins including hnRNP A1 were found to be in complex with an intramolecular quadruplex structure of G-rich motifs in the *KRAS* nuclease hypersensitive promoter region.[112] UP1 or hnRNP A1 were subsequently reported to destabilize two *KRAS* promoter quadruplex structures[113, 114] raising the possibility that they may affect transcription by converting quadruplex structures in critical promoter regions into duplex form.[113] In step with this conjecture, transcription activity was shown to be elevated when two quadruplex elements in a G-rich tract upstream to the *HRAS* TSS were destabilized by a different protein, the MAZ transcription factor.[43]

2.2.3.4 *Quadruplex unwinding DNA helicases*: A number of DNA helicases that were identified in diverse species resolved quadruplex forms different G-rich sequences. Among the enzymes that possessed quadruplex unwinding activity were members of the RecQ family of DNA helicases; Sgs1,[115] WRN,[116] and BLM[115, 117] as well as other helicases including the SV40 large T-antigen,[118] FANCJ,[119] RHAU,[120] Pif1,[121, 122] Dna2,[123] G4 resolvase 1 (G4R1),[120, 124] and DHX9[125] (see [126] for review). Putative roles of tetraplex resolving helicases in gene regulation were indirectly inferred from few global analyses of gene expression and from a direct study G4R1.

Indirect evidence for the involvement of quadruplex unwinding helicases in the regulation of transcription was gained by global comparison of mRNA levels in wild type and in RecQ helicase-deficient cells. Microarray analyses indicated that absence of Sgs1 in yeast cells resulted in preferential suppression of the expression of loci that contained in their transcription units putative quadruplex-forming motifs.[127] By contrast, transcription of quadruplex-containing genes was preferentially upregulated in human cells that were deprived of WRN or BLM helicases.[128] These results suggested that helicase-catalysed unwinding of critically placed quadruplexes may either repress or enhance the transcription of multiple genes in different species or cell types.

Direct evidence for helicase-mediated modulation of transcription was obtained in a study of G4R1. This enzyme, a product of the human gene DHX36 that was purified from HeLa cells bound and resolved DNA and RNA quadruplexes.[120, 124] Unwinding by this helicase of a transcription-blocking quadruplex structure in the YY1 promoter region resulted in upregulation of YY1 expression. Also, gene array analysis of samples derived from 258 breast cancer patients identified positive correlation between the levels of G4R1 and of YY1.[61] These findings were consistent with the idea that G4R1 upregulated the transcription of the YY1 gene by unwinding a transcription-blocking putative quadruplex in its 5'-UTR.

## 3 REGULATION OF TRANSCRIPTION BY NON-PROMOTER G-QUADRUPLEXES

Computerized genomic analyses indicated that putative quadruplex structures that are located in regions other than the 5'-UTR of genes may also contribute to the regulation of transcription. A survey of 2 kb stretches upstream and downstream to the TSS revealed that about 16% of 18,217 human genes contained potential quadruplex forming motifs in the non-template DNA strand downstream to the TSS.[129] Since most of these elements were overrepresented in the first intron of the surveyed genes, it was argued that their guanine richness could not be accounted for by the presence of characteristic upstream G-rich motifs such as SP1 binding sites. Specifically, conserved G-rich intron 1 (GrIn1) elements $d(G_{\geq 3}N_xG_{\geq 3}N_xG_{\geq 3}N_xG_{\geq 3})$ with a potential to form G-quadruplex structures were identified in the 5'termini of the first introns of a large proportion of human genes.[129] Considering the proximity of the first intron to the TSS, quadruplexes formed therein may contribute to transcriptional regulation by affecting the loading of transcription factors or by modulating pre-mRNA splicing.[129] The possible linkage between first intron G-quadruplex motifs and gene transcription was illustrated by a specific case of the expression of the topoisomerase 1 (TOP1) encoding gene. Levels of TOP1 mRNA varied by more than 5.7-fold in a panel of 60 cell lines and the amounts of TOP1 transcripts were negatively correlated to the presence of potential quadruplex elements in the first intron of the TOP1 gene.[130] Interestingly, however, although G-rich tracts were detected in both DNA strands, motifs with the highest probability of generating quadruplex structure were located in the template strand. Thus, in this specific case, template strand G-quadruplexes in the first intron possibly acted as negative regulators of transcription.[130]

Independent bioinformatics analysis detected additional potential non-5'-UTR G-quadruplex structures that may contribute to gene regulation. It was observed that genes that contained in their non-template strand potential quadruplex-forming motifs within 0.5

Kb downstream to the TSS were expressed at higher levels than genes with no such elements.[131] It was suggested that quadruplex formation in the non-transcribed strand prevented its re-annealing to the template strand. As a result the transcribed DNA was maintained in an open conformation and gene transcription was enhanced.[131] A proposed mode of regulation of gene transcription by quadruplex motifs downstream to the TSS linked their formation to transcriptional pausing. The expression of multiple genes is modulated by the pausing of RNA polymerase II (Pol II) downstream to the TSS.[132] Bioinformatics analysis revealed that G-rich sequences were more abundant in genes at which Pol II paused than in non-paused genes. Most interestingly, pausing was correlated with the presence of GrIn1 in the non-template strand at the 5'-end of the first introns of genes.[133] Yet, it is less likely that pausing resulted from direct blocking of Pol II by the quadruplex structures since GrIn 1 sites occurred at around 200 bp after the TSS whereas Pol II pauses 20-50 bases downstream to the TSS.

Direct demonstration of repression of transcription by a potential quadruplex far downstream from the 5'-UTR was obtained for the gene that encodes the transcription factor Signal Transducer and Activator of Transcription 3 (STAT3). A G-rich sequence that potentially formed a G-quadruplex structure was identified in the 3' terminus of the STAT3 gene.[134] Insertion of this sequence downstream to a luciferase reporter gene lowered its transcription in transfected cells. By contrast, a mutated sequence that lost its potential to fold into tetraplex structure did not affect luciferase expression. Also, the ligand cepharanthine that stabilized *in vitro* the STAT3 quadruplex structure inhibited STAT3 expression in cardiomyocytes.[134]

## References

1. H. J. Lipps and D. Rhodes, *Trends Cell Biol*, 2009, **19**, 414-422.
2. S. Millevoi, H. Moine and S. Vagner, *Wiley Interdiscip Rev RNA*, 2012.
3. S. Lattmann, M. B. Stadler, J. P. Vaughn, S. A. Akman and Y. Nagamine, *Nucleic Acids Res*, 2011, **39**, 9390-9404.
4. A. N. Sexton and K. Collins, *Molecular and cellular biology*, 2011, **31**, 736-743.
5. S. Lattmann, B. Giri, J. P. Vaughn, S. A. Akman and Y. Nagamine, *Nucleic Acids Res*, 2010, **38**, 6219-6233.
6. E. P. Booy, M. Meier, N. Okun, S. K. Novakowski, S. Xiong, J. Stetefeld and S. A. McKenna, *Nucleic Acids Res*, 2012, **40**, 4110-4124.
7. C. Schaffitzel, I. Berger, J. Postberg, J. Hanes, H. J. Lipps and A. Pluckthun, *Proc Natl Acad Sci U S A*, 2001, **98**, 8572-8577.
8. C. C. Chang, J. Y. Wu, C. W. Chien, W. S. Wu, H. Liu, C. C. Kang, L. J. Yu and T. C. Chang, *Anal Chem*, 2003, **75**, 6177-6183.
9. C. C. Chang, I. C. Kuo, I. F. Ling, C. T. Chen, H. C. Chen, P. J. Lou, J. J. Lin and T. C. Chang, *Anal Chem*, 2004, **76**, 4490-4494.
10. X. Ji, H. Sun, H. Zhou, J. Xiang, Y. Tang and C. Zhao, *Nucleic Acid Ther*, 2011, **21**, 185-200.
11. A. Bugaut and S. Balasubramanian, *Nucleic Acids Res*, 2012.
12. J. L. Huppert and S. Balasubramanian, *Nucleic Acids Res*, 2005, **33**, 2908-2916.
13. A. K. Todd, M. Johnston and S. Neidle, *Nucleic Acids Res*, 2005, **33**, 2901-2907.
14. J. Eddy and N. Maizels, *Nucleic Acids Res*, 2006, **34**, 3887-3896.
15. Z. Du, P. Kong, Y. Gao and N. Li, *Biochemical and biophysical research communications*, 2007, **354**, 1067-1070.
16. Y. Zhao, Z. Du and N. Li, *FEBS letters*, 2007, **581**, 1951-1956.

17. J. L. Huppert and S. Balasubramanian, *Nucleic Acids Res*, 2007, **35**, 406-413.
18. A. Verma, K. Halder, R. Halder, V. K. Yadav, P. Rawal, R. K. Thakur, F. Mohd, A. Sharma and S. Chowdhury, *J Med Chem*, 2008, **51**, 5641-5649.
19. Z. Du, Y. Zhao and N. Li, *Nucleic Acids Res*, 2009, **37**, 6784-6798.
20. F. Kouzine and D. Levens, *Front Biosci*, 2007, **12**, 4409-4423.
21. T. A. Brooks and L. H. Hurley, *Nat Rev Cancer*, 2009, **9**, 849-861.
22. H. M. Wong and J. L. Huppert, *Mol Biosyst*, 2009, **5**, 1713-1719.
23. K. Halder, R. Halder and S. Chowdhury, *Mol Biosyst*, 2009, **5**, 1703-1712.
24. T. Simonsson, P. Pecinka and M. Kubista, *Nucleic Acids Res*, 1998, **26**, 1167-1172.
25. S. Kendrick and L. H. Hurley, *Pure Appl Chem*, 2010, **82**, 1609-1621.
26. T. A. Brooks and L. H. Hurley, *Genes Cancer*, 2010, **1**, 641-649.
27. V. Gonzalez and L. H. Hurley, *Annu Rev Pharmacol Toxicol*, 2010, **50**, 111-129.
28. K. B. Marcu, S. A. Bossone and A. J. Patel, *Annu Rev Biochem*, 1992, **61**, 809-860.
29. L. M. Facchini and L. Z. Penn, *Faseb J*, 1998, **12**, 633-651.
30. M. D. Cole and S. B. McMahon, *Oncogene*, 1999, **18**, 2916-2924.
31. S. Pelengaris and M. Khan, *Arch Biochem Biophys*, 2003, **416**, 129-136.
32. D. Levens, *Genes Cancer*, 2010, **1**, 547-554.
33. Y. Qin and L. H. Hurley, *Biochimie*, 2008, **90**, 1149-1171.
34. S. Rankin, A. P. Reszka, J. Huppert, M. Zloh, G. N. Parkinson, A. K. Todd, S. Ladame, S. Balasubramanian and S. Neidle, *J Am Chem Soc*, 2005, **127**, 10584-10589.
35. H. Fernando, A. P. Reszka, J. Huppert, S. Ladame, S. Rankin, A. R. Venkitaraman, S. Neidle and S. Balasubramanian, *Biochemistry*, 2006, **45**, 7854-7860.
36. P. S. Shirude, B. Okumus, L. Ying, T. Ha and S. Balasubramanian, *J Am Chem Soc*, 2007, **129**, 7484-7485.
37. D. Sun, K. Guo, J. J. Rusche and L. H. Hurley, *Nucleic Acids Res*, 2005, **33**, 6070-6080.
38. K. Guo, V. Gokhale, L. H. Hurley and D. Sun, *Nucleic Acids Res*, 2008, **36**, 4598-4608.
39. Y. Qin, E. M. Rezler, V. Gokhale, D. Sun and L. H. Hurley, *Nucleic Acids Res*, 2007, **35**, 7698-7713.
40. S. L. Palumbo, R. M. Memmott, D. J. Uribe, Y. Krotova-Khan, L. H. Hurley and S. W. Ebbinghaus, *Nucleic Acids Res*, 2008, **36**, 1755-1769.
41. S. Cogoi and L. E. Xodo, *Nucleic Acids Res*, 2006, **34**, 2536-2549.
42. S. Cogoi, M. Paramasivam, A. Membrino, K. K. Yokoyama and L. E. Xodo, *J Biol Chem*, 2010, **285**, 22003-22016.
43. A. Membrino, S. Cogoi, E. B. Pedersen and L. E. Xodo, *PLoS One*, 2011, **6**, e24421.
44. R. De Armond, S. Wood, D. Sun, L. H. Hurley and S. W. Ebbinghaus, *Biochemistry*, 2005, **44**, 16341-16350.
45. K. Guo, A. Pourpak, K. Beetz-Rogers, V. Gokhale, D. Sun and L. H. Hurley, *J Am Chem Soc*, 2007, **129**, 10220-10228.
46. T. S. Dexheimer, D. Sun and L. H. Hurley, *J Am Chem Soc*, 2006, **128**, 5404-5415.
47. H. Li, Y. Liu, S. Lin and G. Yuan, *Chemistry*, 2009, **15**, 2445-2452.
48. Y. Xu and H. Sugiyama, *Nucleic Acids Res*, 2006, **34**, 949-954.
49. N. Khan, A. Avino, R. Tauler, C. Gonzalez, R. Eritja and R. Gargallo, *Biochimie*, 2007, **89**, 1562-1572.
50. M. Nambiar, G. Goldsmith, B. T. Moorthy, M. R. Lieber, M. V. Joshi, B. Choudhary, R. V. Hosur and S. C. Raghavan, *Nucleic Acids Res*, 2011, **39**, 936-948.
51. R. Basundra, A. Kumar, S. Amrane, A. Verma, A. T. Phan and S. Chowdhury, *Febs J*, 2010, **277**, 4254-4264.

52. A. Yafe, S. Etzioni, P. Weisman-Shomer and M. Fry, *Nucleic Acids Res*, 2005, **33**, 2887-2900.
53. M. C. Hammond-Kosack, M. W. Kilpatrick and K. Docherty, *J Mol Endocrinol*, 1993, **10**, 121-126.
54. A. Lew, W. J. Rutter and G. C. Kennedy, *Proc Natl Acad Sci U S A*, 2000, **97**, 12508-12512.
55. P. Catasti, X. Chen, R. K. Moyzis, E. M. Bradbury and G. Gupta, *J Mol Biol*, 1996, **264**, 534-545.
56. A. Baral, P. Kumar, R. Halder, P. Mani, V. K. Yadav, A. Singh, S. K. Das and S. Chowdhury, *Nucleic Acids Res*, 2012, **40**, 3800-3811.
57. C. Broxson, J. Beckett and S. Tornaletti, *Biochemistry*, 2011, **50**, 4162-4172.
58. A. Siddiqui-Jain, C. L. Grand, D. J. Bearss and L. H. Hurley, *Proc Natl Acad Sci U S A*, 2002, **99**, 11593-11598.
59. J. Seenisamy, E. M. Rezler, T. J. Powell, D. Tye, V. Gokhale, C. S. Joshi, A. Siddiqui-Jain and L. H. Hurley, *J Am Chem Soc*, 2004, **126**, 8702-8709.
60. X. D. Wang, T. M. Ou, Y. J. Lu, Z. Li, Z. Xu, C. Xi, J. H. Tan, S. L. Huang, L. K. An, D. Li, L. Q. Gu and Z. S. Huang, *J Med Chem*, 2010, **53**, 4390-4398.
61. W. Huang, P. J. Smaldino, Q. Zhang, L. D. Miller, P. Cao, K. Stadelman, M. Wan, B. Giri, M. Lei, Y. Nagamine, J. P. Vaughn, S. A. Akman and G. Sui, *Nucleic Acids Res*, 2012, **40**, 1033-1049.
62. S. Ghosh, P. Majumder, S. K. Pradhan and D. Dasgupta, *Biochim Biophys Acta*, 2010, **1799**, 795-809.
63. M. W. Freyer, R. Buscaglia, K. Kaplan, D. Cashman, L. H. Hurley and E. A. Lewis, *Biophys J*, 2007, **92**, 2007-2015.
64. F. X. Han, R. T. Wheelhouse and L. H. Hurley, *J Am Chem Soc*, 1999, **121**, 3561-3570.
65. H. Han, D. R. Langley, A. Rangan and L. H. Hurley, *J Am Chem Soc*, 2001, **123**, 8902-8913.
66. P. Weisman-Shomer, E. Cohen, I. Hershco, S. Khateb, O. Wolfovitz-Barchad, L. H. Hurley and M. Fry, *Nucleic Acids Res*, 2003, **31**, 3963-3970.
67. M. J. Morris, K. L. Wingate, J. Silwal, T. C. Leeper and S. Basu, *Nucleic Acids Res*, 2012, **40**, 4137-4145.
68. C. L. Grand, H. Han, R. M. Munoz, S. Weitman, D. D. Von Hoff, L. H. Hurley and D. J. Bearss, *Mol Cancer Ther*, 2002, **1**, 565-573.
69. R. Halder, J. F. Riou, M. P. Teulade-Fichou, T. Frickey and J. S. Hartig, *BMC Res Notes*, 2012, **5**, 138.
70. L. H. Hurley, D. D. Von Hoff, A. Siddiqui-Jain and D. Yang, *Semin Oncol*, 2006, **33**, 498-512.
71. T. M. Ou, Y. J. Lu, C. Zhang, Z. S. Huang, X. D. Wang, J. H. Tan, Y. Chen, D. L. Ma, K. Y. Wong, J. C. Tang, A. S. Chan and L. Q. Gu, *J Med Chem*, 2007, **50**, 1465-1474.
72. J. Dai, M. Carver, L. H. Hurley and D. Yang, *J Am Chem Soc*, 2011, **133**, 17673-17680.
73. J. N. Liu, R. Deng, J. F. Guo, J. M. Zhou, G. K. Feng, Z. S. Huang, L. Q. Gu, Y. X. Zeng and X. F. Zhu, *Leukemia*, 2007, **21**, 1300-1302.
74. M. Franceschin, L. Rossetti, A. D'Ambrosio, S. Schirripa, A. Bianco, G. Ortaggi, M. Savino, C. Schultes and S. Neidle, *Bioorg Med Chem Lett*, 2006, **16**, 1707-1711.
75. W. J. Zhang, T. M. Ou, Y. J. Lu, Y. Y. Huang, W. B. Wu, Z. S. Huang, J. L. Zhou, K. Y. Wong and L. Q. Gu, *Bioorg Med Chem*, 2007, **15**, 5493-5501.

76. Y. Ma, T. M. Ou, J. Q. Hou, Y. J. Lu, J. H. Tan, L. Q. Gu and Z. S. Huang, *Bioorg Med Chem*, 2008, **16**, 7582-7591.
77. R. V. Brown, F. L. Danford, V. Gokhale, L. H. Hurley and T. A. Brooks, *J Biol Chem*, 2011, **286**, 41018-41027.
78. Y. Qin, J. S. Fortin, D. Tye, M. Gleason-Guzman, T. A. Brooks and L. H. Hurley, *Biochemistry*, 2010, **49**, 4208-4219.
79. D. Sun, W. J. Liu, K. Guo, J. J. Rusche, S. Ebbinghaus, V. Gokhale and L. H. Hurley, *Mol Cancer Ther*, 2008, **7**, 880-889.
80. T. Taka, K. Joonlasak, L. Huang, T. Randall Lee, S. W. Chang and W. Tuntiwechapikul, *Bioorg Med Chem Lett*, 2012, **22**, 518-522.
81. M. Bejugam, S. Sewitz, P. S. Shirude, R. Rodriguez, R. Shahid and S. Balasubramanian, *J Am Chem Soc*, 2007, **129**, 12926-12927.
82. K. I. McLuckie, Z. A. Waller, D. A. Sanders, D. Alves, R. Rodriguez, J. Dash, G. J. McKenzie, A. R. Venkitaraman and S. Balasubramanian, *J Am Chem Soc*, 2011, **133**, 2658-2663.
83. Z. A. Waller, S. A. Sewitz, S. T. Hsu and S. Balasubramanian, *J Am Chem Soc*, 2009, **131**, 12628-12633.
84. A. K. Todd and S. Neidle, *Nucleic Acids Res*, 2008, **36**, 2700-2704.
85. E. A. Raiber, R. Kranaster, E. Lam, M. Nikan and S. Balasubramanian, *Nucleic Acids Res*, 2012, **40**, 1499-1508.
86. P. Kumar, V. K. Yadav, A. Baral, D. Saha and S. Chowdhury, *Nucleic Acids Res*, 2011, **39**, 8005-8016.
87. K. Walsh and A. Gualberto, *J Biol Chem*, 1992, **267**, 13714-13718.
88. J. Shklover, S. Etzioni, P. Weisman-Shomer, A. Yafe, E. Bengal and M. Fry, *Nucleic Acids Res*, 2007, **35**, 7087-7095.
89. A. Yafe, J. Shklover, P. Weisman-Shomer, E. Bengal and M. Fry, *Nucleic Acids Res*, 2008, **36**, 3916-3925.
90. J. Shklover, P. Weisman-Shomer, A. Yafe and M. Fry, *Nucl. Acids Res.*, 2010, gkp1208.
91. S. J. Berberich and E. H. Postel, *Oncogene*, 1995, **10**, 2343-2347.
92. L. Ji, M. Arcinas and L. M. Boxer, *J Biol Chem*, 1995, **270**, 13392-13398.
93. T. S. Dexheimer, S. S. Carey, S. Zuohe, V. M. Gokhale, X. Hu, L. B. Murata, E. M. Maes, A. Weichsel, D. Sun, E. J. Meuillet, W. R. Montfort and L. H. Hurley, *Mol Cancer Ther*, 2009, **8**, 1363-1377.
94. V. Gonzalez, K. Guo, L. Hurley and D. Sun, *J Biol Chem*, 2009, **284**, 23622-23635.
95. V. Gonzalez and L. H. Hurley, *Biochemistry*, 2010, **49**, 9706-9714.
96. M. Borgognone, P. Armas and N. B. Calcaterra, *Biochem J*, 2010, **428**, 491-498.
97. E. F. Michelotti, T. Tomonaga, H. Krutzsch and D. Levens, *J Biol Chem*, 1995, **270**, 9494-9499.
98. D. J. Uribe, K. Guo, Y. J. Shin and D. Sun, *Biochemistry*, 2011, **50**, 3796-3806.
99. Z. Yu, J. D. Schonhoft, S. Dhakal, R. Bajracharya, R. Hegde, S. Basu and H. Mao, *J Am Chem Soc*, 2009, **131**, 1876-1882.
100. A. C. Connor, K. A. Frederick, E. J. Morgan and L. B. McGown, *J Am Chem Soc*, 2006, **128**, 4986-4991.
101. Y. Wang, H. Zhang, L. A. Ligon and L. B. McGown, *Biochemistry*, 2009, **48**, 8189-8194.
102. J. D. Schonhoft, A. Das, F. Achamyeleh, S. Samdani, A. Sewell, H. Mao and S. Basu, *Biopolymers*, 2010, **93**, 21-31.
103. J. Xiao, J. A. Carter, K. A. Frederick and L. B. McGown, *J Sep Sci*, 2009, **32**, 1654-1664.

104. J. Xiao and L. B. McGown, *J Am Soc Mass Spectrom*, 2009, **20**, 1974-1982.
105. G. Sarig, P. Weisman-Shomer, R. Erlitzki and M. Fry, *J Biol Chem*, 1997, **272**, 4474-4482.
106. G. Sarig, P. Weisman-Shomer and M. Fry, *Biochem Biophys Res Commun*, 1997, **237**, 617-623.
107. R. Erlitzki and M. Fry, *J Biol Chem*, 1997, **272**, 15881-15890.
108. P. Weisman-Shomer, E. Cohen and M. Fry, *Nucleic Acids Res*, 2002, **30**, 3672-3681.
109. P. Weisman-Shomer, Y. Naot and M. Fry, *J Biol Chem*, 2000, **275**, 2231-2238.
110. H. Fukuda, M. Katahira, E. Tanaka, Y. Enokizono, N. Tsuchiya, K. Higuchi, M. Nagao and H. Nakagama, *Genes Cells*, 2005, **10**, 953-962.
111. H. Fukuda, M. Katahira, N. Tsuchiya, Y. Enokizono, T. Sugimura, M. Nagao and H. Nakagama, *Proc Natl Acad Sci U S A*, 2002, **99**, 12685-12690.
112. S. Cogoi, M. Paramasivam, B. Spolaore and L. E. Xodo, *Nucleic Acids Res*, 2008, **36**, 3765-3780.
113. M. Paramasivam, A. Membrino, S. Cogoi, H. Fukuda, H. Nakagama and L. E. Xodo, *Nucleic Acids Res*, 2009, **37**, 2841-2853.
114. M. Paramasivam, S. Cogoi and L. E. Xodo, *Chemical communications*, 2011, **47**, 4965-4967.
115. M. D. Huber, D. C. Lee and N. Maizels, *Nucleic Acids Res*, 2002, **30**, 3954-3961.
116. M. Fry and L. A. Loeb, *J Biol Chem*, 1999, **274**, 12797-12802.
117. H. Sun, J. K. Karow, I. D. Hickson and N. Maizels, *J Biol Chem*, 1998, **273**, 27587-27592.
118. N. Baran, L. Pucshansky, Y. Marco, S. Benjamin and H. Manor, *Nucleic Acids Res*, 1997, **25**, 297-303.
119. Y. Wu, K. Shin-ya and R. M. Brosh, Jr., *Mol Cell Biol*, 2008, **28**, 4116-4128.
120. J. P. Vaughn, S. D. Creacy, E. D. Routh, C. Joyner-Butt, G. S. Jenkins, S. Pauli, Y. Nagamine and S. A. Akman, *J Biol Chem*, 2005, **280**, 38117-38120.
121. C. Ribeyre, J. Lopes, J. B. Boule, A. Piazza, A. Guedin, V. A. Zakian, J. L. Mergny and A. Nicolas, *PLoS Genet*, 2009, **5**, e1000475.
122. C. M. Sanders, *Biochem J*, 2010, **430**, 119-128.
123. L. Zheng, M. Zhou, Z. Guo, H. Lu, L. Qian, H. Dai, J. Qiu, E. Yakubovskaya, D. F. Bogenhagen, B. Demple and B. Shen, *Molecular cell*, 2008, **32**, 325-336.
124. S. D. Creacy, E. D. Routh, F. Iwamoto, Y. Nagamine, S. A. Akman and J. P. Vaughn, *J Biol Chem*, 2008, **283**, 34626-34634.
125. P. Chakraborty and F. Grosse, *DNA Repair (Amst)*, 2011, **10**, 654-665.
126. Y. Wu and R. M. Brosh, Jr., *Febs J*, 2010, **277**, 3470-3488.
127. S. G. Hershman, Q. Chen, J. Y. Lee, M. L. Kozak, P. Yue, L. S. Wang and F. B. Johnson, *Nucleic Acids Res*, 2008, **36**, 144-156.
128. J. E. Johnson, K. Cao, P. Ryvkin, L. S. Wang and F. B. Johnson, *Nucleic Acids Res*, 2010, **38**, 1114-1122.
129. J. Eddy and N. Maizels, *Nucleic Acids Res*, 2008, **36**, 1321-1333.
130. W. C. Reinhold, J. L. Mergny, H. Liu, M. Ryan, T. D. Pfister, R. Kinders, R. Parchment, J. Doroshow, J. N. Weinstein and Y. Pommier, *Cancer Res*, 2010, **70**, 2191-2203.
131. Z. Du, Y. Zhao and N. Li, *Genome Res*, 2008, **18**, 233-241.
132. J. Li and D. S. Gilmour, *Curr Opin Genet Dev*, 2011, **21**, 231-235.
133. J. Eddy, A. C. Vallur, S. Varma, H. Liu, W. C. Reinhold, Y. Pommier and N. Maizels, *Nucleic Acids Res*, 2011, **39**, 4975-4983.
134. S. Lin, S. Li, Z. Chen, X. He, Y. Zhang, X. Xu, M. Xu and G. Yuan, *Bioorg Med Chem Lett*, 2011, **21**, 5987-5991.

# THE REALITY OF QUADRUPLEX NUCLEIC ACIDS AS A THERAPEUTIC TARGET

G. N. Parkinson[1]

[1]Department of Pharmaceutical and Biological Chemistry, UCL School of Pharmacy, London WC1N 1AX, UK

## 1 INTRODUCTION

The potential for the formation of G-quadruplex motifs to alter key aspects of regulatory control of the cell appears convincing based on the widespread distribution of putative quadruplex-forming sequences (PQSs) throughout the cell. Some well characterized molecular and functional targets for possible therapeutic intervention involving DNA/RNA G-quadruplex folding[1,2,3] currently include: at the single stranded 3' ends of telomeric DNA, promoter regions of dsDNA, mRNA both introns and exons at 5' and 3' ends and near genes, within TERRA[4] (telomeric repeat-containing RNA), ribosome biogenesis, transcription, translation, and replication. A renewed interest in targeting nucleic acids to control cellular signalling and gene expression came from the discovery and development of RNA interference (RNAi) technologies for controlling posttranscriptional suppression of gene expression for medicinal and biotechnology applications. It is an ability of nucleic acids to form unusual secondary structures and affect the transcriptome[5] that is central to this chapter. The formation of stable quadruplex motifs provides a rational for the development of new molecular targeted therapeutic agents through rational drug design. These targets provide real opportunities for novel therapeutic intervention, particularly as G-quadruplex stabilization has been shown to be readily induced *in-vitro,* by the addition of several classes of small molecule ligands.

In this chapter, we will review the current thinking and explore the aspirations of those involved in quadruplex drug discovery and development. Early on it was postulated and perhaps optimistically hoped, that a simple strategy of G-quadruplex stabilization, through small molecule interactions, could offer a realistic model of translating traditional chemotherapeutic approaches, as developed for dsDNA, to generate new quadruplex specific pharmaceutical agents. The fundamental question addressed here is the reality of quadruplexes as a viable therapeutic target to elicit biologically relevant effects. Are quadruplexes good, bad or indeed irrelevant, and what makes a good therapeutic target? I will review the current situation in targeting these motifs, where the hurdles lie and what, if any, opportunities exist.

## 2 TARGETING DNA/RNA QUADRUPLEXES

### 2.1 Telomeres

Structural biologists were drawn to telomeres with a realization that the single stranded telomeric DNA overhangs at the ends of chromosomes could form unusual secondary structures through guanine-guanine self-associations. The formation or inhibition of these secondary structures appeared to play a role in protection and maintenance of chromosomes. The subsequent identification of the enzyme telomerase and its up regulation in the vast majority of cancers placed telomerase inhibition at the forefront of cancer therapeutic research.[6] Telomerase reverse transcriptase is a ribonucleoprotein, that uses a 11 nucleotide segment of the 451 nucleotide RNA component as a template for the 3' addition of G-rich d(TTAGGG) repeats onto the 3' end of single stranded DNA, to ensure the integrity of chromosomal DNA during repeated cycles of cell division. The concept of preventing hybridization of the 3' end of chromosomal DNA to the RNA template provides a simple strategy for telomerase inhibition.[7] The design of selective ligands that promote and stabilize quadruplex formed from the G-rich ssDNA repeats located at the ends of chromosomes might then provide an avenue for eliciting damage repair pathway responses[8] specific to cancer cells. A simple commercial *in vitro* assay is routinely used to test telomerase enzyme efficiency under a variety of conditions and in the presence of ligands. The assay only requires a priming DNA template, NTP's and cell extracts containing functional telomerase enzymes. The ability of the enzyme to add hexanucleotide repeats to the priming sequence is reduced with the addition of components that stabilize G-quadruplex formation, such as $K^+$ metal ions, or selective ligands quadruplex inducing ligands. This is a commonly used assay with many variants, typically incorporating a PCR amplification step. Using the TRAP (Telomeric Repeat Amplification Protocol) variant, compounds can be quickly assayed to determine their effectiveness in interfering with telomerase's ability to add tandem repeats to the primer sequence. In order to find ligands that selectively induce quadruplex formation and/or stabilization, and not generally inhibit polymerase activity and or act as telomerase substrate binding inhibitor, modified telomerase assays have been developed.[9] Efforts to incorporate high-throughput screen methods by removing the need for time consuming gel electrophoresis and PCR steps will aid in ligand screening. One such technique uses a solid support with chronocoulometry (CC) where the electrochemical signal developed is proportional to the length of newly synthesized DNA, this is achieved by measuring the binding of hexaammineruthenium(III) chloride.[10] Cell lysate extracts are passed over the solid support in the presence of ligands. An advantage is that the inhibition determined in this method can discriminate between telomerase and telomere binding. Currently *in vivo* assays used for measuring inhibition through quadruplex formation still neglect the many associated proteins at the ends of the chromosome, particularly those with high affinity both the double and single stranded telomeric. A proof of concept for targeting the RNA templating region of the RNA component (hTR) comes from the development of GRN163L (Imetelstat, GERON), an hTR template antisence antagonist in phase I/II clinical studies. We need to be cautious as the complexity of the structural, spatial and temporal arrangement of DNA and associated proteins are lost restricting our ability to comprehensively model quadruplex formation and telomere associations.

Since the validation of telomerase inhibition through quadruplex stabilization by specific small molecules ligands,[11] a significant proportion of the compounds identified are positively charged small molecules with polyaromatic and heterocylic rings. These general properties offer certain advantages for drug development; molecules can penetrate the cell, enter the nucleus and readily associate with anionic DNA. Typically these aromatic molecules are planar, stacking externally on the 5' and 3' terminal tetrads and offer opportunities for further design to avoid duplex DNA binding by the addition of three or more substituents around the central core. One important outlier to this simple design concept is an uncharged cyclic polyamine telomestatin, a natural product from the bacteria Streptomyces anulatus. Containing several oxazole rings it is able to inhibit telomerase, prevent the 3' single stranded binding protein (protection of telomere 1 – POT1) associate and selectively stabilize human telomeric DNA.[12] Synthesis of oxazole ring based mimetics of telomestatin[13, 14] offers the opportunity to build a family of selective ligands targeting telomeric DNA without the addition of charged groups. Further identification of new families of ligands to target telomere quadruplexes will come with the development of new PCR and gel electrophoresis free telomerase inhibition assays, allowing the efficient screening of large family of ligands, however drug-like ligand libraries targeting DNA and RNA are currently limited in scope.

Several ligands have been tested for their efficacy in cellulo and in xenograft models including: telomestatin,[15] RHPS4,[16] BRACO19 a 3,6,9-trisubstituted acridine,[17] a series of naphthalene diimides.[18] Activity for RHPS4 in melanoma cells and human transformed fibroblasts triggered specific DNA damage response factors, gamma-H2AX, RAD17, and 53BP1 and worryingly induced the overexpression TRF2 or POT1 as telomere-protective factors. Tumors were shown to be resistant to RHPS4 through either TRF2 or POT1 over-expression.[19] The development of ligands targeting telomere stabilization is currently limited by several factors and need to be developed further before progressing to human trials, however they do show promise and may avoid resistance through alternative lengthening of telomeres (ALT) mechanisms. There is no doubt that telomerase will remain an important target for intervention, even though it is not yet a clinically validated target as it is differentially over-expressed in most human cancers above normal somatic cells.

## 2.2 Promoters

Certain G-rich sequences upstream of transcription sites within promoter regions are particularly susceptible to nuclease activity, and have a potential to form secondary structures creating opportunities for gene regulation. Through a mechanism of G-quadruplex stabilization, they offer pathways to develop molecular-based targeted approaches for the control of gene expression. An analysis of G-rich regions within promoters began in 1996 with the identification of the chicken-β-globin gene.[20] Not until 2005 that a systematic analysis of the human genome for G-rich regions[21, 22] conducted, focusing first on transcription start sites, and later on promoter sites providing tantalizing evidence for a role of G-quadruplex motifs in transcriptional regulation, folding as intramolecular quadruplexes these cis-acting regulatory elements are attractive molecular targets unique in sequence and distinctive in topology. They can be validated readily both *in vitro* and *in vivo* assays by monitoring changes in protein and mRNA expression levels in cells, and assays can be scaled up for high-throughput drug screening. In addition, potential resistance through point mutations is less likely than the gene products, with fewer molecular copies to be targeted. However, several difficulties arise, first the complexity of the systems in which the folding occurs, their transient nature and the variability of topologies induced. The variability of topologies is particularly a problem for promoters as they tend to have several G-rich runs that can containing more than the required minimum of four G-stretches.

Currently there are only five structurally validated molecular targets for rational drug design and therapeutic intervention, c-Myc,[23] c-Kit21 c-Kit87up,[24] hTERT,[25] RET promoter sequence,[26] and Bcl-2[27] all presenting a parallel topology *in vitro*, with three stacked tetrads, but still retaining unique ligand/quadruplex interfaces. A further nine therapeutically relevant promoter sequences, with a focus on proto oncogenes, have been characterized and are awaiting structural elucidation. Support for ligands with the capacity to modulate transcription comes from TMPyP4, a small molecule ligand with little quadruplex selectivity over duplex DNA and its partner sequence c-Myc.[28] Another is a trisubstituted isoalloxazine[29] that alters the expression *in cellulo* of an important proto-oncogene c-kit. A recent determination by NMR methods of a complex between a modified c-MYC NHE III$_1$ sequence MycG4, and two quindolines[30] reveals a stable complex in potassium, with the two crescent-shaped ligands stacked externally over the 5' and 3' tetrads. The authors argued that the flanking residues that form a binding pocket are adaptable and that the ligand first stacks on one guanine of the tetrad, and then positions to exploit groove/loop interactions. A cautionary note is provided by Boddupally et al. & Hurley who used a cellular based assay in a Burkitt's lymphoma cell line (CA46-specific) to show that none of the 3 lead compounds of the 11-substituted quindoline analogues acted through down-regulation c-MYC gene expression, even though all showed *in vitro* transcriptional control of c-MYC expression.[31]

Hitting other G-quadruplexes than those intended may be difficult to control. Crossover activity between promoter and telomeric targets of ligands stabilizing quadruplexes has been documented for the naphthaline diiamide BMSG-SH-3. Although effective as an anti-tumor agent in an *in vivo* pancreatic cancer xenograft model, here reduction in telomerase activity was also accompanied by a reduction of the gene product HSP90, which contains two putative quadruplex forming sequences.[18] The great abundance of PQS identified in the genome does raise questions over the specific roles they may individually play. This may be partly addressed by the possibility of quadruplexes acting *in trans* with zinc finger transcription factors, perhaps acting within a specific cell type and cell cycle context.[32] Significant evidence is amassing in support of promoter as molecular targets, and so this truly is the start of drug discovery for promoter regions.

## 2.3 RNA G-quadruplexes

The potential for the formation of stable quadruplexes in RNA was recognized in 1993[33] with an investigation the self-association of HIV-1NL4-3 RNA via G-rich regions. Furthermore, the Insulin-like growth factor II (IGF-II) mRNAs contained G-rich regions in the 3' region that in the presence of K$^+$ ions *in vitro* blocked reverse transcription.[34] High resolution NMR structural data on a intermolecular quadruplex forming sequence r(UGGGGU)[35], identified in E. coli 5S RNA,[36] confirmed the stability of these motifs, in solution and solid state in 2001.[37] Structural data on RNA oligomers[38] folding as dimeric quadruplex complexes provided some explanation for the associations seen with mRNA-fragile X mental retardation protein (FMRP) binding and HIV-1 genome RNA dimerization. Structural diversity within RNA quadruplex forming sequences was apparent with the observation from NMR data of bulged uridine residues as sites for protein-RNA and RNA-RNA interaction.[39] The recent identification of TERRA transcripts (G-rich telomeric repeat-containing RNAs) from the C-rich telomeric strand of telomeres, which is thought to play a key role in chromatin remodelling, spurred a renewed interest in RNA quadruplex formation. NMR and X-ray methods used to study two r(UUAGGG) repeats[40,41] have shown parallel stranded dimeric quadruplex formation consistent with biophysical data. Structural data on RNA intramolecular folded quadruplexes and *in vivo* data is desperately required to validate current models for the association of longer RNA G-rich sequences. Bioinformatic searches have revealed that, similar to genomic DNA, mRNA sequences also host an increased abundance of PQS as compared to the overall frequency seen in the transcriptone.[42] Further evidence is emerging for the possible roles G-quadruplex formation could play in the regulation of gene expression. *In vitro* luciferase assays have shown that the introduction of G-rich PQS, in several locations 5' of the UTR can reduce translational expression. Furthermore, transfections of human cells by plasmids containing similar reporter elements have also shown translational suppression.[43] Direct RNA-RNA interactions between small RNA and mRNA to control gene expression through quadruplex stabilization has also been established.[44]

We need to keep in mind that targeting RNA secondary motifs is extremely challenging. There have been successes with antibiotics that include erythromycins, tetracyclines, and aminoglycosides that target prokaryotic RNA in the ribosome, and with paromomycin with binding to the 30S tRNA decoding A site. In contrast, the targeting of HIV-1 Trans-activation response element (TAR) RNA has so far not yet yielded an approved therapeutic agent.

## 3 THE SUITABILITY OF QUADRUPLEXES AS A MOLECULAR TARGET

### 3.1 Targeting nucleic acids

With the identification of each new molecular target, it is becoming clearer that significant challenges lay ahead in targeting quadruplexes. Many compounds have been designed that can specifically interact with quadruplexes over dsDNA and other topologies but a fundamental issue to all these studies has been the design of selective ligands that can discriminate between the different folded topologies and loop sequences. To address this issue we can turn towards a history of drug discovery targeting genomic dsDNA and the development of small molecule ligands in an attempt to draw some parallels and address some of the upcoming challenges.

The similarities between B-form dsDNA (B) and four stranded G-quadruplexes (GQ) are quite striking; both topologies display grooves of variable width running along their phosphate backbones, stabilized by ordered water networks. Additionally, the nucleotides stack (GQ = 3.1-3.3 Å, B = 3.4 Å) in a twisted helical arrangement (GQ = 30°, B = 36.7°), adopting rigid rod-like structures, utilizing both Watson-Crick and Hoogsteen base pairing, thereby generating distinct sequence dependent hydrogen bonding patterns of accepters and donors. The presence of nucleophile groups, N7 of adenine and guanine bases, along with a charged phophodiester backbone have offered opportunities for the design of selective ligands. It is then not surprising that small molecules ligands that associate with canonical DNA can readily bind to telomeric DNA. Ligand interactions with major DNA/RNA features comprise of major and minor grooves, the Hoogsteen or Watson-Crick faces, end-stacking and intercalation between stacked basepairs. Similarly, specific ligands for quadruplex interaction have exploited the same interfaces, and although intercalation between the stacked tetrads has yet to be been fully structurally validated, ligands readily stack on 3' and 5' ends and intercalate between stacked quadruplexes. Unlike quadruplexes, canonical DNA has the structural flexibility to untwist and open up to incorporate planer polyaromatic moieties, such as daunomycin, proflavin and amsacrine with a preference to stack between AT basepairs. This mode of binding has not been observed directly for stacked G-tetrads however at the 3' and 5' ends of the stacked tetrads the nucleotides in the connecting loops readily rearrange to incorporate and stabilize polyaromatic ligands by π-π stacking[45] as shown for a series of bimolecular Oxytricha nova complexes[46, 47] and the c-MYC promoter sequence[30]. Combination of intercalation and groove binding observed for actinomycin D mirrored in a mode of binding of the high affinity selective ligands associated with stacked quadruplexes, BRACO-19,[46] an acridine FD-121,[45] and naphthalene diimides.[48] The distortion of dsDNA helical structures upon ligand binding occurs as a common feature for bioactive small molecule agents. Structural data for quadruplex/ligand complexes has not yet shown such distortions in stacked tetrads, in contrast structural evidence has revealed that ligand binding is accompanied by significant rearrangement of external loops. The most effective small molecule stabilizing G-quadruplexes exploit end stacking mechanisms, through either maximizing π-π interactions between the polyaromatic ligands and guanine bases or by using shape complimentarily.

Double stranded DNA and RNA readily associate via anti-parallel phosphate backbone arrangements to generate distinctive major, and minor grooves. The antibiotics distamycin A, and netropsin are dsDNA minor groove binders that exploit both cationic side-chains for non-specific ionic interactions with the phosphate backbone and multiple hydrogen bonding groups to bind in a sequence selective manner, threading deep into the minor grooves of dsDNA. The grooves offer common feature for ligand binding particularly as quadruplexes can adopt a range of groove widths dependent on the mix of phosphate backbone orientations. Biophysical studies on distamycin, and netropsin and their association to several quadruplex DNA topologies suggested external end stacking rather than groove binding with 1:1 binding stoichiometries[49] along with a family of biaryl polyamides based on distamycin.[50] Efforts by experts to exploit minor groove binding have yet to yield tangible results,[51] restricted perhaps by the reduced opportunities for selectively due to limited hydrogen bonding patterns of N2 and N3 in quadruplexes.

A significant class of ligands commonly used as therapeutics are alkylating agents that covalently attach to DNA/RNA bases through the nucleophilic N7 and to lesser extent N3 and N2 atoms. Their activation comes through cytochrome P450, where reactions result in mono-functional and bifunctional agents able to cross-link DNA as well as other biomolecules. This cross-linking ability leads to damage of DNA synthesis and maintenance mechanisms resulting in the death of rapidly dividing cells. Carboplatin, is such a cytotoxic agent, able to cross-link inter and intra molecularly anchoring together DNA with distorted geometries. The idea to combine ligands with general quadruplex affinity to electrophilic moieties that place the alkylating agent adjacent to nucleophiles has recently been developed as a strategy. These hybrid agents[52, 53] were able to show binding to quadruplexes and efficacy against melanoma cells, with effects on telomere function and telomerase expression. Thus, even though designing selectivity for an individual molecular target may not be achievable an element of selectivity against certain cancer cell lines may still be achievable through other factors.

## 3.2 Exploiting novel structural features of G-quadruplexes

To realize opportunities for innovative ligand design the unique structural features need to be characterized. Contribution to our structural understanding of quadruplex topology has come through both X-ray crystallographic and NMR spectroscopy studies. We observed that core stacked tetrads offer a common and stable platform for the selective association of ligands, while the connecting loops of variable nucleotide sequence offer opportunities to extract an element of specificity amongst quadruplexes. The X-ray diffraction structure determination of the intramolecular human telomeric quadruplex in complex with TMPyP4 highlights the flexibility of the TTA linking nucleotides to generate unexpected sites of interaction while retaining the stacked core.[54] Direct end stacking of TMPyP4 onto the external tetrad is sterically hindered, but binding to the DNA quadruplex is achieved by rearranging the TTA loop residues by stacking the adenine from the loops onto the external tetrad, pairing it with a thymine from 5' end to form a dinucleotide pair that forms the ideal ligand binding platform. Conversely, the binding of TMPyP4 to MT3-MMP mRNA, a 20nt purine sequence that sits 5' to the mRNA of MT3-MMP inhibits translation, and destabilizes the RNA quadruplex. In cellulo, using a luciferase based reporter assay in transfected in HeLa cells TMPyP4 actually up-regulates translation. Loop flexibility was again exploited by the disubstituted acridine FD121 to generate a novel binding platform through its associations to TERRA RNA folded as a quadruplex. Here the UUA loops were distorted to create an expanded flat planar tetrad by recruiting the adenines to form a new planar hexad arrangement that maximizes $\pi$-$\pi$ stacking.[45] Loop flexibility makes rational drug design challenging, unless we can build a sufficiently large structural database to construct validated models. Even beyond this, efforts towards rational drug design are further complicated where several distinct quadruplex topologies co-exist only separated by small energy barriers. This problem is highlighted by the human telomeric sequence in the presence of potassium cations, where several unique and stable conformations exist together in solution within a with a flat free-energy landscape. A possible option is to use high throughput *in vivo* screening assays based on FRET[55]. Technology that can exploit the inherent structural flexibility we have observed from both NMR and X-ray analyses to self-select for stabilizing complexes, ideal when searching for new classes of ligands.

## 3.3 Strategies and agents

With the identification of a suitable molecular target, using rational based drug design approach should eventually deliver selective compounds that induce the required response. An alternative strategy is to develop agents that are non-specific to a particular topology to hit a broad range of molecular targets. Pyridostatin is such an agent; long term exposure of cells to pyridostatin, a non-specific G-quadruplex interacting small molecule ligand, induces cell-cycle arrest through DNA-damage checkpoint activation, a mechanism consistent with double strand breaks. While the biochemical data were consistent with general DNA damage, a specific sarcoma proto-oncogene (SRC) was identified as responsive for the reduction of both levels of mRNA and protein in MRC-5-SV40 cells. Analysis of the SRC gene revealed 25 PQSs folded as stable quadruplexes, and in the presence of 2 µM pyridostatin modulates SRC activity in the breast cancer cell line MDA-MB-231. However, many other genes known to contain promoters with high PQS did not respond to the presence of pyridostatin. Screening of compounds through a panel of cancer cell lines is an effective but expensive method for determining ligand specificity. A traditional approach of using an anticancer non-optimized therapeutic index LD50/ED50, where ED50 represents 50% of the maximum effective concentration may be more suitable. The extension of the genome project at the Sanger Institute to sequence the exons and flanking introns of 600 cancer cell lines may help rationalize such an approach.

Quarfloxin (CX-3543), belongs to a class of fluoroquinolone ligands that was once thought to target the c-myc promoter, but subsequently shown to interact with G-rich elements in human ribosomal DNA (rDNA) and invoke *in vivo* antitumor activity. Addition of quarfloxin to yeast cells results in down regulation of rRNA expression and disrupts associations between nucleolin and rDNA.[56] Developed by Cylene Pharmaceuticals for human trials it has completed phase I and should have commenced phase II trials, focusing on carcinoid/neuroendocrine tumors in 2011. BMVC[57] is a carbazole derivative, displaying antitumor activity being able to stabilize quadruplexes. Its ability to fluoresce and stain has led to its application as a cancer cytological diagnosis tool, and is currently under clinical trials. BMVC4, a related carbazole, has the potential to be developed as a chemotherapeutic agent against both telomerase positive and ALT cancer cells, suppressing c-myc expression through quadruplex stabilization and inducing DNA breaks.[58]

Monoclonal antibodies offer an attractive and selective strategy to targeting quadruplexes. Unfortunately, they are restricted in their application as therapeutic agents as the molecular targets reside in the cytoplasm or nucleus and are not accessible on the cell surface. Peptide nucleic acids[59] offer interesting opportunities, being neutral in charge with bases linked together with a polyamide backbone instead of a phosphate backbone, are very stable in a cellular environments, although they tend to bind complimentary DNA only when under torsional stress. However, their ability to bind to the C-rich strand of DNA upstream of the BCL2 gene and promote quadruplex formation provides a new opportunity to target BCL2 gene expression with small molecule ligands.

## 4 CONCLUSIONS AND PERSPECTIVES

The well defined topology of G-quadruplexes make them very attractive as molecular therapeutic targets, offering realistic opportunities for rational drug design using standard tools to progress hit molecules into lead compounds. Furthermore, putative efficacy of the ligands can be readily determined *in vivo* using the vast array of molecular tools already developed and *in cellulo* using robust cell-based reporter assays. G-quadruplexes as molecular targets are stable, easy to synthesize and handle, and are readily amenable to structure based analysis in a variety of conditions. Telomerase based assays and FRET based ligand screening assays are being refined for use in high-throughput formats, along with developments in *in silico* modelling for drug screening.[60, 61] The efficient implementation of these methodologies has enabled researches to characterize and unfortunately reject many ligands as general quadruplex stabilizers. Drug discovery efforts have tended to follow the same rational used for dsDNA, an important target for many cancer cytotoxic therapeutic agents, and assumed quadruplexes could be targeted likewise by small molecule ligands. However, DNA interacting small molecules used as therapeutic agents are generally non-selective and act as cytotoxic agents with low potency. Our expectations from successfully designing small molecule kinase inhibitors such as, nilotinib, dasatinib, imatinib and mesylate with high affinities in the nM range to protein targets may not be comparable to targeting DNA. Many effective anti-cancer drugs target the distinctive grooves of dsDNA, something that has proved elusive for quadruplex targets agents. A move towards the development of a more general cytotoxic agent, but still selective against cancer cell lines may appear to be a regressive step, however at least tools exist to verify the pathway and mechanisms of action. Based on the earlier experiences it is perhaps not surprising the challenges present in designing and developing small molecules to target four-stranded motifs.

An interesting proposition is the targeting several eukaryotic promoter sequences in a concerted manner, thereby altering the expression levels of related set of oncogenes associated with the six hallmarks of cancer. For a detailed review of the targets see Books and Hurley's review.[62] Such a strategy offers a path to develop compounds with fewer side effects but would be difficult to implement if combination therapies are required. Unfortunately, quadruplex formation within promoters and the desired modulation of gene expression by quadruplex stabilizing small molecule ligands still needs to be validated as a viable mechanism. The telomere remains a very attractive molecular target even though it still lacks strong clinical evidence, while a move towards targeting mRNA and TERRA is still in its infancy, but shows significant promise as a tool to target specific gene expression. Structural data for RNA quadruplexes is revealing the dominance of parallel backbone geometries, masking perhaps the unique structural features that could make them ideal targets. The diverse range of RNA quadruplexes loop sequences and sub-structures anticipated from the sequences needs to be explored with an aim to reveal unique binding pockets and aid in maximizing selectivity. Perhaps most importantly a clearer biological understanding is urgently required to define the context for quadruplex formation in a complex cellular environment.

Several experienced groups specializing in medicinal chemistry have devoted significant efforts in designing new ligand scaffolds and expanding those of existing families. Many of the new and existing compounds under development display significant affinity towards quadruplexes, but lack the required selectivity to distinguish effectively between different topologies. Perhaps the diversity of molecular agents targeting quadruplexes is not sufficient. Very few ligand libraries are available that are designed exclusively for targeting DNA/RNA. The introduction of natural product[63, 64] libraries is beginning to address this issue somewhat. As discussed above the dynamic nature of quadruplexes still presents challenges for *in silico* and *in vitro* screening. However, recent work on the related transactivation response element (TAR) from HIV type 1 (HIV-1), was successful when a combined approach of NMR spectroscopy and computational molecular dynamics were used to virtually screen for small molecules.[20]

This chapter is restricted in scope due to space limitations, with many important therapeutically related quadruplexes deserving greater discussion. From a target based drug-development perspective, clearly additional structural data is required, with matching biophysical and ligand binding data. Efforts to provide part of this data as an updateable online resource, such as QuadPredict[65] will help define stable topologies from sequences derived from bioinformatics. What is truly striking is the diversity of approaches taken to target quadruplexes and the variety of regulatory pathways they can affect as well as the diversity of their distribution throughout the cell. These are early days still in the scientific journey of discovery for these unusual, robust structural motifs.

## References

1. M. Folini, L. Venturini, G. Cimino-Reale and N. Zaffaroni, *Expert Opin Ther Targets*, 2011, **15**, 579-593.
2. D. Yang and K. Okamoto, *Future Med Chem*, 2010, **2**, 619-646.
3. A. Bugaut and S. Balasubramanian, *Nucleic Acids Res*, 2012, **40**, 4727-4741.
4. Y. Xu and M. Komiyama, *Methods*, 2012.
5. G. G. Jayaraj, S. Pandey, V. Scaria and S. Maiti, *RNA Biol*, 2012, **9**, 81-86.
6. N. W. Kim, M. A. Piatyszek, K. R. Prowse, C. B. Harley, M. D. West, P. L. Ho, G. M. Coviello, W. E. Wright, S. L. Weinrich and J. W. Shay, *Science*, 1994, **266**, 2011-2015.
7. A. M. Zahler, J. R. Williamson, T. R. Cech and D. M. Prescott, *Nature*, 1991, **350**, 718-720.
8. R. Rodriguez, S. Muller, J. A. Yeoman, C. Trentesaux, J. F. Riou and S. Balasubramanian, *J Am Chem Soc*, 2008, **130**, 15758-15759.
9. C. Y. Chen, Q. Wang, J. Q. Liu, Y. H. Hao and Z. Tan, *J Am Chem Soc*, 2011, **133**, 15036-15044.
10. S. Sato and S. Takenaka, *Anal Chem*, 2012, **84**, 1772-1775.

11. D. Sun, B. Thompson, B. E. Cathers, M. Salazar, S. M. Kerwin, J. O. Trent, T. C. Jenkins, S. Neidle and L. H. Hurley, *J Med Chem*, 1997, **40**, 2113-2116.
12. M. Y. Kim, H. Vankayalapati, K. Shin-Ya, K. Wierzba and L. H. Hurley, *J Am Chem Soc*, 2002, **124**, 2098-2099.
13. D. S. Pilch, C. M. Barbieri, S. G. Rzuczek, E. J. Lavoie and J. E. Rice, *Biochimie*, 2008, **90**, 1233-1249.
14. J. Linder, T. P. Garner, H. E. Williams, M. S. Searle and C. J. Moody, *J Am Chem Soc*, 2011, **133**, 1044-1051.
15. T. Tauchi, K. Shin-ya, G. Sashida, M. Sumi, S. Okabe, J. H. Ohyashiki and K. Ohyashiki, *Oncogene*, 2006, **25**, 5719-5725.
16. P. Phatak, J. C. Cookson, F. Dai, V. Smith, R. B. Gartenhaus, M. F. Stevens and A. M. Burger, *Br J Cancer*, 2007, **96**, 1223-1233.
17. A. M. Burger, F. Dai, C. M. Schultes, A. P. Reszka, M. J. Moore, J. A. Double and S. Neidle, *Cancer Res*, 2005, **65**, 1489-1496.
18. M. Gunaratnam, M. de la Fuente, S. M. Hampel, A. K. Todd, A. P. Reszka, A. Schatzlein and S. Neidle, *Bioorg Med Chem*, 2011, **19**, 7151-7157.
19. E. Salvati, C. Leonetti, A. Rizzo, M. Scarsella, M. Mottolese, R. Galati, I. Sperduti, M. F. Stevens, M. D'Incalci, M. Blasco, G. Chiorino, S. Bauwens, B. Horard, E. Gilson, A. Stoppacciaro, G. Zupi and A. Biroccio, *J Clin Invest*, 2007, **117**, 3236-3247.
20. A. C. Stelzer, A. T. Frank, J. D. Kratz, M. D. Swanson, M. J. Gonzalez-Hernandez, J. Lee, I. Andricioaei, D. M. Markovitz and H. M. Al-Hashimi, *Nat Chem Biol*, 2011, **7**, 553-559.
21. A. K. Todd, M. Johnston and S. Neidle, *Nucleic Acids Res*, 2005, **33**, 2901-2907.
22. J. L. Huppert and S. Balasubramanian, *Nucleic Acids Res*, 2005, **33**, 2908-2916.
23. A. T. Phan, Y. S. Modi and D. J. Patel, *J Am Chem Soc*, 2004, **126**, 8710-8716.
24. A. T. Phan, V. Kuryavyi, S. Burge, S. Neidle and D. J. Patel, *J Am Chem Soc*, 2007, **129**, 4386-4392.
25. K. W. Lim, L. Lacroix, D. J. Yue, J. K. Lim, J. M. Lim and A. T. Phan, *J Am Chem Soc*, 2010, **132**, 12331-12342.
26. X. Tong, W. Lan, X. Zhang, H. Wu, M. Liu and C. Cao, *Nucleic Acids Res*, 2011, **39**, 6753-6763.
27. J. Dai, D. Chen, R. A. Jones, L. H. Hurley and D. Yang, *Nucleic Acids Res*, 2006, **34**, 5133-5144.
28. A. Siddiqui-Jain, C. L. Grand, D. J. Bearss and L. H. Hurley, *Proc Natl Acad Sci U S A*, 2002, **99**, 11593-11598.
29. M. Bejugam, S. Sewitz, P. S. Shirude, R. Rodriguez, R. Shahid and S. Balasubramanian, *J Am Chem Soc*, 2007, **129**, 12926-12927.
30. J. Dai, M. Carver, L. H. Hurley and D. Yang, *J Am Chem Soc*, 2011, **133**, 17673-17680.
31. P. V. Boddupally, S. Hahn, C. Beman, B. De, T. A. Brooks, V. Gokhale and L. H. Hurley, *J Med Chem*, 2012.
32. P. Kumar, V. K. Yadav, A. Baral, D. Saha and S. Chowdhury, *Nucleic Acids Res*, 2011, **39**, 8005-8016.
33. W. I. Sundquist and S. Heaphy, *Proc Natl Acad Sci U S A*, 1993, **90**, 3393-3397.
34. J. Christiansen, M. Kofod and F. C. Nielsen, *Nucleic Acids Res*, 1994, **22**, 5709-5716.
35. C. Cheong and P. B. Moore, *Biochemistry*, 1992, **31**, 8406-8414.
36. J. Kim, C. Cheong and P. B. Moore, *Nature*, 1991, **351**, 331-332.
37. J. Deng, Y. Xiong and M. Sundaralingam, *Proc Natl Acad Sci U S A*, 2001, **98**, 13665-13670.

38. H. Liu, A. Matsugami, M. Katahira and S. Uesugi, *J Mol Biol*, 2002, **322**, 955-970.
39. B. Pan, K. Shi and M. Sundaralingam, *Proc Natl Acad Sci U S A*, 2006, **103**, 3130-3134.
40. H. Martadinata and A. T. Phan, *J Am Chem Soc*, 2009, **131**, 2570-2578.
41. G. W. Collie, S. M. Haider, S. Neidle and G. N. Parkinson, *Nucleic Acids Res*, 2010, **38**, 5569-5580.
42. R. Kostadinov, N. Malhotra, M. Viotti, R. Shine, L. D'Antonio and P. Bagga, *Nucleic Acids Res*, 2006, **34**, D119-124.
43. A. Arora, M. Dutkiewicz, V. Scaria, M. Hariharan, S. Maiti and J. Kurreck, *Rna*, 2008, **14**, 1290-1296.
44. K. Ito, S. Go, M. Komiyama and Y. Xu, *J Am Chem Soc*, 2011, **133**, 19153-19159.
45. G. W. Collie, S. Sparapani, G. N. Parkinson and S. Neidle, *J Am Chem Soc*, 2011, **133**, 2721-2728.
46. N. H. Campbell, G. N. Parkinson, A. P. Reszka and S. Neidle, *J Am Chem Soc*, 2008, **130**, 6722-6724.
47. N. H. Campbell, M. Patel, A. B. Tofa, R. Ghosh, G. N. Parkinson and S. Neidle, *Biochemistry*, 2009, **48**, 1675-1680.
48. G. N. Parkinson, F. Cuenca and S. Neidle, *J Mol Biol*, 2008, **381**, 1145-1156.
49. I. Prislan, I. Khutsishvili and L. A. Marky, *Biochimie*, 2011, **93**, 1341-1350.
50. K. M. Rahman, A. P. Reszka, M. Gunaratnam, S. M. Haider, P. W. Howard, K. R. Fox, S. Neidle and D. E. Thurston, *Chem Commun (Camb)*, 2009, 4097-4099.
51. A. K. Jain and S. Bhattacharya, *Bioconjug Chem*, 2011, **22**, 2355-2368.
52. F. Doria, M. Nadai, M. Folini, M. Di Antonio, L. Germani, C. Percivalle, C. Sissi, N. Zaffaroni, S. Alcaro, A. Artese, S. N. Richter and M. Freccero, *Org Biomol Chem*, 2012, **10**, 2798-2806.
53. M. Di Antonio, F. Doria, S. N. Richter, C. Bertipaglia, M. Mella, C. Sissi, M. Palumbo and M. Freccero, *J Am Chem Soc*, 2009, **131**, 13132-13141.
54. G. N. Parkinson, R. Ghosh and S. Neidle, *Biochemistry*, 2007, **46**, 2390-2397.
55. D. Renciuk, J. Zhou, L. Beaurepaire, A. Guedin, A. Bourdoncle and J. L. Mergny, *Methods*, 2012.
56. D. Drygin, A. Siddiqui-Jain, S. O'Brien, M. Schwaebe, A. Lin, J. Bliesath, C. B. Ho, C. Proffitt, K. Trent, J. P. Whitten, J. K. Lim, D. Von Hoff, K. Anderes and W. G. Rice, *Cancer Res*, 2009, **69**, 7653-7661.
57. C. C. Chang, J. Y. Wu, C. W. Chien, W. S. Wu, H. Liu, C. C. Kang, L. J. Yu and T. C. Chang, *Anal Chem*, 2003, **75**, 6177-6183.
58. F. C. Huang, C. C. Chang, J. M. Wang, T. C. Chang and J. J. Lin, *Br J Pharmacol*, 2012.
59. I. G. Panyutin, M. I. Onyshchenko, E. A. Englund, D. H. Appella and R. D. Neumann, *Curr Pharm Des*, 2012, **18**, 1984-1991.
60. K. M. Rahman, K. Tizkova, A. P. Reszka, S. Neidle and D. E. Thurston, *Bioorg Med Chem Lett*, 2012, **22**, 3006-3010.
61. D. L. Ma, V. P. Ma, D. S. Chan, K. H. Leung, H. J. Zhong and C. H. Leung, *Methods*, 2012.
62. T. A. Brooks, S. Kendrick and L. Hurley, *Febs J*, 2010, **277**, 3459-3469.
63. X. Cui, S. Lin and G. Yuan, *Int J Biol Macromol*, 2012, **50**, 996-1001.
64. Q. Shang, J. F. Xiang, X. F. Zhang, H. X. Sun, L. Li and Y. L. Tang, *Talanta*, 2011, **85**, 820-823.
65. H. M. Wong, O. Stegle, S. Rodgers and J. L. Huppert, *J Nucleic Acids*, 2010, **2010**.

# SCREENING FOR QUADRUPLEX BINDING LIGANDS: A GAME OF CHANCE?

E. Largy and M.-P. Teulade-Fichou

UMR 176 - Conception, Synthesis and Targeting of Biomolecules, Institut Curie, Bat. 110-112, Université Paris-Sud, 91405 Orsay, France. E-mail: mp.teulade-fichou@curie.fr

## 1 INTRODUCTION

The discovery of molecules displaying physical, chemical or biological properties is at the heart of research activities both in the industry and in academic laboratories. Chemical libraries are key components of the intellectual property of these laboratories and are actively exploited by industrials *via* high-throughput screening (HTS) facilities for drug discovery. In the recent years, high-throughput screening for drug discovery in the academic field has been greatly eased through the creation of institutional and publically funded screening centers (95 academic screening facilities are listed on the Society for Laboratory Automation and Screening website[1]) and the better availability of chemical libraries,[2] and has therefore become more prominent or even routine.[3,4]

In an opening speak at Lille University in 1854, Pasteur stated that "*in the field of observation, chance only favors the prepared minds*".[5,6] HTS could be considered has a systematic line of attack of serendipity,[7] where preparation plays indeed a critical role. Academic screening differs from industrial one on various points that should thus be taken into account during this preparation like the choice of the chemical library, the assay, the target(s), the drug-likeness requirements, etc. In this chapter, we will review these points in the context of quadruplex (G4) targeting. We will also discuss the integration of the necessary steps following the screening *stricto sensu*: data reporting, quality control procedures and hit-to-lead optimization through medicinal chemistry (Figure 1).[8,9]

G4 structures are the subject of a growing interest since they are believed to be involved in many biological processes.[10] An extensive number of small molecules able to bind G4 structures have been reported over the past years with the aim to regulate and/or visualize these processes. Most of the G4 ligands have been rationally designed following simple guidelines, i.e. enhancing π-stacking *via* planar aromatic surfaces, and electrostatics interaction with various possibilities like protonable chains or charged metal center.[11] Well-known examples include the bisquinolinium family (Phen-DC3, 360A),[12,13] BRACO-19,[14] PIPER,[15] etc. Nonetheless, other nonstandard categories of ligands can be found, like Telomestatin[16,17] and its derived compounds (HXDV[18] and TOxaPy[19]), which present uncharged oligomeric heteroaryl scaffolds. Until now, these guidelines have been sufficient to provide ligands of high affinity for G4 nucleic acids but also of high selectivity against other nucleic acid structures, most notably duplex (ds) DNA. G4 structures may indeed only exist transiently in the cell and, in any case, will be massively overwhelmed by ds-DNA. G4-interactive small molecules should exhibit a significant binding selectivity over ds-DNA and more largely over other nucleic acid secondary structures in order to be used as drugs or probes. However, rational design has been unable so far to provide small molecules displaying a marked selectivity for a given G4 entity, associated to a single DNA or RNA sequence and therefore to a precise biological process or even disease. It is consequently not surprising to observe an increasing interest in high-throughput screening assays in the G4 field, which could ultimately allow the discovery of highly specific G4 ligands, potentially not obeying the rules dictated by rational design. Over the last ten years, the G4 community has developed a wide range of physical or virtual methodologies to detect and quantify small molecules-G4 interaction,[20,21] and some of them have already been applied to screenings of chemical libraries. We will hereafter examine the use of these methodologies, sometimes in combination, in the context of academic screening with practical examples and discuss their "high-throughput potential".

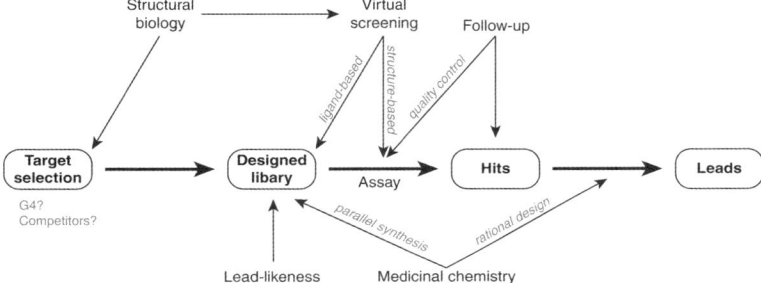

**Figure 1**  *Integrated workflow example. Inspired by Fara et al.*[22]

## 2 PREPARING YOUR MIND

### 2.1 Tool *vs* drugs

It should be kept in mind that academic screenings offer different prospects than industrial screenings, for instance the discovery of compounds directed to a biological target, for mechanistic studies, and not particularly related to an investigated disease.[3] For these *tool compounds*, costs of production, biological half-life or bioavailability are not important parameters.[4] However, high selectivity (no off-targets) is an essential parameter in order to resolve binding and subsequent biological effect mechanisms. Consequently, screening methodologies enabling to quantify structural selectivities (competitive FRET-melting, HT-G4-FID, SPRi, ELISA, etc.; see section 3.1) should be preferred, as they will allow to quickly discarding useless compounds. Drug development is also possible within the academical frame, but pharmacokinetics and drug metabolism will be the major issues to work out.[4]

### 2.2 Functional *vs* binding

*Functional screenings*, i.e. recording small molecules influence on cells is the most common way of screening but is expensive and does not give information on the underlying mechanisms.[4] Hereafter, most of the reviewed methodologies enable *binding screenings*, which are (substantially) less expensive. Functional studies may be performed anyway, but only on few identified hits, which allow discarding innocent ligands (not agonizing or antagonizing the targeted activity). Careful choice of the target(s) is an important factor of binding screening, and may be particularly difficult in the G4 field. As mentioned previously, G4 nucleic acids can be found in various domains, display a high polymorphism, and their structure within cellular environment is still a matter of debate (molecular crowding, chaperone proteins, etc.).[23] Moreover, one of the most difficult challenge is the field is to find ligands able to bind preferentially to a quadruplex over others (G4 vs G4 selectivity), in order to correlate cellular effects and binding. The use of a single G4-forming oligonucleotide as target for screening will limit the possibility of discovering G4 vs G4 selective compounds. In the light of current knowledge, it is thus essential to use a panel of oligonucleotides originating from various domain to bring more structural diversity and thus more selectivity criteria: nucleic acid type (DNA vs RNA), molecularity (monomer vs multimer), strand orientation (parallel, antiparallel, hybrid), loop sequence and size, number of G-quartets, etc. Structural biology is then an invaluable tool to select appropriate sequences. The same comment could be made with competitors (GC-/AT-rich or genomic DNA and RNA). For these reasons, the use of screening assays allowing the use of different oligonucleotide sequences in parallel is recommended. Alternatively, various sequences can be used in follow-up experiments (see for example Hartig et al.[24] where oligonucleotides with different loop composition and topologies have been used).

### 2.3 Chemical libraries

Choosing the right chemical library is a key factor to ensure a success in screening. Focusing on a particular chemical class (*targeted libraries*) through computational techniques such as data mining, or virtual screening may be timesaving (see for instance the screenings performed by Xu[25] or Teulade-Fichou[26] and described below), and has been reviewed by Guy et al.[27] This can also be proven useful if one wants to screen for a drug, as it will allow discarding non drug-like compounds before physically screening.[28]

In order to obtain potent and selective candidate molecules, synthesis of various derivatives is required (hit-to-lead optimization; see section 4.3).[3, 4] For this reason, screening of easily modifiable compound libraries, which is usually not the case of commercial libraries, is advised. Designed libraries, constructed from easy-to-assemble building blocks, are a good alternative. An example of a small library of unnatural polyamides obtained by combinatorial synthesis of aromatic amino acids can be found elsewhere in this chapter (section 3.1.3).[29] However, it should be mentioned that such libraries can be biased towards a certain type of chemical scaffolds (e.g. condensed aromatics) and/or binding mode (e.g. π-stackers) and may consequently reduce the scope of the results. Finally, natural product mixtures represent another alternative to chemical library screening and some examples will be given hereafter (see section 3.1.6).

### 2.4 Physical screening (PS) vs. Virtual screening (VS)

Although depicted as a very profitable technique in the previous sections, high-throughput screening still suffers from some drawbacks. Like in any scientific experiments, errors can occur, notably *via* biases of a given method/technology (a very informative example is given by Wu and coworkers[30]). Also, bigger chemical libraries do not necessarily guarantee increased hit numbers.[31] Virtual screening allows the evaluation of *virtually* unlimited number of molecules, for a reduced cost; a smaller library of selected structures can then be physically assayed. Computer-aided methodologies are useful to i) discard undesired structures (see above section 2.3), ii) identify structures with targeted properties (such as drug-like scaffolds) and iii) help the post-HTS analysis process (decision making notably[32]). In the first two cases, it acts as a filter to reduce the search space in order to accelerate putative findings while saving money. Rather than an alternative to physical screening, virtual screening should thus be considered as a valuable complement, and both strategies can be fully integrated.[22] Finally, it is certainly more advantageous for academic labs, having rather limited resources, to work with high quantities of data (*high-content screening*)[33, 34] than with high number of molecules (high-throughput screening).

## 3 ASSAYS

Numerous analytical assays, aimed at studying small molecule/G4-DNA binding, have been published in the recent years. For the sake of conciseness, the principle of the assays will not be systematically described. For more details on the principle of analytical assays widely used by the community, please refer to Defrancq and Randazzo reviews.[20, 21] Hereafter, we present some of the advantages and drawbacks of the methods developed and/or applied to screening for G4 binders. Examples of screening will be given, with information on the used strategy (preparation, results, and follow-up).

### 3.1 Physical assays

*3.1.1 Fluorescence-based assays.* Various assays based on fluorescence detection have been developed to screen for G4 ligand discovery, most notably the FRET-melting[35] (Förster resonance energy transfer) and HT-G4-FID (High-Throughput G-quadruplex Fluorescent Intercalator Displacement) assays.[36] FRET-based assays are among the few examples of purely physical HTS of chemical libraries (> 1000 molecules) for G4 binder discovery. The first temperature-dependent fluorescence assay-based screening reported was performed exploring the Natural Product Chemistry Institute chemical library (ICSN, France).[37] Among 1000 molecules, two natural compounds (0.2 % hit-rate) inducing a significant stabilization (> 5°C) of the human telomeric G4 were identified and subsequently evaluated on their telomere shortening capabilities notably, to show their drug potentiality.

Recently, the FRET-melting assay was used to screen 2307 compounds, with significant chemical structure variations, from three NCI libraries.[38] The study has been aimed at discovering drug-like G4 binders (*i.e.* following the Lipinski Rule of Five). Fluorescently labeled telomeric and duplex forming sequences were used to study the affinity and selectivity of the molecules, respectively. Thirteen compounds, producing a 7.3°C or higher G4 stabilization at 1 µM and a low duplex stabilization (< 3.5 °C), were found (0.6 % hit-rate). Interestingly, authors gave suggestions for the chemical optimization of some hit compounds in order to improve their properties, as well as the generation of small focused libraries. Worthy of note, the 13 identified compounds have low molecular weights (327—533 Da), ClogP ranged from 0.31 to 5.86, 5 or less H-bond donors and 7 or less H-bond acceptors, consistent with the Lipinski Rule of Five. The FRET-melting assay was also used after different virtual screenings described elsewhere in this chapter (section 3.2.2).[39, 40]

A very interesting automated HTS, based on the FRET effect, has been led by Hartig and coworkers, on an unfocused chemical library of 34 000 compound from the University of Konstanz Screening Facility.[24] In their methodology, the unfolded telomeric repeat, labeled by FAM and TAMRA is mixed with small molecules. G4 ligands induce isothermal folding of the oligonucleotide upon binding and subsequent quenching of FAM fluorescence (Figure 2). Importantly, the *Z'* parameter (see section 4.1) was calculated

(0.88) prior to screen, in order to validate the suitability of the assay. 63 compounds were identified (> 50% activity at 10 µM; 0.1 % hit-rate) with the primary screen, and the already known G4-binder berberine was among the seven most potent compounds, which validated the protocol. Further experiments on the hits involved competitive FRET-melting experiments (G4 vs ds selectivity), as well as polymerase stop assay, circular dichroism and thermal denaturation measurements of three other oligonucleotides with different loop composition and topologies (G4 vs G4 selectivity). This work encompasses many important aspects of the G4 domain (G4 vs ds and G4 vs G4 selectivities notably) as well as some high-throughput requirements (large chemical library, automation, quality control).

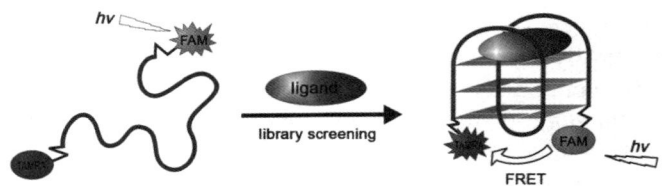

**Figure 2**  *Principle of the isothermal FRET screening assay with human telomeric DNA. Reprinted with permission from Benz et al.[24], copyright 2011 John Wiley and Sons*

The HT-G4-FID assay is a fast, easy to implement and costless procedure, using unmodified oligonucleotides, developed in order to employ the classical FID assay to the discovery of selective G4 ligands by screening.[36] The FID assay has already been performed as a semi-automated procedure for the discovery of ds-DNA binders,[41] thus confirming its feasibility. Single-point and HT-G4-FID titration assays have recently been performed to screen *circa* 8 500 compounds from the Institut Curie chemical library in a multi-step procedure involving a virtual primary screening, and a molecular modeling final step (Figure 3).[26] This physical and virtual assay combination enabled i) the discovery of 4 new structurally-related G4 binders (0.1% hit-rate), displaying G4 vs ds selectivity but also a preference for certain G4 topologies and ii) the understanding of the structure-activity relationship ruling the selectivity of this series. These results could ultimately allow the synthesis of lead compounds. In another study, fifteen published G4 binders were assayed with 22 sequences displaying various secondary structures, underlining the capability of the single-point HT-G4-FID assay to quickly and easily identify compounds with G4 vs ds or G4 vs G4 selectivities.[42]

Other fluorescent assays were developed for screening purpose. Noteworthy, Lacroix *et al.* designed the duplex-quadruplex competition test to screen for specific telomerase RNA G4 ligands, using any kind of unlabeled competitors.[43] The screening of a 169 molecule library (biased towards G4-ligands) afforded 11 excellent hits (6.5% hit-rate) and acts as a proof of concept for further HTS transposition. Other fluorescent-based methods useful for screening includes the *Mix and Measure* assay by Bolton *et al.*[44], a strategy relying on graphene oxide sheets,[45] and a label-free test using G4-based DNAzyme hemin displacement.[46] Finally, a designed peptide library was screened using native polyacrylamide gel electrophoresis with Sybr Gold as stain.[47] Such a methodology cannot be applied to HTS but offers new insights for the identification of peptide specific for a unique G4.

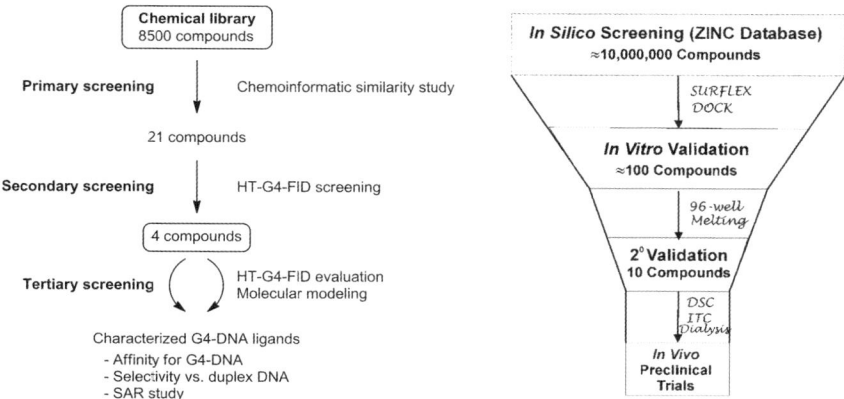

**Figure 3** *Examples of multi-step procedure integrating virtual and physical screening. Left: reprinted from Largy et al.[26], with kind permission from Springer Science and Business Media, copyright 2011. Right: Reprinted with permission from Holt et al.[39], copyright 2010 John Wiley and Sons*

*3.1.2 Surface plasmon resonance imaging (SPRi).* Surface plasmon resonance imaging (SPRi) enables testing up to 1000 compounds in a single run (Figure 4),[48] using SPR technology, therefore providing interaction real-time kinetics. Development of this technique to assay small molecules has been performed recently, using various G4-, ds- and ss-DNA oligonucleotides as target and a set of four G4 binders.[49] Results were in agreement with FRET-melting data from the literature and SPRi successfully discriminated non-selective from selective ligands. SPRi could thus be a very useful technique to screen large chemical libraries.

**Figure 4** *Image obtained from the prism with SPRi–Plex device. SH-modified 25 mers oligonucleotides spotted on a SPRi prism using a quill pin with the spotter Chip Writer Pro from Biorad, in a microarray of 1050 sites with a diameter of 150 μm. Reprinted from Pillet et al.[48], copyright 2010, with permission from Elsevier*

*3.1.3 Enzymatic assays.* In order to screen a library of unnatural polyamides obtained *via* combinatorial (split-and-pool one-bead-one-compound) synthesis against surface immobilized human telomeric G4 DNA, direct ELISA (enzyme-linked immunosorbent assay) with streptavidin-horseradish peroxidase polymer conjugate detection has been chosen.[29] The library is biased towards π-staking G4 ligands as nucleic acid building blocks chosen to construct the polyamides are aromatic (e.g. tyrosine, tryptophan, acridine, etc.), thus considerably lowering the probability of identifying a non-canonical ligand but likely to increase the hit-rate. Specificity of binding of these polyamides was measured as the ratio of ELISA absorbance between G4 and non-G4 forming oligonucleotides. As expected, compounds possessing multiple charges displayed lower selectivities than the single-charged ones, due to unspecific electrostatic interactions with DNA. Eight compounds, identified by tandem mass spectrometry in the initial screening were re-assayed by ELISA titration, which globally confirmed the first results. SPR and FRET-melting were used as alternative validation assays and also confirmed the initial screen results. It is worth noting that the use of two or more independent assays to confirm the initial screen data is particularly important as it allows avoiding bias from a particular methodology.[30] For instance, the authors specify that ELISA is biased towards detection of low off-rate interactions (in order to survive dissociation during washings) whereas SPR enables detecting all binding events simultaneously. ELISA is a rapid and already automated methodology and can thus prove useful for screening selective G4 binders as it enables screening against multiple oligonucleotides in parallel.

A colorimetric assay using a combination of nucleic acid-coated gold nanoparticles (AuNPs) and exonuclease I has been developed recently, and has been calibrated with four compounds so far.[50] Enzymatic cleavage of the nucleic acid chains results in AuNP destabilization, followed by a rapid aggregation characterized by a visible color change. The use of duplex competitor in a parallel experiment (not coated on AuNP) allows selectivity measurements. This assay has various advantages since it can be performed with any G4 sequence or competitor, and could potentially be performed with a microplate absorbance reader.

Finally, the polymerase stop assay was used in combination with virtual screening in various studies, either as a primary screen[51-53] and/or as a follow-up experiment.[24, 53]

*3.1.4    Electrospray Ionization Mass Spectrometry (ESI-MS).* By analogy with studies carried out with ds-DNA, ESI-MS is now performed to analyze small molecule/G4-DNA interactions.[54] This technique is sensitive and enables stoichiometry and binding constant determination.[55] All species are observable (*e.g.* different stoichiometries), which allows to use virtually any sequence, or even a mixture of sequences, to quantify the specificity of binding. However, response factors of the free molecules and complexes are to be considered to this end.[56] On the whole, mass spectrometry is a useful tool as it could be used to screen ligands for a precise target, several targets (selectivity) and even to determine binding constants.

*3.1.5    Circular Dichroism (CD).* Circular dichroism is routinely performed to study nucleic acid conformation[57] and has been e as a primary screening method in a single publication by Chang *et al*.[58] The method takes advantage of the $Cu^{2+}$-induced unfolding of G4 structures by monitoring the disappearance of the 291 nm CD band of the human telomeric sequence in potassium-rich buffer. This spectral modification is delayed in the presence of ligands, as a function of their affinity (the slower, the better). Five BMVC derivatives, as well as reference compounds BRACO-19 and DODCI, have been screened following this protocol. Even if high-throughput CD systems are commercially available, it is unlikely that such an assay be applied to HTS of large libraries since it is rather time consuming (kinetics measurements up to one hour), nucleic acid consuming, and does not give insight into structural selectivities.

*3.1.6    Screening from natural extracts.* Analysis of natural plant extracts is an alternative to chemical library screening. Various methodologies have been applied to this end. Dialysis has been used in combination with NMR to screen from natural plant extracts.[59] After incorporation of G4 within the natural mixture, the high molecular weight ligand-G4 complexes are separated by dialysis and identified by NMR. Natural extracts have also been screened using HPLC, by comparing the peak profiles in presence or absence of G4 DNA.[60] Four compounds were isolated, and then characterized by mass spectrometry and NMR. Their G4 binding properties were finally studied through CD-melting experiments.

A three-step methodology based only on $^1$H-NMR was developed, where the presence of G4 binders is detected using characteristic peaks (exploiting DOSY), and structures of aforementioned ligands are ultimately determined by 2D techniques (e.g. HSQC, HMBC).[61, 62] All these methodologies are certainly useful to fish out ligands from a complex mixture but do not give information on structural selectivity and have a rather low throughput.

*3.1.7 Functional screening.* A functional cell-based screening for the discovery of *c-kit* down-regulating small molecules, using a luciferase gene reporter assay (figure 5), was described recently by Balasubramanian and coworkers.[63] A plasmid containing a *c-kit* promoter, driving expression of the *luc2* gene, was transfected in a cell line from the human gastric carcinoma (HGC-27) in order to evaluate *c-kit* activity. A control promoter that does not contain G4-forming sequences was used as a control. 173 compounds from a partially focused chemical library were assayed at different concentrations (0.2 to 5 µM), leading to the identification of five molecules reducing *c-kit* expression. After statistical analysis of the results (Z-score calculation, notably), three compounds were discarded (1.2 % hit-rate). Reproducibility of the down-regulation of c-*kit* by the 2 remaining compounds was evaluated by quantitative polymerase chain reaction (qPCR). Finally, binding of the two hits to G4 sequences was measured by SPR. Micromolar $K_d$ were found for both compounds with the c-kit2 sequence (2-, 5- and 10-fold lower than with c-myc, the human telomeric and c-kit1 sequences, respectively). Both compounds also displayed a good selectivity against ds-DNA.

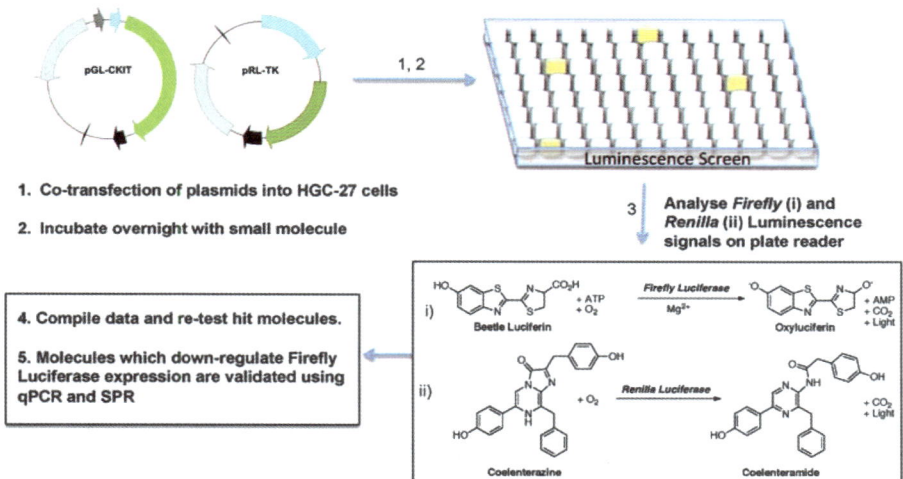

**Figure 5** *Schematic of the luciferase gene reported assay for discovery of* c-kit *down-regulating compounds. Reprinted with permission from McLuckie et al.[63] Copyright (2011) American Chemical Society.*

## 3.2 Virtual assays

*3.2.1 General points.* Virtual screening (VS) is currently considered as a cheaper, faster and more versatile alternative complement to HTS.[64] Usually, VS implies to have knowledge of the 3D structure of the target (although it is not mandatory) in order to assay a 3D ligand database by docking and subsequent scoring. In reality, a wide range of methodologies are used[65] but two general approaches are commonly employed, i.e. *receptor-based* (or *structure-based, i.e.* docking) and *ligand-based* (or *similarity-based*) virtual screenings.[66] Ligand-based VS is relying on physico-chemical similarities with known ligands of a target. Docking may nevertheless be performed with the identified molecules in a second step (case of *forward filtering*).[64] This methodology is obviously biased by the known ligand(s) properties and reduce the hit diversity, but is a fast and reliable way to narrow the chemical space. Ligand-based VS have been thoroughly reviewed by Willett[67, 68] and Bajorath[69, 70], and used recently for G4 binder discovery.[25, 26] Alternatively, in structure-based VS, filters can be applied after docking (*backwards filtering*).[64] In any case, filters are very useful to focus the database on "valuable" structures (drug-likeness notably). By analogy with HTS, target selection is also very important and may strongly impact the output (see section 2.2). Various docking softwares are commercialized but aimed to small molecules-protein binding evaluation. However, examples of nucleic acid related studies using docking software (AutoDock, Surflex) have appeared in the last years.[71, 72] Scoring is another fundamental parameter of VS, whatever docking software is employed, since it determines which molecules can be considered as potential binders. Molecular docking for virtual screening of small molecules binding G4 DNA is a topic exhaustively reviewed by Ma et al.[73-75] Below are briefly described recent applications of virtual screening studies for G4 binder discovery.

*3.2.2 Examples of virtual screenings.* High-throughput docking of a 100 000 drug-like compound database by the ICM method (Molsoft), using an X-Ray structure of the human telomeric G4, led to the identification of a new binder.[40] Its affinity and selectivity was determined by FRET-melting and its putative binding mode was studied by CD and molecular modeling. This docking methodology was performed in following studies to support the discovery, evaluation and optimization of platinum(II) complex families.[51, 52] ICM was also used to perform the virtual screening of a 20 000 natural compound database (AnalytiCon Discovery) for their c-myc G4 binding properties. The 5 best scoring molecules were assayed by polymerase stop assay.[53] Fonsecin B was identified as the most promising candidate and further studied by UV-vis titration for its c-myc G4 affinity as well as selectivity vs. ds- and ss-DNA, molecular modeling, and finally dose-response polymerase stop assay, showing comparable potency than TMPyP4.

Virtual screening enables the study of various targets "simultaneously". An example was published in 2010, where 29 plant extract compounds where docked with six proteins and telomeric G4 DNA, using AutoDock.[76] Two compounds were found to bind G4 DNA, including quercetin, a molecule already known for its G4 binding properties.[77]

Chaires and Trent proposed a strategy fully integrating virtual and physical screening (figure 3).[39] The first steps consisted of an *in silico* screening of the ZINC database (*circa* $11.3 \times 10^5$ drug-like compounds) using Surflex-Dock. Interestingly, the authors decided to use various sequences (G4, ds, triplex) as target to mimic competitive binding sites, and considered different possible binding sites for each target. After ranking, the best 160 compounds were selected for physical HTS, *i.e.* a 96-well plate FRET-melting assay with an antiparallel human telomeric G4, allowing to select a single compound for further studies (differential scanning calorimetry).

Randazzo and coworkers made use of VS to discover groove binding G4 ligands, a category of compounds more difficult to obtain through rational design but displaying a higher potential for G4 *vs* G4 selectivity.[78] A 6000-compound database (Life Chemicals) was screened for their binding to [d(TG$_4$T)]$_4$ using Autodock4. 137 hits were ranked according to their predicted binding free energies and after discarding of some compounds (visual inspection notably), 30 molecules were finally selected (0.5% hit-rate) and further assayed by NMR, molecular modeling and ITC measurements.[79] Groove binding compounds were also discovered through a ligand-based approach based on a series of 38 anthraquinone derivatives exhibiting cytotoxicity against telomerase-positive cells.[25] The pharmacophore model, obtained by generating a set of conformers from the anthraquinone series, was built using the CATALYST software, and used to filter a 9820-compound database. 176 compounds were identified (1.8% hit-rate) and 20 were selected on the basis of their availability, notably. Follow-up experiments (CD-melting and NMR experiments) were used to evaluate the affinity, selectivity and binding mode of 2 non-planar ligands. Another ligand-based screening, associated with the HT-G4-FID assay was published more recently.[26] In this study, 3 well-characterized and potent G4 binders were used as target compounds to filter an 8500 compound database by a similarity criterion. The use of various target compounds allowed obtaining more molecular diversity within the result set. 21 candidates were then assayed by single-point and concentration dependent HT-G4-FID, and molecular modeling (see section 3.1.1).

## 4 QUALITY CONTROL, FOLLOW-UP AND DATA REPORTING

### 4.1 Quality control and data management

Experimental measurements are not error-free, and the variability of results may be an obstacle to decision making and profitable use of results. Few questions should therefore be asked: Is the assay reliable? Is this compound valuable? Quality control encompasses different aspects like chemical library integrity, assay reliability, process monitoring, post-HTS data treatment or validation of hit chemical structures.[2] Statistics can be used to unravel data and make them truly significant through careful analysis of the employed assay, bias and mislead identification, as well as result treatment.[32] Quality control protocols include *Z'* trend monitoring[80] (see the screening by Hartig *et al.*[24] for an example), use of standards[81], plate pattern recognition algorithms[82] and liquid handler and reader performance monitoring.[83] In the same vein, data reporting should be performed methodically and presented in publications to avoid key information loss. Inglese, Shamu and Guy have proposed a 5-step report as follows: assay (controls, sensitivity, etc.), chemical library, HTS process (interpolate controls), post-HTS analysis (follow-up assays, chemical structure validation) and results (Figure 6).[9]

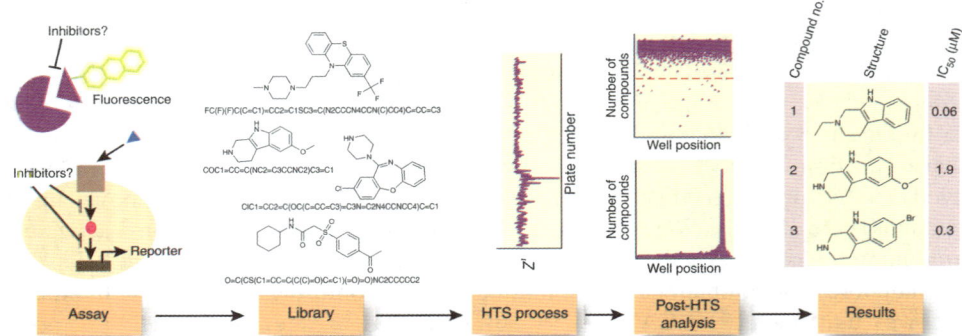

**Figure 6** *The flow of materials and data from assay to reported results in HTS. Reprinted by permission from Macmillan Publishers Ltd: Nature Chemical biology (Inglese et al)[9], copyright 2007*

### 4.2 Follow-up assays

Follow-up assays are to be used to i) confirm the activity of the compounds and ii) give additional information on the hits (selectivity, structure-activity, binding mode, etc.). The use of different assays/technologies is a requirement to avoid biases and therefore to confirm hit activity.[30] Previous examples of G4 binder screenings contained a non-exhaustive list of assays already used to do so, but other analytical methods can be found in the literature.[20, 21]

### 4.3 Medicinal chemistry

As part of the quality control/follow-up assays procedure, the independent synthesis and characterization of hit molecules should be performed to confirm a putative activity.[2] Another important point to take into account before screening is that the resulting hit(s) obtained will generally only be the starting point of a medicinal chemistry campaign aimed at developing an efficient lead small molecule.[3] In other words, the production of one potent compound necessitates the synthesis of a large family of derivatives from the initial hits from the screening, through rational design,[4] as proposed for instance by Thurston and Neidle.[38] Results of follow-up assays giving structure-activity and thermodynamics of binding information are very valuable at this point.[26, 39] As already discussed, the screening of chemical libraries containing easily modifiable compounds is here a substantial advantage.

## 5 CONCLUSION

In an attempt to face the difficult challenge of specifically targeting G4 structures and discover original scaffold with unique properties (binding mode notably), the G4 community has developed a wide range of assays, among which some of them may be applicable to high-throughput screening. Some examples presented in this chapter demonstrate the relevance of screening as they contributed to the discovery of new binders, displaying targeted properties such as non-canonical bindings, G4 *vs* G4 selectivity, or drug-likeness. Combination of virtual and physical methodologies was also successfully employed to discover new scaffolds binding to G4, create constrained libraries (biased toward G4 binder, drug-like molecules, etc.), gain time and give additional insight into small molecules-DNA binding. However, this large number of assays is not sufficient, and it is essential that further screening campaigns systematically implement quality control protocols, wise choose and management of the assay, of the chemical library to screen with, and of the target(s).[84] We indeed believe that i) the G4 community should abide by the standards follow-up procedures that have been developed in the last twenty years, in order to make full profit of the results and ii) medicinal chemistry, structural biology, chemoinformatics (e.g. Lipinski Rule of Five, ADMET properties), physical and virtual screenings should be fully integrated in order to obtain valuable G4-interactive compounds.

## Acknowledgments

We thank N. Saettel for helpful discussions and proofreading of this chapter. E.L. is grateful for the support provided by the CNRS and the Institut Curie.

## References

1. SLAS, 2012, Academic Screening Facilities Directory: http://www.slas.org/screeningFacilities/facilityList.cfm
2. *Nat. Chem. Biol.*, 2009, **5**, 127.
3. *Nat. Chem. Biol.*, 2007, **3**, 433.
4. T. Kodadek, *Nat. Chem. Biol.*, 2010, **6**, 162.
5. *Dans les champs de l'observation, le hasard ne favorise que les esprits préparés.*
6. L. Pasteur, *Oeuvres de Pasteur*, Masson, Paris, 1854.
7. H. Kubinyi, *J. Recept. Signal Transduction Res.*, 1999, **19**, 15.
8. R. Macarron, M. N. Banks, D. Bojanic, D. J. Burns, D. A. Cirovic, T. Garyantes, D. V. Green, R. P. Hertzberg, W. P. Janzen, J. W. Paslay, U. Schopfer and G. S. Sittampalam, *Nat. Rev. Drug Discovery*, 2011, **10**, 188.
9. J. Inglese, C. E. Shamu and R. K. Guy, *Nat. Chem. Biol.*, 2007, **3**, 438.
10. H. J. Lipps and D. Rhodes, *Trends Cell Biol.*, 2009, **19**, 414.
11. D. Monchaud and M.-P. Teulade-Fichou, *Org. Biomol. Chem.*, 2008, **6**, 627.
12. A. De Cian, E. DeLemos, J.-L. Mergny, M.-P. Teulade-Fichou and D. Monchaud, *J. Am. Chem. Soc.*, 2007, **129**, 1856.
13. C. Granotier, G. Pennarun, L. Riou, F. Hoffschir, L. R. Gauthier, A. De Cian, D. Gomez, E. Mandine, J. F. Riou, J. L. Mergny, P. Mailliet, B. Dutrillaux and F. D. Boussin, *Nucleic Acids Res.*, 2005, **33**, 4182.
14. S. M. Gowan, J. R. Harrison, L. Patterson, M. Valenti, M. A. Read, S. Neidle and L. R. Kelland, *Mol. Pharmacol.*, 2002, **61**, 1154.
15. O. Y. Fedoroff, M. Salazar, H. Han, V. V. Chemeris, S. M. Kerwin and L. H. Hurley, *Biochemistry*, 1998, **37**, 12367.
16. K. Shin-ya, K. Wierzba, K. Matsuo, T. Ohtani, Y. Yamada, K. Furihata, Y. Hayakawa and H. Seto, *J. Am. Chem. Soc.*, 2001, **123**, 1262.
17. M.-Y. Kim, H. Vankayalapati, K. Shin-ya, K. Wierzba and L. H. Hurley, *J. Am. Chem. Soc.*, 2002, **124**, 2098.
18. S. G. Rzuczek, D. S. Pilch, E. J. LaVoie and J. E. Rice, *Bioorg. Med. Chem. Lett.*, 2008, **18**, 913.
19. F. Hamon, E. Largy, A. Guédin, M. Rouchon-Dagois, A. Sidibe, D. Monchaud, J.-L. Mergny, J.-F. Riou, C.-H. Nguyen and M.-P. Teulade-Fichou, *Angew. Chem., Int. Ed.*, 2011, **50**, 8745.
20. P. Murat, Y. Singh and E. Defrancq, *Chem. Soc. Rev.*, 2011, **40**, 5293.

21. B. Pagano, S. Cosconati, V. Gabelica, L. Petraccone, S. De Tito, L. Marinelli, V. La Pietra, F. Saverio di Leva, I. Lauri, R. Trotta, E. Novellino, C. Giancola and A. Randazzo, *Curr. Pharm. Des.*, 2012, **18**, 1880.
22. D. C. Fara, T. I. Oprea, E. R. Prossnitz, C. G. Bologa, B. S. Edwards and L. A. Sklar, *Drug Discovery Today: Technol.*, 2006, **3**, 377.
23. L. Petraccone, B. Pagano and C. Giancola, *Methods*, 2012, DOI: 10.1016/j.ymeth.2012.02.011.
24. A. Benz, V. Singh, T. U. Mayer and J. S. Hartig, *ChemBioChem*, 2011, **12**, 1422.
25. Q. Li, J. Xiang, X. Li, L. Chen, X. Xu, Y. Tang, Q. Zhou, L. Li, H. Zhang, H. Sun, A. Guan, Q. Yang, S. Yang and G. Xu, *Biochimie*, 2009, **91**, 811.
26. E. Largy, N. Saettel, F. Hamon, S. Dubruille and M.-P. Teulade-Fichou, *Curr. Pharm. Des.*, 2012, **18**, 1992.
27. A. A. Shelat and R. K. Guy, *Curr. Opin. Chem. Biol.*, 2007, **11**, 244.
28. M. M. Hann and T. I. Oprea, *Curr. Opin. Chem. Biol.*, 2004, **8**, 255.
29. J. E. Redman, S. Ladame, A. P. Reszka, S. Neidle and S. Balasubramanian, *Org. Biomol. Chem.*, 2006, **4**, 4364.
30. M. A. Sills, D. Weiss, Q. Pham, R. Schweitzer, X. Wu and J. J. Wu, *J. Biomol. Screening*, 2002, **7**, 191.
31. S. Fox, H. Wang, L. Sopchak and S. Farr-Jones, *J. Biomol. Screening*, 2002, **7**, 313.
32. I. Coma, J. Herranz and J. Martin, *Methods. Mol. Biol.*, 2009, **565**, 69.
33. J. P. McCoy, Jr., *Nat. Methods*, 2011, **8**, 390.
34. A. Dove, *Nat. Biotechnol.*, 2003, **21**, 859.
35. A. De Cian, L. Guittat, M. Kaiser, B. Sacca, S. Amrane, A. Bourdoncle, P. Alberti, M.-P. Teulade-Fichou, L. Lacroix and J.-L. Mergny, *Methods*, 2007, **42**, 183.
36. E. Largy, F. Hamon and M.-P. Teulade-Fichou, *Anal. Bioanal. Chem.*, 2011, **400**, 3419.
37. B. Brassart, D. Gomez, A. De Cian, R. Paterski, A. Montagnac, K.-H. Qui, N. Temime-Smaali, C. Trentesaux, J.-L. Mergny, F. O. Gueritte and J.-F. Riou, *Mol. Pharmacol.*, 2007, **72**, 631.
38. K. M. Rahman, K. Tizkova, A. P. Reszka, S. Neidle and D. E. Thurston, *Bioorg. Med. Chem. Lett.*, 2012, **22**, 3006.
39. P. A. Holt, R. Buscaglia, J. O. Trent and J. B. Chaires, *Drug Dev. Res.*, 2011, **72**, 178.
40. D.-L. Ma, T.-S. Lai, F.-Y. Chan, W.-H. Chung, R. Abagyan, Y.-C. Leung and K.-Y. Wong, *ChemMedChem*, 2008, **3**, 881.
41. L. S. Glass, A. Bapat, M. R. Kelley, M. M. Georgiadis and E. C. Long, *Bioorg. Med. Chem. Lett.*, 2010, **20**, 1685.
42. P. L. Tran, E. Largy, F. Hamon, M.-P. Teulade-Fichou and J.-L. Mergny, *Biochimie*, 2011, **93**, 1288.
43. L. Lacroix, A. Seosse and J.-L. Mergny, *Nucleic Acids Res.*, 2011, **39**.
44. S. Paramasivan and P. H. Bolton, *Nucleic Acids Res.*, 2008, **36**.
45. H. Wang, T. Chen, S. Wu, X. Chu and R. Yu, *Biosens. Bioelectron.*, 2012, **34**, 88.
46. L. Fu, B. Li and Y. Zhang, *Anal. Biochem.*, 2012, **421**, 198.
47. K. Kobayashi, N. Matsui and K. Usui, *J. Nucleic Acids*, 2011, **2011**, 572873.
48. F. Pillet, C. Thibault, S. Bellon, E. Maillart, E. Trévisiol, C. Vieu, J. M. François and V. A. Leberre, *Sens. Actuators, B*, 2010, **147**, 87.
49. F. Pillet, C. Romera, E. Trevisiol, S. Bellon, M.-P. Teulade-Fichou, J.-M. Francois, G. Pratviel and V. A. Leberre, *Sens. Actuators, B*, 2011, **157**, 304.
50. C. Chen, C. Zhao, X. Yang, J. Ren and X. Qu, *Advanced Materials*, 2010, **22**, 389.
51. P. Wu, D.-L. Ma, C.-H. Leung, S.-C. Yan, N. Zhu, R. Abagyan and C.-M. Che, *Chem.--Eur. J.*, 2009, **15**, 13008.

52. P. Wang, C.-H. Leung, D.-L. Ma, S.-C. Yan and C.-M. Che, *Chem.--Eur. J.*, 2010, **16**, 6900.
53. H.-M. Lee, D. S.-H. Chan, F. Yang, H.-Y. Lam, S.-C. Yan, C.-M. Che, D.-L. Ma and C.-H. Leung, *Chem. Commun.*, 2010, **46**, 4680.
54. J. L. Beck, *Aust. J. Chem.*, 2011, **64**, 705.
55. F. Rosu, E. De Pauw and V. Gabelica, *Biochimie*, 2008, **90**, 1074.
56. V. Gabelica, F. Rosu and E. De Pauw, *Analytical Chemistry*, 2009, **81**, 6708.
57. J. Kypr, I. Kejnovska, D. Renciuk and M. Vorlickova, *Nucleic Acids Res.*, 2009, **37**, 1713.
58. J.-F. Chu, Z.-F. Wang, T.-Y. Tseng and T.-C. Chang, *J. Chin. Chem. Soc.*, 2011, **58**, 296.
59. Q. Shang, J.-F. Xiang, X.-F. Zhang, H.-X. Sun, L. Li and Y.-L. Tang, *Talanta*, 2011, **85**, 820.
60. G. Bai, X. Cao, H. Zhang, J. Xiang, H. Ren, L. Tan and Y. Tang, *J. Chromatogr., A*, 2011, **1218**, 6433.
61. Q. Zhou, L. Li, J. Xiang, Y. Tang, H. Zhang, S. Yang, Q. Li, Q. Yang and G. Xu, *Angew. Chem., Int. Ed.*, 2008, **47**, 5590.
62. Q. Zhou, L. Li, J. Xiang, H. Sun and Y. Tang, *Biochimie*, 2009, **91**, 304.
63. K. I. McLuckie, Z. A. Waller, D. A. Sanders, D. Alves, R. Rodriguez, J. Dash, G. J. McKenzie, A. R. Venkitaraman and S. Balasubramanian, *J. Am. Chem. Soc.*, 2011, **133**, 2658.
64. G. Klebe, *Drug. Discov. Today*, 2006, **11**, 580.
65. G. Schneider, *Nat. Rev. Drug Discovery*, 2010, **9**, 273.
66. S. Ghosh, A. Nie, J. An and Z. Huang, *Curr. Opin. Chem. Biol.*, 2006, **10**, 194.
67. P. Willett, *Drug. Discov. Today*, 2006, **11**, 1046.
68. P. Willett, *Wiley Interdisciplinary Reviews: Data Mining and Knowledge Discovery*, 2011, **1**, 241.
69. H. Geppert, M. Vogt and J. Bajorath, *J. Chem. Inf. Model.*, 2010, **50**, 205.
70. H. Eckert and J. Bajorath, *Drug. Discov. Today*, 2007, **12**, 225.
71. C. G. Ricci and P. A. Netz, *J. Chem. Inf. Model.*, 2009, **49**, 1925.
72. P. A. Holt, P. Ragazzon, L. Strekowski, J. B. Chaires and J. O. Trent, *Nucleic Acids Res.*, 2009, **37**, 1280.
73. D.-L. Ma, V. P.-Y. Ma, D. S.-H. Chan, K.-H. Leung, H.-J. Zhong and C.-H. Leung, *Methods*, 2012, DOI: 10.1016/j.ymeth.2012.02.001.
74. D.-L. Ma, D. S.-H. Chan, P. Lee, M. H.-T. Kwan and C.-H. Leung, *Biochimie*, 2011, **93**, 1252.
75. D.-L. Ma, D. S.-H. Chan and C.-H. Leung, *Chem. Sci.*, 2011, **2**, 1656.
76. N. Phosrithong and J. Ungwitayatorn, *Med. Chem. Res.*, 2010, **19**, 817.
77. H. Sun, Y. Tang, J. Xiang, G. Xu, Y. Zhang, H. Zhang and L. Xu, *Bioorg. Med. Chem. Lett.*, 2006, **16**, 3586.
78. S. Cosconati, L. Marinelli, R. Trotta, A. Virno, L. Mayol, E. Novellino, A. J. Olson and A. Randazzo, *J. Am. Chem. Soc.*, 2009, **131**, 16336.
79. R. Trotta, S. De Tito, I. Lauri, V. La Pietra, L. Marinelli, S. Cosconati, L. Martino, M. R. Conte, L. Mayol, E. Novellino and A. Randazzo, *Biochimie*, 2011, **93**, 1280.
80. J. H. Zhang, *J. Biomol. Screening*, 1999, **4**, 67.
81. I. Coma, L. Clark, E. Diez, G. Harper, J. Herranz, G. Hofmann, M. Lennon, N. Richmond, M. Valmaseda and R. Macarron, *J. Biomol. Screen.*, 2009, **14**, 66.
82. V. Makarenkov, P. Zentilli, D. Kevorkov, A. Gagarin, N. Malo and R. Nadon, *Bioinformatics*, 2007, **23**, 1648.

83. P. B. Taylor, S. Ashman, S. M. Baddeley, S. L. Bartram, C. D. Battle, B. C. Bond, Y. M. Clements, N. J. Gaul, W. E. McAllister, J. A. Mostacero, F. Ramon, J. M. Wilson, R. P. Hertzberg, A. J. Pope and R. Macarron, *J. Biomol. Screen.*, 2002, **7**, 554.
84. W. P. Janzen and P. Bernasconi, eds., *High Throughput Screening Methods and Protocols, Second Edition*, Humana Press, New York, USA, 2009.

# RECOGNITION OF G-QUADRUPLEXES BY METAL COMPLEXES

Kogularamanan Suntharalingam and Ramon Vilar*

Department of Chemistry, Imperial College London, London SW7 2AZ, UK
(r.vilar@imperial.ac.uk)

## 1 INTRODUCTION

The interaction of metal complexes with DNA is an area that has attracted great interest for over four decades. This interest has been largely fuelled by the discovery of cisplatin's anticancer activity by Rosenberg.[1] Since then, a very large number of studies have been published focusing in understanding the interactions between platinum complexes and DNA.[2] Although there is still some debate as to which exact platinum-DNA adducts are more biologically relevant, it is widely accepted that coordination leads to DNA distortions such as unwinding and bending, which impede cellular processes such as transcription and replication. On the other hand, metal complexes that interact non-covalently with DNA have also attracted great interest. In particular, metallo-intercalators (i.e. metal complexes containing ligands that are able to π-π stack in-between base pairs of DNA) have received considerable attention as optical probes and as potential anticancer agents.[3]

Until recently, most studies in this area had focussed on the interaction of metal complexes with B-DNA. However, DNA can assemble in a wide range of different topologies (e.g. Z-DNA, 3-way junctions, triplexes, quadruplexes). With the mounting evidence that several of these non-canonical DNA structures play key biological roles, there has been an increasing interest in understanding their interaction with small molecules – including metal complexes. In particular, over the past few years there have been a considerable number of studies aimed at designing and developing metal complexes that can selectively interact with guanine-quadruplexes (G-quadruplexes).[4] This chapter will provide an overview of the key features metal complexes have that make them attractive as G-quadruplex DNA binders. This will be followed by a review of the main classes of metal complexes that have been shown to interact efficiently with G-quadruplex DNA as well as those complexes that have been studied as optical probes for DNA.

## 2 WHAT MAKES METAL COMPLEXES GOOD G-QUADRUPLEX BINDERS?

G-quadruplexes have unique structural features that can be targeted by small molecules. Their tetra-stranded structure composed of stacks of guanine tetrads (Figure 1) provides a unique structural target for ligand design. Indeed, most quadruplex DNA binders reported to date are based on planar organic heteroaromatic systems that are able to interact via π-π stacking with these tetrads.[5,6] In addition to a π-π stacking core, DNA binders often have other structural features designed to interact with quadruplexes' distinct loops and grooves.

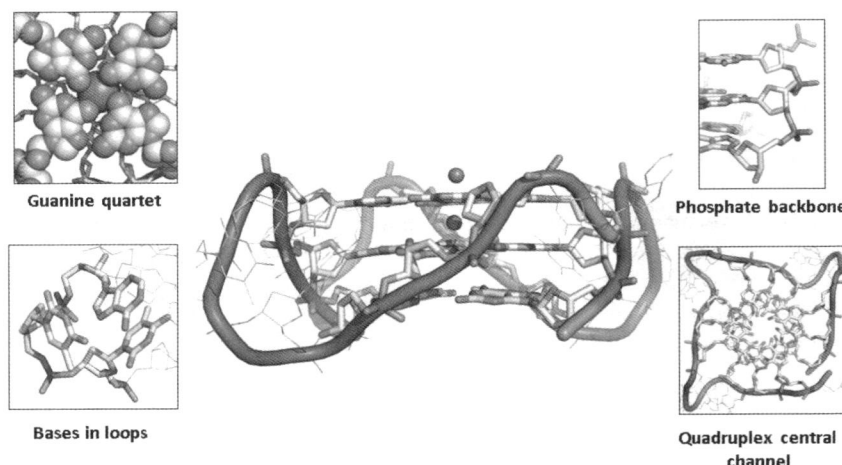

**Figure 1** *G-quadruplex DNA has several structural features that can be targeted by small molecules including the guanine quartets, oligonucleotide loops and grooves and the quadruplex central channel. The figure (adapted from reference 4) was generated with PyMol using crystallographic data deposited in the PDB (1KF1).*

Metal complexes have a very broad range of structural and electronic properties that can be successfully exploited when designing quadruplex DNA binders. In addition, they often have interesting optical, magnetic or catalytic properties which can be exploited for the development of G-quadruplex probes and cleaving agents. Metal centers can be considered structural templates with the ability to organize ligands in specific geometries and relative orientations for optimal quadruplex binding. A wide range of metal complexes can be prepared readily from a small collection of ligands with suitable properties for DNA interactions allowing for the efficient generation of libraries of related compounds. Variations can be easily introduced by modifying the ligands (retaining the coordination geometry) or by changing the metal center (yielding complexes of different geometries). In addition, the electron-withdrawing properties of metal centers reduce the electron density on coordinated aromatic ligands, yielding electron-poor systems which are able to π-π stack more efficiently on top of the guanine tetrads. Also, the metal center in the complex can be positioned so that it provides extra electrostatic stabilization by interacting with the guanines' oxygen atoms which point to the middle of the G-tetrads. Metal complexes also have the possibility of interacting with quadruplexes by direct coordination to the bases of the phosphate backbone.

## 3 BINDING VIA NON-COVALENT INTERACTIONS

In most examples of metal complexes that bind to quadruplex DNA the metal centre acts as a template to organise the ligand(s) in a square planar arrangement, allowing the corresponding complex to non-covalently stack on top of the guanine quartet. However, the metal ion can also play a spectator role, whereby it coordinates to an intrinsically planar ligand (e.g. polyaromatic macrocycles). In addition metal ions can coordinate planar ligands to yield non-planar structures (e.g. octahedral ruthenium complexes) capable of stabilising quadruplex DNA via their interactions with the rim of the quartet or the loops and grooves of quadruplexes. A comprehensive review of this area was recently published[4] and therefore, in this chapter, we aim to highlight only the most noteworthy and recent examples of metal complexes as quadrulex DNA binders.

## 3.1 Metal complexes with macrocyclic ligands

Metalloporphyrins were the first class of metal-containing compounds to be studied for their quadruplex DNA binding properties.[7] These complexes bind largely through π-stacking interactions with the external tetrad.[8] Many of these metalloporphyrins have been modified to include cationic meso groups to provide additional electrostatic interactions with the DNA backbone. Meso-methylpyridinium and methylquinolinium side arms are a popular choice and have been successfully employed to stabilised quadruplex structures (especially the human telomeric sequence). The metal ion residing in the central cavity also contributes to stabilisation through electrostatic interactions.

Metal derivatives (with $Cu^{II}$, $Zn^{II}$, $Mn^{II}$, $Ni^{II}$) of the prototypical $TMPyP_4$ (meso-pyridinium-substituted) porphyrin ligand are the most thoroughly studied metalloporphyrins (Figure 2a). The binding mode of this series is heavily dependent on the metal centre. The zinc(II) and copper(II) complexes, which adopt square planar and square pyramidal geometries respectively, are proposed to bind via the external G-tetrad.[9-11] The manganese(III)-$TMPyP_4$ complex (with two axial ligands) on the other hand, is thought to bind exclusively via the quadruplex grooves, yielding the best selectivity in the $TMPyP_4$ series, a 10-fold preference for quadruplex DNA (over duplex DNA).[12,13] Similarly, another porphyrin based complex, a highly cationic manganese(III) species (designed with extended side chains, Figure 2b) was reported to selectively recognise the grooves of quadruplex DNA and as a result exhibit an extraordinary selectivity of 10000-fold.[14] More recently, porphyrin complexes have utilised phenol based linkers to attach electron deficient planar head groups to intercalate between tetrads and to prepare metalloporphyrin dimers.[15] Although these second generation metalloporphyrins provide diversity in structural design, their quadruplex DNA binding properties have proved quite disappointing (one order of magnitude selectivity at best). Recently, Monti reported a hetero-nuclear $Zn^{II}$-$Pt^{II}$ porphyrazine complex (Figure 2c) which was shown to interact strongly with human telomeric quadruplex DNA.[16] Interestingly this complex displayed encouraging photosensitizing properties, making it therapeutically attractive.

Metal phthalocyanines were investigated on the basis that they have larger polyaromatic surfaces than porphyrins, hence would display more favourable π-π stacking interactions with the external G-tetrad. Indeed, nickel(II) and zinc(II) phthalocyanines display higher affinity for quadruplex DNA compared to similar metalloporphyrin complexes.[17,18] Functionalization of the phthalocyanines' core with positively charged quaternary ammonium groups yields even better quadruplex DNA binders, which also have the ability to induce quadruplex formation and conversion between different quadruplex topologies. Remarkably, one member of this family, a zinc(II) cationic phthalocyanine complex with guanidinium side arms (Zn-DIGP, Figure 2d), was shown by Luedtke to display the strongest binding interaction reported for a small molecule and quadruplex DNA with $K_d < 2$ nM for c-Myc.[19] It is proposed to end-stack, with the guanidinium side arms participating in groove interactions. The emissive properties of Zn-DIGP were also used to evaluate their cellular uptake and localisation. Fluorescence confocal microscopy showed that the complex is readily internalised in several living and fixed cells (including HeLa, MCF7, B16F10, SH-SY5Y, E. coli BL-21, and SK-Mel-28).

Interestingly, it was shown that there is very little non-specific staining of duplex DNA with Zn-DIGP, as revealed by co-staining experiments using Hoechst 33342.

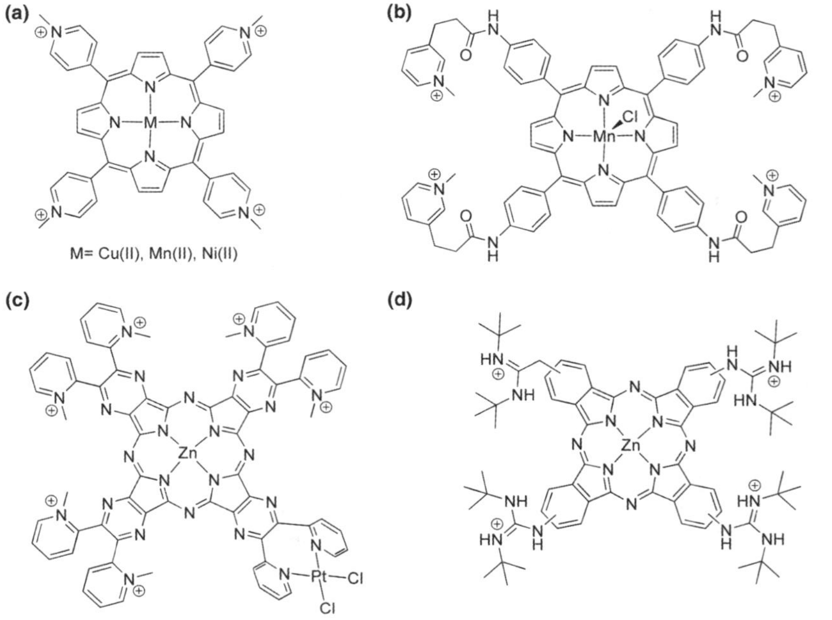

**Figure 2** *Examples of metal complexes with planar macrocyclic ligands that have been reported to interact with quadruplex DNA. The metal ion plays a non-structural role, whereby it occupies the central cavity*

The major drawback of many (with notable exceptions) of the macrocyclic based metal complexes described above is their poor selectivity for quadruplex DNA. Their planar structures usually facilitate intercalation between nucleotides in duplex DNA as well as stacking atop the guanine tetrads in quadruplex DNA. The addition of positively charged side arms has enhanced affinity towards quadruplex DNA, however due to non-specific electrostatic interactions to duplex DNA, selectivity has not significantly improved. In an attempt to address this issue, Miyoshi and Sugimoto have reported a copper(II)-phthalocyanine complex featuring negatively charged sulfonate substituents.[20] The complex displayed improved selectivity, although affinity toward quadruplex DNA was modest ($K_a \approx 10^4$ M$^{-1}$). In contrast, a similar zinc(II)-phthalocyanine complex bearing negatively charged carboxylate substituents did not display measurable affinity for quadruplex DNA,[21] suggesting that the position and number of negative substituents are key to the binding profile.

### 3.2 Metal complexes with non-cyclic ligands

More recently metal complexes involving non-cyclic polypyridyl planar ligands have attracted great interest as potential quadruplex stabilizers. Unlike complexes with macrocyclic ligands, in this type of complexes the metal plays an important structural role. The metal ions can position non-planar ligands into planar geometries ideal for G-quartet binding. The majority of these complexes have also been adapted with protonatable amine side arms to facilitate supplementary interactions with the loops and grooves of quadruplex

DNA. One notable exception is the gold(III) bis-pyrazolylpyridine complex reported by Lippert and Vilar (the first gold complex reported to interact with quadruplex DNA) which displayed very high affinity for quadruplex DNA (evidenced by FID, SPR and VTCD studies) even though it does not possesses protonatable side arms (Figure 3a).[22]

Metal-salphen and salen complexes have been thoroughly studied by Neidle and Vilar.[23,24] Specifically, the nickel(II)- and copper(II)-salphen complexes functionalised with cyclic amine side chains have been established as excellent quadruplex binders (Figure 3b). Electron withdrawing groups have also been utilised to reduce the electron density of the aromatic group with the aim of promoting more effective π-π stacking interactions. This family of complexes displayed some of the highest degrees of stabilization recorded by the FRET assay (at best $\Delta T_m = 33°C$ at 1 µM). They also showed some selectivity over duplex DNA (50-fold). Furthermore, based on data from the TRAP-LIG assay, Neidle and Vilar reported that several of the complexes are very good telomerase inhibitors (low micromolar $^{tel}EC_{50}$ values reported).[25] Very recently the first X-ray crystal structure of quadruplex DNA interacting with a metal complex was reported.[26] The study found the nickel(II)- and copper(II)-salphen complexes to π-π stack atop quadruplex DNA with the amine side arms occupying the DNA grooves. This structural study confirmed the binding mode proposed several years ago for these complexes.

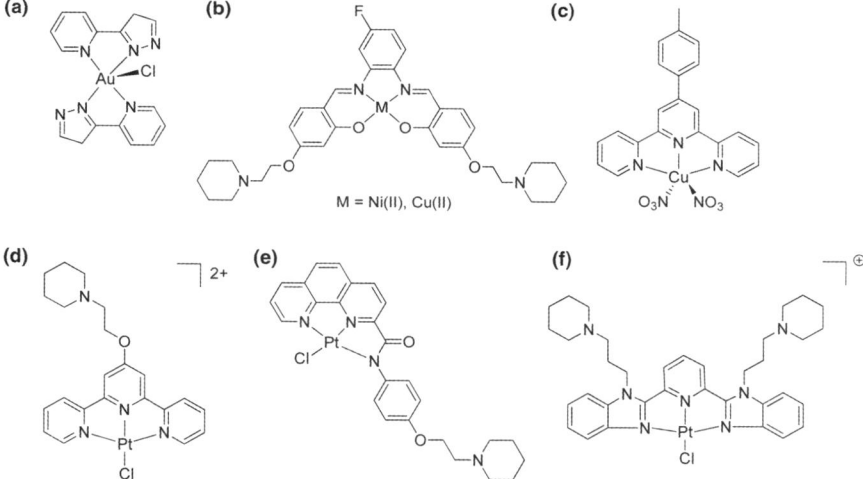

**Figure 3** *Examples of non-cyclic polypyridyl metal complexes which have been shown to stabilise quadruplex DNA.*

Binding of metal terpyridine complexes (especially with platinum) to duplex DNA has long been established. However only within the last 5 years, have metal terpyridine complexes been recognised as good quadruplex binders. Teulade-Fichou reported that square planar platinum(II) and square pyramidal copper(II)-terpyridine complexes (Figure 3c) have high affinity for quadruplex DNA ($\Delta T_m$, 1µM = around 16 °C from FRET studies and $^{G4}DC_{50}$ = < 0.5µM from FID studies).[27] In contrast, trigonal bipyramidal zinc(II) and octahedral ruthenium(III)-terpyridine complexes where shown to have very low affinity towards quadruplex DNA. This study highlighted the necessity for quadruplex stabilisers to have at least one planar surface to display effective π-π stacking interactions with the external tetrads of G-quadruplexes. In other studies reported by Vilar it was shown that addition of side chains comprising of cyclic amine head-groups to platinum(II) terpyridines (Figure 3d), leads to moderate enhancements in binding.[28]

Other poly-pyridyl metal complexes have also shown great promise as potential quadruplex stabilisers. Platinum(II) phenanthroline complexes are one example. Early work conducted by Vilar showed that platinum(II) phenanthrolines, modified with pendant cyclic amine or pyridine side arms through single amide linkages (Figure 3e), are capable of stabilising telomeric quadruplex DNA ($\Delta T_m$, 1 μM = 20 °C as determined by FRET studies) and inhibiting telomerase ($^{tel}EC50$ = 49.5 μM from TRAP assays).[29,30] Control tests performed with the free ligands found limited interactions, demonstrating the important role played by metals in quadruplex binding. Subsequent studies undertaken by Sissi also highlighted the significance of metals.[31,32] A range of metal cations ($Cu^{2+}$, $Ni^{2+}$, $Mn^{2+}$, $Zn^{2+}$ and $Mg^{2+}$) were shown to assemble functionalised and bridged phenanthroline ligands into optimal geometries that were then able to recognise specific quadruplex conformations. The uncoordinated ligands did not have the capacity to do likewise. Further evidence for binding of phenanthroline-based complexes (with platinum, nickel and ruthenium) to quadruplex DNA was provided by Ralph using electrospray ionisation mass spectrometry (ESI-MS) and circular dichroism spectroscopy (CD).[33]

Other acyclic metal complexes shown to stabilise quadruplex DNA through π-π stacking interactions are those reported by Che. Platinum(II) complexes containing dipyridophenazine, phenanthroimidazol and C-deprotonated phenylpyridine ligands were shown to bind via the expected external end-stacking mode with affinities in the region of $10^7$ $M^{-1}$ (determined by a host of techniques including UV/Vis spectroscopy, competition dialysis, NMR spectroscopy and emission titration).[34] More recently the same group, with the aid of computational docking studies, reported a series of square planar platinum(II) complexes with 2,6-bis(benzimidazol-2-yl)-pyridine and 2,6-bis-(pyrazol-3-yl)pyridine ligands (Figure 3f) that interact strongly and selectively with c-Myc quadruplex DNA.[35]

## 3.3 Non-planar metal complexes and supramolecular assemblies

As well as planar molecules, several non-planar metal complexes such as octahedral ruthenium(III) complexes containing polyaromatic ligands, have been shown to be very good quadruplex DNA binders. These complexes are generally made up of bidentate aromatic nitrogen ligands and it is these ligands that are proposed to interact with quadruplex DNA (via π-π stacking with parts of the external G-quartet and intercalating between neighbouring G-tetrads). The metal itself is not considered to contribute to these interactions, however due to its positive charge it is proposed to engage in electrostatic interactions with the loops and grooves.

A series of luminescent bimetallic ruthenium(II) complexes with the ditopic ligands, tetrapyridophenazine and tetraazatetrapyridopentacene (Figure 4a) were reported by Thomas and Haq to bind to different quadruplex DNA sequences.[36] This series of complexes were proposed to bind via "end-pasting" or threading through the lateral loops, while partial intercalation remained a possibility considering the highly dynamic nature of quadruplex structures. Interestingly, binding of these di-ruthenium complexes to quadruplex DNA is accompanied by a blue-shifted "light-switch" effect. The enhancement of the emission was more pronounced in the presence of quadruplex than with duplex DNA. Subsequently, Batagglia and Thomas reported the use of these di-ruthenium complexes for the direct imaging of DNA in living cells.[37] It was shown that the bipyridine analogue is readily taken up by cells and acts as an *in cellulo* nuclear DNA stain.

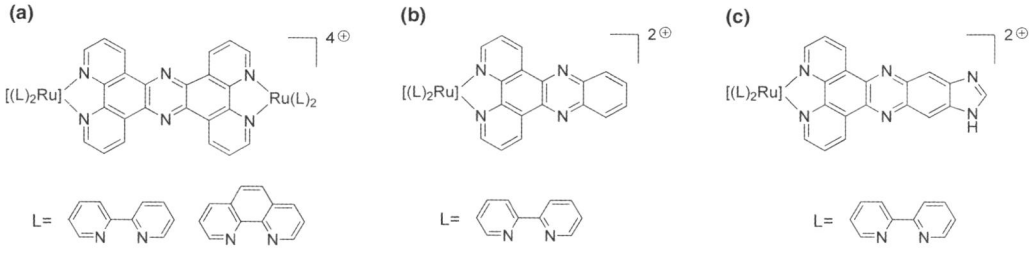

**Figure 4** *Examples of non-planar ruthenium(II) complexes which have been shown to stabilise quadruplex DNA. These complexes interact via ligand π-stacking and groove/loop electrostatic interactions*

Shi has reported the interaction of analogous mono-ruthenium complexes (see Figure 4b and 4c) with quadruplex DNA. In one of these studies, it was shown that [Ru(bpy)$_2$(dppz)]$^{2+}$ induces telomeric DNA to fold into an intramolecular antiparallel G-quadruplex. Furthermore, it was shown that upon addition of telomeric DNA (in either K$^+$ or Na$^+$ buffered solutions) the emission intensity of the ruthenium(II) complexes was markedly enhanced indicating that this complex can also act as light switch for quadruplex DNA. Similarly, [Ru(bpy)$_2$(dppzi)]$^{2+}$ (see Figure 4c) was shown to interact with telomeric quadruplex DNA and to act as a light switch. The emission of this complex in buffered solution is quenched by [Fe(CN)$_6$]$^{4-}$.[38] Interestingly, upon addition of G-quadruplex DNA to this mixture, the emission of the complex is restored and enhanced. This process can be repeated upon further addition of [Fe(CN)$_6$]$^{4-}$ to the solution allowing to cycle the on/off properties of this light switch by addition of G-quadruplex DNA or [Fe(CN)$_6$]$^{4-}$.

Polymetallic supramolecular assemblies have also been shown to interact with G-quadruplexes. For example, triple-helical chiral cylinders made up of three ditopic ligands and two metal ions (either nickel or iron, Figure 5a), originally shown by Hannon to bind to three-way junctions,[39,40] were shown by Qu to selectively recognise telomeric quadruplex DNA.[41] S1 nuclease cleavage assays suggest that the supramolecular complex protects the Htelo sequence from cleavage at the two ends. The authors propose that the cylinders stack atop the quadruplex through its extensive hydrophobic exterior and at the same time the metal centres engage in ionic interactions with the loops and grooves. More recent studies on this system found the P-enantiomer to induce Htelo quadruplex formation in salt deficient conditions.[42] Formation was reported to be fast, efficient and loop sequence dependent. Furthermore the complex was found to exhibit selectivity for Htelo DNA over duplex DNA as well as over other quadruplex forming sequences (namely *c-Myc* and *c-kit* DNA).

A self-assembled metallo-supramolecular complex made up of four platinum-ethylene diamine components held together by bipyridine linkers (Figure 5b), was found by Sleiman to bind strongly to quadruplex DNA ($\Delta T_m$ at 0.75 μM = 34.5 °C using the FRET assay) and at the same time inhibit telomerase activity ($^{tel}IC_{50}$ = *ca.* 0.2 μM using a modified TRAP assay).[43] Binding to quadruplex DNA was attributed to a combination of factors including the square arrangement of the four bipyridine units, the overall electropositive nature of the complex and the hydrogen bonding ability of the Pt-ethylene diamine components. Molecular dynamics simulations found that the bipyridine groups were non-planar with respect to the tetrad, hence π-π stacking is not likely to be the main binding mode. Electrostatic and hydrogen bonding interactions of the terminal platinum atoms and the ethylene diamine groups with the phosphate backbone were proposed to be the predominant modes of binding.

A cubic metallo-supramolecular cage was reported by Therrien and Vilar.[44] This assembly was designed in an attempt to hinder intercalation between base pairs in duplex DNA and thereby improve the selectivity towards quadruplex DNA. The complexes were made up of two zinc(II) tetrapyridinoporphyrins with meso-substituted nitrogen donors, held in place by bridging 2,5-dihydroxy-1,4-benzoquinonato ligands and ruthenium(II)-arene anchors at each corner (Figure 5c). The hetero-metallic complexes were found to bind very strongly to quadruplex DNA (Htelo and *c-Myc*), however selectivity was moderate. This was attributed to the highly cationic nature of the cubes, which led to non-specific electrostatic interactions.

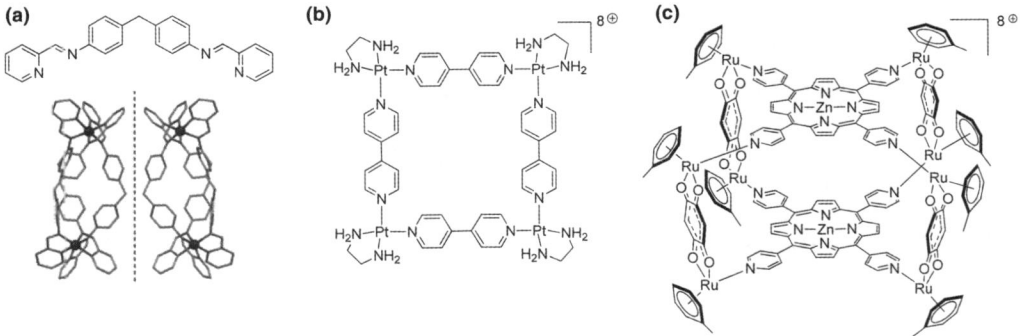

**Figure 5** *Examples of supramolecular assemblies that interact with quadruplex DNA.*

## 4 DIRECT COORDINATION OF METAL COMPLEXES TO QUADRUPLEXES

Several studies have been reported aimed at understanding the interactions of cisplatin (and some of its analogues) with quadruplex-forming sequences. As indicated above, platinum(II) can coordinate to nucleobases, particularly to the N7 atom of guanines. In G-quartets, the N7 of guanines is involved in hydrogen bonding therefore is less accessible for coordination by metal centres. Initial studies by Chottard showed that when the *Tetrahymena* quadruplex-forming sequence $(T_2G_4)_4$ was incubated with $[Pt(NH_3)_3(H_2O)]^{2+}$, *cis*-$[Pt-(NH_3)_2(H_2O)_2]^{2+}$ or *trans*-$[Pt(NH_3)_2(H_2O)_2]^{2+}$, platination of selected guanines was indeed observed.[45] In the case of the tri-amine complex $[Pt(NH_3)_3(H_2O)]^{2+}$, mono- and poly-platinated adducts with four of the sixteen guanines were obtained. These four guanines are not involved in forming the G-quartets – consistent with the hypothesis that if N7 is involved in hydrogen bonding as part of a quartet it will be less accessible for coordination to a metal centre. As per the diamine complexes *cis*-$[Pt-(NH_3)_2(H_2O)_2]^{2+}$ and *trans*-$[Pt(NH_3)_2(H_2O)_2]^{2+}$, both showed to have the ability to cross-link two free guanines at the ends of the quadruplex structure. Interestingly, *trans*-$[Pt(NH_3)_2(H_2O)_2]^{2+}$ is better at crosslinking while *cis*-$[Pt-(NH_3)_2(H_2O)_2]^{2+}$ reacts faster with guanines favouring multiplatination. Molecular dynamics calculations indicated that for both the mono- and bis-cross-linked DNA adducts, there is no major perturbation of the quadruplex structures. These studies were later extended to the human telomere sequences $AG_3(T_2AG_3)_3$ and $(T_2AG_3)_4$.[46] It was shown that *cis*-$[Pt-(NH_3)_2(H_2O)_2]^{2+}$ and *trans*-$[Pt(NH_3)_2(H_2O)_2]^{2+}$ coordinated to four adenines of each sequence as well as four out of the twelve guanines. The observed cross-linking pattern indicated that the quadruplex structure present in solution corresponded to the parallel conformation. Similar studies were carried out with the di-platinum complexes [{trans-PtCl(NH$_3$)$_2$}$_2$NH$_2$(CH$_2$)$_n$NH$_2$]Cl$_2$ (n = 2 or 6) which, unexpectedly, showed that the major cross-link is between two guanines belonging to the same G-quartet.[47] More recent studies have also shown that under certain conditions cisplatin can destabilize quadruplexes by coordinating to N7 of guanines – and hence disrupt the hydrogen bonding interactions within quartets.[48]

In addition to studies aimed at understanding the interaction of cisplatin and its analogues with G-quadruplexes, there have been some reports of quadruplex DNA binders based on platinum complexes containing aromatic tethers (see Figure 6). In this type of complex, the aromatic moiety is designed to interact via π-π stacking with the guanine tetrad while the platinum complex contains labile ligands that favour the coordination of the platinum(II) centre with DNA. For example, Bierbach reported that Pt-ACRAMTU (see Figure 6a) has the ability to interact with telomeric quadruplex DNA (via π-π stacking) and platinate adenine.[49] It was shown that coordination to a given adenine was kinetically favoured over adduct formation with guanine. This unprecedented reactivity suggested that platinum(II) complexes containing polyaromatic tethers could be used to target the telomere for selective chemotherapeutic intervention. A similar approach to targeting quadruplex DNA, where a platinum complex was linked to a quinacridine tether (see Figure 6b), was reported by Teulade-Fichou.[50] In this study it was shown that two very stable adducts between the complex and 22AG (an oligonucleotide that mimics the human quadruplex DNA structure) formed and specific guanines where identified as the main platination targets. Interestingly, it was shown that this Pt-quinacridine complex traps preferentially the antiparallel structure of the 22AG quadruplex.

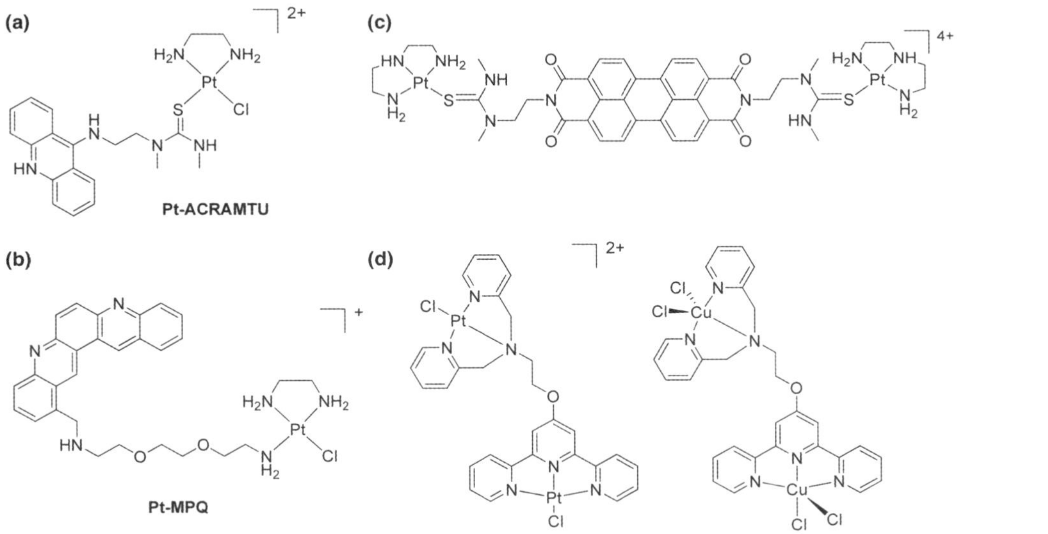

**Figure 6** *Examples of compounds with a planar core (suitable for π-π stacking on to guanine quartets) with metal-based substituents*

Other examples where planar cores (designed to stack on top of the guanine tetrads) have been functionalised with metal-containing substituents are those shown in Figures 6c and 6d. The di-platinum perylene derivative reported by Bierbach (Figure 6c) interacts with telomeric DNA with high affinity and selectivity.[51] The proposed mode of interaction involves π-π stacking of the perylene core on top of the guanine tetrad, with the platinum-containing substituents aiding in improving the selectivity and binding affinity of the compound (even though no direct coordination is proposed to be involved). On the other hand, Vilar has shown that the dimetallic systems based on a terpyridine-DPA ligand (Figure 6d), have high affinity towards *c-Myc* and Htelo quadruplex DNA structures.[52] The di-copper complex showed a remarkable enhanced affinity for G-quadruplexes as compared to the mono-copper counterpart. Although it has not been unambiguously established, it is likely that the high affinity between these complexes and G-quadruplexes involves direct coordination of one of the metal centres to the phosphates or nucleobases.

Bombard and Teulade-Fichou have reported that terpyridine-based platinum(II) complexes can interact with quadruplex DNA by direct coordination. Their platination ability was shown to be influenced by the extension of the aromatic surface surrounding the metal centre.[53] These complexes (particularly with the ttpy ligand, Figure 7) were shown to selectively coordinate to adenine residues of the quadruplex's loops. The same group has reported that the nature of the metal used in this type of complex has an important influence in the binding mode towards DNA. A series of copper(II), palladium(II) and platinum(II) complexes with terpyridine, ttpy, BisQ, tBisQ and tMebip (Figure 7) were reported to interact with quadruplex DNA. The affinities of the palladium(II) and platinum(II) complexes towards quadruplex DNA were considerably different – even though the complexes of both metals are structurally analogous. These differences were proposed to be partly a consequence of the different binding modes of the complexes: the binding of the platinum(II) derivatives takes place mainly via non-covalent interactions while in the case of the corresponding palladium(II) complexes, direct coordination of the metal centre to nucleobases prevails even at short incubation times.[54]

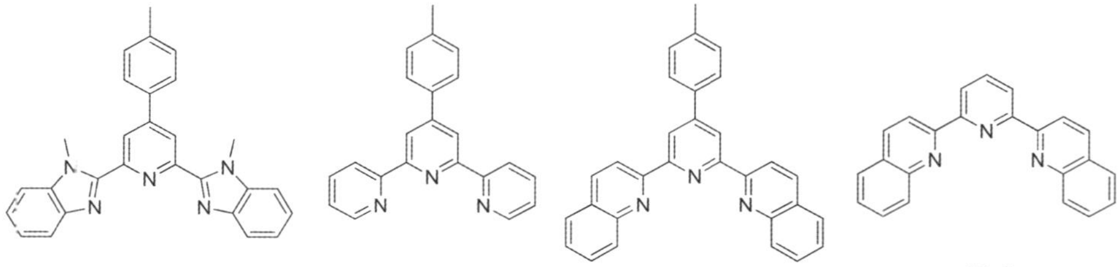

**Figure 7** *Ligands used for the synthesis of a series of plaadium(II) and platinum(II) complexes that displayed distinct binding modes towards G-quadruplex DNA*

## 5 CONCLUDING REMARKS

Since the seminal work by Hurley and Needle in 1997 where they showed that small aromatic molecules could target tetra-stranded structures in telomeric DNA and inhibit telomerase,[55] the number of molecules reported to interact with quadruplexes has grown rapidly. While initially metal complexes featured rarely as quadruplex DNA binders, the past 5 years have seen a growing number of studies that show the great ability metal complexes have to bind and stabilise G-quadruplexes. Considering the large number of sequences in the human genome that are guanine-rich and can potentially form G-quadruplexes, there is growing interest in developing new molecules with enhanced affinity and, more importantly, selectivity for a given G-quadruplex DNA structure. Metal complexes have provided new families of compounds for achieving this aim emerging as increasingly important compounds in the search for novel quadruplex DNA binders – and consequently provide new potential leads for the development of anticancer drugs as well as optical probes for detection and cellular imaging.

REFERENCES

1   B. Rosenberg, L. VanCamp, J. E. Trosko and V. H. Mansour, *Nature*, 1969 **222** 385.
2   R. C. Todd and S. J. Lippard, *Metallomics*, 2009 **1** 280.
3   H.-K. Liu and P. J. Sadler, *Acc. Chem. Res.*, 2011 **44** 349.
4   S. N. Georgiades, N. H. Abd Karim, K. Suntharalingam and R. Vilar, *Angew. Chem., Int. Ed.*, 2010 **49** 4020.
5   A. Arola and R. Vilar, *Curr. Top. Med. Chem.*, 2008 **8** 1405.
6   D. Monchaud and M.-P. Teulade-Fichou, *Org. Biomol. Chem.*, 2008 **6** 627.
7   E. Izbicka, R. T. Wheelhouse, E. Raymond, K. K. Davidson, R. A. Lawrence, D. Y. Sun, B. E. Windle, L. H. Hurley and D. D. Von Hoff, *Cancer Res.*, 1999 **59** 639.
8   D.-F. Shi, R. T. Wheelhouse, D. Sun and L. H. Hurley, *J. Med. Chem.*, 2001 **44** 4509.
9   S. E. Evans, M. A. Mendez, K. B. Turner, L. R. Keating, R. T. Grimes, S. Melchoir and V. A. Szalai, *J. Biol. Inorg. Chem.*, 2007 **12** 1235.
10  L. R. Keating and V. A. Szalai, *Biochemistry*, 2004 **43** 15891.
11  J. Pan and S. Zhang, *J. Biol. Inorg. Chem.*, 2009 **14** 401.
12  A. Maraval, S. Franco, C. Vialas, G. Pratviel, M. A. Blasco and B. Meunier, *Org. Biomol. Chem.*, 2003 **1** 921.
13  C. Vialas, G. Pratviel and B. Meunier, *Biochemistry*, 2000 **39** 9514.
14  I. M. Dixon, F. Lopez, A. M. Tejera, J. P. Esteve, M. A. Blasco, G. Pratviel and B. Meunier, *J. Am. Chem. Soc.*, 2007 **129** 1502.
15  I. M. Dixon, F. Lopez, J.-P. Estève, A. M. Tejera, M. A. Blasco, G. Pratviel and B. Meunier, *ChemBioChem*, 2005 **6** 123.
16  I. Manet, F. Manoli, M. P. Donzello, C. Ercolani, D. Vittori, L. Cellai, A. Masi and S. Monti, *Inorg. Chem.*, 2011 **50** 7403.
17  L. Ren, A. Zhang, J. Huang, P. Wang, X. Weng, L. Zhang, F. Liang, Z. Tan and X. Zhou, *ChemBioChem*, 2007 **8** 775.
18  L. Zhang, J. Huang, L. Ren, M. Bai, L. Wu, B. Zhai and X. Zhou, *Bioorg. Med. Chem.*, 2008 **16** 303.
19  J. Alzeer, B. R. Vummidi, P. J. C. Roth and N. W. Luedtke, *Angew. Chem., Int. Ed.*, 2009 **48** 9362.
20  H. Yaku, T. Murashima, D. Miyoshi and N. Sugimoto, *Chem. Commun.*, 2010 **46** 5740.
21  J. Alzeer and N. W. Luedtke, *Biochemistry*, 2010 **49** 4339.
22  K. Suntharalingam, D. Gupta, P. J. Sanz Miguel, B. Lippert and R. Vilar, *Chem. Eur. J.*, 2010 **16** 3613.
23  J. E. Reed, A. A. Arnal, S. Neidle and R. Vilar, *J. Am. Chem. Soc.*, 2006 **128** 5992.
24  A. Arola-Arnal, J. Benet-Buchholz, S. Neidle and R. Vilar, *Inorg. Chem.*, 2008 **47** 11910.
25  J. Reed, M. Gunaratnam, M. Beltran, A. P. Reszka, R. Vilar and S. Neidle, *Anal. Biochem.*, 2008 **380** 99.
26  N. H. Campbell, N. H. A. Karim, G. N. Parkinson, M. Gunaratnam, V. Petrucci, A. K. Todd, R. Vilar and S. Neidle, *J. Med. Chem.*, 2012 **55** 209.
27  H. Bertrand, D. Monchaud, A. De Cian, R. Guillot, J.-L. Mergny and M.-P. Teulade-Fichou, *Org. Biomol. Chem.*, 2007 **5** 2555.
28  K. Suntharalingam, A. J. P. White and R. Vilar, *Inorg. Chem.*, 2009 **48** 9427.
29  J. E. Reed, S. Neidle and R. Vilar, *Chem. Commun.*, 2007 4366.

30  J. E. Reed, A. J. P. White, S. Neidle and R. Vilar, *Dalton Trans.*, 2009 2558.
31  S. Bianco, C. Musetti, A. Waldeck, S. Sparapani, J. D. Seitz, A. P. Krapcho, M. Palumbo and C. Sissi, *Dalton Trans.*, 2010 **39** 5833.
32  C. Musetti, L. Lucatello, S. Bianco, A. P. Krapcho, S. A. Cadamuro, M. Palumbo and C. Sissi, *Dalton Trans.*, 2009 3657.
33  J. Talib, C. Green, K. J. Davis, T. Urathamakul, J. L. Beck, J. R. Aldrich-Wright and S. F. Ralph, *Dalton Trans.*, 2008 1018.
34  D.-L. Ma, C.-M. Che and S.-C. Yan, *J. Am. Chem. Soc.*, 2009 **131** 1835.
35  P. Wang, C. H. Leung, D. L. Ma, S. C. Yan and C. M. Che, *Chem. Eur. J.*, 2010 **16** 6900.
36  C. Rajput, R. Rutkaite, L. Swanson, I. Haq and J. A. Thomas, *Chem. Eur. J.*, 2006 **12** 4611.
37  M. R. Gill, J. Garcia-Lara, S. J. Foster, C. Smythe, G. Battaglia and J. A. Thomas, *Nature Chem.*, 2009 **1** 662.
38  S. Shi, J. Zhao, X. Gao, C. Lv, L. Yang, J. Hao, H. Huang, J. Yao, W. Sun, T. Yao and L. Ji, *Dalton Trans.*, 2012 **41** 5789.
39  J. Malina, M. J. Hannon and V. Brabec, *Chem. Eur. J.*, 2007 **13** 3871.
40  I. Meistermann, V. Moreno, M. J. Prieto, E. Moldrheim, E. Sletten, S. Khalid, P. M. Rodger, J. C. Peberdy, C. J. Isaac, A. Rodger and M. J. Hannon, *Proc. Natl. Acad. Sci.*, 2002 **99** 5069.
41  H. Yu, X. Wang, M. Fu, J. Ren and X. Qu, *Nucleic Acids Res.*, 2008 **36** 5695.
42  C. Zhao, J. Geng, L. Feng, J. Ren and X. Qu, *Chem. Eur. J.*, 2011 **17** 8209.
43  R. Kieltyka, P. Englebienne, J. Fakhoury, C. Autexier, N. Moitessier and H. F. Sleiman, *J. Am. Chem. Soc.*, 2008 **130** 10040.
44  N. P. Barry, N. H. Abd Karim, R. Vilar and B. Therrien, *Dalton Trans.*, 2009 10717.
45  S. Redon, S. Bombard, M.-A. Elizondo-Riojas and J.-C. Chottard, *Biochemistry*, 2001 **40** 8463.
46  S. Redon, S. Bombard, M.-A. Elizondo-Riojas and J.-C. Chottard, *Nucleic Acids Res.*, 2003 **31** 1605.
47  I. Ourliac-Garnier, M.-A. Elizondo-Riojas, S. Redon, N. P. Farrell and S. Bombard, *Biochemistry*, 2005 **44** 10620.
48  I. Ourliac-Garnier, A. Poulet, R. Charif, S. Amiard, F. Magdinier, K. Rezai, E. Gilson, M.-J. Giraud-Panis and S. Bombard, *J. Biol. Inorg. Chem.*, 2010 **15** 641.
49  L. Rao and U. Bierbach, *J. Am. Chem. Soc.*, 2007 **129** 15764.
50  H. Bertrand, S. Bombard, D. Monchaud and M.-P. Teulade-Fichou, *J. Biol. Inorg. Chem.*, 2007 **12** 1003.
51  L. Rao, J. D. Dworkin, W. E. Nell and U. Bierbach, *J. Phys. Chem. B*, 2011 **115** 13701.
52  K. Suntharalingam, A. J. P. White and R. Vilar, *Inorg. Chem.*, 2010 **49** 8371.
53  H. Bertrand, S. Bombard, D. Monchaud, E. Talbot, A. Guedin, J.-L. Mergny, R. Grunert, P. J. Bednarski and M.-P. Teulade-Fichou, *Org. Biomol. Chem.*, 2009 **7** 2864.
54  E. Largy, F. Hamon, F. Rosu, V. Gabelica, E. De Pauw, A. Guedin, J.-L. Mergny and M.-P. Teulade-Fichou, *Chem. Eur. J.*, 2011 **17** 13274.
55  D. Sun, B. Thompson, B. E. Cathers, M. Salazar, S. M. Kerwin, J. O. Trent, T. C. Jenkins, S. Neidle and L. H. Hurley, *J. Med. Chem.*, 1997 **40** 2113.

# Chapter 5

## Applications in Bioanalytics, Therapy and Molecular Electronics

Chapter Editor: **Wolfgang Fritzsche**

The unique properties of G4-based nucleic acids provide the base for a variety of applications.

A whole set of biomedical as well as bioanalytical applications is based on the ability of G-quartets to stabilize defined three-dimensional nucleic acid structures that exhibit a high affinity to a target molecule. Such aptamers can therefore show similar binding properties as antibodies with comparable applications in diagnostics and therapy, but exhibit striking advantages like ex-vivo synthesis and increased physicochemical stability. Moreover, they can even show catalytic behavior that can be utilized for bioanalytical purposes. The chapter contains examples for applications in these fields.

The structural properties described in previous chapters are the base for applications in molecular nanotechnology and –electronics. Nucleic acids represent the most promising materials in these fields, and here G4 structures show even outstanding mechanical stability and length control from the nano- into the micrometer range. An important step on the way to respective applications is the integration of G4-based nanostructures into technical environments such as microelectrodes. Here electrical field-based approaches – e.g. dielectrophoresis DEP – represent the most promising technique, which has been demonstrated for the integration and subsequent characterization of even single G4 structures. These techniques have also been used to enable the characterization of electrical properties of G4 assemblies as described in this chapter.

In conclusion, by presenting various biomedical as well as nanobiotechnological demonstrations this chapter demonstrates the great application potential of G4-based nanostructures.

# CATALYTIC G-QUADRUPLEXES

Dipankar Sen

Department of Molecular Biology & Biochemistry, Simon Fraser University, Burnaby, British Columbia V5A 1S6, Canada

## 1 INTRODUCTION

The discovery that the nucleic acids, RNA and DNA, can be catalytic (reviewed in 1,2), has opened up a large field of research, both to investigate the range of inherent catalytic potential of the nucleic acids, and to exploit such potential for practical (notably, biomedical) utility. As polymers with chemical functionalities, RNA and DNA are somewhat lacking, relative to proteins that have a rich repertoire of side-chain functional groups. Indeed, the catalytic RNAs (ribozymes) found in nature, though sophisticated catalysts, catalyze a relatively narrow range of chemical reactions (phosphoester hydrolysis and transfer; and, peptide synthesis). The question set forth by this observation is: is the catalytic potential of nucleic acids necessarily limited to these reactions? In recent years powerful molecular biological methodologies, most notably *in vitro* selection (SELEX), have enabled both the isolation ands study of not only novel ribozymes that are not found in nature, but also catalytic DNAs (deoxyribozymes or DNAzymes), which catalyze a wide range of both the familiar and often unexpected reactions. One device that even naturally occurring protein enzymes utilize for broadening their catalytic repertoire is by co-opting the chemical and catalytic functionalities of a variety of small molecule cofactors, including metal ions, the different vitamins, and heme. *In vitro* selection-generated ribozymes and DNAzymes have likewise been found to be capable of co-opting such cofactors (most notably, heme) for their catalytic function.

*In vitro* selection (SELEX)[3-5] is carried out as follows: large, random-sequences libraries ($10^{13}$-$10^{15}$ different sequences) of single-stranded DNA or RNA are allowed to fold into complex shapes, and then subjected to a selection screen for the ability to either bind a target compound or catalyze a specified reaction. DNA/RNA sequences successful at the desired task are then amplified and subjected to several further rounds of selection, until a highly enriched pool is obtained. At this point the individual binding/catalytic sequences (aptamers and DNAzymes/ribozymes, respectively) can be isolated and characterized by cloning and sequencing of the final enriched pool (reviewed by Eckstein[6]).

Of all the aptamers and catalytic DNAs and RNAs obtained by SELEX to date, a significant number (especially, aptamers) incorporate G-quadruplexes within their folded/active structures. None of the naturally occurring ribozymes incorporate G-quadruplexes; however, a significant number of SELEX-generated DNAzymes, as well as artificial ribozymes rationally designed from pre-existing ones, incorporate G-quadruplexes. These catalysts can be classified into two broad classes: (a) those that intrinsically incorporate G-quadruplexes (and of which, a subset *require* the G-quadruplex directly for their catalytic function), and (b) those rationally designed variants of naturally occurring ribozymes, in which the G-quadruplex plays a controlling function.

## 2. RESULTS

### 2.1 Ribozymes and DNAzymes that intrinsically incorporate G-quadruplexes

*2.1.1 DNAzymes that catalyze porphyrin metallation.* In 1996, Li *et al.*[7] selected DNA aptamers that specifically and tightly bound a distorted porphyrin, N-methylmesoporphyrin (NMM), with a dissociation constant in the nanomolar range. NMM is known to be a mimic of the reaction transition state for porphyrin metallation by ferrochelatase enzymes, whose substrates are biological porphyrins such as protoporphyrin IX (PPIX). Li and Sen (1998)[8] reasoned that if DNA aptamers could be found that bound the distorted NMM preferentially over planar porphyrins such as PPIX and mesoporphyrin IX (MPIX), then such aptamers should catalyze the metallation of these latter porphyrins. From their SELEX experiment, these authors found a 33-nucleotide, guanine-rich DNA aptamer, PS5.ST1, did indeed efficiently catalyze metallation of MPIX by $Zn^{2+}$ and $Cu^{2+}$ ions in an aqueous buffer.[8] PS5.ST1, and two shorter sequence variants derived from it, PS5.M and PS2.M, obeyed Michaelis-Menten kinetics for copper insertion into MPIX. Under optimal conditions, the following catalytic parameters were measured: $k_{cat}$: 1.3 $min^{-1}$; $K_M$: 40 $\mu$M.

Li and Sen[9] further showed that with PS5.M, an interesting result of the binding of substrate porphyrin to its G-quadruplex was a notable enhancement of the porphyrin's basicity. It was proposed that this shift likely underlay PS5.M's observed catalysis of porphyrin metallation.[10] At the same time as the above results were reported, a catalytic RNA (also G-rich) for porphyrin metallation was also reported.[11] It appeared likely, though not shown directly, that this guanine-rich ribozyme also formed a G-quadruplex, and was able to catalyze metallation using strategies similiar to its DNA counterparts.[12]

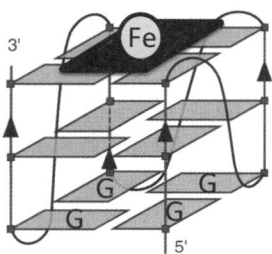

**Figure 1** *A schematic diagram of a parallel-stranded G-quadruplex, such as PS2.M, complexed with hemin (shown in black).*

*2.1.2 DNA and RNA G-quadruplexes that bind hemin and show enhanced peroxidase and peroxygenase activity.* Hemin [Fe(III)-heme] was found to be an excellent competitive inhibitor of porphyrin metallation by the DNAzymes PS2.M and PS5.M.[8] Such strong binding of hemin by these two G-quadruplex DNAzymes implied that the latter could be regarded as aptamers for binding hemin (Figure 1). A question then arose: could such hemin-DNA complexes possibly show any of the catalytic oxidative properties of many hemoproteins? Travascio *et al.* (1998) investigated whether the abovementioned DNA-hemin complexes, in the presence of hydrogen peroxide, showed peroxidase (1-electron oxidation) activity above the background, i.e. the activities of hemin alone or hemin mixed in the presence of non-binding, control DNA.[13] Surprisingly, both aptamer-hemin complexes showed notably enhanced peroxidase activity compared to the two controls. Under optimized reaction conditions, PS2.M-hemin showed ~250-fold higher peroxidase activity than the controls.[13]

The UV-vis spectra of the aptamer-hemin complexes showed striking differences from the spectra of disaggregated hemin in the absence of aptamer, including a ~2-fold hyperchromicity of hemin's intense Soret absorption, as well as important changes in the visible spectrum.[13] Overall, the spectra of the aptamer-hemin complexes closely resembled those of hemoproteins such as metmyoglobin (which also shows a notable peroxidase activity). Two important mechanistic insights were obtained about the above hemin-aptamer complexes: (a) First, the RNA version of PS2.M ("rPS2.M"), complexed to hemin, also showed comparable peroxidase activity as PS2.M-hemin.[14] (b) Binding to DNA/RNA appeared to enhance the basicity (reduce the acidity) of a water ligand axially coordinated to the heme iron, and this alone was a significant reason for the enhancement of peroxidation catalysis, at close to neutral pH, by the hemin-binding DNA and RNA aptamers.[15]

Recently, these peroxidase DNAzymes and ribozymes have also been found to catalyze 2-electron oxidations (peroxygenase[16] and NADH oxidase[17] activities) over and above the 1-electron oxidation (peroxidase) activity. A significant body of work by numerous groups has now established that parallel-stranded G-quadruplexes in general, whether made of DNA or RNA, appear to bind hemin strongly and also to catalyze the above categories of oxidative reactions (reviewed in 18).

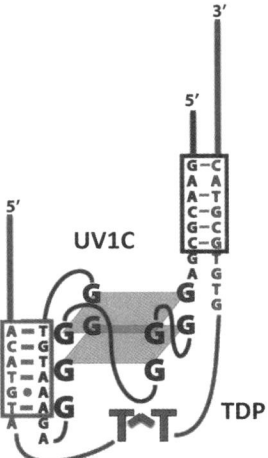

**Figure 2** *A schematic drawing of the UV1C photolyase DNAzyme, bound to its DNA substrate, TDP.*

*2.1.3 A G-quadruplex DNAzyme with photolyase activity.* In 2004, Chinnapen and Sen[19] reported the use of DNA SELEX to select for a 42-nt DNA, UV1C, that was capable of using UV light to photo-reactivate a cyclobutane (CPD) thymine dimer placed within a single-stranded substrate DNA, TDP (Figure 2). No extraneous cofactor was needed for this catalysis, which occurred most optimally ($k_{cat}/k_{uncat}$ = 25,000) at 305 nm wavelength, at the edge of DNA's UV absorption spectrum. The folded UV1C-TDP complex was found to incorporate a G-quadruplex, whose formation was key to UV1C's catalytic activity.[19] It was proposed that the G-quadruplex served as an antenna for harvesting light in the 300-310 nm wavelength range, and that one or more photo-excited guanine within the quadruplex transferred an electron to the substrate's thymine dimer, leading to the latter's repair to thymine bases.[19] Subsequent work on this DNAzyme has included contact crosslinking experiments, which have clearly established that within the folded DNAzyme-substrate complex the thymine dimer is indeed located very close in space to key guanines of the G-quadruplex. Such spatial closeness is necessary for efficient electron transfer to occur between the quadruplex and the thymine dimer.[20,21]

*2.1.4 A self-capping G-quadruplex DNAzyme.* In 2000, Breaker and colleagues reported the *in vitro* selection of DNAzymes that were able to react with ATP to "cap" their own 5' terminal phosphates.[22] This remarkable catalytic activity resulted in the generation of a 5'-5' linked pyrophosphate caps at these sites. Of the different classes of capping DNAzymes selected, the Class I sequence, a 41-nt DNAzyme, was particularly noteworthy in that it folded into a G-quadruplex (Figure 3). Extensive mutagenesis as well as chemical protection studies on this DNAzyme led to a structural model for it, in which the capping site, the 5'-most nucleotide of the sequence (a guanine), appeared to be positioned in an incomplete G-quartet made up of three guanines and a vacant site. It remains a tantalizing possibility that this vacant site may function as the ATP-binding site of this DNAzyme.

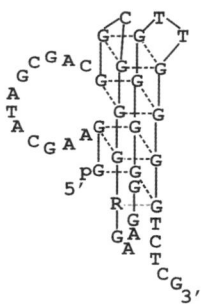

**Figure 3** *Folded structure of a self-capping G-quadruplex DNAzyme*

*2.1.5 A kinase G-quadruplex DNAzyme.* McManus and Li (2008) reported the selection of a family of kinase DNAzymes, Dk1-Dk5.[23] Dk-1 to Dk-4 all required the manganese(II) cation for their catalytic activity, and were able to utilize only one nucleotide, either ATP or GTP. Dk-5, however, had notably different characteristics—it could make use of a variety of metal ions as catalytic cofactors (including calcium, magnesium, manganese(II) and copper(II)), and it could utilize both ATP and GTP as substrates.[23] Structural studies on Dk-5 showed that it was also structurally very different from the other four DNAzymes, and that it formed a complex, pseudoknotted G-quadruplex.[23] To what extent this unique folded structure of the DNAzyme correlates with its distinctive substrate and cofactor utilization capabilities, will require further studies.

*2.1.6 Self-cleaving G-quadruplexes.* A remarkable and unanticipated property of a number of intramolecularly folded G-quadruplexes was reported by Tianhu Li and colleagues.[24,25] An artificial (non-biological) DNA G-quadruplex was originally designed to have an external loop positioned very proximal to the oligonucleotides's terminal 3'-OH group. An expectation was that perhaps the 3'-OH (if activated as a nucleophile by the presence of magnesium ions and histidine) might cleave within the DNA loop. Indeed, such a cleavage activity in the loop was found, and it depended on the presence of both magnesium and histidine, and took place at moderate temperatures (< 45° C).[24] A similar DNA backbone cleavage was subsequently reported by the authors to occur in the mixed-topology G-quadruplex formed by a tetra-repeat of the human telomeric DNA sequence, i.e. [d(TTAGGG)$_4$]. In this latter case, however, cleavage took place not in a loop but within a single-stranded flank to the quadruplex.[25] These DNA self-cleavage reactions required low millimolar concentrations of both magnesium ions and L-histidine, and proceeded at 20°-30° over a period of hours. Although interesting, this self-cleavage activity remains mechanistically somewhat mysterious. Cleavage has been reported to date only *in cis*, i.e. in intramolecular fashion. It would be particularly interesting if such cleavage could also be observed *in trans*, for that would surely lead to a better understanding of this surprising catalysis.

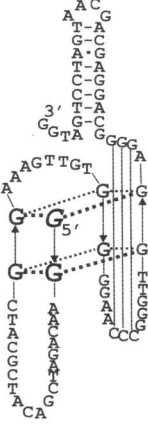

**Figure 4** *The pseudoknotted G-quadruplex structure of the Dk-5 kinase DNAzyme.*

## 2.2. Ribozymes under the control of grafted G-quadruplex motifs.

*2.2.1. Hammerhead ribozymes controlled by G-quadruplexes.* The hammerhead self-cleaving RNA motif is the smallest and perhaps the best studied of all naturally occurring catalytic RNA motifs. In its natural state, the ribozyme is not subject to allosteric control. However, in a series of classic studies Breaker and colleagues[30] reported the design of *allosteric* hammerhead ribozymes, fashioned from the physical and functional linkage of the ribozyme to various ligand-binding aptamers. The idea here was that ligand-binding-induced conformational changes in the aptamer domain could be structurally communicated to the ribozyme domain, to either enhance or diminish the latter's catalytic activity. Following this general theme, two groups, those of Hartig[26] and Katahira[27], have designed hammerhead ribozymes whose activity is controlled by the formation (or lack of formation) of two distinct kinds of G-quadruplexes.

Wieland and Hartig (2006) set out to create allosteric, G-quadruplex-containing hammerhead ribozymes that were either activated or inhibited by G-quadruplex binding ligands, such as the cationic porphyrin, TMPyP4 (meso-5,10,15,20-tetrakis-(N-methyl-4-pyridinio)porphine).[26] Figure 5 shows the design of their key construct, the GQP-HHR, which is structurally destabilized relative to a 'wild-type' hammerhead ribozyme (wt-HHR) by both the shortening (to 3 base pairs) and destabilization (introduction of G•A base pairs) of one of the latter's three constituent stems.[26] A $G_4UG_3UG_3UG_3$ loop was then grafted to the end of this truncated arm (in the GQP-HHR construct).[26] Curiously, TMPyP4 was found to be a very potent inhibitor of even the unmodified, wt-HHR ribozyme. However, this same compound, added to GQP-HHR, powerfully resurrected the latter's activity, albeit not to the level of the optimally functioning wt-HHR.[26]

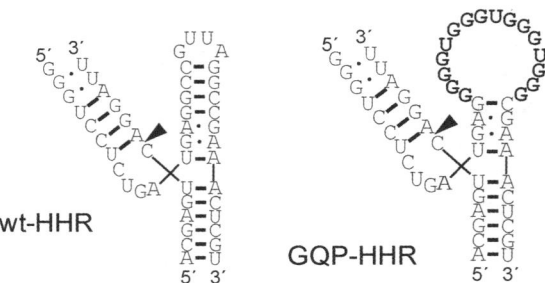

**Figure 5** *Sequences and secondary structures of the wt-HHR and GQP-HHR ribozymes.*

In a parallel study, Katahira and colleagues[27] used the sequence $(AGG)_4$, which forms a highly unusual, dimerized, G-quadruplex with a dimerization interface formed between two G-and A-rich base hexads. Figure 6 shows that one of their $(AGG)_4$-modified ribozymes, which is catalytically inactive in the absence of potassium (Figure 6, *b*), folds (as well as dimerizes) in the presence of potassium to give a catalytically active dimeric ribozyme. Other variants of this general design show the opposite behaviour, i.e. are repressed by potassium and G-quadruplex formation. These "intelligent" ribozymes are therefore uniquely responsive to potassium, and provide a paradigm for the control of hammerhead ribozyme activity by this metal ion.[27]

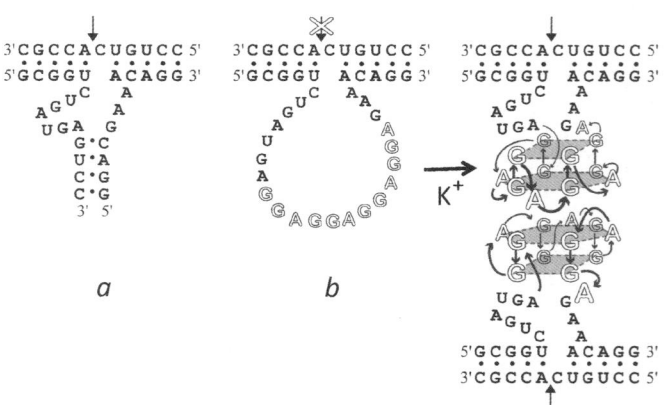

**Figure 6** *Sequences and secondary structures of a wild-type hammerhead ribozyme (a) and an "intelligent", potassium-responsive, hammerhead ribozyme (b).*

*2.2.2 HDV ribozymes controlled by G-quadruplexes.* While the hammerhead ribozyme has a relatively simple folded structure, that readily lends itself to manipulation and substitution such that allosteric variants of the ribozyme can be readily generated, the naturally occurring, self-cleaving, HDV ribozyme presents a more challenging target in terms of such manipulations. This is not least because HDV forms a tightly packed folded structure. Also, it does not display multiple folding pathways, which property could have been exploited to generate allosteric HDV ribozymes. Nevertheless, Beaudoin and Perreault (2008) attempted to structurally weaken one of the stems of the folded HDV ribozyme and graft a G-quadruplex-capable RNA loop onto it.[28] Astonishingly, such a successful ribozyme was constructed. It was found to be catalytically inactive *in trans*, in the absence of potassium, but its catalysis could be restored (in the presence of the mechanistically required magnesium) by the presence of potassium, stabilizer of G-quadruplexes. Activation was even more optimal with the simultaneous presence of both potassium and TMPyP4 (which were found synergistically to stabilize the grafted G-quadruplex). The authors proposed that the *inactive* structure of this so-called "G-quartzyme" likely contained catalytically detrimental pairings between key elements of the ribozyme and the grafted G-quadruplex forming loop.[28]

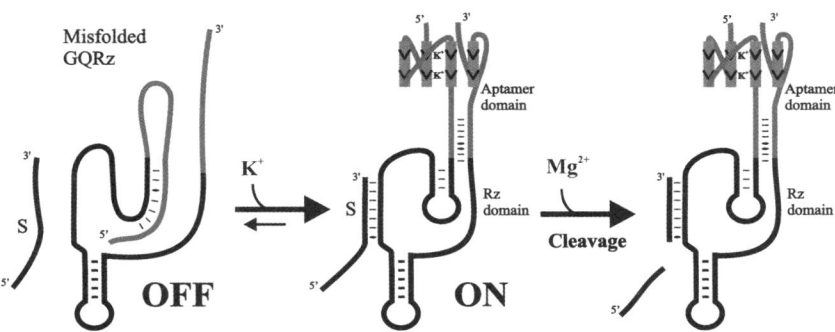

**Figure 7** *The various folded and functional states of the G-quartzyme (GQz).*

3. CONCLUSIONS

In a thoughtful review on the range and diversity of tertiary structures adopted by known DNAzymes, McManus and Li note that while numerous catalytic DNA G-quadruplexes have been obtained by SELEX, no natural ribozyme nor any SELEX-derived ribozyme has been shown to adopt this particular fold.[29] McManus and Li go on to propose that this may be related to the significantly higher structural polymorphism available to DNA G-quadruplexes relative to their RNA counterparts (RNA has been found to fold only to parallel-stranded quadruplexes).[29] However, as summarized above, G-quadruplexes are by no means incompatible with RNA and catalysis. Indeed, the examples summarized in this review testify to the high utility of grafted RNA G-quadruplex domains in controlling the activity of even naturally occurring ribozymes.

## 4. ACKNOWLEDGMENTS

I wish to thank all the colleagues who sent me figures, modified and amended from their original publications, for use in this review.

**References**

1. J.A. Doudna and T.R. Cech, *Nature*, 2002, **418**, 222.
2. C. Höbartner and S.K. Silverman, *Biopolymers*, 2007, **87**, 279.
3. A.D. Elllington and J.W. Szostak, *Nature*, 1990, **346**, 818.
4. C. Tuerk and L. Gold, *Science*, 1990, **249**, 505.
5. D.L. Robertson and G.F. Joyce, *Nature*, 1990, **344**, 467.
6. F. Eckstein, *Expert Opin. Biol. Ther.*, 2007, **7**, 10121.
7. Y. Li, C.R. Geyer, D. Sen, *Biochemistry,* 1996, **35**, 6911.
8. Y. Li and D. Sen, *Nat. Struct. Biol.*, 1996, **3**, 743.
9. Y. Li and D. Sen, *Biochemistry,* 1997, **36**, 5589.
10. Y. Li and D. Sen, *Chem. Biol.*, 1998, **5**, 1.
11. M.M. Conn, J.R. Prudent and P.G. Schultz, *J. Am. Chem. Soc.*, 1996, **118**, 7012.
12. D. Sen, C.R. Geyer, *Curr. Opin. Chem. Biol.*, 1998, **2**, 680.
13. P. Travascio, Y. Li and D. Sen, *Chem. Biol.*, 1998, **5**, 505.
14. P. Travascio, A.J. Bennet, D.Y. Wang and D. Sen, *Chem. Biol.,*1999, **6**, 779.
15. P. Travascio, D. Sen, A.J. Bennet, *Can. J. Chem.*, 2006, **84**, 613.
16. L.C. Poon, S.P. Methot, W. Morabi-Pazooki, F. Pio, A.J. Bennet and D. Sen, *J. Am. Chem. Soc.*, 2011, **133**, 1877.
17. E. Golub, R. Freeman and I. Willner, *Angew. Chem. Int. Ed. Engl.*, 2011, **50**, 1.
18. D. Sen and L.C. Poon, *Crit. Revs. Biochem. Mol. Biol.*, 2011, **46**, 478.
19. D.J. Chinnapen and D. Sen, *Proc. Natl. Acad. Sci. USA*, **101**, 65.
20. D.J. Chinnapen and D. Sen, *J. Mol. Biol.*, 2007, **365**, 1326-1336.
21. G.S. Sekhon and D. Sen, *Biochemistry*, 2009, **48**, 6335.
22. Y. Li, Y. Liu and R.R. Breaker, *Biochemistry*, 2000, **39**, 3106.
23. S.A. McManus and Y. Li, *J. Mol. Biol.*, 2008, **375**, 960.
24. X. Liu, X. Li, T. Zhou, Y. Wang, M.T. Ng, W. Xu, W. and T. Li, *Chem. Comm.* 2008, **3**, 380.
25. X. Li, M.T. Ng, Y. Wang, T. Zhou, S.T. Chua, W. Yuan and T. Li, *Bioorg. Med. Chem. Lett.*, 2008, **18**, 5576.
26. M. Wieland and J.S. Hartig, *Angew. Chem. Int. Ed. Engl.*, 2006, **45**, 5875.
27. T. Nagata, Y. Sakurai, Y. Hara, T. Mashima, T. Kodaki and M. Katahira, *FEBS J.*, 2012, **279**, 1456.
28. J.-D. Beaudoin and J.-P. Perreault, *RNA*, 2008, **14**, 1018.
29. S.A. McManus and Y. Li, *Molecules*, 2010, **15**, 6269.
30. R.R. Breaker, *Nature*, 2004, **432**, 838.

# CATALYTIC G-QUADRUPLEXES FOR THE DETECTION OF TELOMERASE ACTIVITY

Joanna Kosman, Bernard Juskowiak

Faculty of Chemistry, Adam Mickiewicz University, Grunwaldzka 6, 60-780 Poznan, Poland

## 1 INTRODUCTION

Telomeres – the ends of eukaryotic chromosomes – play a very important role in protection and prolongation of cell life. Those structures consist of DNA with a specific sequence and specialized proteins. The necessity of existence of telomeres is dictated by the imperfection of the DNA replication mechanism. During every replication event, the lagging strand is shortened due to incomplete replication. To prevent the loss of information stored in coding regions of chromosome, the ends contain non-coding tracks $(TTAGGG)_n$, which are involved in telomere formation. However, those repeats is also are shortened upon replication and when their length approaches a critical point, the cell undergoes apoptosis. For that reason, the telomeres are called the clock of cell live. The number of allowed division is specified for every kind of cell by Hayflick limit and for human being is around 55 cycles [1]. Still, there are some cells for which division is the main function (e.g., stem cells), so they need to bypass this limitation. Stem cells produce enzyme telomerase, which can add 5'-GGTTAG-3' repeats to the chromosome ends [2]. This enzyme is also present in 85% of cancer cells [3], thus the study concerning telomerase are really important in order to understand the mechanism of diseases, and to develop new diagnostic tools and potential therapeutics.

Telomerase is a ribonucleoprotein that consists of telomerase reverse transcriptase (TERT) and an integral telomerase RNA (TR) unit. The protein part is the location of catalytic site, which is responsible for DNA synthesis using the TR template [4]. The telomerase enzyme operates in two stages. First phase consists in the synthesis of a single 6-nucleotide repeat. The second stage involves regeneration of the template RNA through separation of the DNA/RNA hybrid, rearrangement of RNA in relation to DNA primer and recreation of DNA/RNA hybrid [5]. The described above mechanism of synthesis is processive. Telomerase can also stop after first addition of the single repeat and instead of template regeneration, it may dissociate from the DNA. The synthesis of oligonucleotide by telomerase is the same as for every polymerase, but TR regeneration and processivity is unique for telomerase [6]. It is also possible that several telomerase molecules can operate on one telomere elongation. This type of action is distributive and was artificially achieved by shortening of telomeres (during telomerase inhibition) or by telomerase overexpression [7].

The telomerase enzyme can be linked with many diseases. telomerase activity is one of the main causes that are responsible for cancer cells immortality [8]. It was also proven that mutations in the six telomerase components are associated with human telomere-mediated disorders: aplastic anemia (AA), idiophatic pulmonary fibrosis (IPF) [9] and dyskeratosis congenita (DC) [10]. The association of telomerase with many diseases caused the great impetus for developing sensitive diagnostic tools. The most common assay for telomerase activity was described by Kim et al in 1994 [11]. Telomeric repeat amplification protocol (TRAP) is based on the PCR technique. The telomerization process is conducted on TS primer (5'-TTA GGC AGC TCG TCT CAA-3'). The product is amplified by PCR technique with the TS primer and a second primer complementary to it. The results are visualized by polyacrylamide gel electrophoresis or other techniques. This method is however expensive, requires special equipment and is time consuming. Additionally, PCR amplification can be inhibited by cells extract components [12]. All of these drawbacks caused the search of novel telomerase assay, which for the one hand will avoid all of those defects and for the other one will be even more sensitive. The quest met on their road the new systems, peroxidase mimicking DNAzymes, which were featured with great potential in bioanalysis. This chapter will highlight the potential of DNAzymes in telomerase activity detection.

## 2 PEROXIDASE-MIMICKING DNAZYMES

The peroxidase- mimicking DNAzymes were discovered by Sen and co-workers in 1998 [13], who found out that hemin, the cofactor of every enzyme with peroxidase activity, can itself catalyze this reaction. The scientists searched for biomolecule, which could enhance this activity and thanks to SELEX technique (Systematic Evolution of Ligands by Exponential Enrichment), they found such specific DNA oligonucleotides. The guanine-rich DNA sequences can form four-stranded structures called G-quadruplexes. Those structures proved to be important from biological point of view [14, 15]. It turns also out that G-quadruplexes can have bioanalytical potential as well. The G-quadruplex selected by SELEX appeared to be an aptamer for molecule of hemin and this complex could perform catalytic reaction between hydrogen peroxide and organic substrate [15]. The DNAzymes possess many advantages over conventional protein enzymes (thermal stability, simple synthesis and purification, hybridization), which make them better replacement of protein peroxidase in most of analytical applications. Thanks to hybridization phenomenon it is possible to develop a simple assay for any genomic sequence and also other molecules using adequate aptamers. The most interesting assays involving peroxidase-mimicking DNAzymes were described in our review of bioanalytical potential of this system [16]. In bioassays based on peroxidase-mimicking DNAzymes, two indicator techniques: colorimetry and chemiluminescence are commonly used. For the reaction with spectrophotometric readout, the ABTS (2,2'-azino-bis(3-ethylbenzthiazoline-6-sulphonic acid) or TMB (3,3',5,5'-tetramethylbenzidine) are used. Luminol is most commonly used in the Assays that involved chemiluminescence measurements. the. The molecule of hemin binds to G-quadruplex quartets by end-stacking interaction (Fig. 1). Each G-quadruplex posses two external quartets, so one molecule can bind two molecules of hemin. This protoporphirin-iron complex itself possesses low peroxidase activity but after binding to the G4 DNA, the activity level rises significantly that enables development of bioassays [17]. There is evidenced relationship between the quadruplex structure and activity of DNAzyme. Shangguan et al reported that the highest activity exhibited unimolecular parallel G-quadruplexes like PS2.M (5'GTGGGTAGGGCGGGTTGG3')

[18]. This oligonucleotide is the most frequently employed in bioanalytical assays as an enzymatic label for detection of DNA targets and small molecules using DNAzyme approach.

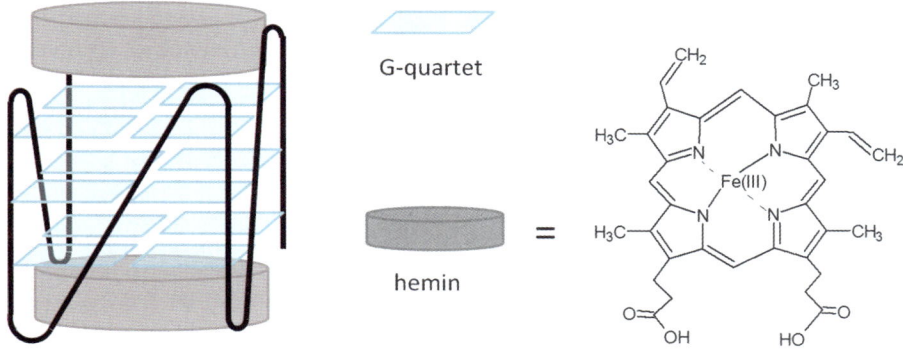

Figure 1. *Model of hemin binding by a parallel G-quadruplex.*

The G-quadruplex structure depends mainly on oligonucleotide sequence, but environmental conditions can also have an impact. This is an important issue in case of DNA sequences, which naturally form structures that exhibit low peroxidase activity like telomeric sequence. By tuning the environmental conditions, one can affect the structure of telomeric sequences and thus enhance peroxidase activity of DNAzyme based on this sequence. Such attempts we have undertaken with quite positive results [19]. On the other hand it was reported that DNAzymes from telomeric DNA are not really a good approach for telomerase assay [20]. Authors noticed the non-linear relationship between telomeric strand length and peroxidase activity. The neighboring external quartets in two consecutive G-quadruplexes form hydrophobic environment, which binds hemin stronger, and therefore, exhibit higher peroxidase activity [20]. However, the experiments with longer oligonucleotides that can form larger number of G-quadruplexes, indicated satisfactory relationship between telomere length and peroxidase activity. Probably for longer strands one can neglect contribution from the weak-activity external sites since the high-activity internal sites dominate. With the elongation of telomeric strand, the number of internal active sites increases and the total activity of the system increases proportionally. Shangguan and others reported that sites of hemin binding located between two G-quadruplexes could be less accessible to hydrogen peroxide, which is able to degrade hemin molecule and to inhibit DNAzyme [21]. This is another confirmation to the hypothesis that internal sites of hemin binding are connected with higher peroxidase activity of multiple G-quadruplexes.

3 Telomerase activity detection

Two approaches can be distinguished in developing assays for telomerase activity that are based on peroxidase DNAzymes. The DNAzyme can be formed on telomeric strands (direct approach) or can be added to the system as an external hybridization probe (indirect approach). As it was mentioned earlier, the telomeric strands self-assemble to form antiparallel or hybrid structures that exhibit rather low peroxidase activity. Therefore, telomerase assays are mainly based on indirect approach with probes bearing PS2.M sequence (the aptamer which exhibit the highest activity). The first report about telomerase detection by peroxidase DNAzymes has been published in 2004. Willner et al presented an assay based on catalytic nucleic acid beacon [22]. The main principle of their detection format was based on generation of the DNAzyme after beacon opening caused by hybridization with a target telomeric sequence. The hairpin structure consisted in two domains: one, which can form G-quadruplex and exhibit catalytic activity (I) and other one that is complementary to the telomeric sequence (II). The G-quadruplex, however, cannot be form in the absence off target because the G-rich sequence is involved in the stem duplex of hairpin structure. However the addition of telomerase that elongates the primer located at the end of domain II, causes hairpin rearrangement and hybridization between domain II and newly generated telomeric sequence and release of domain I. The hemin can now create complex with just created G-quadruplex by domain I and resulting DNAzyme catalyzes the reaction between $H_2O_2$ and ABTS. Authors reported that this assay allowed for the detection of telomerase in cell extracts down to 500 cancer cells.

The same group reported also the heterogeneous, chemiluminescence telomerase assay [23]. The scheme of this approach is depicted in Figure 2. The Au-coated glass plate was modified by telomerase primer (5'-TTT AAA TCC GTC GAG CAG AGT-3'). The telomerase enzyme in the presence of dNTPs elongated the primer strand by adding TTAGGG repeats. In the next step, an external probe was added. This probe consisted of the peroxidase DNAzyme part and a sequence complementary to the telomeric repeats. After hybridization, the substrates for peroxidase reaction were added (luminol, hydrogen peroxide) and the chemiluminescence signal was measured. This approach includes two amplification steps: first, the hybridization of several probes to the single telomeric strand and second, the peroxidase DNAzyme catalyzed luminol reaction that generates many photons. The detection limit for this amplification approach was established at the level of 1000 HeLa cells in the sample.

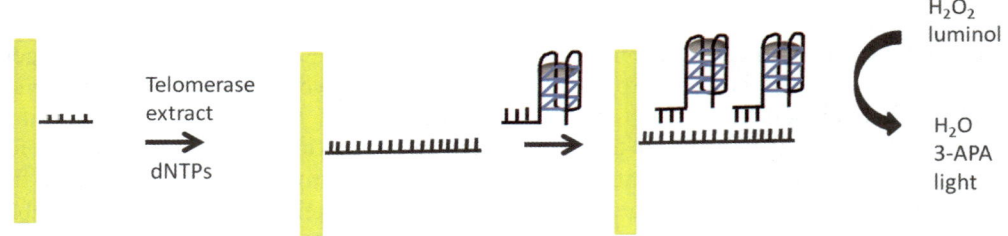

Figure 2. *A chemiluminescence assay of telomerase detection on a gold surface with the use of external DNAzyme probe.*

There have been also developed two other assays, which involved the use of gold nanoparticles: one, with the involvement of external probe and the other one based on the direct approach. The indirect format was proposed by Willner group [24] and the principle of detection was similar to the chemiluminescence assay described previously. The difference consisted in the type of the probe used in the method. Instead of predesigned oligonucleotide labels, authors used DNA-functionalized gold nanoparticles. Au-NPs were modified by oligonucleotide consisting of two domains: one, complementary to the telomeric unit sequence and second, the peroxidase DNAzyme domain. Each telomeric strand could bind gold nanoparticle modified with many DNAzyme labels. The detection limit for this approach corresponded to 1000 HeLa cells in the analyzed solution.

Li et al proposed the method illustrated in Figure 3. The first step of this approach included preparation of magnetic microbeads functionalized with gold nanoparticles (MB-AuNPs). In the next step AuNPs were modified by an oligonucleotide complementary to the TS primer (S1). Separately, the telomerase extract was used to elongate TS primer in the presence of dNTPs. Hybridization between TS domain of elongated telomeric strands and S1 capturing sequences led to immobilization of telomeric DNAs while their TTAGGG repeats remained free to self-assemble. After hemin and potassium ions addition, the peroxidase DNAzymes were formed and chemiluminescence signal could be measured from the reaction of luminol with $H_2O_2$. This method allowed for the detection of telomerase in extracts from 100 HeLa cells.

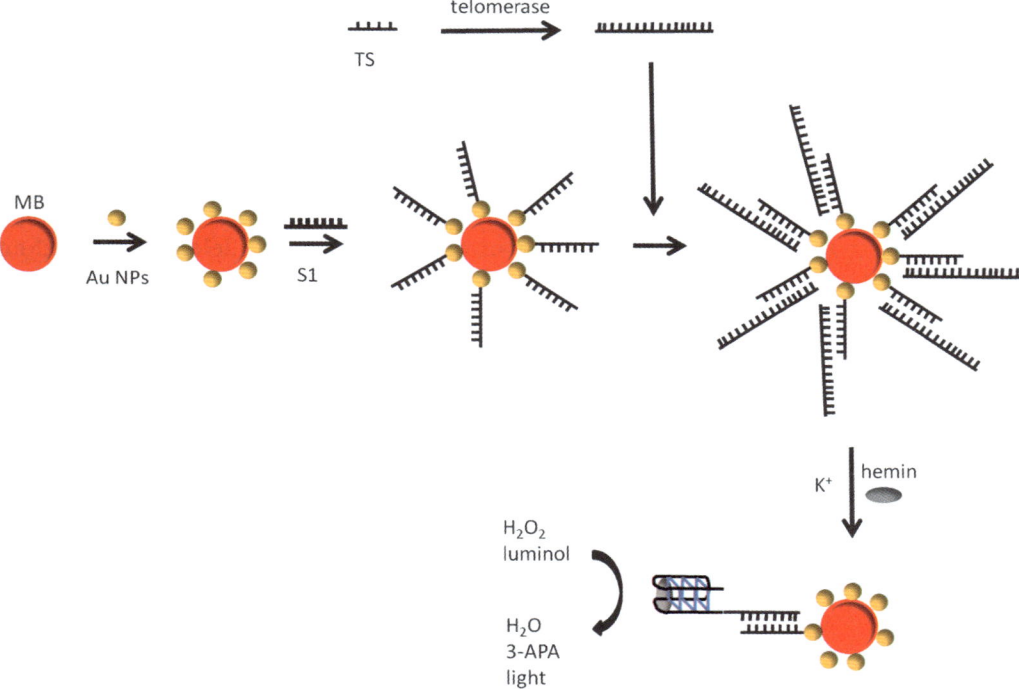

Figure 3. *Direct telomerase hybridization assay on magnetic microbeads functionalized with gold nanoparticles (MB-AuNPs).*

The simple direct approach of telomerase activity detection by peroxidase-type DNAzymes was reported by Willner group [26]. The scheme of this homogeneous detection method is depicted in Figure 4. The telomerase enzyme (cell extract) extends the TS primer by adding TTAGGG repeats. The elongated strand in the presence of hemin and potassium ions can form G-quadruplexes that exhibit peroxidase activity in the TMB oxidation reaction and colorimetric readout. This method in comparison with presented above is characterized by great simplicity and beside telomerase primer, it does not require the use of other DNA labels or nanoparticles. The reported detection limit was reported to approach 200 cells/µl.

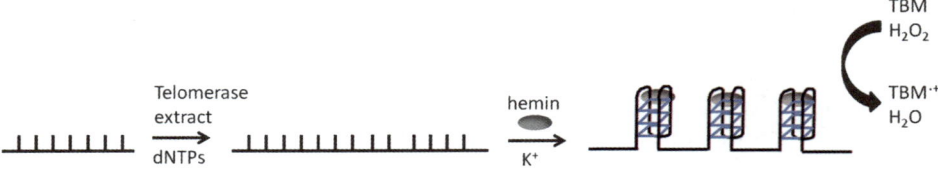

Figure 4. *Direct approach of telomerase detection by DNAzyme based on G-quadruplexes self-assembled on the telomeric strand.*

## 4 Conclusions

The telomerase enzyme is considered to be a cancer marker, therefore, any attempt to develop new diagnostic tools for detection of its activity, has great importance. Approaches reviewed here utilized peroxidase DNAzymes as a sensor platform. The greatest advantage of those systems is the amplification of the analytical signal without PCR technique, which is expensive and time-consuming. Presented examples of assays for telomerase based on DNAzyme system illustrate how great is the bioanalytical potential of these G-quadruplex structures. It should be mention that different approaches to exploit quadruplex structures for telomerase detection have been attempted. Takenaka et al. reported the use of ferrocenyl naphthalene diimide intercalator that gave electrochemical signal after binding telomeric tetraplex motives, which enabled telomerase activity detection [27]. Willner et al. detected telomerase activity exploiting the hemin binding to self-assembled G-quadruplexes using its quenching effect on the fluorescence of quantum dots or applying surface plasmon resonance (SPR) technique [28]. Despite the fact that so many methods in this area were already presented there still is the space for new approaches. The key factor is the sensitivity limit. Reduction of detection limit can be achieved in two ways. Firstly, through choosing the technique with higher intrinsic sensitivity, like for example, fluorescence. Secondly, through optimization of environmental conditions in order to transform the low activity antiparallel structures to the parallel quadruplexes with high activity..

## References

1. L. Hayflick, *Exp. Cell Res.*, 1996, **37**, 3, 614–636.
2. A.M. Olovnikov, *J. Theor. Biol.*, 1973, **41**,1 ,181–90
3. G.B. Morin, *Cell*, 1989, **59**, 521-529
4. D. Shippen-Lentz and L.H. Blackburn, *Science*, 1990, **247**, 546-552.
5. J.D. Podievsky and J.J.-L. Chen, *Mutat. Res.*, 2012, **730**, 3-11.
6. J.-L. Chen and C.W. Greider, *EMBO J.*, 2003, **22**, 304-314.
7. Y. Zhao, E. Abreu, J. Kim, G. Stadler, U. Eskiocak, M.P. Terns, R.M. Terns, J.W. Shay and W.E. Wright, *Mol. Cell*, 2011, **42**, 297-307.
8. C. Autexier and C.W. Greider, *Biochem.*, 1998, **21**, 387-391.
9. N.D. Nielson and A.A. Bertuch, *Mutat. Res.*, 2012, **730**, 43-51.
10. M. Armanios, *Mutat. Res.*, 2012, **730**, 52-58.

11. N.W. Kim, M.A. Piatyszek, K.R. Prowse, C.B. Harley, M.D. West, P.L. Ho, G.M. Coviello, W.E. Wright, S.L. Weinrich and J.W. Shay, *Science*, 1994, **266**, 2011-2015.
12. N.W. Kim and F. Wu, *Nucleic Acids Res.*, 1997, **25**, 2595-2597.
13. P. Travascio, Y. Li, D. Sen, *Chem. Biol.*, 1998, **5**, 505-517.
14. J.L. Huppert, *Biochimie*, 2008, **90**, 1140-1148.
15. W. Zhou, N.J. Brand and L. Ying. *J. Cardiovasc. Tran. Trans.*, 2011, **4**, 256-270.
16. J. Kosman and B. Juskowiak, *Anal. Chim. Acta*, 2011, **707**, 7-17.
17. Y. Li and D. Sen, *Biochemistry*, 1997, **36**, 5589-5599.
18. X. Cheng, X. Liu, T. Bing, Z. Cao and D. Shangguan, *Biochemistry*, 2009, **48**, 7817-7823.
19. J. Kosman and B. Juskowiak, *Cent. Eur. J. Chem.*, 2012, **10**, 368-372.
20. L. Stefan, F. Denat and D. Monchuad, *J. Am. Chem. Soc.*, 2011, **133**, 20405-20415.
21. X. Yang, C. Fang, H. Mei, T. Chang, Z. Cao and D. Shangguan, *Chem. Eur. J.*, 2011, **17**, 14475-14484.
22. Y. Xiao, V. Pavlov, T. Niazov, A. Dishon, M. Kotler and I. Willner, *J. Am. Chem. Soc.*, 2004, **126**, 7430-7431.
23. V. Pavlov, Y. Xiao, R. Gill, A. Dishon, M. Kotler and I. Willner, *Anal. Chem.*, 2004, **76**, 2152-2156.
24. T. Niazov, V. Pavlov, Y. Xiao, R. Gill and I. Willner, *Nano Lett.*, 2004, **4**, 1683-1687.
25. Y. Li, X. Li, X. Ji and X. Li, *Biosens. Bioelectron.*, 2011, **26**, 4095-4098.
26. R. Freeman, E. Sharon, C. Teller, A. Henning, Y. Tzfati and I. Willner, *ChemBioChem*, 2010, **11**, 2362-2367.
27. S. Sato, H. Kondo, T. Nojima and S. Takenaka, *Anal. Chem.*, 2005, **77**, 7304-7309.
28. E. Sharon, R. Freeman, M. Riskin, N. Gil, Y. Tzfati and I. Willner, *Anal. Chem.*, 2010, **82**, 8390-8397.

# G-QUADRUPLEX FORMING OLIGONUCLEOTIDES WITH TAILOR-MADE MODIFICATIONS AS EFFECTIVE APTAMERS FOR POTENTIAL THERAPEUTIC APPLICATIONS

Domenica Musumeci and Daniela Montesarchio

Dipartimento di Scienze Chimiche, Università di Napoli Federico II, via Cintia 4, Complesso Universitario di Monte Sant'Angelo, I-80126, Napoli, Italy

## 1 INTRODUCTION

Nucleic acid-based aptamers are synthetic oligonucleotides able to bind biomedically relevant proteic targets with affinity and specificity only comparable to antibodies.[1,2] Such a dramatic avidity and specificity of recognition is typically due to the peculiar three-dimensional architectures these oligomers can adopt, mainly involving unusual DNA or RNA conformations as bulges, hairpins or multi-stranded structures. With respect to antibodies, aptamers exhibit remarkable advantages in terms of size, non-immunogenicity and wide synthetic accessibility, which render them attractive tools for potential therapeutic applications.

The research in this field has received a considerable stimulus with the approval by the US FDA, in December 2004, of pegaptanib sodium (Macugen), a pegylated, modified 28-mer RNA acting as anti-vascular endothelial growth factor (anti-VEGF) for the treatment of neovascular age-related macular degeneration (AMD).[3,4] Since then, a wide number of aptamers with improved pharmacological properties have been developed, in some cases also undergoing advanced clinical trials.[5]

Among the combinatorially selected aptamers endowed with significant bioactivity, many are G-rich oligomers. These compounds, still sharing a common structural feature, *i.e.* the ability to fold into stable G-quadruplex conformations under physiological conditions, can recognize very different proteic targets.[6,7] This apparent contrast can be explained considering the extremely high polymorphism of G-quadruplex structures: indeed, G-quadruplex-based aptamers with even very similar molecular skeletons can result into a large variety of different conformations, thus accounting for the diversity of biological effects produced.

In the design of potential therapeutic agents, oligonucleotides generally require chemical modifications imparting higher enzymatic resistance, efficient cell uptake and a more favourable pharmacokinetic profile. Most aptamers are usually unstable in the bloodstream due to nuclease digestion, with half-lives measured in few minutes. Remarkably, in some cases G-quadruplex foldings can result into intricate and compact structures, subjected to almost negligible enzymatic degradation. As a consequence, within the cell G-quadruplex forming aptamers can show residence times sufficient to display their functions even in the form of unmodified phosphodiester oligomers.[8] On the other hand, to ensure full druggability, G-rich oligomers have to include suitable modifications in their skeletons to overcome the other typical drawbacks associated with the *in vivo* usage of oligonucleotides of both the ribo- and deoxyribo- series – *i.e.* poor cell uptake and unfavourable pharmacokinetic properties.[9] In this frame, the wide repertoire of chemical modifications well optimized in the context of antisense/antigene strategies, useful to convert natural oligonucleotides into biomedically viable tools, can be conveniently exploited also to enhance the conformational stability and pharmacological performance of G-quadruplex structures.[10,11]

In this contribution, several examples of G-quadruplex forming oligonucleotides designed for therapeutic applications are analyzed. Particular emphasis is given to tailor-made structural modifications enhancing the *in vivo* efficacy of G-rich aptamers endowed with antiviral, anticancer or anticoagulant activity.

## 2 ANTI-HIV ACTIVE G-QUADRUPLEX FORMING OLIGONUCLEOTIDES

In the last two decades, many synthetic G-rich oligonucleotides have been identified as promising anti-HIV drugs *in vitro*.[12,13] Particularly, several G-quadruplex forming aptamers interacting with HIV-1 proteins, such as reverse transcriptase, HIV protease, Tat protein and HIV-integrase, have been developed by either rational design or SELEX techniques. Inhibition of HIV-1 replication may be achieved through a variety of mechanisms, but essentially two strategies have been pursued: i) inhibition of virus binding and entry to the target cell, and ii) inhibition of virus integration.

The first G-quadruplex forming oligonucleotide identified as potent anti-HIV agent ($IC_{50}$ = 0.3 μM) was the phosphorothioate 8-mer d($^{5'}$TTGGGGTT$^{3'}$), forming a tetramolecular, parallel-stranded G-quadruplex, which was shown to bind to the V3 loop of the envelope protein gp120 and to inhibit virus adsorption and cell fusion.[14] The tetrameric G-quadruplex structure provides a rigid, compact complex, which strongly interacts with the cationic V3 loop due to its highly anionic character. Useful modification within this sequence, enhancing both thermal and enzymatic stability, proved to be the replacement of dG residues with 2'-deoxy, 2'-fluoro-D-arabinofuranosyl nucleic acid (2'F-ANA) monomers, able to stabilize G-tetrads requiring guanines in anti conformations.[15]

Interesting anti-HIV activity was also found in the sequence d($^{5'}$GGGTTTTGGG$^{3'}$), forming a bimolecular G-quadruplex, able to inhibit HIV-1-induced syncytium formation and virus production in peripheral blood mononuclear cells. The antiviral activity of this oligonucleotide increased when the phosphodiester linkages were replaced with phosphorothioate bonds. *In vivo* data showed that phosphorothioate d($^{5'}$GGGTTTTGGG$^{3'}$) was capable of blocking the interaction between gp120 and CD4, specifically inhibiting the entry of T-cell line-tropic HIV-1 into cells.[16]

HIV-1 integrase is the target of the second type of G-rich aptamers.[17,18] Within this family of aptamers, one of the most potent compounds discovered to date is the unmodified 16-mer designated as *93del*, of sequence $^{5'}$d($G_4TG_3AG_2AG_3T$)$^{3'}$, *in vitro* inhibiting both the processing and strand transfer functions of HIV-1 integrase at low nanomolar concentrations and exhibiting a very good selectivity index (>1000).[19] Detailed NMR investigations showed that, in the presence of $K^+$ ions, this 16-mer adopts an unusual dimeric interlocked parallel-stranded G-quadruplex architecture.[20] Using computational methods, including protein-DNA docking and molecular dynamics simulation in explicit solvent, the binding of *93del* to HIV-1 integrase was modelled, allowing insight into the interactions of this aptamer with key residues of the catalytic loops of the viral protein, thus stabilized in catalytically inactive conformations.[21] Recent studies have elucidated the *ex-vivo* mechanism of action of this oligonucleotide, proving that *93del*, identified as a strong *in vitro* inhibitor of HIV-integrase, is able to efficiently enter cells - with enhanced uptake if in the presence of HIV-1 (acting as a vector) - and to affect the very early steps of HIV-1 replication cycle.[22,23]

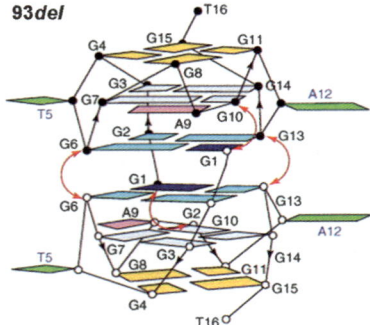

**Figure 1**  *Schematic representation of the 93del three-dimensional folding (adapted from ref. 12, with permission from Elsevier, copyright 2005).*

A remarkable HIV-1 integrase inhibition, with $IC_{50}$ values in the nanomolar range, has been discovered previously also in the sequences d($^{5'}GTG_2TG_3TG_3TG_3T^{3'}$), named as T30177,[17] and d($^{5'}G_3TG_3TG_3TG_3T^{3'}$), known as T30695.[18] To investigate the structure-activity relationships and to improve inhibition of HIV-1 integrase activity, a series of analogs of T30695 carrying positively charged residues or large hydrophobic groups have been synthesized.[24] In the mini-library of derivatives investigated, T residues in the loop domains were replaced with 5-amino dU or with 5-propynyl dU, and dG residues in the loop domains were substituted with T. From the analysis of the melting temperatures ($T_m$), inhibition of HIV-1 integrase activity ($IC_{50}$), and inhibition of HIV-1 replication in cell culture ($EC_{50}$), a relationship between thermal stability of the G-quadruplex structures and ability to inhibit HIV-1 proliferation in cell culture was proposed.

An in-depth NMR study - carried out on T30695 in comparison with a set of extended analogs - has recently demonstrated that this oligomer forms a dimeric structure stabilized by the stacking of two propeller-type parallel-stranded G-quadruplex subunits, in which all the guanine residues participate to the G-tetrad core formation.[25] In parallel, also T30177 has been shown to form a dimeric G-quadruplex structure, with six G-tetrad layers involving the stacking of two propeller-type parallel-stranded G-quadruplex subunits at their 5'-end; all twelve guanines in the sequence participate in G-tetrad formation, with an interruption in the first G-tract due to a thymine forming a bulge between two adjacent G-tetrads.[26]

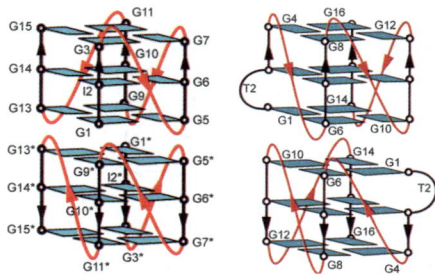

**Figure 2**  *Schematic representation of T30965 (left, from ref. 25) and T30177 (right, from ref. 26) three-dimensional folding in $K^+$ solution (reproduced with permission from Oxford University Press, copyright 2011).*

The generally poor cellular uptake of oligonucleotides in principle determines high extracellular concentration of G-quadruplex-based candidate drugs. Therefore, independently from the results obtained from *in vitro* studies, most G-quadruplex forming oligonucleotides are thought to exert their antiviral activity *in vivo* primarily through inhibition of HIV adsorption into host cells, and particularly through binding to viral gp120 protein. This effect, analyzed in *93del*, cannot be excluded also for the other known G-quadruplex-based HIV integrase inhibitors, and has renovated the interest towards G-rich aptamers *ad hoc* designed to target envelope protein gp120.

Following the studies carried out by Wyatts *et al.* on d($^{5'}$TTGGGGTT$^{3'}$),[14] promising anti-HIV activity was found in several modified G-rich oligonucleotides, investigated by Hotoda and coworkers. In the course of their studies, they demonstrated that various G-rich oligomers, bearing suitable substituents at their 5'-end, are non antisense anti-HIV active compounds. d($^{5'}$TGGGAG$^{3'}$) was selected as the *lead* sequence, found to be active against HIV-1 at submicromolar concentration only when conjugated at the 5' position with bulky aromatic moieties. This behaviour was first interpreted as a result of improved cellular uptake due to the presence of large lipophilic groups attached to the oligonucleotide chain. In successive articles, the authors considered the possibility that large aromatic groups at the 5'-end of the native sugar-phosphate backbone could produce stabilizing hydrophobic effects. Their final conclusion was that both the G-quadruplex forming oligonucleotide backbone and the cluster of 5'-end substituents contribute to the interaction of the 6-mers with viral gp120.[27] The most potent anti-HIV *in vitro* activity was observed in the analog identified as R-95288, bearing the 3,4-dibenzyloxybenzyl (DBB) and 2-hydroxyethylphosphate residues, respectively, at the 5' and 3' ends.[27,28,29] Modification of the guanines in this sequence through N2-methylation enhanced both thermal stability of the resulting G-quadruplex complexes and *in vitro* anti-HIV activity.[30] On the contrary, replacement of dG residues with 8-aza-3-deaza-2'-deoxyguanosine monomers led to a general decrease in antiviral activity.[31]

In order to better elucidate the structure-activity relationships in G-quadruplex forming oligonucleotides endowed with antiviral activity, 6-mer d($^{5'}$TGGGAG$^{3'}$) was chosen as a useful model system in studies carried out by Montesarchio *et al.*; thus, some representative examples of Hotoda's anti-HIV active sequences – carrying, respectively, the 4,4'-dimethoxytriphenylmethyl (DMT), *tert*-butyldiphenylsilyl (TBDPS) and 3,4-dibenzyloxybenzyl (DBB) groups at the 5' end - were synthesized and examined in a systematic physico-chemical characterization by DSC, CD and molecular modeling analyses, carried out in comparison with the unmodified oligonucleotide.[32] The obtained results showed that large aromatic groups at the 5'-end of d($^{5'}$TGGGAG$^{3'}$) play a decisive role in favoring the G-quadruplex formation processes. The unmodified sequence is indeed able to adopt a tetrameric, parallel stranded G-quadruplex structure, although it is not thermodynamically stable at physiological temperature and its formation is very slow. Conversely, the aromatic groups at the 5'-end of d($^{5'}$TGGGAG$^{3'}$) dramatically enhance both equilibrium and rate of formation of the quadruplex complexes, with *Tm* values higher than 70 °C. Interestingly, the overall stability of the investigated G-quadruplexes well correlate with the IC$_{50}$ data: these findings suggest a strict relationship between stability and rate of formation of the G-quadruplexes, on one side, and anti-HIV activity of 5'-modified aptamers, on the other, thus providing a quantitative evidence in support to the hypothesis that the G-quadruplexes are the ultimate active species, responsible for bioactivity.[32]

In an effort to develop novel, effective antivirals, a small library of d($^{5'}$TGGGAG$^{3'}$) derivatives, conjugated with mono- or disaccharides (glucose, mannose and sucrose, respectively) at the 3' or 5'-end, was then prepared, exploiting a fully automated, on-line phosphoramidite-based strategy. Data on the thermal stability of the resulting quadruplexes, determined by CD-melting analysis, and on the anti-HIV properties of the novel, conjugated oligonucleotides revealed significant bioactivity in those compounds generating the most stable G-quadruplex complexes.[33]

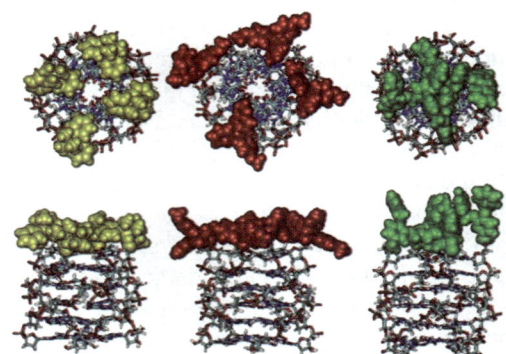

**Figure 3**  *Top and lateral views of the molecular models of the G-quadruplexes generated by 5'-TBDPS-d($^{5'}$TGGGAG$^{3'}$), 5'-DMT-d($^{5'}$TGGGAG$^{3'}$) and 5'-DBB-d($^{5'}$TGGGAG$^{3'}$), from left to right (from ref. 32, reproduced with permission from American Chemical Society, copyright 2007).*

The successive drug optimization process has been therefore based on the assumption that the kinetically and thermodynamically favored formation of the G-quadruplex complex is a pre-requisite for efficient anti-HIV activity. New d($^{5'}$TGGGAG$^{3'}$) derivatives, inserting a variety of different large aromatic groups at the 5'-phosphate end, have been synthesized, then subjected to complete antiviral evaluation, showing the analog carrying the (4-benzyloxy)phenylphosphate residue 6-fold more active than Hotoda's most active R-95288 (IC$_{50}$ = 0.061 vs. 0.37 µM, respectively).[34]

The concept of favoring the G-quadruplex formation process under physiological conditions through suitable chemical modifications of the oligonucleotide backbones has been explored also from a different perspective, i.e. the conversion of the tetrameric G-quadruplex complex generated from d($^{5'}$TGGGAG$^{3'}$) into a unimolecular, constrained folding. This was realized by chemically connecting the 3'-ends of four d($^{5'}$TGGGAG$^{3'}$) strands using bunchy oligonucleotides,[35] built on a solid support incorporating a suitable dendritic spacer. IC$_{50}$ values in the nanomolar range (0.082 µM) were observed for the compound designed as 1L, bearing TBDPS groups at the 5'-end of the bunchy G-rich sequences.[36]

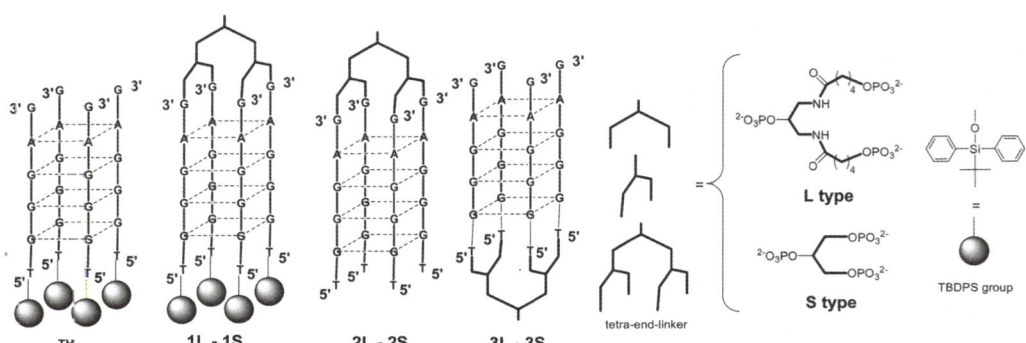

**Figure 4**  *Schematic representation of the bunchy oligonucleotides carrying the Hotoda's sequence investigated in ref. 36.*

Hydrophobic TBDPS groups were also inserted at the 5-position of the terminal thymine residues of the Hotoda's sequence through linkers of different length, resulting also in this case into a general increase of the anti-HIV activity compared to 5'-DMT-d($^{5'}$TGGGAG$^{3'}$).[37]

A remarkable increase of the *in vitro* anti-HIV activity of both the Hotoda's sequence and T30177 has been observed also upon backbone modifications - replacement of natural phosphodiester with locked nucleic acid (LNA) residues - and conjugation with (R)-1-O-(pyren-1-ylmethyl)glycerol (intercalating nucleic acid, INA) or (R)-1-O-[4-(1-pyrenylethynyl)phenylmethyl]glycerol (twisted intercalating nucleic acid, TINA). Incorporation of LNA or INA/TINA monomers produces up to a 8-fold improvement of the anti-HIV-1 activity of these G-rich oligomers, also in this case found to form more thermally stable G-quadruplex complexes.[38]

## 3 ANTI-CANCER G-QUADRUPLEX FORMING OLIGONUCLEOTIDES

The ability of certain oligonucleotide sequences to form G-quadruplex structures has been also exploited to develop novel anti-cancer agents which, binding to critical G-quadruplex binding proteins in tumor cells, could promote activation of apoptosis.[39] In the late '90s, Bates, Miller, Thomas and Trent discovered that various G-rich oligonucleotides (named with the acronym *GROs*) were able to strongly inhibit proliferation in a number of human tumor cell lines. Interestingly, the biological activity of these oligonucleotides correlated with binding to nucleolin, a multifunctional phosphoprotein highly overexpressed in cancer cells, both intracellularly and on the cell surface, whose levels are related to the rate of cell proliferation.[40] Subsequently, a set of derivatives of the most potent GRO investigated, *i.e.* GRO29A, a 3'-aminoalkyl modified 29-mer carrying the sequence d($^{5'}$TTTGGTGGTGGTGGTTGTGGTGGTGGTGG$^{3'}$), were evaluated in their anti-proliferative activity. Remarkable activity, similar to the parent GRO29A, was observed in the corresponding phosphorothioate (PS29A) and mixed DNA/2'-O-methyl RNA (MRdG29A) oligonucleotide analogs, able to significantly inhibit proliferation in a number of tumor cell lines, while the 2'-O-methyl RNA (MR29A) analog was not significantly effective, probably for the differences in the grooves of the resulting G-quadruplex, as evidenced by molecular modeling.[41] Successively, it was found that both the 3'-modification and the 5'-terminus "tail" d($^{5'}$TTT$^{3'}$) of GRO29A were not essential for activity, and the subsequent preclinical and clinical studies were carried out on a truncated 26-mer sequence, d($^{5'}$GGTGGTGGTGGTTGTGGTGGTGGTGG$^{3'}$), which became known as AGRO100 while being developed by Aptamera, and then as AS1411, after acquisition by Antisoma in 2005.[42] Most notably, AS1411 – now in Phase II - is the first oligonucleotide-based aptamer to reach advanced clinical trials for the treatment of cancer,[43] and currently some other GROs, forming stable G-quadruplex structures, are being evaluated as potential anticancer agents.

In this frame, Jing and coworkers explored G-rich oligonucleotides able to form intramolecular G-quadruplex (GQ) structures which resulted potent inhibitors of two molecular targets important in cancer therapy: 1) the signal transducer and activator of transcription (STAT) 3, a critical mediator of oncogenic signaling in many cancer forms, including the majority of prostate and breast cancers, and 2) the hypoxia-inducible factor-1α subunit (HIF-1α), a basic helix-loop-helix transcription factor involved in angiogenesis and required for cancer growth. In particular, the unmodified 16-mer d($^{5'}$GGGCGGGCGGGCGGGC$^{3'}$), designated as T40214, demonstrated ability to inhibit *in vitro* STAT3 DNA-binding activity,[44] especially when intracellularly delivered with an effective system such as polyethyleneimine.[45] Furthermore, T40214 dramatically suppressed the growth of xenografts of prostate and breast cancer cells in which Stat3 is constitutively activated *in vivo*, thus opening the way to a novel class of chemotherapeutic agents holding promise for the systemic treatment of many forms of metastatic cancer.[46] On the other hand, the lead compounds JG243 and JG244, carrying respectively the sequences d($^{5'}$GGCGGGCAGGCGGG$^{3'}$) and d($^{5'}$GGCGGGTAGGCGGG$^{3'}$), were found to selectively target HIF-1α, decrease the levels of both HIF-1α and HIF-2α ($IC_{50}$ < 2 µmol/l), inhibit the expression of HIF-1-regulated proteins and suppress the growth of prostate, breast, and pancreatic tumor xenografts.[47] These oligomers, designed on the basis of structural considerations of the target and a computational approach, were able to form intramolecular parallel G-quadruplex structures similar to T40214 (Figure 5).[47] Recently, it was reported that the combination of the anti-cancer agents T40214 (the STAT3 inhibitor) and JG244 (the HIF-1α inhibitor) showed increased therapeutic efficacy and reduced drug resistance in prostate cancer therapy.[48]

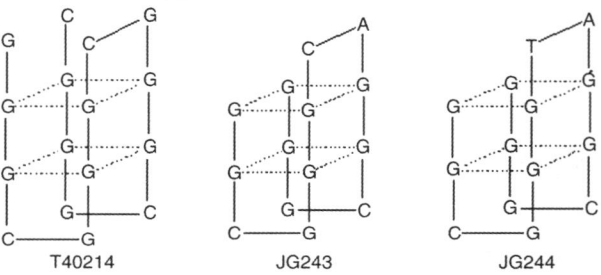

**Figure 5** *Schematic representation of the intramolecular G-quadruplex structures of T40214, JG243 and JG244.*

Other potential anti-cancer GROs, whose activity is correlated with their capacity to adopt a G-quadruplex conformation, were the telomeric G-tail oligonucleotides, TG18, TG24, and TG48 ODNs, which exhibited potent growth inhibitory activity against many tumor cells in culture,[49] and the homooligomer of 2'-deoxyguanosine G20, that induced apoptosis specifically in the malignant esophageal cell line, OE19, *in vitro*.[50]

## 4 ANTI-COAGULANT G-QUADRUPLEX FORMING OLIGONUCLEOTIDES

In 1992 Bock *et al.*[51] identified by SELEX single-stranded DNA sequences able to inhibit clot formation through high affinity binding with thrombin, one of the major target for anticoagulation and cardiovascular diseases therapy. These synthetic oligonucleotides, named thrombin binding aptamers (TBAs), contained a 14-17 consensus sequence with eight highly conserved guanine residues which form a central core composed of two guanine quartets. Among the TBAs, the 15-mer d($^{5'}$GGTTGGTGTGGTTGG$^{3'}$) ($TBA_{15}$, Figure 6) has been the object of numerous studies, since it has the strongest anticoagulant activity acting at nanomolar concentration on the two procoagulant functions of thrombin:[52] the activation of fibrinogen and the platelet aggregation interacting with the fibrinogen-binding exosite of the protein[53] (exosite I, Figure 6). According to detailed NMR and X-ray structural studies, $TBA_{15}$, in the presence of specific cations such as $K^+$ or of thrombin, forms an intramolecular, antiparallel G-quadruplex with a chair-like conformation consisting of two G-quartets connected *via* a TGT loop and two TT loops.[54]

In 1997 longer oligonucleotide aptamers (29-, 27- or 31-mers), denoted as $TBA_{29}$, $TBA_{27}$ and $TBA_{31}$, were found to bind thrombin with 20- to 50-fold higher affinity ($K_d$ =

0.5 nM) than TBA$_{15}$.[55] These aptamers recognise the heparin binding exosite (exosite II, Figure 6) and for this reason have only moderate effects on the activation of fibrinogen. TBA$_{29}$, TBA$_{27}$ and TBA$_{31}$ contain both a G-quadruplex and a duplex motif in their conformation, with a 15-nucleotide "core" sequence that has a close structural similarity to TBA$_{15}$.

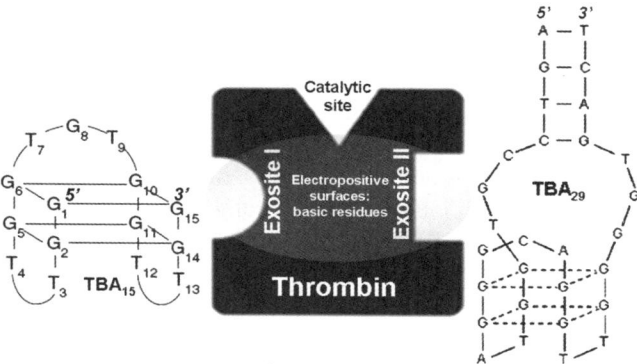

**Figure 6** *Interaction of TBA$_{15}$ and TBA$_{29}$ with the thrombin exosites I and II, respectively.*

TBA$_{15}$ was evaluated in Phase I clinical trial as an anticoagulant for potential use in acute cardiovascular settings, and showed a rapid dose-related anticoagulant response. However, the high amount of the aptamer needed to achieve the desired effect, resulting in a sub-optimal dosing profile, convinced the company Archemix Corp. and Nuvelo Inc., which conducted these experiments, to discontinue the subsequent development of TBA$_{15}$ itself.[56] Nevertheless, TBA$_{15}$, for its simple composition and stable G-quadruplex structure, has been extensively used as a model compound for structural and diagnostic investigations: a variety of novel chemical modifications of this aptamer are being proposed, giving in many cases promising TBA analogs on which, in our opinion, clinical trials could be re-opened in the next future.

A number of different approaches have been adopted to improve the enzymatic degradation resistance of TBA$_{15}$ as well as its anti-thrombin effect for potential *in vivo* applications. In various strategies TBA$_{15}$ maintained its natural backbone and was only modified in the sequence by the replacement of several bases in the loops and/or in the G-quartets, or by inserting 5'- or 3'-extensions. Many research groups developed TBA$_{15}$ variants by inserting non-natural modifications of the nucleobases, or of the nucleotide backbone. Studies on the structure, stability and biological properties of these chemically modified TBA$_{15}$ derivatives have been extensively surveyed in two recently appeared reviews.[6,57]

Most of the chemical modifications afforded an increase of the overall stability of the G-quadruplex structure and/or of the affinity for thrombin, as discussed in the contribution by A. Aviñó and R. Eritja (*Synthesis and properties of oligonucleotides forming G-quadruplexes*) in Chapter 2 of this book. However, only in very few cases anticoagulant derivatives more effective than the natural counterpart were discovered. In particular, an increased inhibitory activity towards thrombin was obtained with 3' terminal elongations realized with the attachment to the 3' end of TBA$_{15}$ in a case of a poly(dA) tail (dA$_{15}$),[58] which afforded a modified aptamer with also a survival time in human plasma about 3-fold higher compared to the original aptamer, and in another case of an abasic-site linker (ab$_4$),[59] which resulted in a TBA$_{15}$ variant 1.5-fold more potent than the unmodified TBA$_{15}$. The TBA$_{15}$-(ab)$_4$ is even more active than a second generation exosite I targeting aptamer, known as NU172, having the sequence d($^{5'}$AATACTCCGGGATGGAAGAGAGGACCGGT$^{3'}$), that is currently in Phase 2 clinical trials as an anticoagulant.[60]

As far as the nucleobase modifications are concerned, the introduction of a naphthylmethyl group into $N_2$ of $G_6$ increased both *in vitro* and *in vivo* the thrombin inhibitory activity by ca. 60%.[61] An increase in activity was also observed upon introduction of a benzyl group into $N_2$ of $G_6$ and $G_{11}$ and of relatively small groups, such as methyl and propynyl, into the $C_8$ positions of $G_1$, $G_5$, $G_{10}$, and $G_{14}$.[61] The replacement of four T residues in $TBA_{15}$ with 4-thio-dU resulted in significant improvements in anticoagulant activity in a position-dependent manner.[62] Compared to $TBA_{15}$, the analog with the sequence d($^{5'}$GG(dU$^{4\text{-thio}}$)TGG(dU$^{4\text{-thio}}$)G(dU$^{4\text{-thio}}$)GGT(dU$^{4\text{-thio}}$)GG$^{3'}$) showed a 2-fold increased inhibition in thrombin-catalyzed fibrin clot formation and platelet aggregation. In relation to the inhibition of thrombin-induced fibrin formation from purified fibrinogen and activation of washed platelets, the TBA thio-analog was 3-fold and 12-fold more effective than the original aptamer, respectively.[62]

As far as the backbone modifications are concerned, a $TBA_{15}$ containing one unlocked nucleic acid (UNA) monomer in position 7 was identified as more potent in blood clotting than the unmodified counterpart.[63] On the other hand, infusion into monkeys of an oligonucleotide containing four formacetal groups replacing the phosphodiester linkage showed an increased *in vivo* anticoagulant effect and an extended *in vivo* half-life compared to the unmodified aptamer.[64]

In many cases it emerged that for the thrombin aptamers, anti-thrombin activity is not dictated only by the thermal stability of the G-quadruplex structure or by the affinity for thrombin. Indeed, several examples of modified $TBA_{15}$ demonstrated that the most stable G-quadruplex structure was not the most active anticoagulant, as in the case of: i) the analog containing a 5'-5' inversion site, d($^{3'}$GGT$^{5'-5'}$TGGTGTGGTTGG$^{3'}$), which showed increased thermal stability and thrombin-affinity, while possessed reduced anticoagulant activity compared to unmodified TBA,[65,66] or ii) the LNA-containing TBAs for which the less stable aptamer with LNA at G2 showed higher activity with respect to the most stable aptamers with LNA in position G5, T7, or G8.[67]

Different and very efficient approaches, explored in order to improve both TBA activity and stability, reside in the realization of multivalent-TBAs which, exploiting polyvalent interactions, have showed a powerful activity toward thrombin (even 2 orders of magnitude higher than unmodified TBAs) as well as high stability to enzymatic degradation. These systems were realized through: i) the construction of bivalent or dendritic TBAs; ii) the integration of TBAs on nanoparticles, essentially gold and iron oxide ones, or iii) the assembly of TBAs on DNA nanostructures. For examples, $TBA_{15}$ (or a modified derivative with a pyrene pendant) and $TBA_{29}$ were connected through a poly-dA (dA$_{15}$) linker, affording a bivalent ligand that recognizes thrombin with high affinity, binding both regulatory exosites simultaneously (Figure 7).[68-70] The bivalent aptamer is more active than the single precursors of which it is composed, with more than 30-fold higher anticoagulant effect. In another approach, dendritic TBAs were assembled by linking multiple copies of $TBA_{15}$ (dendritic part) - through optimized polyethylene glycol linkers (16.8 nm) - to $TBA_{27}$, defined as the anchor domain since it allows the proximity of the dendritic part to thrombin thanks to its highest affinity for the protein.[71] It was demonstrated that the double units $TBA_{15}$-dendrimer assembly (DualTBA$_{15}$) showed improved anticoagulation properties with respect to the $TBA_{15}$ alone, as well as to the linear assembly $TBA_{15}$-PEG-$TBA_{27}$ and the Triple-$TBA_{15}$.

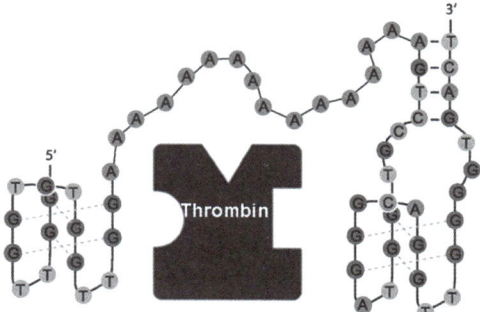

**Figure 7** *Schematic representation of the binding to thrombin of the bivalent anticoagulant aptamer formed by $TBA_{15}$ (on the left) and $TBA_{29}$ (right side), connected through a poly-dA linker.*

The integration of anticoagulant TBA aptamers into nanoparticles, such as gold[72-76] and iron oxide ones,[77,78] provided new hybrid nano-systems which exhibit higher biological performance due to polyvalent interactions with the target thrombin protein, extraordinary stability in plasma, as well as unique optical, electrochemical and magnetic properties. Various TBAs-conjugated nanoparticles showed enhanced anticoagulant potency as a result of multivalent interactions and steric blocking effects. Here we mention: 1) $TBA_{29}$-conjugated gold nanoparticles ($TBA_{29}$-Au NPs), resulting into an anticoagulant two orders of magnitude stronger than free $TBA_{29}$;[72] 2) $TBA_{15}$ and $TBA_{29}$ with proper flexible linkers attached to Au NPs either through Watson-Crick hybridization (self-assembling approach),[73] or covalently linked through disulfide bonds.[74-76] In particular, optimizing the linker ($dT_{15}$) and introducing stem pairs ($P_8$) between the aptamers and the nanoparticles a covalent conjugate ($TBA_{15}/TBA_{29}$–Au NPs) was obtained which possessed high binding affinity toward thrombin ($K_d = 8.86 \times 10^{-12}$ M, two orders of magnitude higher than free $TBA_{29}$), anticoagulant potency 75-fold stronger than $TBA_{15}$, and a high resistance to nuclease digestion in plasma.[75] A further optimized TBAs-Au NPs was realized through a molecularly imprinted (MP) approach, exploiting the presence of thrombin (Thr), thus acting as a template, during the conjugation reaction between $TBA_{15}$-$dT_{15}$-SH and $TBA_{29}$-$dT_{15}$-SH and 13 nm Au NPs (Figure 8).[74] Thrombin molecules were removed from Au NPs surfaces after the conjugation. The nanosystem with a density of ca. 42 $TBA_{15}$ and 42 $TBA_{29}$ molecules per Au NP showed enhanced binding affinity toward thrombin ($K_d = 5.2 \times 10^{-11}$ M) and anticoagulant potency 8 times higher compared to $TBA_{15}/TBA_{29}$-$T_{15}$-Au NPs prepared in the absence of thrombin.

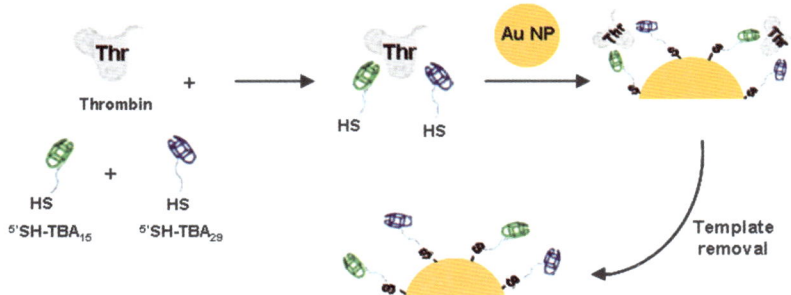

**Figure 8** *Schematic representation of molecularly imprinted approach for the realization of optimized $TBA_{15}/TBA_{29}$-conjugated gold nanoparticles.*

## 5 CONCLUSIONS

In the last decade a variety of nucleic acid-based aptamers able to fold into stable G-quadruplex structures have been developed and evaluated in their therapeutic potential. Most relevant results concern G-rich oligonucleotides designed to inhibit HIV-1, or to display anticancer and anticoagulant activity.

As far as anti-HIV G-rich oligomers are concerned, potent HIV-1 integrase inhibitors, with $IC_{50}$ values in the low nanomolar range, were found in 16-mers known as *93del*, T30177 and T30965, which in-depth NMR analyses proved to be folded in unusual dimeric G-quadruplex structures, providing compact, thermally and enzymatically stable three-dimensional arrangements. Another interesting example is the Hotoda's 6-mer d($^5$'TGGGAG$^{3'}$), forming tetramolecular, parallel-stranded G-quadruplex complexes, with activity in the submicromolar range if 5'-conjugated with large aromatic groups.

As a general remark, anti-HIV active G-rich oligomers are thought to primarily interact with viral gp120 in the form of G-quadruplex complexes. Several results, from our laboratories and from other research groups, corroborate this hypothesis, demonstrating a good correlation between thermal stability of the corresponding G-quadruplexes and antiviral activity, and showing the kinetic parameters as relevant as the thermodynamic ones in influencing the overall G-quadruplex formation process. In all the investigated cases, backbone (*e.g.*, replacement of natural phosphodiester with phosphorothioate, LNA or 2'F-ANA monomers), nucleobase modifications (*e.g.*, N2-methylation of the guanine residues) or conjugation (*e.g.*, with large aromatic groups or flexible linkers chemically connecting the extremities of the four G-rich strands) producing enhanced thermal stability of the G-quadruplex structures also resulted into significantly increased biological activity.

As far as anticancer and anticoagulant G-rich oligomers are concerned, interestingly, many of the most active sequences are unmodified, or minimally modified, oligonucleotides, thus suggesting that, in the case of monomolecular G-quadruplex structures, very compact conformations are realized, which display high thermal stability and marked resistance to nuclease digestion, as well as enhanced cellular internalization. Thus, extensive chemical modifications within the sugar-phosphate backbone may not be necessary for G-quadruplex-forming aptamers to be used for these *in vivo* applications. Notably, certain cell types (essentially cancer cells and some immune cells) preferentially internalize G-quadruplex-forming oligonucleotides, compared to unstructured sequences. However, the mechanism of their internalization remains unclear.

If all the research efforts focused in this field have not resulted into any approved drug to date, it has to be noted that, however, a significant number of G-quadruplex forming oligonucleotides are undergoing preclinical or even advanced clinical trials. In addition, it may be expected that the synergy between combinatorial methods - allowing the rapid discovery of selective aptamers – and the rational search for tailored modifications - to be inserted in the oligonucleotide backbone in a successive drug optimization process - will produce novel, promising candidate drugs in a very near future.

**Acknowledgments**

AIRC, Italian Association for Cancer Research (grant n. 11947) is gratefully acknowledged.

## References.

1. X. Ni, M. Castanares, A. Mukherjee and S.E. Lupold, *Curr. Med. Chem.*, 2011, **18**, 4206.
2. A.D. Keefe, S. Pai and A. Ellington, *Nat. Rev. Drug Discov.*, 2010, **9**, 537.
3. C. Campa and S.P. Harding, *Curr. Drug Targets*, 2011, **12**, 173.
4. E.W.M. Ng and A.P. Adamis, *Ann. NY Acad. Sci.*, 2006, **1082**, 151.
5. D.H.J. Bunka, O. Platonova and P.G. Stockley, *Curr. Opin. Pharmacol.*, 2010, **10**, 557.
6. W.O. Tucker, K.T. Shum and J.A. Tanner, *Curr. Pharm. Des.*, 2012, **18**, 2014.
7. B. Gatto, M. Palumbo and C. Sissi, *Curr. Med. Chem.*, 2009, **16**, 1248.
8. V. Đapić, V. Abdomerović, R. Marrington, J. Peberdy, A. Rodger, J.O. Trent and P.J. Bates, *Nucl. Acids Res.*, 2003, **31**, 2097.
9. R. E. Wang, H. Wu, Y. Niu and J. Cai, *Curr. Med. Chem.*, 2011, **18**, 4126.
10. B. Saccà, L. Lacroix and J.-L. Mergny, *Nucleic Acids Res.*, 2005, **33**, 1182.
11. For general methods in the chemical synthesis of G-rich oligonucleotides, see the chapter by A. Aviñó and R. Eritja, *Synthesis and properties of oligonucleotides forming G-quadruplexes*, in this book.
12. S.H. Chou, K.-H. Chin and A.H.-J. Wang, *Trends Biochem. Sci.*, 2005, **30**, 231.
13. J.J. Turner, M. Fabani, A.A. Arzumanov, G. Ivanova and M.J. Gait, *Biochim. Biophys. Acta*, 2006, **1758**, 290.
14. J.R. Wyatt, T.A. Vickers, J.L. Roberson, R.W. Buckheit, Jr., X. Klimkait, E. De Baets, P.W. Davis, B. Rayner, J.L. Imbach and D.J. Ecker, *Proc. Natl. Acad. Sci USA*, 1994, **91**, 1356.
15. C.G. Peng and M.J. Damha, *Nucleic Acids Res.*, 2007, **35**, 4977.
16. J. Suzuki, N. Miyano-Kurosaki, T. Kuwasaki, H. Takeuchi, G. Kawai and H. Takaku, *J. Virol.*, 2002, **76**, 3015.
17. A. Mazumder, N. Neamati, J.O. Ojwang, S. Sunder, R.F. Rando and Y. Pommier, *Biochemistry*, 1996, **35**, 13762.
18. N. Jing and M.E. Hogan, *J. Biol. Chem.*, 1998, **273**, 34992.
19. V.R. de Soultrait, P.Y. Lozach, R. Altmeyer, L. Tarrago-Litvak, S. Litvak and M.-L. Andreola, *J. Mol. Biol.*, 2002, **324**, 195.
20. A.T. Phan, V. Kuryavyi, J.-B. Ma, A. Faure, M.-L. Andreola and D.J. Patel, *Proc. Natl. Acad. Sci. USA*, 2005, **102**, 634.
21. M. Sgobba, O. Olubiyi, S. Ke and S. Haider, *J. Biomol. Struct. Dyn.*, 2012, **29**, 863.
22. M. Métifiot, A. Faure, V. Guyonnet-Duperat, P. Bellecave, S. Litvak, M. Ventura and M.-L. Andréola, *Oligonucleotides*, 2007, **17**, 151.
23. A. Faure-Perraud, M. Métifiot, S. Reigadas, P. Recordon-Pinson, V. Parissi, M. Ventura and M.-L. Andreola, *Antiv. Ther.*, 2011, **16**, 383.
24. N. Jing, E. De Clercq, R.F. Rando, L. Pallansch, C. Lackman-Smith, S. Lee and M.E. Hogan, *J. Biol. Chem.*, 2000, **275**, 3421.
25. N.Q. Do, K.W. Lim, M.H. Teo, B. Heddi and A.T. Phan, *Nucleic Acids Res.*, 2011, **39**, 9448.
26. V.T. Makundan, N.Q. Do and A.T. Phan, *Nucleic Acids Res.*, 2011, **39**, 8984.
27. H. Hotoda, M. Koizumi, R. Koga, M. Kaneko, K. Momota, T. Ohmine, H. Furukawa, T. Agatsuma, T. Nishigaki, J. Sone, S. Tsutsumi, T. Kosaka, K. Abe, S. Kimura and K. Shimada, *J. Med. Chem.*, 1998, **41**, 3655.

28  M. Koizumi, R. Koga, H. Hotoda, T. Ohmine, H. Furukawa, T. Agatsuma, T. Nishigaki, K. Abe, T. Kosaka, S. Tsutsumi, J. Sone, M. Kaneko, S. Kimura and K. Shimada, *Bioorg. Med. Chem.*, 1998, **6**, 2469.

29  M. Koizumi, M. Kaneko, T. Ohmine, H. Furukawa and T. Nishigaki, 1999, Pat. JP98/04625; PCT/WO99/19474.

30  M. Koizumi, K. Akahori, T. Ohmine, S. Tsutsumi, J. Sone, T. Kosaka, M. Kaneko, S. Kimura and K. Shimada, *Bioorg. Med. Chem. Lett.*, 2000, **10**, 2213.

31  S. Jaksa, B. Kralj, C. Pannecouque, J. Balzarini, E. De Clercq and J. Kobe, *Nucleosides Nucleotides Nucleic Acids*, 2004, **23**, 77.

32  J. D'Onofrio, L. Petraccone, L. Martino, G. Di Fabio, L. De Napoli, C. Giancola and D. Montesarchio, *Bioconjugate Chem.*, 2007, **18**, 1194.

33  J. D'Onofrio, L. Petraccone, L. Martino, G. Di Fabio, A. Iadonisi, J. Balzarini, C. Giancola and D. Montesarchio, *Bioconjugate Chem.*, 2008, **19**, 607.

34  G. Di Fabio, J. D'Onofrio, M. Chiapparelli, B. Hoorelbeke, D. Montesarchio, J. Balzarini and L. De Napoli, *Chem. Commun.*, 2011, **47**, 2363.

35  G. Oliviero, N. Borbone, A. Galeone, M. Varra, G. Piccialli and L. Mayol, *Tetrahedron Lett.*, 2004, **45**, 4869.

36  G. Oliviero, J. Amato, N. Borbone, S. D'Errico, A. Galeone, L. Mayol, S. Haider, O. Olubiyi, B. Hoorelbeke, J. Balzarini and G. Piccialli, *Chem. Commun.*, 2010, **46**, 8971.

37  W. Chen, L. Xu, L. Cai, B. Zheng, K. Wang, J. He and K. Liu, *Bioorg. Med. Chem. Lett.*, 2011, **21**, 5762.

38  E.B. Pedersen, J.T. Nielsen, C. Nielsen and V.V. Filichev, *Nucleic Acids Res.*, 2011, **39**, 2470.

39  C.R. Ireson and L.R. Kelland, *Mol. Cancer Ther.*, 2006, **5**, 2957.

40  P.J. Bates, J.B. Kahlon, S.D. Thomas, J.O. Trent and D.M. Miller, *J. Biol. Chem.*, 1999, **274**, 26369.

41  V. Dapić, P.J. Bates, J.O. Trent, A. Rodger, S.D. Thomas and D.M. Miller, *Biochemistry*, 2002, **41**, 3676.

42  P.J. Bates, D.A. Laber, D.M. Miller, S.D. Thomas and J.O. Trent, *Exp. Mol. Pathol.*, 2009, **86**, 151.

43  P.J. Bates, E.W. Choi and L.V. Nayak, *Methods Mol. Biol.*, 2009, **542**, 379.

44  N. Jing, Y. Li, X. Xu, W. Sha, P. Li, L. Feng and D.J. Tweardy, *DNA Cell Biol.*, 2003, **22**, 685.

45  P. Weerasinghe, Y. Li, Y. Guan, R. Zhang, D.J. Tweardy and N. Jing, *Prostate*, 2008, **68**, 1430.

46  N. Jing, Y. Li, W. Xiong, W. Sha, L. Jing and D.J. Tweardy, *Cancer Res.*, 2004, **64**, 6603.

47  Y. Guan, K.R. Reddy, Q. Zhu, Y. Li, K. Lee, P. Weerasinghe, J. Prchal, G.L. Semenza and N. Jing, *Mol. Ther.*, 2010, **18**, 188.

48  K.R. Reddy, Y. Guan, G. Qin, Z. Zhou and N. Jing, *Prostate*, 2011, **71**, 1796.

49  H. Qi, C.P. Lin, X. Fu, L.M. Wood, A.A. Liu, Y.C. Tsai, Y. Chen, C.M. Barbieri, D.S. Pilch and L.F. Liu, *Cancer Res.*, 2006, **66**, 11808.

50  T.R. Schwartz, C.A. Vasta, T.L. Bauer, H. Parekh-Olmedo and E.B. Kmiec, *Oligonucleotides*, 2008, **18**, 51.

51  L.C. Bock, L.C. Griffin, J.A. Latham, E.H. Vermaas, and J.J. Toole, *Nature*, 1992, **355**, 564.

52  L.C. Griffin, G.F. Tidmarsh, L.C. Bock, J.J. Toole and L.L. Leung, *Blood*, 1993, **81**, 3271.

53  M. Tsiang, A.K. Jain, K.E. Dunn, M.E. Rojas, L.L. Leung and C.S. Gibbs, *J. Biol. Chem.*, 1995, **270**, 16854.
54  J.A. Kelly, J. Feigon and T.O. Yeates, *J. Mol. Biol.*, 1996, **256**, 417.
55  D.M. Tasset, M.F. Kubik and W. Steiner, *J. Mol. Biol.*, 1997, **272**, 688.
56  S.M. Nimjee, C.P. Rusconi and B.A. Sullenger, *Annu. Rev. Med.*, 2005, **56**, 555.
57  A. Aviñó, C. Fàbrega, M. Tintoré and R. Eritja, *Curr. Pharm. Des.*, 2012, **18**, 2036.
58  S. Uehara, N. Shimada, Y. Takeda, Y. Koyama, Y. Takei, A. Ando, S. Satoh, A. Uno and K. Sakurai, *Bull. Chem. Soc. Jpn.*, 2008, **81**, 1485.
59  M.C.R. Buff, F. Schäfer, B. Wulffen, J. Muller, B. Potzsch, A. Heckel and G. Mayer, *Nucleic Acids Res.*, 2010, **38**, 2111.
60  J. Wagner-Whyte, S.F. Khuri, J.R. Preiss, J.C. Kurz, K. Olson, P. Hatala, R.M Boomer, J.M. Fraone, N. Brosnan, A. Makim, S.D. Lewis, L. Cai, T. McCauley, R. Hutabarat, C. Horvath, W.D. Funk, S.R. Deitcher, H. Tratte, B. Hussaini, P. Treanor, J.B. Rottman and J.L. Diener, *J. Thromb. Haemost.*, 2007, **5**, P-S-067.
61  G.X. He, S.H. Krawczyk, S. Swaminathan, R.G. Shea, J.P. Dougherty, T. Terhorst, V.S. Law, L.C. Griffin, S. Coutre and N. Bischofberger, *J. Med. Chem.*, 1998, **41**, 2234.
62  S. Mendelboum Raviv, A. Horvath, J. Aradi, Z. Bagoly, F. Fazakas, Z. Batta, L. Muszbek and J. Harsfalvi, *J. Thromb. Homeost.*, 2008, **6**, 1764.
63  A. Pasternak, F.J. Hernandez, L.M. Rasmussen, B. Vester and J. Wengel, *Nucleic Acids Res.*, 2011, **39**, 1155.
64  G. X. He, J.P. Williams, M.J. Postich, S. Swaminathan, R.G. Shea, T. Terhorst, V.S. Law, C.T. Mao, C. Sueoka, S. Coutre and N. Bischofberger, *J. Med. Chem.*, 1998, **41**, 4224.
65  L. Martino, A. Virno, A. Randazzo, A. Virgilio, V. Esposito, C. Giancola, M. Bucci, G. Cirino and L. Mayol, *Nucleic Acids Res.*, 2006, **34**, 6653.
66  B. Pagano, L. Martino, A. Randazzo and C. Giancola, *Biophys. J.*, 2008, **94**, 562.
67  L. Bonifacio, F.C. Church and M.B. Jarstfer, *Int. J. Mol. Sci.*, 2008, **9**, 422.
68  J. Müller, D. Freitag, G. Mayer and B. Pötzsch, *J. Thromb. Haemostasis*, 2008, **6**, 2105.
69  J. Müller, B. Wulffen, B. Pötzsch and G. Mayer, *ChemBioChem.*, 2007, **8**, 2223.
70  F. Rohrbach, M.I. Fatthalla, T. Kupper, B. Pötzsch, J. Müller, M. Petersen, E.B. Pedersen and G. Mayer, *ChemBioChem.*, 2012, **13**, 631.
71  Y. Kim, D.M. Dennis, T. Morey, L. Yang and W. Tan, *Chem. Asian J.*, 2010, **5**, 56.
72  Y.-C. Shiang, C.-C. Huang, T.-H. Wang, C.-W. Chien and H.-T. Chang, *Adv. Funct. Mater.*, 2010, **20**, 3175.
73  Y.-C. Shiang, C.-L. Hsu, C.-C. Huang and H.T. Chang, *Angew. Chem. Int. Ed. Engl.*, 2011, **50**, 7660.
74  C.L. Hsu, H.T. Chang, C.T. Chen, S.C. Wei, Y.C. Shiang and C.C. Huang, *Chem. Eur. J.*, 2011, **17**, 10994.
75  C.L. Hsu, S.C. Wei, J.W. Jian, H.T. Chang, W.H. Chen and C.C. Huang, *RSC Adv.*, 2012, **2**, 1577.
76  Y.J. Liao, Y.C. Shiang, C.C. Huang and H.T. Chang, *Langmuir,* 2012, **28**, 8944.
77  D. Musumeci, G. Oliviero, G.N. Roviello, E.M. Bucci and G. Piccialli, *Bioconjug. Chem.*, 2012, **23**, 382.
78  K. Taira, K. Abe, T. Ishibasi, K. Sato and K. Ikebukuro, *J. Nucleic Acids*, 2011, 103872.

# DEP-BASED INTEGRATION OF G-QUADRUPLEX STRUCTURES

Christian Leiterer[1], Andreas Kopielski[1], Irit Lubitz[2], Alexander Kotlyar[2], Antti-Pekka Eskelinen[3], Päivi Törmä[3] and Wolfgang Fritzsche[1]

[1]Department of Nanobiophotonics, Institute of Photonic Technology, 07745 Jena, Germany
[2]Department of Biochemistry, Tel Aviv University, Ramat Aviv 69978, Israel
[3]Department of Applied Physics, Aalto University, P.O. Box 15100, 00076 AALTO, Finland

## 1 INTRODUCTION

DNA represents a very potent tool in nanotechnology and nanoengineering. Beginning with the research of Seeman et al.[1] DNA was used to assemble nanoscale 2D and 3D structures. DNA origami introduced by Rothemund demonstrates the power of DNA self-assembly to construct impressive nanostructures in 2D[2] and 3D[3] or for using it as a breadboard to form nanostructures from different materials like gold nanoparticle[4,5] or carbon nanotubes[6,7]. These entire DNA self-assembly techniques are based on simple Watson-Crick base pairing forming DNA double strands. In addition 3 and 4 strands can associate by means Hoogsteen base pairing resulting in the formation of triple and quadruple strands structures.

This article mainly deals with quadruplex structures and their nanotechnological applications. We will report synthesis of these synthesize these structures from short G-rich oligonucleotides (GGGGTTGGGG) of various lengths and characterize the assembled structures with AFM. Furthermore we will discuss the possibility to integrate these short oligonucelotide based structures and longer monomolecular G-wires structures into electrode gaps using dielectrophoresis (DEP) an electrical field based integration approach.

## 2 METHOD AND RESULTS

### 2.1 Synthesis of G-quadruplex structures from oligonucleotides

G-Quadruplex structures can be synthesized through a hybridization reaction of G-containing oligonucleotides (GGGGTTGGGG). In this reaction the oligonucleotides form a superstructure by self-hybridization which can result in micrometer long Quadruplex structures also termed G-wires[8]. A schematic of the wire formation is shown in **figure 1.**

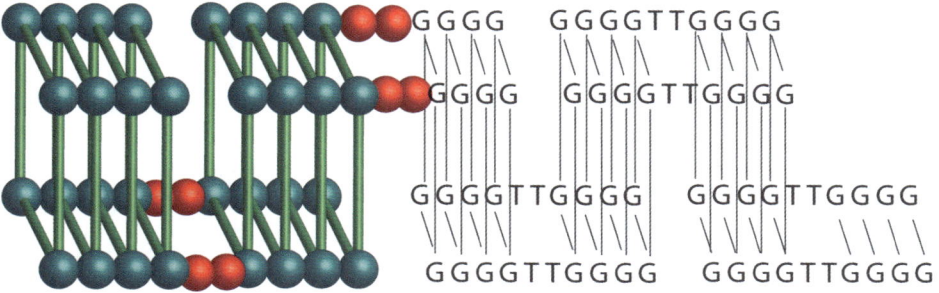

**Figure 1** *Schematic illustration: short oligonucleotides assembling into long G-quadruplex superstructures.*

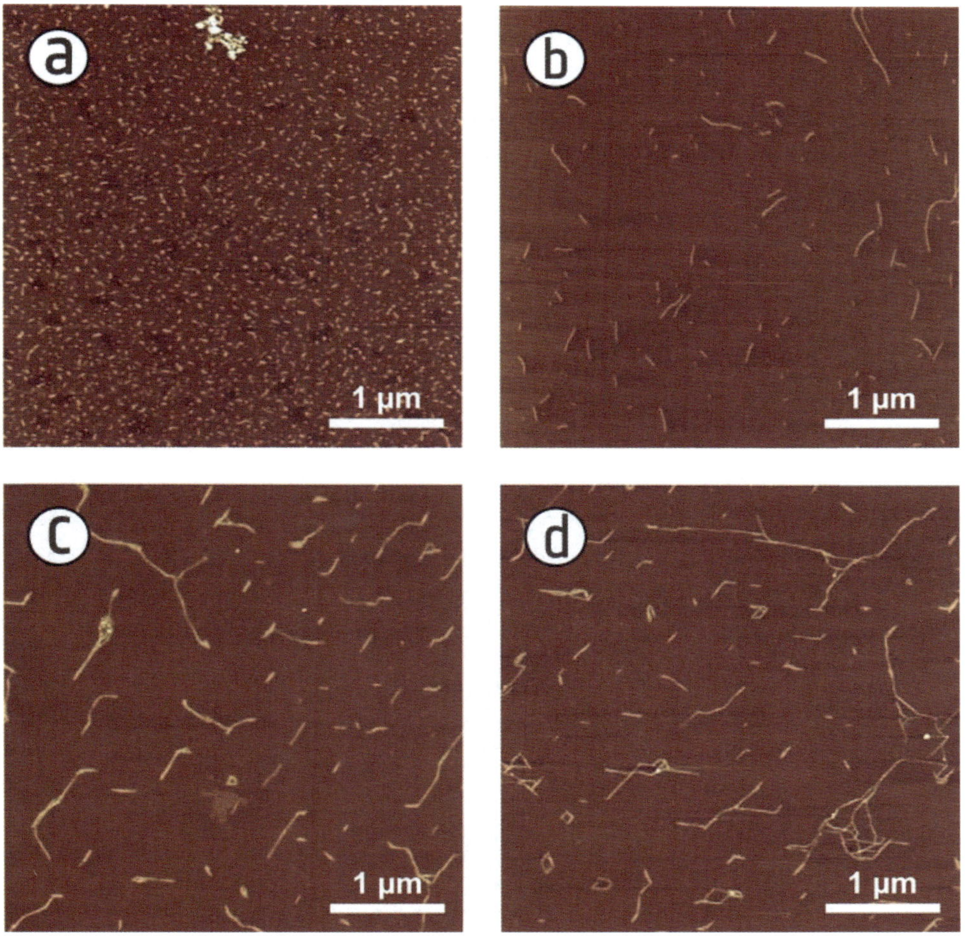

**Figure 2** *AFM images of oligonucleotide-based G-quadruplex wires on mica. The structures were assembled in 50 ng/μl solution of GGGGTTGGGGG in 20 mM Tris-HCl (pH 7 – 7.5), 4 mM $MgCl_2$ (a) in presence of 50 mM NaCl (b) 2 M (c) 3 M NaCl (d) Notice the apparent orientation of the structures in (c) and (d) along the crystal axis of mica*[9].

In the following experiments the goal was to synthesize G-Quadruplex structures for the integration into electrode gaps via dielectrophoresis. Therefore the assembled structures should have the length of several micrometers, since the gap size is 2 µm. In order to achieve such structures, the synthesis of the G-quadruplex wires was conducted as follows: 50 ng/µl oligonucleotide (GGGGTTGGGG) were dissolved in 20 mM Tris-HCl (pH 7 – 7.5), 4 mM $MgCl_2$ and NaCl varying from 50 mM to 3 M). All samples were heated in a thermocycler for 110 min (10 min 95°C // 20 min 90°C // 20 min 85°C // 30 min 80°C // 30 min // 75°C) to promote the hybridization of the oligonucleotides. By varying NaCl concentration one can synthesize oligonucleotide based G-Quadruplex structures of different length (up to 2 µm). The molecular morphology of the wires was characterized on a mica surface by AFM (Nanoscope III, Dimension®-3100, DI, Santa Barbara, CA) (**figure 2**).

The G-quadruplex structures synthesized as described above can not be used for the DEP (cf. section 2.2) due to the high ion concentration in the sample. Therefore, G-quadruplex structures that are stable in low salt concentrations had to be prepared.

These G-quadruplexes were prepared from GGGGTTGGGG (6 ng/µl) in 2.5 mM NaCl, 12 mM Tris-HCl (7 – 7.5) and 1.5 mM $MgCl_2$. These G-quadruplex wires were shorter tan those formed at high NaCl concentrations and have a length of up to 1 µm on mica and ~200 nm on oxidized silicon surfaces respectively (see **figure 3**). These structures were used for the DEP integration (see 2.2).

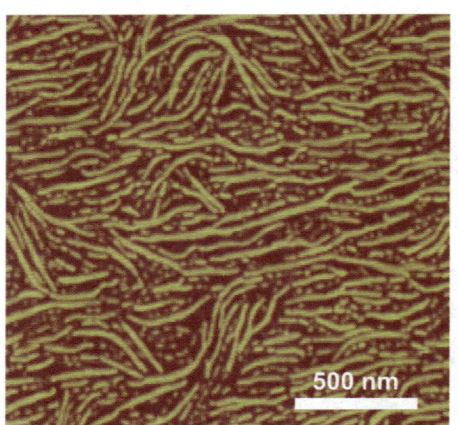 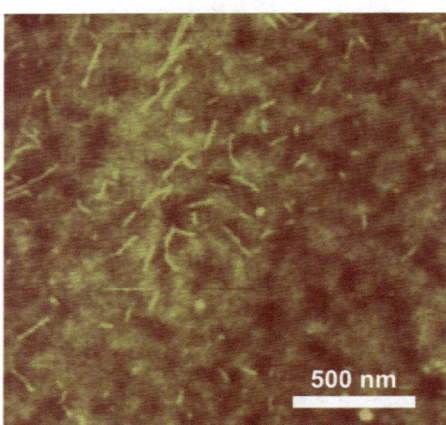

**Figure 3**  *AFM images of oligonucleotide-based G-quadruplex wires on mica (a) and oxidized silicon (b) surfaces. The structures were assembled in 6 ng/µl solution of GGGGTTGGGG in: 2.5 mM NaCl, 12 mM Tris-HCl (pH 7-7.5) and 1.5 mM $MgCl_2$.*

## 2.2 Integration of G-wire using dielectrophoresis

In order to integrate G-quadruplex structures into an electrical environment a method is needed, which does not require any post sputtering or metal evaporation techniques such as photo- or electron-beam-lithography which will most likely destroy the DNA during the integration procedure. Therefore here we employed a technique which does not affect biological materials in such a harsh way. Moreover, it allows for a rather parallel approach of integration. This technique is based on an AC electrical field applied to polarizable objects. The objects experience a force, which either pull them to or from the electrodes, as soon as voltage is applied. This technique is well known for its ability to realize marker-free cell sorting[10], but it has been shown that it can be also used to integrate nanostructures like semiconducting nanowire[11], carbon nanotubes[12], metal nanoparticle[13] or DNA[14,15].

### 2.2.1 Electrode layout and experimental configuration.

For the DEP integration process of the G-quadruplex wires both an electron beam (e-beam) lithographic and a photolithographic chip layout were realized using standard sputtering and lift-off technique **(figure 4)**. The photolithographic chip consists of two individual electrodes (2 µm gap size), and the other of 50 parallel contacted electrode pairs (~50 nm gap size).

**Figure 4**  *SEM image of the electrode chips utilized for DEP. The photolithographic structured electrodes (right) consist of two individual tapered electrodes with 2 µm distance made from gold on a titanium adhesive layer. These electrodes are used for the DEP of the oligonucleotide based G-quadruplex structures. The e-beam patterned electrodes (left) consist of 50 electrode pairs which can be all addressed in parallel. The electrodes are made from gold as well and have an average gap size is about 50 nm.*

For the DEP procedure the electrodes were connected to a function generator (33220A, Agilent, Santa Clara, CA) supporting high frequent (10 kHz – 1 MHz) voltage for the DEP process. As soon as aqueous G-quadruplex solution was applied to the chips the DEP force starts to pull molecules in the electrode gap. An average DEP process in our experimental configuration took about 10min. Subsequently the chips were rinsed with dH$_2$O and dried using compressed air for optical and AFM imaging. The setup is schematically illustrated in **figure 5**.

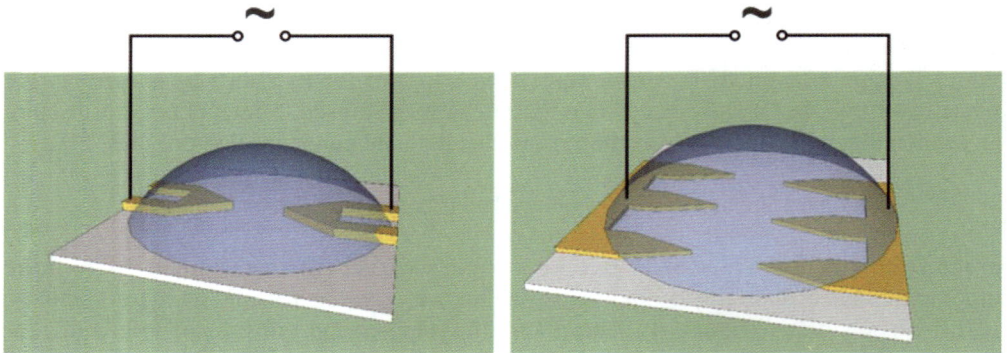

**Figure 5**  *Schematic illustrations of the used electrode layouts and the experimental configuration. The DEP process took place in a droplet that contained the aqueous G-Quadruplex solution covering the whole chip. The electrodes were connected to an AC power supply providing up to 2.5 V at 10 – 1000 kHz for the setup.*

### 2.2.2 DEP integration of oligonucleotide G-Quadruplex structures.

By varying voltage, frequency and duration of the process, working parameters were determined. It has been shown that a voltage of 2.5 V at a frequency range from 100 kHz to 1 MHz is working well to trap G-quadruplex structures and immobilize them on the electrodes. The process duration of 10 min seemed to be appropriate for these parameters. In the following **figure 6** AFM measurement from the trapped G-quadruplexes at three different frequencies (10 kHz, 100 kHz, 1 MHz) are shown and compared to DEP trapped dsDNA (48.502 bp).

As it can be seen in **figure 6a**, dsDNA can be stretched across the electrode gap. Several single DNA strands have been stretched. The oligonucleotide-based G-quadruplex structures can also be trapped and aligned at the tips of the electrodes under similar conditions like dsDNA (2.5 V, 100 kHz – 1 MHz). Single G-quadruplex structures (marked by triangle arrow) as well as agglomerated ones are immobilized in the electrode gap (**figure 6, b – d**). The G-quadruplex structures also seem to orientate along the electrical field when using frequencies from 100 to 1000 kHz (**figure 6, c, d**). At frequencies lower than 10 kHz no orientation along the field lines was observed (**figure 6, d**). Similar results have been obtained for dsDNA (not shown here). Additional problem is the strong electrolysis occurring in salt-containing solutions at frequencies below 100 kHz. This leads to electrode damage and higher ionic based movements. Also an increased pH-value near the electrodes surface caused by electrolysis may affect the biomolecules.

### 2.2.3 Integration of long mono-molecular G-quadruplex structures.

In the second attempt smaller, e-beam structured electrode gaps were fabricated for the DEP (**figure 7 a**). Also longer and more stable assembled G-quadruplex structures were used. These structures are made from one ssDNA molecule containing only guanines[16].

The previous used DEP parameters were (1 MHz, 1 V, 10min) used, but the voltage was lowered due to the reduced gap size.

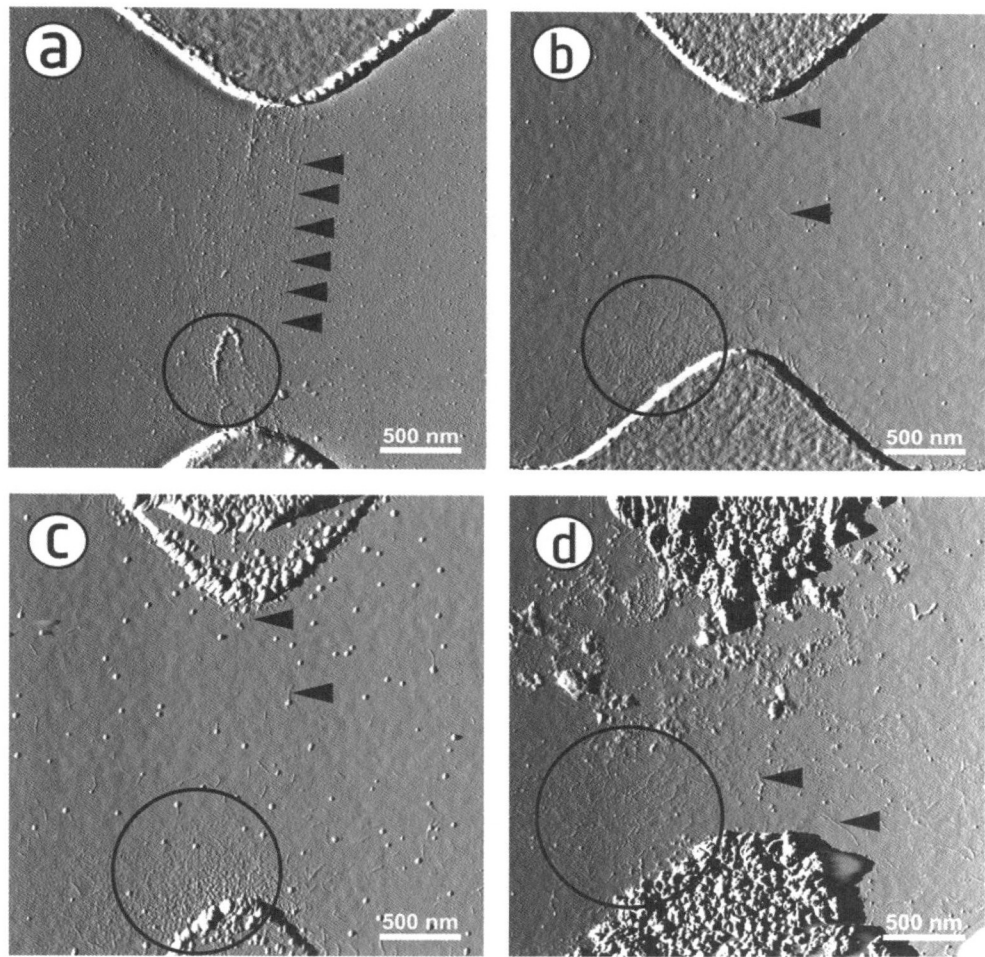

**Figure 6** *DEP trapping of G-quadruplex structures assembled from short G-rich oligonucleotides (GGGGTTGGGG) in comparison with trapped dsDNA. Triangles refer to single structures, circle marks area the agglomerated structures. (a) dsDNA trapped and stretched across the gap at 1.5 V, 1 MHz. (b) G-quadruplex structures trapped in between the electrodes at 2.5 V, 1 MHz, no electrode damage is noticed. (c) G-quadruplex structures trapped in between the electrodes at 2.5 V 100 kHz, electrolysis starts to damage the electrodes. (d) G-quadruplex structures trapped at 2.5 V 10 kHz, strong electrode damage due to electrolysis occurs, G-quadruplex structures are hardly visible.*

The G-quadruplex structures were fluorescently stained using YOYO-1 iodide (Invitrogen, Karlsruhe, Germany) to verify the successful integration and fixation of the G-quadruplex structures in the electrode gap by DEP. The following image shows the bare electrode structures imaged by dark field microscopy are shown in **figure 7a.** The fluorescence microscopy image (see **figure 7b**) of the gap shows the DEP fixated G-quadruplex structures. The fluorescence signal only appears inside and near the electrode gaps where the DEP force was present.

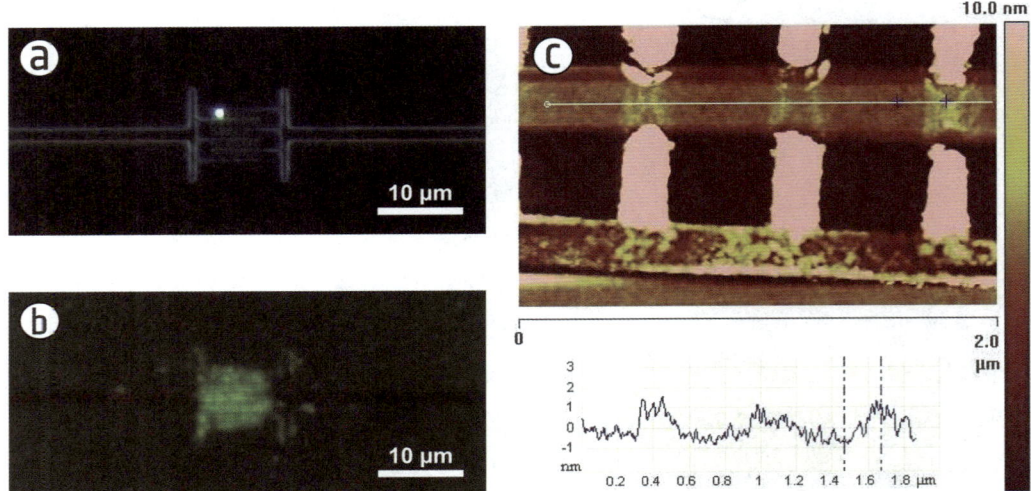

**Figure 7** *DEP trapped G-quadruplex structures in nano electrode gaps (~50 nm). (a) dark field microscopic image of the 50 parallel addressed electrode pairs before the DEP process (empty) (b) fluorescence microscopic image of the YOYO-1 stained G-quadruplex structures after the DEP process. Fluorescence signal appears near the electrode gaps, showing the successful trapping of the G-quadruplex structures. (c) AFM measurement across 3 of the 50 electrode gaps showing the trapped G-quadruplex structures with an average height of about 1.8 nm. There integrity of the structure seems to be affected either by intercalating fluorescence dye and/or the strong electrical field.*

Taking a closer look on the gaps by AFM-measurements (**figure 7c**), structures with an average height of about 2 nm which corresponds to that of G-quadruplexes can be detected in the electrode gap. However, the G-quadruplex could not be visualized as uniform structures. This may be due to damaging of the structures induced by the strong electrical as well as the intercalation of the fluorescence dye that is known to change the DNA structure.

3 CONCLUSION

We demonstrated integration of the G-quadruplex structures in an electrode setup using DEP. Higher DEP frequencies (~1 MHz) resulted in more defined integration and reduces adverse effects caused by electrolysis or strong ionic movement. This technique allows simple preparation of samples for electrical measurement on single biological molecules, including G-Quadruplex assemblies. We demonstrated that the length of the used structures, the electrode geometry and the nature and concentration of the buffer, strongly affect the integration process. Optimizing conditions for integration of biomolecules will enable further investigations of charge migration in dsDNA and G-Quadruplexes, which has been widely discussed by experimentalists and theoreticians over many years.

*Acknowledgements:*
Financial support by BMBF project NAWION (FKZ: 16SV5386K, V4MNI014) is gratefully acknowledged. We thank Academy of Finland for financial support (project numbers 134630, 135000, 141039).

[1] Seeman, N. C. Scientific American 2004, 290, 64–75.
[2] Rothemund, P. W. K. Nature 2006, 440, 297–302.
[3] Doerr, A. Nature Methods 2011, 8, 454–454.
[4] Ding, B.; Deng, Z.; Yan, H.; Cabrini, S.; Zuckermann, R. N.; Bokor, J. Journal of the American Chemical Society 2010, 132, 3248–3249.
[5] Kuzyk, A.; Schreiber, R.; Fan, Z.; Pardatscher, G.; Roller, E.-M.; Högele, A.; Simmel, F. C.; Govorov, A. O.; Liedl, T. Nature 2012, 483, 311–314.
[6] Maune, H. T.; Han, S.; Barish, R. D.; Bockrath, M.; Iii, W. A. G.; Rothemund, P. W. K.; Winfree, E. Nature Nanotechnology 2009, 5, 61–66.
[7] Eskelinen, A.-P.; Kuzyk, A.; Kaltiaisenaho, T. K.; Timmermans, M. Y.; Nasibulin, A. G.; Kauppinen, E. I.; Törmä, P. Small 2011, 7, 746–750.
[8] Marsh, T. C.; Vesenka, J.; Henderson, E. Nucleic Acids Research 1995, 23, 696–700.
[9] Vesenka, J.; Bagg, D.; Wolff, A.; Reichert, A.; Moeller, R.; Fritzsche, W. Colloids and Surfaces B: Biointerfaces 2007, 58, 256–263.
[10] Pethig, R.; Markx, G. Trends in Biotechnology 1997, 15, 426–432.
[11] Lao, C. S.; Liu, J.; Gao, P.; Zhang, L.; Davidovic, D.; Tummala, R.; Wang, Z. L. Nano Letters 2006, 6, 263–266.
[12] Li, J.; Zhang, Q.; Yang, D.; Tian, J. Carbon 2004, 42, 2263–2267.
[13] Kretschmer, R.; Fritzsche, W. Langmuir 2004, 20, 11797–11801.
[14] Washizu, M.; Suzuki, S.; Kurosawa, O.; Nishizaka, T.; Shinohara, T. IEEE Transactions on Industry Applications 1994, 30, 835–843.
[15] Wolff, A.; Leiterer, C.; Csaki, A.; Fritzsche, W. Front. Biosci 2008, 13, 6834–6840.
[16] Borovok, N.; Molotsky, T.; Ghabboun, J.; Porath, D.; Kotlyar, A. Analytical Biochemistry 2008, 374, 71–78.

# CONDUCTIVE BEHAVIOR OF G4-DNA-SILVER NANOPARTICLE STRUCTURES

T. Parviainen,[1] G. Eidelshtein,[2] A. Kotlyar,[2] and J.J. Toppari[1]

[1]Nanoscience Center, Department of Physics, P.O. Box 35, FI-40014 University of Jyväskylä, Finland
[2]Department of Biochemistry, George S. Wise Faculty of life Sciences and The Center of Nanoscience and Nanotechnology, Tel Aviv University, Ramat Aviv 69978, Israel

## 1 INTRODUCTION

DNA has been proven to be a very flexible and versatile molecule within a wide variety of fields in nanotechnology.[1] Many DNA-based templates and constructs have been introduced,[2-4] as well as various approaches to implement molecular devices for bionanotechnology and nanoelectronics. DNA conjugates may exhibit electrically conductive or semi-conductive behaviour and thus, serve as elements in nanoelectronic devices and circuits. However, electronic transport measurements on double stranded (ds) DNA have so far yielded very controversial results.[5]

It is known that besides the classical double helix, certain sequences can adopt four-stranded conformations. This is especially true for relatively short (16–32 bases)[6-9] and long (thousands of bases)[10] G-rich sequences. These structures (termed G4-DNA) are comprised of stacked tetrads; each of the tetrads arising from the planar association of four guanines by Hoogsteen hydrogen bonding.[11] G4-DNA seems like a better candidate for electrical purposes since they are stiffer compared to dsDNA and made solely of G bases, characterized by the lowest ionization potential among the DNA bases, thus promoting more efficient charge migration along DNA. In addition, we have already demonstrated a clear electrostatic polarizability of G4-wires which is indicative of possible electrical conductivity.[12]

For demonstration of the electrical conductivity of G4-DNA, one needs to position the molecule between nanoscale electrodes. Dielectrophoresis (DEP), consisting of the movement of any polarizable particle within a non-uniform electric field,[13] has been widely exploited in controllable manipulation and positioning of a diverse variety of objects, including micron-[13-16] and nanoscale[16-21] DNA. Due to its polarizability DNA is ideally suited for the DEP manipulation.

Here we report preparation, trapping and subsequent electrical conductivity measurements of individual conjugates composed of three 20 nm silver nanoparticles (AgNP) connected to each other by means of 20 tetrad G4-DNA linkers. We have demonstrated that the conjugate trapped between the nanoscale electrodes exhibits highly conductive behaviour (R ~ 1 kΩ) at voltages in the range from -0.7 V to 0.7 V. However, at voltages below – 0.7 V and above +0.7 V the conjugates were non-conductive.

## 2 EXPERIMENTAL PROCEDURES

### 2.1 Preparation of the conjugates

Unless otherwise stated, reagents were obtained from Sigma-Aldrich (USA) and were used without further purification.

*2.1.1 DNA samples.* The deoxyoligonucleotides, were purchased from Alpha DNA (Montreal, Canada). Approximately 1 mg of oligonucleotide powder was dissolved in 200 µL of 0.1M LiOH and incubated for 30 min at room temperature. The solution was subsequently passed at room temperature through a pre-packed Sephadex G-25 DNA-Grade column (Amersham-Biosciences) equilibrated with 2 mM Tris-Ac (pH 8.5).

*2.1.2 Synthesis of AgNPs.* AgNPs with 20 nm diameter were prepared by $AgNO_3$ reduction in the presence of borohydride ($NaBH_4$) as follows: 0.45 mL of freshly prepared $AgNO_3$ was added into 180 mL of deionized/filtered water precooled to 4°C. Then, 0.90 mL of 50 mM sodium citrate solution and 0.75 mL of 0.6 M $NaBH_4$ were added to the reaction mixture while vigorously stirring in the melting ice bath. The solution developed a clear yellow color. After that the solution was kept at 4°C over night during which it turned to dark yellow. 0.72 mL of 2.5 M LiCl was added to the solution under vigorous stirring at ambient temperature. The mixture was incubated for 15 min and centrifuged at 11,000 rpm for 1.2 h at 20°C in a Sorval SS-34 Rotor. A clear supernatant was carefully discarded and the pellet was suspended in 4-5 mL of residual supernatant; OD of the particle suspension should be approximately equal to 100 at 400 nm. 40 µL of 1 mM $a_{10}A_5$ ("a" refers to phosphorothioated adenosine and "A" to a regular adenosine) solution in deionized water was added to 4 mL of the above particle suspension followed by the addition of 100 µL of 1 M NaCl. The mixture was incubated for 1 h at 50°C and was subsequently loaded onto a Sepharose 6B-CL 16/40 column equilibrated with 10 mM Na-Pi buffer (pH 7.4) at a flow rate of 0.6 mL/min. The particles were isocratically eluted with 40-50 mL of the buffer after a dark-blue fraction containing nanoparticle aggregates. The particles completely separate from the non-covalently bound $a_{10}A_5$ (eluted 20 mL after the particle fraction) and the particle aggregates were collected from the column and concentrated by centrifugation at 14,000 rpm for 25 min on Eppendorf table 5424 centrifuge.

The resulting nanoparticles were screened for their size and uniformity by TEM, revealing an average diameter of 20±3 nm. The UV−Vis spectra showed a characteristic absorption peak at 397 nm. Concentration of the particles was calculated using an extinction coefficient ($\varepsilon$) of $3 \times 10^9$ $M^{-1}cm^{-1}$ at 390 nm.[22]

*2.1.3 Gel Electrophoresis.* The DNA-NP samples were loaded onto 2% agarose gel 7x7 cm, and electrophoresed at 4°C by 130 V for 40 min. Tris-Acetate-EDTA (TAE) buffer, containing 40 mM Tris-Acetate and 1 mM EDTA, in addition to being used to prepare the agarose, also served as the running buffer.

*2.1.4 HPLC separation.* The size-exclusion chromatography was conducted on a TSK-gel G-5000-PW HPLC column (7.8 × 300 mm) from TosoHaas (Japan) connected to an Agilent 1100 HPLC system with a photodiode array detector. The molecules were eluted in 20 mM Tris–Ac (pH 8.5), for 30 min at a flow rate of 0.5 mL/min. Peaks were identified from their retention times obtained from the absorbance at 260 and 400 nm for DNA and DNA-AgNP conjugates respectively.

*2.1.5 TEM imaging.* TEM samples were prepared by dropping 2.5 µL of the sample solution in 20 mM Tris-Ac (pH 8.5) onto a carbon coated grid (400 mesh). The grids (before depositing) were negatively glow discharged using Emitech K100X glow discharger. After incubated for 5 min at ambient temperature, the excess solution was removed by blotting with filter paper. TEM imaging was performed on a TEM JEM model 1200 EX instrument operated at an accelerating voltage of 120 kV.

## 2.2 Trapping and electrical measurements

*2.2.1 Fabrication of nanoelectrodes.* The nanoelectrodes were fabricated by conventional electron-beam-lithography on a slightly *p*-doped silicon chip covered with a ~300 nm thick insulating $SiO_2$ layer. Before patterning the chip was cleaned by hot acetone and subsequently by sonication in isopropanol (IPA) followed by drying in nitrogen flow. A thin film of 2% 950 PMMA A resist (MicroChem Corp, USA) was spin coated onto the surface of the chip at 2,000 rpm, followed by baking on a hot plate at 160°C for 5 minutes. The exposure was done by a Raith e-Line electron beam lithography system. After the exposure the sample was treated for 30 s with a development solution containing methyl isobutyl ketone (MIBK) and IPA mixed at 1:3 volume ratio. The development process was stopped by rinsing the chip with pure IPA. The revealed $SiO_2$ surface was finally cleaned from the undeveloped residues of the resist by 15 W oxygen flash in Oxford Instruments Plasmalab 80 Plus reactive ion etching (RIE) chamber for 10 s.

Metallization of the patterned chip with a 2 nm layer of titanium (to increase adhesion of gold to silicon) and 25 nm layer of gold was done by electron beam evaporation in an ultrahigh vacuum chamber. For the lift-off the chip was left in acetone overnight and finally sonicated in acetone for 10 seconds. Just prior to the DEP trapping the chip was exposed to oxygen plasma (100 W) for 1 min in a RIE chamber in order to remove residues left from the lift-off and create a hydrophilic layer on the surface.

*2.2.2 Dielectrophoretic trapping.* The DEP trapping procedure was done inside a sealed metallic box with bottom compartment filled with water in order to prevent drying of the sample during the trapping. The chip was placed on a nylon stage covered with epoxy, and equipped with sharp metallic probes which were placed carefully on the contact pads of the electrodes. Probes were connected to a signal generator (Agilent 33120A arbitrary waveform generator). Frequencies and voltages of the applied signal were varied between 1 kHz and 10 MHz and between 2 and 6 V, respectively. 5 µL of a G4-DNA-AgNP conjugate solution was dropped on the electrodes, the box was sealed and the AC voltage was set. Trapping time was varied from 5 to 12 min depending on the concentration of the particles. After trapping the chip was rinsed with double-distilled water and dried by a nitrogen flow. The voltage was turned off and the chip was removed from the stage.

*2.2.3 AFM imaging.* The presence of nanoparticle structures between the electrodes after trapping was verified by imaging with atomic force microscope (AFM) Veeco Dimension 3100. The imaging was done by tapping mode with 300 kHz standards tapping mode tips.

*2.2.4 Electrical measurements.* All electrical measurements were carried out at ambient temperature (~21°C) inside an electromagnetically shielded room with isolated noiseless grounding. The chip was placed on the same nylon probe stage used for trapping, and the stage was placed inside an electrically insulated, specially grounded metallic box. Humidity and temperature inside the measuring box was measured by a HIH-3602-A Honeywell Humidity sensor. Voltage between the electrodes was applied using a homemade battery-powered DAC circuitry which was controlled by specially programmed Labview-program. High voltages and any quick changes in voltage were avoided in order to prevent disintegration or discharging of the sample. The current and voltage were measured using battery-powered preamplifiers DL-Instruments Model 1211 and Model 1201, respectively. In the first measurement, the voltage was swept by finite steps of ~5 mV from -0.5 to 0.5 V. In the subsequent measurements the range was gradually increased usually up to ±1.2 V. The positive and negative voltages were swept either separately or within the same cycle. The current and voltage signals were collected by a computer through a data acquisition card DAQ, National Instruments PXI-10311. Each current-reading was acquired by averaging 1,000 individual samplings at the sampling rate of 10,000 scans/s, thus yielding an effective integration time of 0.1 s. In addition, after each voltage change the system was given at least 300 ms to settle before the measurement.

Samples were measured at ambient humidity and often also at high relative humidity (up to RH ~ 90%). Additional measurements at high relative humidity (up to RH ~ 90%) were also performed. In the latter case the humidity was slowly increased by flowing steam into the sample box while constantly measuring I-Vs.

*2.2.5 Statistical analysis of the data.* All the measured I-Vs were converted into first derivative by numerical derivation to obtain conductance values. These values were further statistically analyzed.

## 3 RESULTS

### 3.1 Synthesis and characterization of G4-DNA-AgNP conjugates

Short G-rich oligonucleotides are known to assemble into high-molecular weight multistrand aggregates.[23,24] These aggregates form spontaneously upon dissolving of commercial preparations of G-rich oligonucleotide. They, however, can be disassembled at pH higher than 12; addition of 0.1 M LiOH to the oligonucleotide solution leads to dissociation of the aggregates into single oligonucleotide strands. The dissociation is driven by deprotonation of G-bases at the N1-site at pH higher than pKa (9.4) and is caused by strong electrostatic repulsion of the deprotonated strands.

*3.1.1 Preparation of G-DNA.* We dissolved an oligonucleotide containing 20 central G-base fragment ($G_{20}$) flanked by two runs of phosphorothioated adenines ($a_5$) on both sides, i.e. $a_5G_{20}a_5$, in 0.1 M LiOH. The pH was subsequently reduced during chromatography of the alkaline oligonucleotide solution on a Sephadex G-25 column equilibrated with 2 mM Tris-Ac (pH 8.5). The oligonucleotide eluted from the column is characterized by a weak CD signal (data not presented), indicating that the strands do not fold into a quadruple helix during the chromatography stage. Addition of Tris-Ac buffer (pH 8.5) to a final concentration of 100 mM to the eluate (OD >100 at 260 nm) induces a strong positive CD band at 258 nm in the spectrum of the oligonucleotide (see Figure 1, solid curve). The CD spectrum of the DNA is slightly affected by K-ions (see Figure 1, dashed curve). The CD spectrum and the effect of the cation on the spectrum are similar to those reported by us for tetra-molecular G4-wires.[10] We thus suggest that the association of four strands takes place upon formation of G-quadruplexes.

*3.1.2 Conjugation.* To prepare G4-DNA-AgNP conjugates we used 20 nm silver particles. The citrate-capped nanoparticles formed during the reduction of silver ions in the presence of citrate, are unstable and tend to agglomerate at salt concentrations higher than 20 mM. In addition, the particles interact with phosphate groups of the DNA backbone. In order to increase the stability and to avoid the backbone-mediated interactions we have coated the particles with a 15-mer oligonucleotide, $a_{10}A_5$. This oligonucleotide is composed of ten phosphorothioate adenine (a) residues that can tightly and covalently anchor the strand to the surface of a silver nanoparticle and five regular adenine (A) bases that form a protective coating layer on the particle surface. Incubation of the particles with

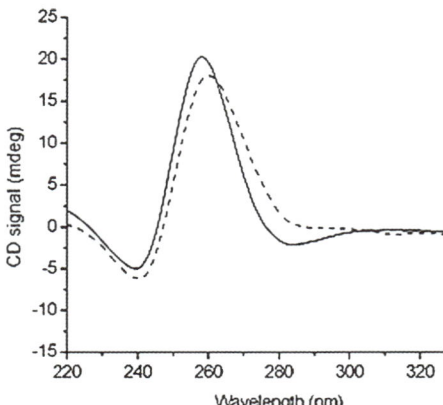

**Figure 1**  CD spectra of G4-quadruplexes formed by $a_5G_{20}a_5$ oligonucleotide before (solid line) and after (dashed line) the addition of 10 mM KCl. Tris-Ac (pH 8.5) was added to $a_5G_{20}a_5$ eluted from a Sephadex G-25 column to a final concentration of 100 mM. The sample (OD at 260 nm ~ 100) was incubated at room temperature for 30 min and was then diluted 100 times with 100 mM This-Ac (pH 8.5). Each spectrum is an average of three scans.

phosphorothioate-functionalized G4-DNA structures, for 2 hours in 10 mM Na-Pi buffer (pH 7.4) containing 250 mM NaCl yielded a mixture of conjugates bearing a discrete number of nanoparticles. The conjugates can be efficiently separated by electrophoresis in a 2% agarose gel (see Figure 2). The yellow bands indicated by arrows and marked by numbers in Figure 2 (lane 2) correspond to conjugates comprising one, two and three particles respectively. To obtain individual conjugates, each band was cut out of the gel with a razor blade. The yellow gel slices were placed in dialysis bags containing TAE buffer; each sample was then electro-eluted into a bag and collected.

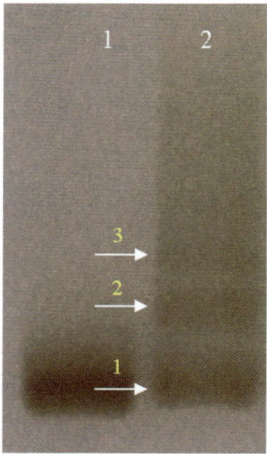

**Figure 2**  Electrophoretic separation of G4-DNA-Ag-NP conjugates. Lane 1 – Ag-NPs (OD at 400 nm is equal to 200); Lane 2 – Ag-NPs were incubated with G4-structures formed in $a_5G_{20}a_5$ solution (see Figure 1) in 10 mM Na-Pi buffer (pH = 7.4) containing 250 mM NaCl for 2 hours at 50°C. 20 µL of each sample were loaded onto a 2% agarose gel and electrophoresis at 130 V for 40 min at 4°C.

*3.1.3 TEM characterization.* The conjugates extracted from each band were characterized by transmission electron microscope (TEM). As seen in Figure 3, conjugates from the gel slices corresponding to bands 1, 2 and 3 composed of 1, 2 and 3 nanoparticles, respectively. The conductance of conjugates electro-eluted from the band 3 and composed of 3 particles was measured as follows.

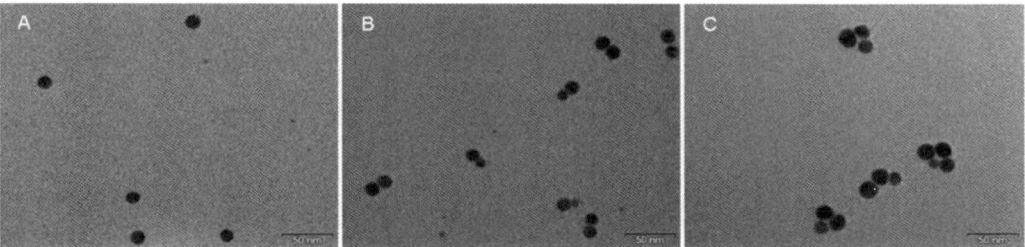

**Figure 3** *TEM images of G4-DNA-AgNP conjugates. Molecules were electro-eluted from gel slices corresponding to bands 1(A), 2(B) and 3(C) of the gel (Figure 2), deposited on 400 mesh copper carbon grids and visualized by TEM.*

## 3.2 Trapping of conjugates

The conjugates were first positioned between the two lithographically fabricated gold-nanoelectrodes (see Figure 4a) by the DEP trapping procedure (see above), and the same electrodes were subsequently used for measuring the conductance of the trapped sample. The trapping method has high efficiency and yield.[19-21]

Prior to the trapping the conjugate solution was diluted with double-distilled water in order to reduce the concentration of the buffer present in the DNA-NP sample. High conductivity of the buffer leads to high currents and, as a result, heating of the gap area, which can further induce convectional flows preventing the trapping,[25] or even destroy the electrodes. The yield of the trapping procedure strongly depends on the voltage and frequency of the AC signal. By choosing optimal parameters one can place a single, conjugate between the electrodes. We have shown that at frequencies lower than ~100 kHz the trapped conjugates are mainly lying on the electrodes and not in the gap between them, whereas at relatively high frequencies (higher than 5 MHz) the conjugates neither bind to the electrodes nor bridge the interelectrode gap.[19-21,26,27] The voltage and the trapping time mainly affected to the amount of gathered conjugates.

Trapping parameters have been tested by us on various DNA- and nanoparticle-based structures.[19-21,26,27] The parameters we selected for the deposition of the G4-DNA-AgNP trimers eluted from the gel band 3 (band 3 in Figure 2 ) are: 3 MHz and 3.5 $V_{pp}$ (peak-to-peak AC voltage). Trapping time was varied from 5 to 12 min depending on the concentration of the conjugate. The presence of conjugate(s) between the electrodes was verified by AFM. As clearly seen in Figures 4c and 4d, we succeeded to position small number (one or two) G4-DNA-AgNP conjugates in between the electrodes. Trapping was not that successful in the case of a sample shown in Figure 4b due to the non-optimal parameters used. The electrode surface is totally covered with the particles, while the gap remains empty.

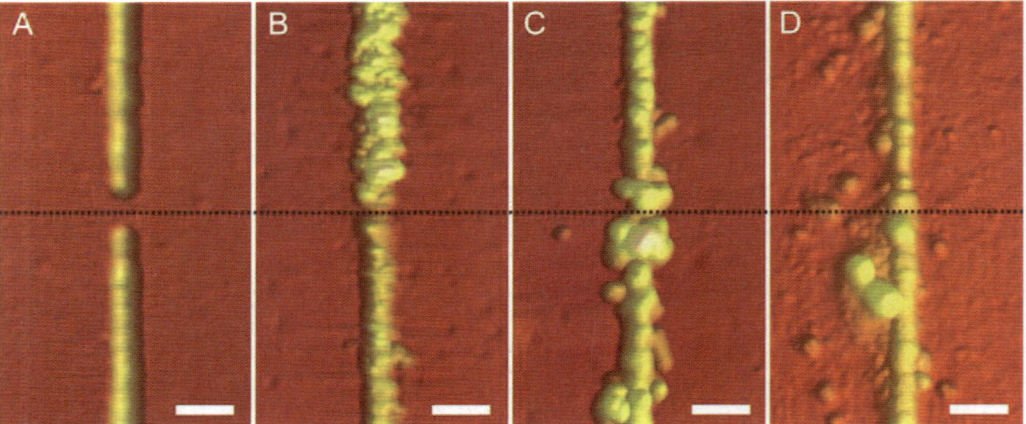

**Figure 4** *AFM-images of empty lithographically fabricated gold nanoelectrodes (A) and electrodes with trapped G4-DNA-AgNP trimers (B-D). B demonstrates unsuccessful trapping of the conjugates, while C and D present successful result. Scale bars are 100 nm and the dotted line visualizes the position of the gap in each image. Samples C and D were used in conductivity measurements.*

### 3.3 Conductance behaviour

Resistance of bare electrodes in ambient conditions is very high (in a 1 TΩ range),[26,27] and it is very strongly reduced upon bridging the interelectrode gap by the conjugates.

*3.3.1 I-V curves.* The I-V characteristics of samples C and D (see Figure 4) are presented in Figure 5 (left and right panels respectively). Both samples show linear behaviour in the range between -0.7 V and 0.7 V. However, the increase of the absolute value of the voltage above 0.7 V results in dramatic increase of the resistance and decrease of the current. The transition between the conductive (R ~ 1 kΩ) and non-conductive (R ~ 1 TΩ) states is very sharp and reversible. Because no change in the I-V pattern in the -0.7 - 0.7 V region was observed upon application of potentials up to ±1.2 V we conclude that no irreversible changes in the nanoparticle-DNA system occur at high absolute values of the applied voltage. It is worth noting that the observed switching from low to high conductivity did not happen every single cycle. In some cases reducing the voltage below 0.7 V was not associated with an increase of the current. The I-V pattern, however, was always restored after one or several cycles. This was also supported by the fact that transition to the low conductivity state never happened if the voltage range did not exceed 0.7 V during sweeping. On the other hand, after switching to the high conductivity state, it was never switched off while sweeping only within voltages of absolute value below 0.7 V. This unusual I-V behaviour was repeated totally on 5 different samples.

*Applications in Bioanalytics, Therapy and Molecular Electronics* 321

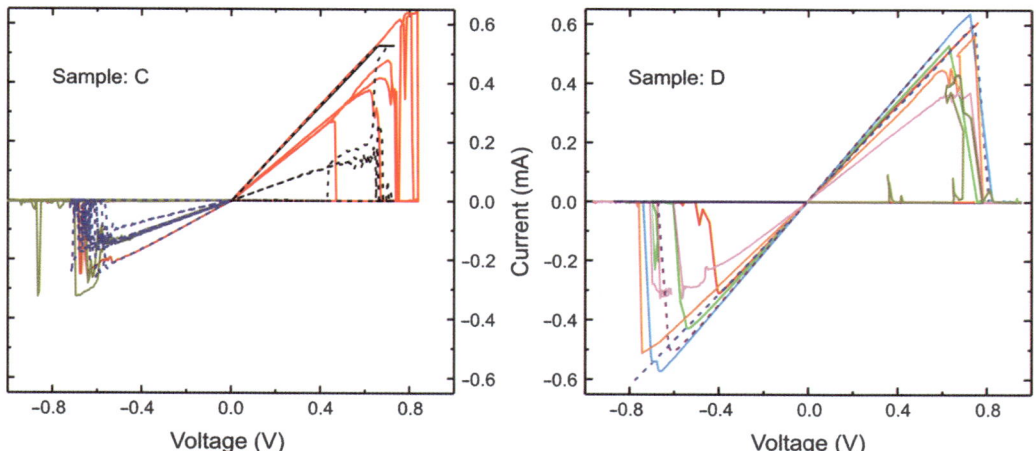

**Figure 5**   *I-V characteristics measured in ambient conditions on samples C and D (see Figure 4). The high conductive parts of all I-V curves correspond to resistances lower than 1 k$\Omega$. Clear switching from conductive to a non-conductive state is observed at voltages of ~ 0.7 V and -0.7 V.*

*3.3.2 Conductance of the high conductive state.* As evident from Figure 5, the conductance of the sample in the range between -0.7 and 0.7 V is changing slightly from cycle to cycle in a random fashion. The conductance within a single cycle is always constant between -0.7 V and 0.7 V. However, switching to a non-conductive mode and getting back to a conductive one slightly changes the conductive characteristics. Figure 6 presents a statistical analysis of the conductance data obtained on 3 different samples.

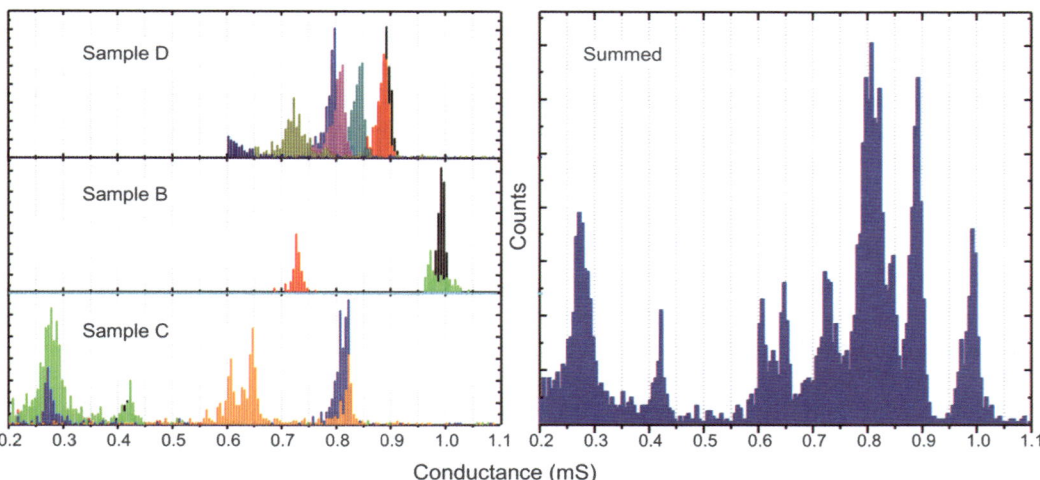

**Figure 6**   *The statistical analysis of conductance data collected from samples B, C and D. Derivative analyses of all I-V curves allowed estimation of possible discrete conductance values for samples D, B and C (left panels). Right panel summarizes statistical data obtained on the three above samples.*

As one can see, the statistical data in Figure 6 show that the conductance of the system is varied and that discrete conductive states exist. One can speculate that the states correspond to electron transfer through particular G4-DNA molecules which orientation and nature can vary from cycle to cycle. We also can suggest that the average difference in the conductance values corresponding to different peaks (approximately ~0.05 mS) corresponds to the conductivity of a single G4-DNA molecule connecting the particles.

## 4 DISCUSSION

One can speculate that the reason for the unusual electrical behaviour demonstrated here can be associated also with detachment of the nanoparticles from the electrodes or from each other at high absolute values of voltage. As NP-DNA conjugates and metal particles can move towards cathode at relatively high values of the applied potential, movement along the electrode plane might physically detach the conjugates from the electrodes leading to transition of the system to the non-conductive state. However, it is most likely that changing of the conjugate position with respect to the electrodes would irreversibly abolish the conductance, which is not the case here.

We therefore think that the "switching" phenomenon is most likely associated with changes of the G4 structure induced by the electrical field. Many studies have shown that the DNA conformation has a major impact on its electrical conductivity.[28,29] The conformational changes can be driven by stretching the DNA during the conjugate movement along the electrode axes in the electrical field or by chemical changes of the G-basis at high redox potentials. It is well known that the G-base characterized by relatively mild values of ionization potential, can undergo reversible redox transition at potentials higher than +0.7 V or lower than -0.7 V. [30-33] The perturbation of the DNA structure induced by oxidation or reduction can lead to a sharp drop of the current at relatively low or high applied potentials.

We have also shown that the switching is affected by the relative humidity (RH). The switching is more robust under high humidity compared to ambient conditions. The reversible "switching" behaviour of a sample showing only insulating characteristics at the ambient conditions, was induced by increasing the relative humidity (RH). This observation also points out for the essential role of G4-DNA conformation for mediating electron transfer between the nanoparticles and electrodes.

## 5 CONCLUSION

Here we have demonstrated the unusual ohmic behaviour, i.e. potential driven switching between high and low conductive states under ambient conditions, of a nanoparticle-DNA conjugate composed of three 20 nm silver nanoparticles connected to each other by means of 20 tetrad G-quadruplex DNA linkers. The conjugate was trapped between gold-nanoelectrodes and its conductivity was probed using a conventional setup for DC conductivity measurements. Clear switching from high (~1 mS) to low (~1 pS) conductive state was observed at potential values of -0.7 V and 0.7 V. We suggest that the transitions between the high to the low conductive states is due to changing of the conformation of the G4-DNA linker induced by oxidation and reduction of the G-bases at potentials of 0.7 and -0.7 V, respectively. This would also imply that the almost equally spaced discrete states observed in the conductance values could yield a conductivity of a single G4-DNA. However, further investigations are needed in order to better understand the mechanism of the observed effect and its importance for nanoelectronics.

## References

1. T.H. LaBean and H.Li, *Nano Today*, 2007, **2**, 26.
2. N.C. Seeman, *Nature*, 2003, **421**, 427.
3. N.C. Seeman, *Nano Lett.*, 2010, **10**, 1971.
4. S.M. Douglas, H. Dietz, T. Liedl, B. Högberg, F. Graf and W.M. Shih, *Nature*, 2009, **459**, 414.
5. R.G. Endres, D.L. Cox and R.R.P. Singh, *Rev. Mod. Phys.* 2004, **76**, 195.
6. G.N. Parkinson, M.P. Lee, and S. Neidle, *Nature*, 2002, **417**, 876.
7. W.J. Qin and L.Y. Yung, *Nucleic Acids Res.* 2007, **35**, e111.
8. S. Sheikholeslami, Y.W. Jun, P.K. Jain and A.P. Alivisatos, *Nano Lett.* **10**, 2655.
9. J. Yu, S. Choi, C.I. Richards, Y. Antoku and R.M. Dickson, *Photochem. Photobiol.* 2008, **84**, 1435.
10. N. Borovok, N. Iram, D. Zikich, J. Ghabboun, G.I. Livshits, D. Porath and A.B. Kotlyar *Nucleic Acids Res.* 2008, **36**, 5050.
11. J.T. Davis, *Angew. Chem., Int. Ed. Engl.* 2004, **43**, 668.
12. H. Cohen, T. Sapir, N. Borovok, T. Molotsky, R. Di Felice, A.B. Kotlyar, D. Porath, *Nano Lett.* 2007, **7**, 981.
13. H.A. Pohl, Dielectrophoresis: the Behaviour of Neutral Matter in Nonuniform Electric Field, Cambridge University Press, Cambridge, UK, 1978.
14. M. Washizu and O. Kurosawa, *IEEE Trans. Ind. Appl.*, 1990, **26**, 1165.
15. M. Washizu, O. Kurosawa, I. Arai, S. Suzuki and N. Shimamoto, *IEEE Trans. Ind. Appl.*, 1995, **31**, 447.
16. R. Hölzel and F. F. Bier, *IEE Proc.: Nanobiotechnol.*, 2003, **150**, 47.
17. L. Ying, S. S. White, A. Bruckbauer, L. Meadows, Y. E. Korchev and D. Klenerman, *Biophys. J.*, 2004, **86**, 1018.
18. A. Wolff, C. Leiterer, A. Csaki and W. Fritzsche, *Front. Biosci.*, 2008, **13**, 6834.
19. A. Kuzyk, B. Yurke, J.J. Toppari, V. Linko and P. Törmä, *Small*, 2008, **4**, 447.
20. S. Tuukkanen, A. Kuzyk, J.J. Toppari, H. Häkkinen, V.P. Hytönen, E. Niskanen, M. Rinkiö and P. Törmä, *Nanotechnology,* 2007, **18**, 295204.
21. S. Tuukkanen, A. Kuzyk, J.J. Toppari, V.P. Hytönen, T. Ihalainen and P. Törmä, *Appl. Phys. Lett.*, 2005, **87**, 183102.
22. J. Yguerabide and E.E. Yguerabide, *Anal. Biochem.* 1998 **262**, 137.
23. V.M. Marathias and P.H. Bolton, *Biochemistry*, 1999, **38**, 4355.
24. J. Gros, F. Rosu, S. Amrane, A. De Cian, V. Gabelica, L. Lacroix and J.L. Mergny, *Nucleic Acids Res.*, 2007, **35**, 3064.
25. R. Hölzel, N. Calander, Z. Chiragwandi, M. Willander and F.F. Bier, *Phys. Rev. Lett.* 2005, **95**, 128102.
26. V. Linko, S.-T. Paasonen, A. Kuzyk, P. Törmä, and J.J. Toppari, *Small*, 2009, **5**, 2382.
27. V. Linko, J. Leppiniemi, S.-T. Paasonen, V.P. Hytönen and J.J. Toppari, *Nanotech.*, 2011, **22**, 275610.
28. A. Yu. Kasumov, D. V. Klinov, P.-E. Roche, S. Gueron & H. Bouchiat, *Appl. Phys. Lett.*, 2004, **84**, 6.
29. X. Guo, A. Gorodetsky, J. Honez, J. Barton and C. Nuckdolls, *Nature* 2008, **3**, 163.
30. D. Douki, *et al. Top. Curr. Chem.*, 2004, **236**, 1.
31. J. Cadet, T. Douki, D. Gasparutto and J.-L. Ravanat, *Mutat. Res.*, 2003, **531**, 5.
32. A. Abbaspour and A. Noori, *Analyst*, 2008, **133**, 1664.
33. R. Hallaj and A. Salimi, *Anal. Methods*, 2011, **3**, 911.

# NOVEL MATERIALS FOR MOLECULAR ELECTRONICS - SYNTHESIS AND CHARACTERIZATION OF LONG G4-DNA

Dvir Rotem[1], Gennady Eidelshtein[2], Alexander Kotlyar[2*] and Danny Porath[1*]

[1]Institute of Chemistry and The Harvey M. Kreuger Family Center for Nanoscience and Nanotechnology, The Hebrew University of Jerusalem, Jerusalem 91904, Israel
[2]Department of Biochemistry, George S. Wise Faculty of Life Sciences and Center for Nanotechnology, Tel Aviv University, Ramat Aviv, 69978, Israel

Corresponding authors: Danny Porath (porath@chem.ch.huji.ac.il) and Alexander Kotlyar (s2shak@post.tau.ac.il)

## 1 INTRODUCTION

Two decades ago it was predicted that the microelectronics industry will not be able to follow the miniaturization trend of Moore's law, due to lithography limitations. Nowadays, wires with 22 nm pitch are used in boards in production and with smaller pitch in research laboratories. Nevertheless, the distance between functional devices is still tens of nanometers using the current production architectures, the production cost is very high and upon miniaturization quantum effects become more and more relevant. At that time, two decades ago, molecular electronics was the "big promise" and there were expectations that future molecule based computers and components will play a central role in computing. In spite of a dramatic increase in our knowledge about molecules we are not there yet and we do not even well understand the conduction mechanisms in one-dimensional polymers that are longer than ~10 nm. The central reason is the huge difficulty in performing measurements on single long molecules, e.g., due to molecule to molecule variability and extreme dependence of the results on the experimental and environmental conditions. Having said that, the brain presents a prime example for the most sophisticated computing and information transfer machine, which is entirely based on cell and molecules. Therefore, with this knowledge in the background the challenge remains to produce and measure molecules that will enable to effectively transfer information, electronic in particular, in single one-dimensional polymers, which are not one-dimensional crystals like carbon nanotubes, and elucidate the electron transfer mechanisms. This was the central objective of our research on DNA and its G-quadruple derivatives.
The dramatic development of the electronic industry in the last decades was based on a progress in reducing the size of devices, which are made of denser and denser integrated circuits. These circuits are currently fabricated with complementary-metal-oxide-semiconductor (CMOS) transistors. Higher transistor density on a single chip means faster circuit performance. However, the trend towards higher integration is restricted by the limitations of the current lithography technologies, by heat dissipation

and by capacitive coupling between different components. In addition, the down-scaling of individual devices to the nanometer range brings their behavior to a new regime, where the fundamental physical laws of quantum mechanics dominate. In conventional silicon-based electronic devices the information is carried by mobile electrons within a band of allowed energies according to the semiconductor band structure. When the electronic devices dimensions shrink to the nanometer scale, these bands turn into discrete energy levels and quantum correlation effects induce localization. Therefore a novel technology, which would exploit the pure quantum mechanical effects that rule at the nanometer scale, is desired to push the miniaturization of integrated circuits further. The search for efficient molecular devices, that would be able to perform operations currently done by silicon transistors, is pursued within this framework. The basic idea of molecular electronics is to use individual molecules to perform as wires, switches, rectifiers and memories[1-3]. Another conceptual idea that is advanced by molecular electronics is the switch from a top-bottom approach, where the devices are extracted from a single large-scale building block, to a bottom-up approach in which the whole system is composed of small basic building blocks with recognition, structuring and self-assembly properties.

DNA is a prospective candidate for molecular electronic devices. In this frame DNA possesses two unique and appealing properties: recognition and a special structuring that suggests its use for self-assembly. Molecular recognition describes the capability of a molecule to form selective bonds with other molecules or with substrates, based on the information stored in the structural features of the interacting partners. The DNA molecular recognition that is based mainly on Watson-Crick base pairing can be easily engineered. Self-assembly, which is the capability of molecules to spontaneously organize themselves in supramolecular aggregates under suitable experimental conditions[4], may drive the design of well-structured systems. A promising development has been recently achieved in controlling the self-assembly of DNA into complex structures, as DNA origami structures [5,6] and in coupling molecules to metal contacts [7]. However, there is still a great controversy around the understanding of the electrical behavior of DNA and of the mechanisms that might control charge mobility through its structure [8]. Several approaches and techniques were developed for the measurement of charge transport through native and synthetic DNA molecules. These methods, each with its own unique advantages, difficulties and drawbacks, yielded a wide range of results that fueled a scientific debate over the conduction properties of DNA molecules. Generally, the data that were accumulated from these experiments suggests that it is possible to transport charge carriers along short single double-stranded DNA (dsDNA) molecules, in bundles of molecules and in networks, although the conductivity is rather poor. However, transport through long single dsDNA molecules (>40 nm) that are attached to the surface is apparently blocked. It may be due to the surface force field that induces many defects in the molecules and blocks the current or any additional reason[7-11].

Charge transport through some other DNA derivatives might be better than through a random sequence dsDNA. However, normally these molecules are synthesized as short oligonucleotides or as part of long DNA molecules. Here we present the development and characterization of long (100's-1000's nm scale) poly(dG)-poly(dC) dsDNA and G-quadruplex DNA molecules that were made by us, in the context of their potential applications in molecular electronics.

The common model for electron transfer through DNA is based on overlap between $\pi$ orbitals in adjacent base pairs [12]. Irregular base pair sequences may lead to localization of charge carriers and reduce the transfer rate of electrons [13,14]. A structure containing a single type of base pair such as poly(dG)–poly(dC) dsDNA may therefore provide a better conditions for π overlap. In addition, guanines, which have the lowest ionization potential among DNA bases, promote charge migration through the DNA. Experimental demonstration of the conducting behavior in short poly(dG)–poly(dC) dsDNA[12,15] and the results of theoretical calculations showing that poly(dG)–poly(dC) dsDNA exhibits better conductance than poly(dA)–poly(dT) dsDNA[16], support an idea of possible application of poly(dG)–poly(dC) dsDNA in molecular electronic devices.

DNA sequences containing stretches of guanines (dG) can form G-quadruplex structures. These structures, commonly named G4-DNA, are comprised of stacked tetrads. Each of them arises from the planar association of four guanines by Hoogsteen hydrogen bonding (Figure 1A). G-quadruplex structures have polymorphism and are characterized by various molecularity, topology and strand orientation[17]. They may be formed from one, two and four separate DNA strands, thus termed mono-, bi- and tetra-molecular G4-DNA, correspondingly[18-23] (Figure 1B). The potential biological role of these structures stimulated studies of their physical and chemical properties. Most of these studies have been performed using short (16–32 bases) G-rich telomeric oligonucleotides that are stable only in the presence of stabilizing cations[24-28]. Interestingly, short G4 oligonucleotides were shown to assemble spontaneously into long molecular wires in the presence of proper monovalent cations[29-31]. However, these wires are comprised of G-oligonucleotide fragments that are not covalently connected. G4-wires are comprised of stacked guanine tetrads, thus providing better conditions for π overlap compared to the base-pairs of the canonical dsDNA. In addition, the fact that these wires are made solely from guanine, that have the lowest ionization potential among the DNA bases, also increases the probability of charge migration through them[32-34].

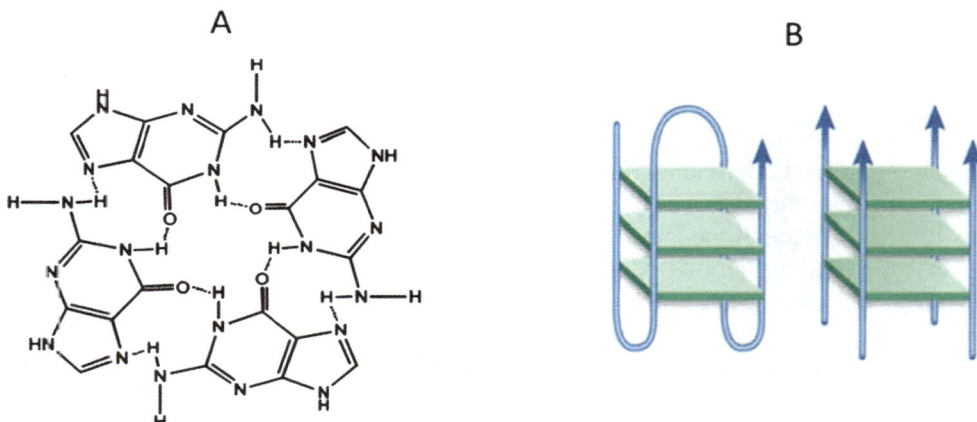

**Figure 1** *G4-DNA (A) A schematic representation of the G-tetrad building block. (B) Mono-molecular G-quadruplex (G4-DNA) is created from a single strand that is 4-folded to create anti-parallel G4-DNA structures (left). Tetra-molecule G4-DNA is formed from four parallel G-strands (right).*

Here we describe the methods that were used for production of long poly(dG)–poly(dC) dsDNA molecules and G4-wires and their morphologic and electric characterization that were done with several single molecule microscopy and spectroscopy techniques.

## 2 RESULTS AND DISCUSSION

### 2.1 Long poly(dG)–poly(dC) dsDNA

Single stranded DNA (ssDNA) do not have continuous and tight π staking along the molecule, which is essential for possible charge transport, especially when deposited on a hard surface. dsDNA, e.g., poly(dG)–poly(dC) molecules, may transport charges only if they are double stranded along all their length, since breaks in their backbone are likely to strongly reduce the π staking and therefore the ability of the polymer to transport current. Molecules produced in the past by commercial kits were found to be characterized by a broad size distribution, nicks and single-stranded fragments along the DNA[35]. These disadvantages stimulated the development of a novel enzymatic procedure for production of poly(dG)–poly(dC), composed of G- and C-homopolymers having equal lengths and lacking breaks along the strands[35].

These molecules were morphologically characterized by atomic force microscopy (AFM) (Figure 2A) and scanning tunneling microscopy (STM)[36]. The STM is a powerful tool for obtaining both structural and electrical information on the submolecular level. The STM is based on the concept of quantum tunneling: When a conducting tip is in proximity to a surface, a bias (voltage difference) applied between the two can allow electrons to tunnel through the vacuum between them. This tunneling current is dependent on the tip position, the applied voltage, and the local density of states of the sample. Monitoring the current changes as the tip scans across the surface may enable imaging at the atomic level. STM imaging of DNA was quite attractive since the STM invention[37-42].

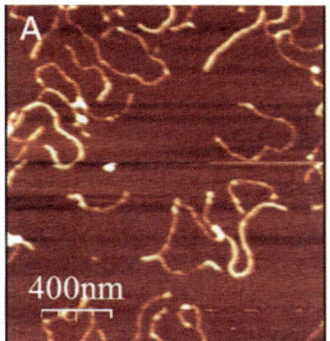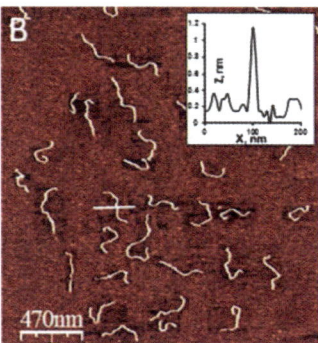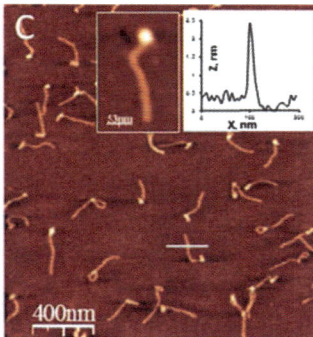

**Figure 2** *AFM images of poly(dG)-poly(dC) dsDNA molecules (A); mono-molecular G4-DNA wires (B); and tetra-molecular G4-DNA wires (C) on mica. The graphs show cross sections of the marked molecules.*

For STM imaging, poly(dG)–poly(dC) molecules were deposited on a highly smoothed gold (111) under ultra high vacuum conditions (Figure 3B). The results of high resolution STM reveal a periodic structure seen as repeating "bulbs" along the molecules. The molecular height obtained from a cross section over the molecule is 0.8 nm. This agrees rather well with results demonstrating lower height of dsDNA than the molecular diameter of the molecule as it was characterized in solutions and crystals (e.g.[43]). A periodic structure is visualized along the molecule, with "bulbs" corresponding to the helix pitches and an average length of 4 nm. This pitch length is longer than the reported pitch length, as extracted from X-ray structure of dsDNA (3.4

Å for the B-form[44]). Although it is not possible to extract the exact structure that the DNA adopts on the surface, it is clearly stretched with respect to its relaxed structure in solution. This can originate either from the deposition method or from the ultra high vacuum conditions (or both) and may change when using other deposition parameters.

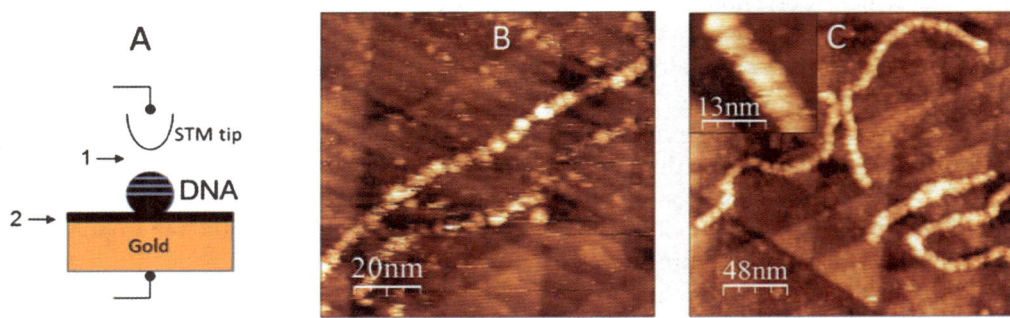

**Figure 3**   STM of DNA molecules (A) Schematic diagram of the double-barrier tunnel junction configuration. The first tunnel junction is formed between the STM tip and the molecule (1). The second junction (2) is between the molecule and the gold surface. (B) STM image of poly(dG)–poly(dC) DNA molecules on a gold (111) surface. (c) STM image of a mono-molecules G4-DNA wires deposited on a gold (111) surface. The inset shows a high resolution image of a G4-DNA wire.

The intrinsic electronic structure of molecules, such as poly(dG)–poly(dC), is one of the factors that determine its response to an external electric field. Hence, it is attractive to get this knowledge in order to reveal these molecules suitability to support electrical currents and the viable transport mechanisms. Scanning tunneling spectroscopy (STS) was chosen for this purpose, since it can measure the electronic density of states (DOS) of single molecules, as was demonstrated for various carbon nanostructures[45], molecular objects[46] and inorganic nanoparticles[47] deposited on substrates. STS is an electrical measurement through a tunnel junction, in which the current (I) between the STM tip and the sample is characterized as a function of the voltage (V) and providing information on the energy levels of the molecule. The measurement configuration on the dsDNA is a double-barrier tunnel junction (Figure 3A), meaning that when the STM tip is mounted above the molecule while measuring, two tunnel junctions are formed: one between the tip and the molecule and the other between the molecule and the conductive surface. I-V curves were measured for single poly(G)-poly(C) dsDNA molecules[43]. The corresponding conductance curves (dI/dV–V) clearly reveal the existence of an excitation gap and emphasize the details of the peak structures (Figure 4A). The average gap width is 2.5 eV at 78 K. The peak-energy distribution, that contain information about the electronic structure of dsDNA molecules when deposited on the surface, shows six distinguishable groups (Figure 4a): three in the negative and three in the positive bias range. Each group in this peak distribution differs energetically from its neighboring groups by at least 0.2 eV, which is beyond the uncertainty limit in these results due to noise. The experimental I–V characteristics are fairly reproducible at 78 K. The gap also shows reproducibility at 300 K and the average gap value does not change within the temperature range 78-300 K. These results are the only reliable and reproducible STS reported so far on long dsDNA molecules that provide direct information on their energy levels on the single molecule level. Note, however, that they are measured in non-physiological conditions: vacuum and attached to a hard surface.

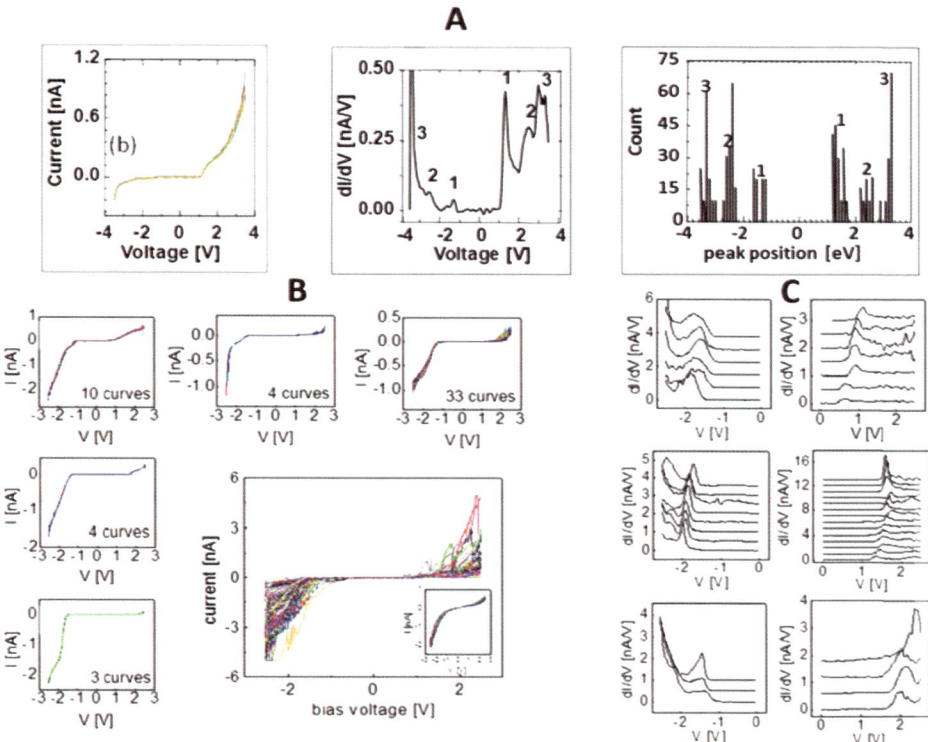

**Figure 4** *STS of DNA molecules (A) Representative STS measurements of poly(dG)–poly(dC) DNA molecules at 78 K presented as I–V curves (left) and averaged conductance curve (dI/dV–V) (middle). Statistical analysis of the peak-energy values taken from 180 experimental curves (right). (B) A data set of 400 STS mono-molecule-G4-DNA wires at 4.5 K (middle) and its clustering to reproducible groups (the rest), presented as I–V. The application of a sorting algorithm to the curve set of panel a results in its splitting to 15 groups, each containing at least three curves. Five of the extracted groups are displayed. (C) Representative conductance curves (dI/dV–V) of mono-molecule-G4-DNA wires obtained by differentiating average I–V curves from sorted groups are shown shifted along the vertical axis. Patterns found in the negative-bias (left) positive-bias (right) regimes are shown.*

## 2.2 Mono-molecular G4-wires

Short G4-DNA oligonucleotides can be spontaneously assembled into long G4-DNA wires. However, these G4-DNA wires are non-uniform polymers having gaps between the oligonucleotides fragments along the wires. A novel procedure for the synthesis of long monomolecular G4-DNA wires was developed (Figure 5A)[49]. In this procedure, the long poly(dG)-poly(dC) molecules, described above, are separated into their poly(dG) and poly(dC) strands at high pH. These strands are isolated from each other by high-pressure liquid chromatography (HPLC). The pH of the poly(dG) molecules is then reduced to 7. As a result, the molecules are folded into monomolecular G4-DNA wires, as demonstrated by absorption and circular dichroism (CD) spectra [49] and by AFM (Figure 2B)[49] and STM imaging (Figure 3C)[50]. The AFM imaging[49] shows that the apparent height of the G4-wires on a mica surface is lower than these molecules nominal diameter in solution (2.8 nm[26,28]). This probably results from surface forces and AFM tip pressure applied to them. However the G4-wire height is about twice the height of dsDNA molecules (Figure 6B), in spite their similar diameters in solution (2.1 nm for B-from of dsDNA[51]). This indicates higher stiffness and resistance to surface and tip forces for the G4-DNA wires. They show also resistance to heat treatment and to DNase[49]. In addition, the strand folding and the stability of the G4-DNA wires are independent of the presence of cations in the solution[49], in contrast to the short G4-DNA structures. For STM imaging G4-DNA wires were deposited onto a highly smoothed gold(111) surface achieved by flame annealing[50]. The G4-wires were 750 tetrads long and contained 40 phosphothioated G-nucleotides at the 5' end of the G-strand that contributed to the immobilization of the molecules on the gold substrate[52]. High-resolution STM scans demonstrate the clear periodic structure of the G4-DNA, with bulbs corresponding to the helix pitch with an average length of ~3.5 nm (Figure 3C). The molecular height is ~1.5 nm, similar to that measured by AFM.

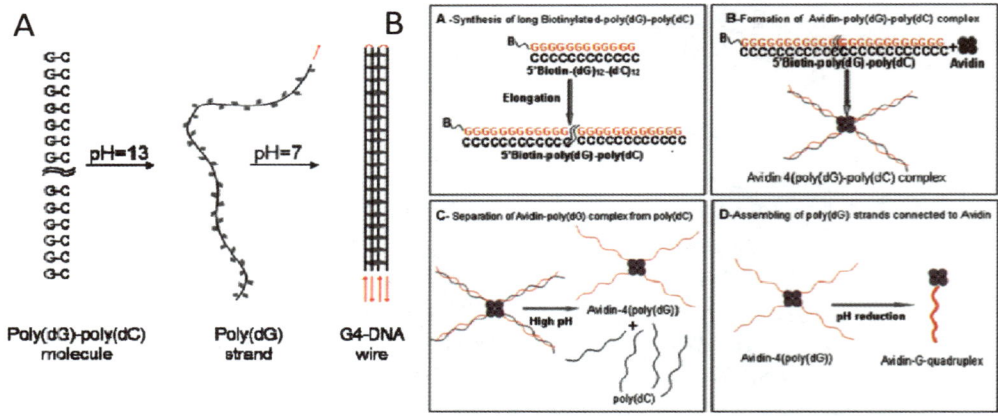

**Figure 5** *Schematic representation of the formation procedures of mono-molecular (A) and tetra-molecular (B) G4-wires.*

As was mentioned before, long (>40 nm) single dsDNA molecules attached to hard surfaces showed neither conductivity nor polarizability[53,54]. It was suggested that deformations induced by surface forces impaired the conductivity of dsDNA in the latter experiments [8,43]. G4-DNA, due to its higher structural rigidity [18,49] is expected to be less influenced by this problem. Moreover, it has a high "surface area" in each tetrad (the molecule cross section) for π-stacking and rich guanine content. These properties make G4-DNA a good candidate for molecular nanowires in nanoelectronics.

The quality of the contacts between the molecule and the electrodes strongly affects the efficiency of charge injection from the electrode to the molecule[55-57]. Therefore, techniques that allow probing the inherent electrical properties of molecules in the absence of contacts are beneficial for initial characterization of molecular wires. One versatile and powerful technique is electrostatic force microscopy (EFM) [58,59] as was demonstrated on carbon nanotubes and dsDNA[54]. EFM is a type of dynamic non-contact AFM (it means that the AFM cantilever is oscillating far from the surface during the scan and is not in surface force range from with the sample) where the electrostatic force is probed. This force arises due to the attraction or repulsion of separated charges on the tip and the sample. Since it is a long-ranged force, it can be detected even 100 nm from the sample. One of the EFM methods that were used to determine the G4-DNA wires polarization is the "plane mode" method. In the "plane mode" operation, the measurement proceeds in two "passes" (Figure 6A). During the "first pass" a topography image of the sample surface is recorded in non-contact dynamic mode. From this image the tip−sample surface plane parameters are obtained. During the "second pass", immediately after the topography imaging, the feedback loop is disconnected, the tip−sample separation is increased to a predefined height above the surface, beyond the van der Waals interactions range, and a phase shift image is measured on the same area while the tip is moving on a plane parallel to the surface using the average slope calculated from the "first pass". In order to measure the electrostatic interaction between the tip and the sample, a bias voltage is applied to the tip during the "second pass". When the tip is sampling a region of the surface (e.g., a molecule) with higher polarizability than that of the substrate background, the interaction changes the resonance frequency of the cantilever and consequently shifts the phase of the oscillations. Therefore, the detected spatial map of the phase shift is expressive of molecular electrostatic polarizability or of charging. "Plane mode" EFM measurements were done on a mica surface that was covered with molecules of two types, dsDNA and a G4- wires (Figure 6B). The measurement reveals a clear pattern of negative phase shift signal at the position of the G4-wires when bias voltage was applied to the tip and no observable signal at 0 V (Figure 6B). If the phase shift is negative (hence reflecting attraction) for both positive and negative voltages, the electrostatic interaction is due to polarizability of the G4-wires. In contrast, a complete absence of the signal in the dsDNA molecules is evident when there is voltage bias. This suggests that the double stranded molecules are not polarized under the experiment conditions. Although the polarizability is only a qualitative index of charge mobility, a viable hypothesis based on the EFM results and on structural traits is that transport experiments may reveal a higher conductivity for G4-wires than for dsDNA, thus identifying a new potential molecular wire. Interestingly, G4-wires that were prepared in the absence of KCl during the folding step of the synthesis failed to show an EFM signal[60]. This suggests that cations play a key role in making G4-wires electrically active.

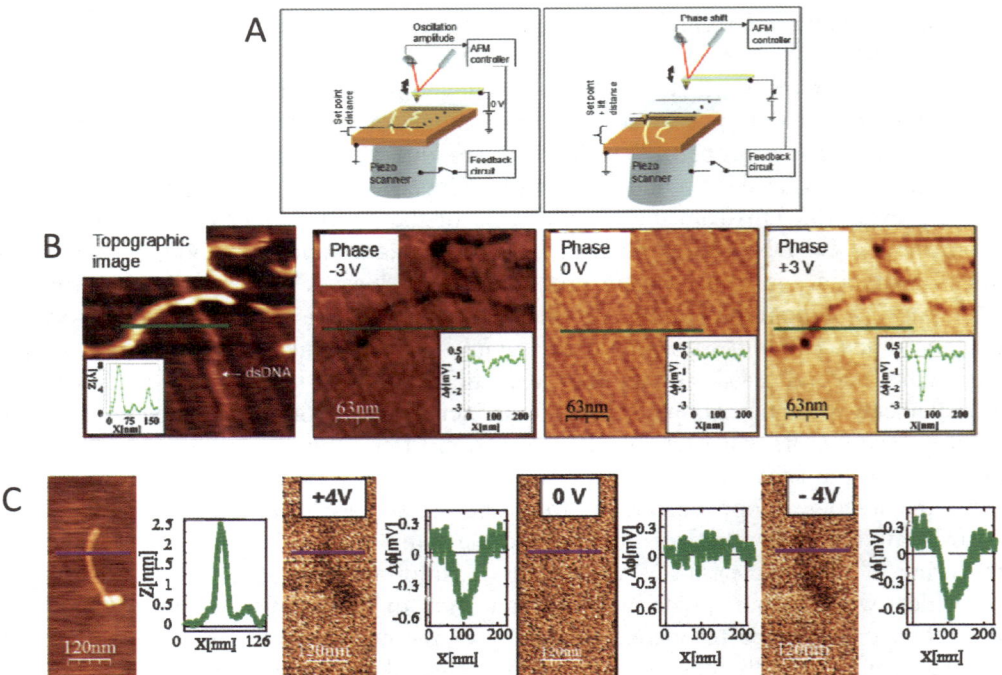

**Figure 6** *EFM of DNA molecules (A) Schematic views of the "plane mode" EFM method. A topography image is acquired and its plane parameters are recorded (left). Then, the tip is lifted and the phase shift is measured with disconnected feedback at lifted height parallel to the previous image plane, with various bias voltages applied to the tip (right). (B) AFM topography image of co-deposited mono-molecular G4-wires and dsDNA. The Cross section demonstrates the G4-wires greater height. (left). Phase shift images of the same area in "plane mode" EFM at −3 V, 0 V, and +3 V, showing clearly that the phase signal shifts only above the location of the G4-wires and only when applying bias voltage. The tip was lifted by 14 nm above the set point value, which was 20 nm above the surface, as extracted from the force−distance calibration. The negative phase shift indicates a decrease in the frequency of the tip oscillations. Line profiles show the magnitude of the signals at the different voltages. (C) Topology image (left) and phase shift images of tetra-molecular G4-wire in "plane mode" EFM at −4 V, 0 V, and +4 V. Line profiles show the magnitude of the signals at the different voltages.*

STS measurements were used for further examination of the G4-wires electrical properties and the energy level spectra[61]. Despite the low measurement temperature (4 K), the series of measured STS curves are characterized by a considerable variability in the I−V curve patterns (Figure 4B). In a careful examination of the data, however, similar curves patterns appeared in different sets, which were taken on different molecules. The origin of this switching is not conclusively clear. Nevertheless, it can be speculated that these changes in the curves may reflect transitions of the molecule between its structural equilibrium states or changes in its energy state or ionization state, stimulated by various possible energy sources. The peak structures in conductance curves (dI/dV−V) should correspond to the electronic energy levels of the G4-DNA wires (Figure 4C). A clustering algorithm was used for extracting the reproducible curve patterns hidden in quite dispersed data sets in order to resolve the energy-level structure of the G4-wires[61]. There is a wide distribution in the position of this peak in the different patterns and it is ranged between -1 V to -2.5 V in the most reproducible curve patterns. The situation in the positive-bias range, which is associated to the unoccupied orbitals, is even more complex and it is more difficult to recognize a set of well-defined patterns similar to those found in the negative-bias range. Nevertheless, some general trends can be outlined. The first peak that appears at positive-bias voltage should represent the helical feature that derives from the guanine lowest unoccupied molecular orbital (LUMO) (Figure 4C). In the curves patterns the current onset is followed by a clear peak that appears between 0.7 and 2.15 V. Generally speaking, the data shows that the first unoccupied-level peak appears at three different energies. In most cases, the dominant peak is accompanied by others, less obvious ones. The statistics of the fundamental energy gap requires an analysis of the curve patterns in the whole voltage range. By taking into account all the measurements at different temperatures (4.5, 78, ~300 K), the gap width manifests a wide distribution, with a maximum at 2.55 V and a mean value of $2.66 \pm 0.89$ V.

The STS data of the G4-wires exhibit a much higher variability than those measured in poly(dG)–poly(dC) molecules (see above[48]), at odds with the intuition that stiffer molecules, the G4-wires, should give more pronounced and clear-cut characteristics. The variability is likely induced by the interplay between intra-plane H-bonding and inter-plane $\pi$-stacking in the determination of each quadruplex DOS peak, whereas $\pi$-stacking is dominating the peak positions of dsDNA. The concurrence of different interactions and the redundancy of energy levels due to multiplets in each tetrad, responsible for significant energy shifts in different dynamical conformations and peak splitting, may in fact render the whole molecular electronic structure more dependent on the mechanical flexibility in G4-DNA.

## 2.3 Tetra-molecular parallel G4-wires

The mono-molecular G4-wire presented above is composed of a single self-folded poly(dG) strand containing hundreds of stacked tetrads[49]. Since in this molecule the poly(dG) strand is folded four times, each two strand fragments are running in opposite directions (Figure 1B, on the left). Tetra-molecular G4-DNA wires composed of four long parallel G-strands (Figure 1B, on the right) might possess different, possibly improved, electric properties with respect to mono-molecular G4-wires. However, uniform inter-molecular G4-wires are not spontaneously assembled in solution, since at high concentrations, which are required for the tetra-molecular folding; G-strands tend to aggregate into bundles comprised of a large number of strands. To avoid aggregation and enable only tetra-molecular G-strand folding, a novel approach has been developed[62]. The basic idea behind this approach is to use avidin-tetramer in order to bring together four G-strands labeled with biotin and therefore enable their folding into a 4-stranded structure. Avidin is a glycoprotein, consisting of four identical subunits, each capable of binding tightly a biotin molecule[63]. The interaction between avidin and biotin is the strongest non-covalent interaction known in biology and it is stable under harsh conditions. The procedure of the tetra-molecular G4-wires production consists of four main stages (Figure 5B)[62]. In the first step, 5′-biotinylated-poly(dG)-poly(dC) molecules are synthesized. Later, a complex between the avidin and four biotinylated poly(dG)-poly(dC) molecules is formed. Next, the poly(dC) strands are separated from poly(dG)-strands connected to the avidin in high pH conditions. The fraction of the four

G-strands connected to avidin is purified using a size-exclusion HPLC. In the final stage, after the pH is reduced, the four G-strands connected to avidin are assembled into tetra-molecular G4-DNA structures.

The morphology of these structures is characterized by AFM imaging (Figure 2C) that shows that each molecule is composed of a linear segment corresponding to the DNA and a brighter (higher) sphere corresponding to the avidin[62]. A comparison between the morphology of the mono- and the tetra-molecular G4-wires shows that the average height, under the same experimental conditions, of the tetra-molecular G4-wires is noticeably larger than that of the mono-molecular ones. This indicates higher stiffness of the tetra-molecular G4-wires. Also the differences in the CD spectra of these wires suggest that they have different structural organization[62]. G4-wires are stable in the absence of cations, similarly to the 4-folded G4-wires.

EFM measurements demonstrated that also the tetra-molecule G4-wires show clear polarizability (Fig 6C)[62]. This is indicative of their possible electrical conductivity. The electrical polarizability together with their high resistance to mechanical deformations makes these wires very promising for nanoelectronic applications.

## 3 CONCLUSION

Here we report our joint progress over the past decade in the synthesis and characterization of poly(dG)-poly(dC) dsDNA molecules and two types of G4-wires. We present their morphological characterization as well as results of STS and polarizability measurements. We find that these molecules offer an attractive and unique alternative to dsDNA towards investigation of electrical transport in long one-dimensional polymers and possibly for implementation in molecular electronic devices.

## References

1  A. Aviram, M. A. Ratner, Engineering Foundation (U.S.) *Molecular electronics: science and technology*, New York Academy of Sciences: New York, 1998.
2  A. Aviram, M. A. Ratner, V. Mujica, United Engineering Foundation (U.S.) *Molecular electronics II*, New York Academy of Sciences: New York, 2002.
3  C. Joachim, J. K. Gimzewski, A.Aviram, *Nature,* 2000, **408**, 541.
4  J. M. Lehn, *Angewandte Chemie International Edition in English,* 1990, **29**, 1304.
5  P. W. K. Rothemund, *Nature,* 2006, **440**, 297.
6  N. C.Seeman, *Annual Review of Biochemistry,* 2010, **79**, 65.
7  E. Braun, Y.Eichen, U.Sivan, G.Ben-Yoseph, *Nature,* 1998, **391**, 775.
8  D. Porath, G. Cuniberti, R. D. Felice. *Long-Range Charge Transfer in DNA II,* 2004, **183**.
9  P. J. de Pablo, F. Moreno-Herrero, J. Colchero, J. Gomez-Herrero, P. Herrero, A. M .Baro, A. M. Ordejon, J. M. Soler, E. Artacho, *Physical Review Letters,* 2000, **85**, 4992.
10  R. G. Endres, D. L. Cox, R. R. P.Singh, *Reviews of Modern Physics*, 2004, **76**, 195.
11  A. J. Storm, J. Van Noort, S. De Vries, C. Dekker, *Applied Physics Letters,* 2001, **79**, 3881.
12  D. Porath, A. Bezryadin, S. de Vries , C.Dekker. *Nature,* 2000, **403**, 635.
13  M. A. Reed, C. Zhou, C. J. Muller, T. P. Burgin, J. M. Tour, *Science,* 1997, **278**, 252.

14  R. Rinaldi, A. Biasco, G. Maruccio, R. Cingolani, D. Alliata, L. Andolfi, P. Facci, F. De Rienzo, R. Di Felice, E. Molinari, *Adv. Mater,* 2002, **14**, 16.
15  J. S. Hwang, K. J. Kong., D,Ahn, G. S. Lee, D. J. Ahn, S. W. Hwang, *Applied Physics Letters,* 2002, **81**, 1134.
16  D. Hennig, E. B. Starikov, J. F. R. Archilla, F. Palmero. *Journal of Biological Physics,* 2004, **30**, 227.
17  S. M. Kerwin, *Current Pharmaceutical Design,* 2000, **6**, 441.
18  J. T. Davis, *Angewandte Chemie International Edition,* 2004, **43**, 668.
19  J. Gros, F. Rosu, S. Amrane, A. De Cian, V. Gabelica, L. Lacroix, J. L. Mergny, *Nucleic Acids Research,* 2007, **35**, 3064.
20  V. M. Marathias, P. H, Bolton. *Biochemistry* , 1999, **38**, 4355.
21  D. Sen, W. Gilbert, *Methods in Enzymology,* 1992 ,**211**, 191.
22  T. Simonsson , *Biological Chemistry,* 2005, **382**, 621.
23  Y. Wang , D. J. Patel, *Structure,* 1994, **2**. 1141.
24  S. Burge, G. N. Parkinson, P. Hazel, A. K.Todd, S. Neidle, *Nucleic Acids Research,* 2006, **34**, 5402.
25  J. Gu, J. Leszczynski, *The Journal of Physical Chemistry A,* 2002, **106**, 529.
26  G. Laughlan, A. I. Murchie, D. G. Norman, M. H.Moore, P. C. Moody, D. M.Lilley, B. Luisi, *Science,* 1994, **265**, 520.
28  K. Phillips, Z. Dauter, A. I. H. Murchie, D. M. J. Lilley, B. Luisi, *Journal of Molecular Biology,* 1997, **273**, 171.
29  T. C. Marsh, J. Vesenka, E. Henderson, *Nucleic Acids Research,* 1995, **23**, 696.
30  E. Protozanova, R. B.Macgregor Jr., *Biochemistry,* 1996, **35**, 16638.
31  D. Sen, W. Gilbert, *Biochemistry,* 1992, **31**, 65.
32  A. Calzolari, R. Di Felice, E. Molinari, A. Garbesi, *Applied Physics Letters,* 2002, **80**, 3331.
33  R. Di Felice, A.Calzolari, H. Zhang. *Nanotechnology,* 2004, *15*, 1256.
34  X.Yang , X. B.Wang , E. R.Vorpagel, L. S.Wang, *PNAS,* 2004, **101**, 17588.
35  A. B. Kotlyar, N. Borovok, T.Molotsky, L. Fadeev, M. Gozin, *Nucleic Acids Research*, 2005, **33**, 525.
36  E. Shapir, H.Cohen, N. Borovok, A. B. Kotlyar, D. Porath, *The Journal of Physical Chemistry B,* **2006**, *110*, 4430.
37  D. P. Allison, L. A.Bottomley, T. Thundat, G. M. Brown, R. P. Woychik, J. J.Schrick, K. B. Jacobson, R. J. Warmack, *PNAS,* 1992, **89**, 10129.
38  M. Amrein, A. Stasiak, H. Gross , E. Stoll, G. Travaglini, *Science,* 1988, **240**, 514.
39  T. P. Beebe Jr, T. E.Wilson, D. F.Ogletree, J. E. Katz, R. Balhorn, M. B. Salmeron, W. J. Siekhaus, *Science,* 1989, **243**, 370.
40  D. D. Dunlap , C. Bustamante, *Nature,* 1989, **342**, 204.
41  S. M. Lindsay, B. Barris, *Journal of Vacuum Science & Technology A: Vacuum. Surfaces. and Films,* 1988, **6**, 544.
42  S. M. Lindsay, T. Thundat, L. Nagahara, U. Knipping, R. L.Rill, *Science,* 1989, **244**, 1063.
43  A. Y. Kasumov, D. V. Klinov, P. E. Roche, S. Gueron, H. Bouchiat, *Applied Physics Letters,* 2004, **84**, 1007.
44  R. R. Sinden , *DNA Structure and Function*, Academic Press, New York. 1994.
45  D. Porath, Y. Levi, M. Tarabiah, O. Millo, *Physical Review B,* 1997, **56**, 9829.
46  J. G. Hou, K.Wang, *Pure and Applied Chemistry*, 2006, **78**, 905.
47  U. Banin, O. Millo, *Annual Review of Physical Chemistry*, 2003, **54**, 465.

48  E. Shapir, H. Cohen, A. Calzolari, C. Cavazzoni, D. A. Ryndyk, G. Cuniberti, A. B. Kotlyar, R. Di Felice, D.Porath , *Nature Materials*, 2007, **7**, 68.
49  A. B. Kotlyar, N. Borovok, T. Molotsky, H. Cohen, E. Shapir, D. Porath, *Advanced Materials,* 2005, **17**, 1901.
50  E. Shapir, L. Sagiv, N. Borovok, T. Molotski, A. B. Kotlyar, D. Porath, *The Journal of Physical Chemistry B,* 2008, **112**, 9267.
51  J. D.Watson , F. H. C.Crick, *Nature*, 1953, **171**, 737.
52  J. Ghabboun, M. Sowwan, H. Cohen, T. Molotsky, N. Borovok, B. Dwir, E. Kapon, A. B. Kotlyar, D. Porath, *Applied Physics Letters,* 2007, **91**, 173101.
53  M. Bockrath, N. Markovic, A. Shepard, M. Tinkham, L. Gurevich, L. P. Kouwenhoven, M. W. Wu, L. L. Sohn, *Nano Letters,* 2002, **2**, 187.
54  C. Gomez-Navarro, F. Moreno-Herrero, P. J.de Pablo, J.Colchero , J.Gomez-Herrero, A. M.Baro, *Proceedings of the National Academy of Sciences* , 2002, **99**, 8484.
55  H. Cohen, C. Nogues, D. Ullien, S. Daube, R. Naaman, D.Porath, *Faraday Discussions,* 2006, **131**, 367.
56  T. Dadosh , Y. Gordin, R. Krahne, I. Khivrich, D. Mahalu, V. Frydman, J. Sperling, A.Yacoby, I. Bar-Joseph, *Nature,* 2005, **436**, 677.
57  A. Salomon, D. Cahen, S. Lindsay, J. Tomfohr, V. B. Engelkes, C. D. Frisbie, *Advanced Materials,* 2003,**15**, 1881.
58  C. Schonenberger , *Physical Review B,* 1992, **45**, 3861.
59  C. Staii, A. T. Johnson , N. J. Pinto, *Nano Letters*, 2004, **4**, 859.
60  H. Cohen , T.Sapir, N. Borovok, T. Molotsky, R. Di Felice, A. B. Kotlyar, D.Porath, *Nano letters*, 2007, **7**, 981.
61  E. Shapir, L. Sagiv, N. Borovok, T. Molotsky, A. B. Kotlyar, D. Porath, *The Journal of Physical Chemistry B,* 2008, **112**, 9267–9269
62  N. Borovok, N. Iram, D. Zikich, J. Ghabboun, G. I.Livshits, D. Porath, A. B. Kotlyar *Nucleic Acids Research* 2008, **36**, 5050.
63  M. Wilchek, E. A. Bayer, *Methods in Enzymology,* 1990, **184**, 5.

# Subject Index

# Subject Index

| | |
|---|---|
| 1,8-naphthyridine | 198 |
| 3-methylxanthine (3MX) | 185, 186, 187, 191 |
| 3,6,9-trisubstituted acridine | 239 |
| 5'-GMP | 3-15 |
| 9-methylguanine (G) | 181, 182, 184, 185 |
| 9-methyluric acid (Ua) | 181, 182, 183, 184, 185, |
| 9-methylxanthine (Xa) | 181, 182, 183, 184, 185 |
| $\alpha/\gamma$ torsional term | 196 |
| $\pi$ | |
|     assistance | 187, 188 |
|     electron(s) | 187, 188 |
|     interaction(s) | 188 |
|     virtual(s) | 188 |
| abasic backbone | 170 |
| Accelrys AFFINITY program | 205 |
| acridine | 198, 200, 205 |
| AFM | 79, 102-108, 111-119, 157, 160, 161, 307-308, 311, 330 |
| AMBER | 196, 197, 198, 200 |
|     AMBER, 11 | 204 |
|     AMBER, 99$\phi$ | 203 |
| Amsterdam Density Functional (ADF) | 180 |
| anion(s) | 181, 184, 185, 191 |
| anti | 167, 168, 170, 171, 172, 173, 174, 175, 176 |
| anti-cancer | 194, 297-298 |
| anti-HIV | 203 |
| anti-HIV agent | 293-297 |
| aptamer | 195, 198, 203, 204, 277-278, 292, |
| assembly(ies) | 179, 184, 185, 186, 187 |
| ATRX | 219 |
| Autodock v4 | 206 |
| Azatrux | 196 |
| backbone | 167, 170, 172, 196, 197, 198, 199, 202, 205 |
| basket type | 168 |
| binding | |
|     binding, energy | 182, 185 |
|     binding, site(s) | 198, 206 |
| bispyridinium derivative(s) | 198 |
| bisquinolinium | 198 |
| BLM helicase | 217, 219 |
| BMVC4 | 244 |
| bond energy | 180, 182, 183, 185, 187 |
| BRACO19 | 239, 248 |
| Brewster angle microscopy (BAM) | 156, 157, 158 |

| | |
|---|---|
| c-MYC gene | 217, 225, 226 |
| C2'-endo | 167 |
| C3'-endo | 167 |
| CEB1 minisatellite | 216 |
| CHARMM 27 force field | 196 |
| chemical libraries | 250 |
| Circular Dichroism (CD) | 148, 149, 150, 176, 177 |
| cluster(s) | 203 |
| computational method(s) | 194, 203 |
| conductor-like Screening Model (COSMO) | 180 |
| continuum solvent | 200 |
| cooperativity | 180, 181, 187-192 |
| coordination | 194, 195, 202 |
| Cornell force field | 196 |
| counterion(s) | 202 |
| crosspeak(s) | 173 |
| crystal structure(s) | 194, 198, 199, 202 |
| cutoff | 196, 203 |
| CX-3543 | 243 |
| cyclic arrangement | 194 |
| cytosine(s) | 198 |
| | |
| D4S43 | 216 |
| Density Functional Theory (DFT) | 180, 202 |
| dielectric constant | 186 |
| dielectrophoresis (DEP) | 307-310, 312, 316, 319 |
| differential pulse voltammetry | 100-108 |
| Diffusion Ordered Spectroscopy (DOSY) | 186, 187 |
| dimer(s) | 182, 183, 184, 191 |
| dimeric | 194, 196, 199, 200 |
| direct coordination to quadruplexes | 270-272 |
| dispersion correction | 180, 185 |
| DNAzymes | 278, 279-280, 286, 290 |
| DSC | 64 |
| dynamic docking | 205 |
| dynamic light scattering (DLS) | 121-133 |
| | |
| EFM | 331-332, 334 |
| Eigen | |
|     Value(s) | 199 |
|     Vector(s) | 199 |
| Elastic Network Models (ENM) | 204 |
| electrophoresis | 318-319 |
| electrostatic | |
|     force(s) | 194, 196, 202 |
|     interaction(s) | 180, 187 |
|     repulsion | 182, 183, 184, |
| Energy Decomposition Analysis (EDA) | 180, 192 |
| ESI-MS | 200 |
| ESP charge(s) | 197 |

*Subject Index*

| | |
|---|---|
| exosite | 204 |
| explicit solvent | 195, 200, 202, 204, 205 |
| Extended Transition State (ETS) | 180 |
| | |
| FANCJ | 219 |
| FD-121 | 242 |
| first intron | 219, 231 |
| folding | 172, 173, 177, 201, 206 |
| Fonsecin B | 205 |
| force field(s) (ff) | 195, 196-198, 202-204 |
| fragile X syndrome | 216 |
| fragmentation scheme | 182, 184 |
| Free Energy Pertubation (FEP) | 201 |
| | |
| G-quadruplex interacting proteins | 228-231 |
| G-residue(s) | 170 |
| G4 helicases | 220, 230 |
| G4-DNA wires | 52, 116-119, 307-308, 327, 330, 333 |
| gas phase | 185, 188-191, 200, 201 |
| Generalised AMBER Force Field (GAFF) | 196 |
| Generalised Born (GB) | 201 |
| genomic instability | 219 |
| geometric formalism | 167, 173, 177 |
| glycosidic | |
|     Bond Angle (GBA) | 167-175, 176, 177 |
|     conformation | 194 |
| gold nanoparticles | 288-289, 301 |
| GRN163L | 238 |
| GRO supramolecular polymers | 49 |
| GROMACS | 203, 204 |
| GROMOS force field | 195 |
| groove(s) | 198, 202, 206 |
|     medium | 169-172 |
|     narrow | 169, 170, 173, 177 |
|     wide | 169, 173, 177 |
|     width(s) | 169, 171-173 |
| guanine riboswitch | 196 |
| guanylic acid | 3-14 |
| | |
| hammerhead ribozymes | 196, 281-282 |
| HDV ribozyme | 283 |
| helical cavity | 194 |
| hemin | 278, 286-287, 289 |
| high-throughput screening | 248 |
| HIV-1 integrase | 293-294 |
| HIV1-Integrase (HIV1-IN) | 204 |
| HMBC | 186, 187 |
| human telomeric DNA | 198 |
| hybrid-2 | 168 |
| hydrogen-bond(s) | 179-183, 185-188, 191, 197, 202 |

| | |
|---|---|
| Hoogsteen | 194 |
| | |
| i-motif(s) | 198 |
| I-U curve | 320-321, 328-329, 332 |
| immunoglobin switch regions | 194, 217 |
| in vivo | 215-217 |
| indole | 205 |
| inhibitory activity | 204 |
| insulin-like hypervariable repeat | 216, 226 |
| interaction(s) | 195-202, 204-206 |
|     energy | 180-184, 188-190 |
| intercalation | 198, 200 |
| ion(s) | 195-198, 200, 202, 203, 206 |
|     channel | 195, 197 |
| ITC measurement(s) | 206 |
| | |
| Langmuir Blodget (LB) film | 155, 156 |
| lead compound(s) | 205 |
| ligand(s) | 196, 198, 200, 201, 205, 206 |
| liponucleosides | 155 |
| lipophilic guanosines | 28-39 |
| localised Enhanced Sampling (LES) | 199 |
| loop(s) | 170-172, 194-205 |
|     diagonal | 170-172, 177, 196, 199 |
|     lateral | 170-173, 177, 199 |
|     propeller | 170-172, 177, 196 |
| Low Barrier Hydrogen Bonds (LBHB) | 180-184, 191 |
| | |
| macrocyclic ligands | 265 |
| matrix(ces) | 199 |
| MEP charge(s) | 197 |
| metal complexes | 264-273 |
| MM-PBSA | 201 |
| modified oligonucleotides, synthesis | 92-96 |
| modified TBA (mTBA) | 204 |
| molecular electronics | 314, 325 |
| molecular docking | 205, 206 |
| Molecular Dynamics (MD) | 195-199, 201, 203- 205 |
| MutSα | 217 |
| | |
| N-methylmesoporphyrin | 278 |
| N. gonorrheae | 219 |
| nanopatterning | 40-47 |
| NMR | 181, 186, 187, 191, 203-206 |
| NOE | 187 |
|     connectivities | 173 |
| NOESY | 173, 186, 187 |
| non-covalent interactions | 264 |
| non-cyclic ligands | 266 |
| non-planar complexes | 268-270 |

## Subject Index

| | |
|---|---|
| octad(s) | 185, 186 |
| oligonucleotides, synthesis | 89-92 |
| optical tweezers | 74 |
| | |
| P53 gene | 221 |
| parm | |
|     94 | 196, 197, 199 |
|     98 | 196, 197, 199, 204 |
|     99 | 196, 200, 203 |
| parm, bsc0 | 196, 197, 198, 202, 203, 204 |
| Particle Mesh Ewald (PME) | 196, 197 |
| pattern(s) | 173, 176 |
| periodic boundary conditions | 197 |
| peroxidase activity | 279, 286, 288, 289 |
| perylene derivative (Tel03) | 200 |
| photolyase activity | 279 |
| PIPER | 248 |
| Platinum (II) Schiff base | 205 |
| point mutations | 226 |
| Poisson Boltzmann (PB) | 200, 201 |
| polyacrylamide del electrophoresis (PAGE) | 148,149 |
| polymer(s) | 187 |
| porphyrin | 198, 201, 205 |
| Potential of Mean Force (PMF) | 201 |
| preparation energy | 180 |
| Principal Component Analysis (PCA) | 199, 200, 204 |
| promoter region(s) | 194 |
| PS2.M | 278 |
| PS5.M | 278 |
| PS5.ST1 | 278 |
| pseudo-planar tetrad | 168 |
| | |
| quadruplex binders | 264 |
| quality control | 257 |
| Quantum Mechanics/Molecular Mechanics (QM/MM) | 202 |
| quantum-chemical calculation | 191 |
| Quantum-regions Interconnected by Local Descriptions (QUILD) | 180 |
| | |
| regulatory regions | 218 |
| Resonance-Assisted Hydrogen Bonding (RAHB) | 187, 188, 192 |
| Restrained Electrostatic Potential Charge (RESP) | 200 |
| RHPS4 | 239 |
| ribbon motif | 30 |
| rise | 202 |
| rmsd | 202, 203 |

| | |
|---|---|
| RNA | 196, 198, 200, 203, 205 |
| RNA quadruplexes | 240 |
| screening | |
|     assays | 251-257 |
|     physical | 250 |
|     virtual | 204-206, 250 |
| screening assay | |
|     Circular Dichroism | 254 |
|     enzymatic | 253 |
|     ESI-MS | 254 |
|     fluorescense | 251-253 |
|     functional | 255 |
|     natural extracts | 254 |
|     surface plasmon resonance | 253 |
|     virtual | 256, 257 |
| SELEX | 277, 286 |
| self-cleavage | 281 |
| silver nanoparticles | 314-315 |
| simulation | 199-204 |
| Small Angle X-ray Scattering (SAXS) | 137-140 |
| solvent phase | 200 |
| stabilizing agent(s) | 198 |
| stacking | 179, 182, 185, 188, 189, 191, 194, 198, 202, 203, 205 |
| stem(s) | 168-174, 176, 177, 194-96, 197-203 |
| STM | 40-48, 327-328, 330 |
| stoichiometry | 198 |
| Strand(s) | 194, 196, 198, 199, 203 |
| STS | 329, 332 |
| sugar pucker(s) | 167-169 |
| Surface Area (SA) | 201 |
| switching between assemblies | 28-39 |
| syn | 167, 168, 171-173, 175, 176 |
| targeting nucleic acids | 241 |
| targets of regulation | 220 |
| TBA | 298-300 |
| telomerase | 285-290 |
| telomere | 179, 189, 191, 192, 194, 217, 285 |
| telomestatin | 239, 248 |
| TEM | 315 |
| temperature gradient gel electrophoresis | 149, 150 |
| TERRA | 237, 240, 243, 245 |
| tetrad(s) | 168-173, 179-186, 188, 190, 191 |
| tetramer(s) | 179-185, 189, 191 |
| tetrameric | 194, 198 |
| tetramolecular | 194 |
| thermal difference | 176 |
| thermal stability | 203 |

*Subject Index*

| | |
|---|---|
| thermodynamic Integration (TI) | 201 |
| thrombin | 298-299 |
| Thrombin Binding Aptamer (TBA) | 173, 195, 198, 201-204 |
| thymine | 200 |
| TMPyP4 | 240 |
| topology | 198, 202 |
| trajectory(ies) | 195-201, 203 |
| transcription by non-promoter quadruplexes | 231 |
| transcription modulation | 225-230 |
| TRAP protocol | 238 |
| triad(s) | 169 |
| truaryl-substituted imidazole derivative TSIZ01 | 205 |
| twist | 202 |
| under selection | 217, 218 |
| unfolding | 201 |
| unimolecular quadruplex | 170, 171 |
| untranslated regions | 219, 224, 229, 231-232 |
| uric acid | 179-182 |
| UV-VIS | 176, 177 |
| VNTRs | 216 |
| Voronoi Deformation Density (VDD) | 181, 184, |
| Watson-Crick | 167, 168, 202 |
| WRN | 219 |
| X-ray | 203-206 |
| X-ray diffraction (of condensed phases) | 140-145 |
| xanthine | 179-181, 183, 185-188, 190, 191 |